CATEGORICAL AND NONPARAMETRIC DATA ANALYSIS

Featuring in-depth coverage of categorical *and* nonparametric statistics, this book provides a conceptual framework for choosing the most appropriate type of test in various research scenarios. Class tested at the University of Nevada, the book's clear explanations of the underlying assumptions, computer simulations, and *Exploring the Concept* boxes help reduce reader anxiety. Problems inspired by actual studies provide meaningful illustrations of the techniques. The underlying assumptions of each test and the factors that impact validity and statistical power are reviewed so readers can explain their assumptions and how tests work in future publications. Numerous examples from psychology, education, and other social sciences demonstrate varied applications of the material. Basic statistics and probability are reviewed for those who need a refresher. Mathematical derivations are placed in optional appendices for those interested in this detailed coverage.

Highlights include the following:

- *Unique coverage of categorical and nonparametric statistics* better prepares readers to select the best technique for their particular research project; however, some chapters can be omitted entirely if preferred.
- *Step-by-step examples* of each test help readers see how the material is applied in a variety of disciplines.
- *Although the book can be used with any program, examples of how to use the tests in SPSS and Excel* foster conceptual understanding.
- *Exploring the Concept boxes* integrated throughout prompt students to review key material and draw links between the concepts to deepen understanding.
- *Problems* in each chapter help readers test their understanding of the material.
- *Emphasis on selecting tests that maximize power* helps readers avoid "marginally" significant results.
- *Website* (www.routledge.com/9781138787827) features datasets for the book's examples and problems, and for the instructor, PowerPoint slides, sample syllabi, answers to the even-numbered problems, and Excel data sets for lecture purposes.

Intended for individual or combined graduate or advanced undergraduate courses in categorical and nonparametric data analysis, cross-classified data analysis, advanced statistics and/or quantitative techniques taught in psychology, education, human development, sociology, political science, and other social and life sciences, the book also appeals to researchers in these disciplines. The nonparametric chapters can be deleted if preferred. Prerequisites include knowledge of t tests and ANOVA.

E. Michael Nussbaum is a professor of educational psychology at the University of Nevada, Las Vegas.

CATEGORICAL AND NONPARAMETRIC DATA ANALYSIS

Choosing the Best Statistical Technique

E. Michael Nussbaum

Routledge
Taylor & Francis Group

NEW YORK AND LONDON

First published 2015
by Routledge
711 Third Avenue, New York, NY 10017

and by Routledge
27 Church Road, Hove, East Sussex BN3 2FA

Routledge is an imprint of the Taylor & Francis Group, an informa business

Library of Congress Cataloging in Publication Data
A catalog record for this book has been requested

ISBN: 978-1-84872-603-1 (hbk)
ISBN: 978-1-138-78782-7 (pbk)
ISBN: 978-0-203-12286-0 (ebk)

Typeset in Stone Serif
by Apex CoVantage, LLC

Printed and bound in the United States of America by
Edwards Brothers Malloy on sustainably sourced paper

This book is dedicated to my mentors who taught me secrets of statistics: Edward H. Haertel, Henry E. Brady, and David R. Rogosa. It is also dedicated to my ever patient wife, Lorraine.

BRIEF CONTENTS

CONTENTS

PREFACE

When scientists perform research, they often collect data that do not come in a form that is suitable for traditional methods, such as analysis of variance (ANOVA) or ordinary linear regression. Categorical and nonparametric data analysis are useful for data that are nominal or ordinal, or when assumptions of traditional metric tests have been violated, for example, when the sample is small. Finally, even when all the assumptions of traditional tests have been met, alternative tests are sometimes more statistically powerful, meaning that those tests are more likely to find a statistically significant result when there is a real effect to be found. There is nothing more frustrating than to spend a year or two on a research project, only to have your results come in as not quite significant (e.g., $p = .07$).

This book was developed both out of my efforts to advise students on dissertations as well as from my own research in the field of argumentation. I found that while much of what I was taught about statistics in graduate school was useful at times, it was also somewhat limited. It was as if I was only taught half the story. In fact, one of my statistics teachers at Stanford University was upset that coverage of categorical data analysis (CDA) was no longer required, because there was too much material already crammed into the statistics core. It was only later, after several years conducting research with actual messy data sets, that I discovered the immense value of categorical and nonparametric data analysis.

It is important to know when to use these techniques and when to use traditional parametric analysis. One major goal of the book is to provide a conceptual framework for choosing the most appropriate type of test in a given situation. One has to consider the underlying assumptions of each test and the factors that impact each test's statistical power. One also has to be able to explain these assumptions (and how the test works) conceptually to both oneself and others (including the audience of a journal article or dissertation). This allows one to make statistically based arguments that can serve as a rationale for using that test. Therefore, another major goal of this book is to provide readers with a conceptual framework that underlies the statistical methods examined and, to some extent, traditional parametric methods as well.

Intended Audience

The primary intended audience is researchers and students in the social sciences (particularly psychology, sociology, and political science) and in related professional domains (such as education). Readers should have some prior knowledge of *t* tests and ANOVA. It is preferable for readers to also have some prior knowledge of linear regression, although bright students can also pick up the basics from reading Chapter 9.

This book is fairly unique in covering both nonparametric statistics and CDA in one volume. With the exception of contingency tables, most textbooks address either one or the other. In the one-semester course that I teach on this topic, I successfully cover both and—as noted above—provide a framework for choosing the best statistical technique. Using the framework requires knowledge of both CDA and nonparametrics. Although it is a challenging course, given students' financial and time constraints, it is often not practical for graduate students who are not specializing in statistics to take separate courses in nonparametric statistics and CDA. Instructors who can provide a two-semester course can of course cover more topics and provide students with more extensive practice; however, a one-semester course can still provide students with some familiarity with the topics and a framework for choosing a statistical methodology for particular research projects. (Students can then get more practice and mastery of that particular technique, as the best way to learn a technique is to use it in the context of applied research.) The book could also be used for a course in only CDA or nonparametrics by just focusing on the applicable chapters, leaving the rest of the material for optional, supplementary reading. Finally, the book is also suitable for researchers and graduate students who are not necessarily enrolled in a course but who desire some knowledge of these alternative techniques and approaches for enhancing statistical power.

Unique Features of the Book

The book is closely tied to those techniques currently available in SPSS and provides direction on using SPSS. (Incorporation into SPSS was in fact a major criterion for choosing what topics would be covered.) Although the book can be used with other statistical packages, SPSS is widely used and there is a need for a text that addresses that audience. Some of the examples and problems in the book also use Excel as a pedagogical tool for building conceptual understanding. Also available is a website for the book at www.Routledge.com/9781138787827 that contains selected data sets for the book's examples and problems and, for the instructor, PowerPoint slides for each chapter, answers to even-numbered problems, and the author's course syllabus.

An important feature of the text is its conceptual focus. Simple computer simulations and the inclusion of *Exploring the Concept* boxes are used to help attach meaning to statistical formulas. Most homework problems were inspired by actual research studies; these problems therefore provide authentic and meaningful illustrations of the techniques presented. Mathematical derivations have been kept to a minimum in the main text but are available in appendices at the end of each chapter.

Content

The book is structured as follows. Chapters 1 through 3 cover basic concepts in probability—especially the binomial formula—that are foundational to the rest of the book. Chapters 4 and 5 address the analysis of contingency tables (i.e., analyzing the relationship between two nominal variables). Chapters 6 through 8 address nonparametric tests involving at least one ordinal variable. (Chapter 8 presents contemporary techniques for testing nonparametric interaction effects, a topic omitted from many other texts.) The book then turns to situations that involve at least one metric variable. Chapter 9 reviews some concepts from linear regression, such as exponential growth and dummy variables, as well as the concept of generalized linear models. All of these concepts are foundational to CDA, which is the focus of the remaining portion of the book. Chapters 10 and 11 cover various types of logistic, ordinal, and Poisson regression. Chapter 12 overviews log-linear models, and Chapter 13 presents the general estimating equations (GEE) methodology for measuring outcomes measured at multiple time points. Chapter 14 covers estimation methods, such as Newton-Raphson and Fisher scoring, for readers desiring a deeper understanding of how the various CDA techniques work. The chapter provides preparation for reading more advanced statistical texts and articles. Finally, Chapter 15 summarizes the various factors that need to be taken into consideration when choosing the best statistical technique.

Overall, the book's organization is intended to take the reader on a step-by-step journey from basic statistical concepts into more advanced terrain.

Acknowledgments

I am truly thankful to all those who assisted me in reviewing chapter drafts. Special thanks to Lori Griswold for reviewing the entire book for clarity and typographical mistakes. Sections of the book were also reviewed by my graduate assistants: Ivan Ivanovich, Jason Boggs, and Marissa Owens. I would also like to thank my editor, Debra Riegert, for her assistance, and Mark Beasley and Hossein Mansouri, both of whom gave valuable input on Chapter 8. I would also like to acknowledge the following reviewers who were solicited by the publisher: Randall H. Rieger, West Chester University; Margaret Ross, Auburn University; Sara Tomek, University of Alabama; Haiyan Wang, Kansas State University; and three anonymous reviewers. Finally, the book would not be possible without all the encouragement and feedback provided by the graduate students who completed my course in categorical/ nonparametric data analysis.

ABOUT THE AUTHOR

Dr. E. Michael Nussbaum is a professor of educational psychology at the University of Nevada, Las Vegas. He holds a PhD from Stanford University and a MPP from the University of California, Berkeley. He is the author of numerous research publications and serves on the editorial boards of the *Journal of Educational Psychology, Contemporary Educational Psychology,* and the *Educational Psychologist.*

CHAPTER 1

LEVELS OF MEASUREMENT, PROBABILITY, AND THE BINOMIAL FORMULA

Categorical and nonparametric data analysis is designed for use with nominal or ordinal data, or for metric data in some situations. This chapter therefore reviews these different levels of measurement before turning to the topic of probability.

Levels of Measurement

In statistics, there are four basic types of variables: (a) nominal, (b) ordinal, (c) interval, and (d) ratio.

A *nominal* variable relates to the presence or absence of some characteristic. For example, an individual will be either male or female. Gender is a dichotomous nominal variable. In contrast, with a multinomial nominal variable, cases are classified in one of several categories. For example, ethnicity is multinomial: Individuals can be classified as Caucasian, African American, Asian/Pacific Islander, Latino, or Other. There is not any particular ordering to these categories.

With an *ordinal* variable, there is an ordering. For example, an art teacher might look at student drawings and rank them from the most to the least creative. These rankings comprise an ordinal variable. Ordinal variables sometimes take the form of ordered categories, for example, *highly creative, somewhat creative,* and *uncreative.* A number of individuals may fall into these categories, so that all the drawings classified as highly creative would technically be tied with one another (the same for the moderately creative and uncreative categories). With ranks, on the other hand, there are typically few if any ties.

With an *interval* or *ratio* variable, a characteristic is measured on a scale with equal intervals. A good example is height. The scale is provided by a ruler, which may be marked off in inches. Each inch on the ruler represents the same distance; as a result, the difference between 8 and 10 inches is the same as between 1 and 3 inches. This is not the case with an ordinal variable. If Drawing A is ranked as more creative than Drawing B, we do not know if Drawing A is just slightly more creative or significantly more creative; in fact, the distances are technically undefined. As a result, we need to use different statistics and mathematical manipulations for ordinal variables than we do for interval/ratio variables. (Much of this book will be devoted to this topic.)

As for the difference between an interval variable and a ratio variable, the defining difference is that in a ratio variable a score of zero indicates the complete absence of something. Height is a ratio variable because zero height indicates that an object has no height and is completely flat (existing in only two dimensions). Counts of objects are also ratio variables. The number of people in a classroom can range from zero on up, but there cannot be a negative number of people. With an interval variable, on the other hand, there can be negative values. Temperature is a good example of something measured by an interval scale, since 0° Celsius is just the freezing point of water, and negative temperatures are possible. However, for the tests discussed in this book, it will usually not be necessary to differentiate between ratio and interval variables, so we will lump them together into one level of measurement. The distinction will only become important when we consider the analysis of count data with Poisson regression. For ease of exposition, in this book I will use the term *metric variable* to refer to those at the interval or ratio levels of measurement.

Metric variables are often also referred to as *continuous,* but this usage fails to recognize that some metric variables are discrete. For example, a count cannot have fractional values; for example, it would be incorrect to say that there are 30.5 people enrolled in a class.

Figure 1.1 shows the three levels of measurement. The metric level is shown on top because it is the most informative. Metric data can always be reduced to ordinal data by using the numerical values to rank the data (e.g., ranking people from the tallest to the shortest based on their heights). Likewise, ordinal data can be reduced to nominal data by performing a median split and classifying cases as *above* or *below* the median. Transforming data from a higher level to a lower level is known as *data reduction.* Data reduction throws away information; for example, knowing that Marie is taller than Jennifer does not tell one how much taller Marie is. Nevertheless, data reduction is sometimes performed if the assumptions of a statistical test designed for metric data are not met. One might then reduce the data to ordinal and perform a statistical test that is designed for ordinal data. One can also reduce ordinal (or metric) data to nominal. One cannot move from a lower level to a higher level in the figure because that requires information that is missing.

Categorical and nonparametric statistics is concerned with statistical methods designed for ordinal- and nominal-level data. However, these methods are often used with metric data when sample sizes are small (and therefore some of the assumptions of *t* tests, ANOVA, and linear regression are not met) or when the data

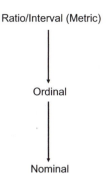

Ratio/Interval (Metric)

Ordinal

Nominal

Figure 1.1 Three levels of measurement. The figure shows that metric data can be reduced to ordinal data, which can in turn be reduced to nominal data. Metric data are the most informative because they carry information on how different the cases are on a variable in quantitative terms. Nominal data are the least informative because they contain no information regarding order or ranks.

are skewed or otherwise highly abnormal. In the latter cases, standard methods may not be as statistically powerful as categorical and nonparametric ones. In reading this book, it is very important to remember the definition of statistical power:

> **Statistical power** refers to the ability to reject the null hypothesis and find a "result." (To be more technically precise, it is the probability that one will reject the null hypothesis when the alternative hypothesis is true.)

Because conducting a study requires a large amount of work, one usually wants to use the most powerful statistical methods. (Obtaining a p value of .06 or .07 is not sufficient to reject the null hypothesis and therefore can be very disappointing to researchers.) That is why, in planning a study, one wants to use the most valid and statistically powerful methods one can.

Probability

All statistical methods—including categorical/nonparametric ones—require an understanding of probability. In the remainder of this chapter, I review the basic axioms of probability and use them to derive the binomial formula, which is the foundation of many of the tests discussed in this book.

In this section, I use a canonical example of tossing coins up in the air and asking questions about the probability of a certain number of them coming up heads. Note that whether a coin comes up heads or tails is a nominal outcome (it either happens or it doesn't)—which is why understanding probability is essential to categorical data analysis.

The Meaning of Probability

If one flips a coin, what is the probability that the coin will come up heads? You may reply "one-half," but what exactly does this statement mean?

One definition of probability is given by the following equation:

$$Probability = \frac{Number\ of\ favorable\ possibilities}{Number\ of\ total\ possibilities},$$ (Eq. 1.1)

assuming that all possibilities are equally likely and mutually exclusive.

So a probability of ½ means that there is one favorable possibility (heads) out of two (heads or tails). However, this example assumes that the coin is fair and not biased, meaning that the possibilities are equally likely. A biased coin might have a little more metal on one side so that heads results 60% of the time. Such coins have been created to cheat at games of chance.

Eq. 1.1 is often useful but because of the restrictive equal-probability assumption, a more general definition of probability is needed. The probability of some event A occurring is the proportion of time that A will occur (as opposed to not-A) in the limit as n approaches infinity, that is:

$$\text{Prob}(A) = \lim (n \to \infty)\frac{f(A)}{n},$$ (Eq. 1.2)

where $f(A)$ is the frequency of A. Thus, the meaning of the statement "The probability that the coin will come up heads is one-half" is that over a large number of flips, about half the time the coin will come up heads. The amount of error decreases as n (the number of flips) increases, so that the proportion will approach Prob(A) as n approaches infinity. Now to assess the probability, one might flip the coin 1,000 times and gauge the relative proportion that the coin comes up heads as opposed to tails. This procedure will only give one an estimate of the true probability, but it will give one a pretty good idea as to whether the coin is fair or biased. Basically, what we are doing is taking a random sample of all the possible flips that could occur.

The definition in Eq. 1.2 reflects a *frequentist* view of probability. Technically, Eq. 1.2 assigns probabilities only to general statements, not to particular facts or events. For example, the statement that 80% of Swedes are Lutherans is a meaningful probability statement because it is a generalization; but the statement that "the probability that John is Lutheran, given that he is Swedish, is 80%" is not meaningful under this definition of probability, because probability applies to relative frequencies, not to unique events.

This position, however, is extreme, given that we apply probability to unique events all the time in ordinary discourse (maybe not about Lutherans, but certainly about the weather, or horse racing, etc.). In my view, probability statements can be meaningfully applied to particular cases if one makes the appropriate background assumptions. For example, if one randomly selects one individual out of the population (of Swedes), then one can meaningfully say that there is an 80% chance that she or he is Lutheran. The background assumption here is that the selection process is truly random (it may or may not be). The existence of the background assumptions means that probability values cannot be truly objective because they depend on whether one believes the background assumptions. Nevertheless, these assumptions are often rational to make in many situations, so in this book I will assign probability values to unique events. (For further discussion of subjectivist, Bayesian notions of probability, which view probability statements as measures of certainty in beliefs, see Nussbaum, 2011).

Probability Rules

Probability of Joint Events

What is the probability that if one flips two coins, both will come up heads? Using Eq. 1.1, it is ¼, because there is one favorable possibility out of four, as shown below.

H   **H**
H   T
T   H
T   T

Another way of calculating the joint probability would be to use the following formula:

$$\text{Prob}(A \ \& \ B) = \text{Prob}(A) * \text{Prob}(B) \ [\text{if Prob}(A) \text{ and Prob}(B)$$
$$\text{are statistically independent}]. \tag{Eq. 1.3}$$

Here, A represents the first coin coming up heads and B represents the second coin coming up heads; the joint probability is $\frac{1}{2} * \frac{1}{2} = \frac{1}{4}$. The formula works because A represents one-half of all possibilities and of these, one half represent favorable possibilities, where B also comes up heads. One-half of one-half is, mathematically, the same as $\frac{1}{2} * \frac{1}{2}$.

An important background assumption, and one that we shall return to repeatedly in this book, is that A and B are statistically independent. What this means is that A occurring in no way influences the probability that B will occur. This assumption is typically a reasonable one, but we could imagine a scenario where it is violated. For example, suppose someone designs a coin with an electrical transmitter, so that if the first coin comes up heads, this information will be transmitted to the second coin. There is a device in the second coin that will tilt it so that it will always comes up heads if the first coin does. In other words: $\text{Prob}(B \mid A) = 1$. This statement means that the probability of B occurring, if A occurs, is certain. We will also assume the converse: Prob(not-B | not-A) = 1. There are therefore only two total possibilities:

H H
T   T

The joint probability of two heads is therefore one-half. The more general rule for joint probabilities (regardless of whether or not two events are statistically independent) is:

$$\text{Prob}(A \ \& \ B) = \text{Prob}(A) * \text{Prob}(B \mid A). \qquad \text{(Eq. 1.4)}$$

The joint probability in the previous example is ½ * 1 = ½.

EXPLORING THE CONCEPT

If two events (A and B) are statistically independent, then Prob(A) = Prob(A | B) and Prob(B) = Prob(B | A). Can you use this fact to derive Eq. 1.3 from Eq. 1.4?

Note that in the rigged coin example, the joint probability has increased from ¼ (under statistical independence) to ½ (under complete dependence). This outcome reflects a more general principle that the probability of more extreme events increases if cases are not statistically independent (and positively correlated). For example, if I throw 10 coins up in the air, the probability that they will all come up heads is

$$\tfrac{1}{2} * \tfrac{1}{2} * \tfrac{1}{2} * \tfrac{1}{2} * \tfrac{1}{2} * \tfrac{1}{2} * \tfrac{1}{2} * \tfrac{1}{2} * \tfrac{1}{2} * \tfrac{1}{2} = (\tfrac{1}{2})^{10} = .001,$$

or 1 in 1,000. But if the coins are programmed to all come up heads if the first coin comes up heads, then the joint probability is again just one-half. If the first coin coming up heads only creates a general tendency for the other coins to come up heads [say Prob($B \mid A$) = 0.8], then the joint probability of two coins coming up heads would be 0.5 * 0.8 = 0.4 (which is greater than ¼th, assuming independence). If 10 coins are flipped, the probability is $0.5 * (0.8)^9 = .067$. This probability is still far greater than the one under statistical independence.

A violation of statistical independence is a serious problem, as it is a violation of the first axiom of statistical theory. For example, my own area of research is on how students construct and critique arguments during small-group discussions. In small-group settings, students influence one another, so, for example, if one student makes a counterargument, it becomes more likely that other students will do so as well, due to modeling effects and other factors. The probability that Student

(A) makes a counterargument is therefore not statistically independent from the probability that Student (B) will, if they are in the same discussion group. Analyzing whether some intervention, such as the use of a graphic organizer, increases the number of counterarguments, without adjusting for the lack of statistical independence, will make the occurrence of Type I errors much more likely. That is because the probability of extreme events (i.e., lots of students making counterarguments) goes way up when there is statistical dependence, so p values are likely to be seriously inflated. There are statistical techniques, such as multilevel modeling, that address the problem, but the technique's appropriateness in argumentation research is still being debated (given that certain sample sizes may be required). Further discussion of multilevel modeling is beyond the scope of this book, but the example illustrates why an understanding of basic probability theory is important.

EXPLORING THE CONCEPT

(a) Suppose one randomly selects 500 households and sends a survey on political attitudes toward female politicians to all the adult males in each household, and sends the same survey to all the adult females (e.g., one to a husband and one to a wife). Would the individual responses be statistically dependent? How might this fact affect the results? (b) How might sampling without replacement involve a violation of statistical independence? For example, what is the probability that if I shuffle a complete deck of 52 cards, the first two cards will be spades? Use Eq. 1.4 to calculate this probability.

Probabilities of Alternative Events

What is the probability that if I select one card from a full deck of 52 cards the card will be spades or clubs? This question relates to alternative events occurring (A or B) rather than joint events (A and B). The probability of alternative events is given by

The Prob(A or B) = Prob(A) + Prob(B) [if Prob(A) and
Prob(B) are mutually exclusive]. (Eq. 1.5)

Applying Eq. 1.5 to our question, we find that the Prob(spades *or* clubs) = Prob(spades) + Prob(clubs) = $\frac{1}{4} + \frac{1}{4} = \frac{1}{2}$. This result makes sense because half the cards will be black. The reason Eq. 1.5 works is because Event A (card being a spade) represents ¼ of the favorable possibilities, and Event B (card being a club) represents another ¼ of the favorable possibilities—so together, the probability that the selected card will be black is the sum of all the favorable possibilities (which is why we add). The background assumption is that the card cannot be both spades and clubs, that is, A and B cannot both occur—the events are mutually exclusive. If the events are not mutually exclusive, one should use Eq. 1.6:

$$\text{Prob}(A \text{ or } B) = \text{Prob}(A) + \text{Prob}(B) - \text{Prob}(A \text{ \& } B). \tag{Eq. 1.6}$$

For example, if one flips two coins, the probability that one or the other (or both) will come up heads is $\frac{1}{2} + \frac{1}{2} - \frac{1}{4} = \frac{3}{4}$, which is the correct answer. If one were to erroneously use Eq. 1.5 to calculate the probability, one would double count the possibility where both coins come up heads.

Summary

In summary, remember that if joint events are involved (involving *ands*), one should multiply (assuming statistical independence), and where alternative events are involved (involving *ors*), one should add (assuming the events are mutually exclusive).

Conditional Probabilities

The last type of probability to be considered is *conditional* probability. These take an IF . . . THEN form; for example, if a student is male, then what is the probability that the student will complete high school? Condition probabilities will be discussed in Chapter 4.

Some Probability Thought Experiments

We shall now progressively consider cases of increasing complexity.

Probability Distribution of the Two-Coin Example

In tossing two coins there are four possible outcomes.

			Number of Heads (H)
(1) H	H		2
(2) H	T		1
(3) T	H		1
(4) T	T		0

Each of the four outcomes is called a *permutation*. If one counts the number of hits (i.e., heads), outcomes (2) and (3) both reflect one head. Although these two outcomes reflect different permutations, they reflect the same *combination*, that is, a

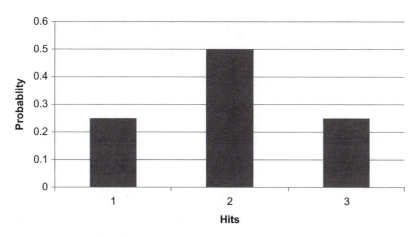

Figure 1.2 Probability histogram for the two-coin problem.

combination of one head and one tail. Combinations reflect the number of heads, but whether the head comes up on the first trial or the second does not matter. The order does matter with permutations.

Let us now make a histogram (frequency graph) of the different combinations (see Figure 1.2).

This figure is shape like an inverted U, except that it is not smooth. Many probability distributions—like the normal curve—have this upward concave characteristic. This is because there are more permutations corresponding to the combinations in the middle than at the extremes, making the more central possibilities "more likely."

One Hundred–Coin Example

Let us take a more extreme case. Suppose one throws 100 coins in the air. This resulting histogram is shown in Figure 1.3. (We have not yet covered the formula that was used in making this figure.)

The most extreme positive case is that all the coins will come up heads. This corresponds to only one permutation. The probability of each head is ½, and these are independent events, so the probability of their joint occurrence, as given by the multiplication rule, is $.5^{100}$. Not surprisingly, this number is extremely small:

.00000000000000000000000000000788861.

On the other hand, the probability of obtaining 50 heads out of 100 is much larger (8%).

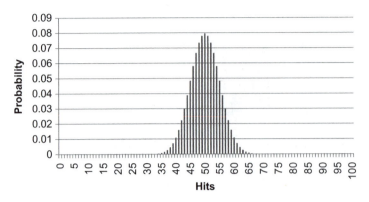

Figure 1.3 Probability histogram for the 100-coin problem.

This is because there are a lot of ways of getting a combination of 50 coins out of 100 (i.e., there are a lot of corresponding permutations). The first 50 coins could come up heads and the second 50 could come up tails; the first 50 coins could come up tails and the second 50 could come up heads; every other coin could come up heads, and so forth. There are also a lot more "random-looking" permutations, such as

    H H H T T T H T T H . . . ,

which is a random sequence that I generated in EXCEL. In fact, the number of permutations comes out to be $1.00891 * 10^{29}$. How this number is derived will be explained later in the chapter. Notice that in comparing Figures 1.2 and 1.3, the one with 100 coins is skinnier, if one thinks of the extremes in percentage terms (100 heads is 100% hits, tails is 0% hits). This result occurs because the more extreme outcomes become less likely as the number of trials increases. For example, the probability of 100% hits in the two-coin case is 25%, not $.5^{100}$ (as calculated above). Later we will see that this is the reason why larger samples provide more precise estimates.

The Binomial Formula

The binomial formula is used to measure the probability of a certain number of hits in a series of yes/no trials. It is a complicated formula, so I will introduce it bit by bit. In my example, I will use a biased rather than a fair coin to make the mathematics easier to follow. The Greek letter pi (π) denotes the probability of a hit. (Notice that the words *pi* and *probability* both start with the letter *p*. Statisticians often choose Greek letter for concepts that, in English, start with the same letter.) Let us suppose that the probability of this particular biased coin coming up heads is 70% ($\pi = 70\%$), and ask: What is the probability of tossing 10 coins (with $\pi = .7$) and obtaining 8 heads?

This question can be broken down into two subparts:

1. What is the probability of a *permutation* involving 8 heads out of 10 coins?
2. How many permutations make up a *combination* of 8 heads out of 10 coins?

Subquestion 1: What is the Probability of One "Permutation"?

Let us consider one permutation, where the first 8 coins come up heads: H H H H H H H H T T. Assuming the tosses are independent, the joint probability is 70% ∗ 70% ∗ 70% ∗ 70% ∗ 70% ∗ 70% ∗ 70% ∗ 70% ∗ 30% ∗ 30% $= .7^8 * .3^2 = \pi^8 * (1 - \pi)^2$. If we define k as the number of hits (in this case $k = 8$), and n as the number of trials (in this case $n = 10$), then the calculation becomes

$$\pi^k * (1 - \pi)^{n-k}. \tag{Eq. 1.7}$$

For the permutation considered previously, the probability is $.7^8 * .3^2 = 0.5\%$. Note that all the permutations associated with 8 coins out of 10 coming up heads are equally likely. For example, the permutation "H H H T H H H H H T" has the probability 70% ∗ 70% ∗ 70% ∗ 30% ∗ 70% ∗ 70% ∗ 70% ∗ 70% ∗ 30% $= .7^8 * .3^2 = 0.5\%$. The only thing that has changed is the order in which the coins come up heads or tails.

Remember that our overall goal is to find the probability of a *combination* of 8 hits out of 10. We shall see that this combination is associated with 45 different permutations, including the two shown above. The probability of one or another of these 45 different permutations occurring can be calculated by adding the individual probabilities together. Because the permutations are mutually exclusive, we can use the addition formula (Eq. 1.5). Because the permutations are equally likely, the calculation reduces to

$$45 * \underline{\pi^k * (1 - \pi)^{n-k}}. \tag{Eq. 1.8}$$

The underlined portion is the probability of one of the permutations occurring (from Eq. 1.7). What I have yet to address is how the 45 is calculated.

Subquestion 2: How Many Permutations Make up a Combination?

How many possible permutations are associated with a combination of 8 hits out of 10? The applicable formula, which is derived conceptually in Appendix 1.1, is

$$\binom{n}{k} = \frac{n!}{k!(n-k)!}. \tag{Eq. 1.9}$$

Here, $\binom{n}{k}$ is the notation for k hits out of n trials. On the right-hand side of the equation, the exclamation mark signifies a factorial product (e.g., $4! = 4 * 3 * 2 * 1$).

Substituting the values from our example yields

$$\frac{10!}{8!(10-8)!} = \frac{3,628,000}{40,320 * 2} = 45.$$

Putting Eqs. 1.9 and 1.7 together yields the *binomial formula*, which gives the probability of a certain number of k hits out of n trials:

$$Prob(k \text{ hits out of } n \text{ trials}) = \frac{n!}{k!(n-k)!} * \pi^k * (1-\pi)^{n-k}. \qquad \text{(Eq. 1.10)}$$

In our example, the terms are $45 * 0.5\% = 23\%$. The first term represents the number of permutations associated with a combination of 8 heads out of 10 coin tosses, and the second term reflects the probability of each permutation.

The calculation can also be performed in EXCEL with the following command:

$= \text{BINOMDIST}(k, n, \pi, 0),$

with the 0 indicating a noncumulative calculation (explained below). So $= \text{BINOMDIST}(8, 10, 0.7, 0) \rightarrow 23\%$. There is almost a one-quarter chance of obtaining exactly 8 heads with this particular biased coin.

EXPLORING THE CONCEPT

Concepts can often be better understood when applied to very simple examples. Use the binomial formula to find the probability of obtaining exactly one head in a toss of two fair coins.

Cumulative Versus Noncumulative Probabilities

Now suppose we wanted to calculate the chance of obtaining, out a toss of 10 coins, 8 heads *or less*. This probability could be derived by calculating the chance of obtaining *exactly* 8 heads or 7 heads or 6 heads or . . . 1 head or 0 heads. These are mutually exclusive outcomes, so according to Eq. 1.5 we can just add the probabilities: $23\% + 27\% + 20\% + 10\% + 4\% + 1\% + 0\% + 0\% + 0\% = 85\%$. (The 0% values are not perfectly equal to zero but are so small that these round to zero.) I calculated the individual probabilities using the EXCEL command "$=\text{BINOMDIST}(k, n, \pi, 0)$", but an easier way is to calculate the cumulative probability by changing the zero in the expression to one. (A one in the last place of the EXCEL command tells EXCEL to use the cumulative probabilities.) So we have

= BINOMDIST(8, 10, 0.7, 1) ➔ 85%.

Now I am going to change the problem a little. The question is, "What is the probability of obtaining 9 heads or more?" (again, out of a toss of 10 coins, with $\pi = 70\%$). This is calculated by subtracting the 85% from 100% (= 15%). Note that the probabilities of obtaining (a) 8 coins or less and (b) 9 coins or more must sum to 100%. This is because the two possibilities are exhaustive of all possibilities and mutually exclusive; one or the other must happen.

Now, instead of asking, "What is the probability of obtaining 9 hits or more?" we ask, "What is the probability of obtaining 8 hits or more?" This value is 100% minus the probability of obtaining 7 hits or less. The EXCEL syntax is

1 − BINOMDIST(7, 10, 0.7, 1) ➔ 38%.

Some students err by putting an 8 rather than a 7 in the formula, because the question asks about the probability of getting 8 heads or more, yielding an incorrect answer of 15%. This approach is incorrect because it is tantamount to calculating the probability of 8 heads or more plus 8 heads or less. These are not mutually exclusive possibilities, because if one obtains exactly 8 heads, both scenarios occur. According to the formula for alternative probabilities, one has to subtract out the probability of getting exactly 8 heads so as to not double count it (see Eq. 1.6). We previously calculated the probability of obtaining eight heads or less at 85% and for eight heads or more at 38%. The two possibilities sum to 123%, which is not a legal probability. We have double counted the probability of obtaining exactly 8 heads (which we calculated above at 23%).

This fact is so important that one should spend a minute now making a mental rule to remember that when you are asked cumulative probability problems that involve *X or more* (rather than *X or less),* you need to use $X - 1$ in the EXCEL formula. EXCEL will only give you cumulative probabilities for *X or less* problems.

The Bernoulli and Binomial Distributions

A probability distribution gives the probability of each and every mutually exclusive outcome of an event. The simplest probability distribution is the Bernoulli distribution, named after the 18th-century mathematician, Daniel Bernoulli. It is the probability distribution associated with a single, discrete event occurring. Table 1.1 presents the Bernoulli distribution for the situation where the probability of a biased coin coming up heads is 70%.

Table 1.1 A Bernoulli Distribution

Heads or Tails	Probability
Heads ($X = 1$)	70%
Tails ($X = 0$)	30%
Sum	100%

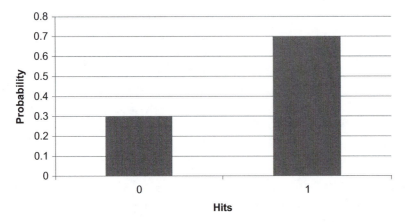

Figure 1.4 Bernoulli distribution for $\pi = 0.7$.

Probability distributions always sum to 100%. Because the different outcomes listed are exhaustive, one of them must happen, so the probability that one or another will happen is found by just adding all the probabilities up (the possibilities are mutually exclusive).

Charting the above Bernoulli distribution yields Figure 1.4. With a Bernoulli distribution, we just toss one coin. With the binomial distribution, we toss more than one coin. So my previous examples using 2, 10, or 100 coins are all associated with binomial distributions. With the binomial distribution, the tosses of the coins must be independent. The formal definition of the binomial distribution is "the distribution resulting from an independent series of Bernoulli trials."

We shall refer to this definition in later chapters, but for now it is just a fancy way of saying that there is more than one nominal event (e.g., we toss up more than one coin).

The shape of a binomial distribution will depend on the value of π (probability of a hit). A binomial distribution can be generated in EXCEL using the binomial formula, with the EXCEL syntax =BINOMDIST($k, n, \pi, 0$).

Figure 1.5 (Panel A) shows a binomial distribution when using a fair coin ($\pi = 50\%$) for different values of k with $n = 10$ (in other words, when we toss 10 fair coins up in the air). The distribution is symmetric. However, with our biased coin ($\pi = 70\%$), the distribution is skewed.

(A)

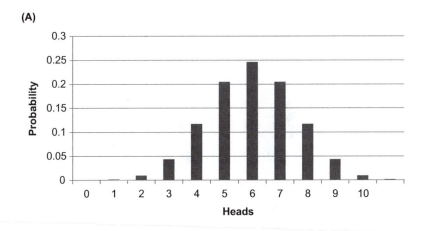

(B)

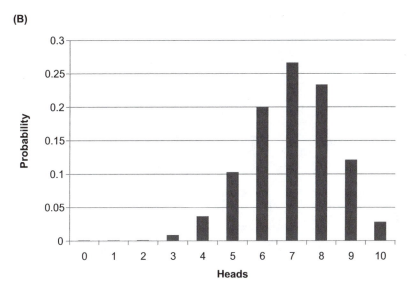

Figure 1.5 Binomial distribution ($n = 10$) for different probabilities of a hit. (A) $\pi = 50\%$. (B) $\pi = 70\%$.

Now we are going to make something interesting and important happen. As the number of trials increase, the distribution becomes more and more symmetric. Figure 1.6 shows that with $n = 100$ (Panel B), the distribution is extremely symmetric. (The distribution is in fact normal except for the fact that the variable "number of hits" is discrete rather than continuous.) Even at $n = 20$ (Panel A) the distribution is somewhat symmetric. Remember that as n increases, the probability of extreme outcomes decreases. So does the probability of somewhat extreme values, although not at the same rate. So in Figure 1.6 (with $n = 20$), consider the probability of moderately extreme values (such as $k = 8$, that is 40% heads), a value which contributes to the skew in Figure 1.5 [specifically, see Panel B at Prob($k = 4$)

(A)

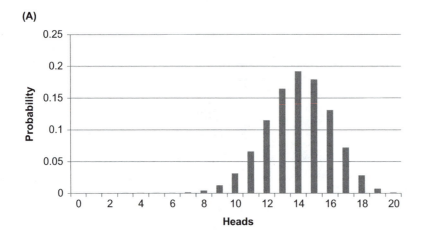

(B)

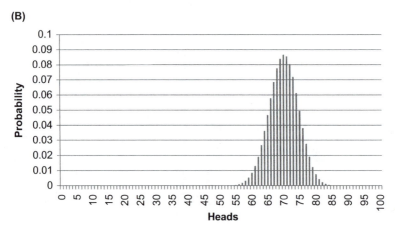

Figure 1.6 Binomial distribution for $\pi = 70\%$ for different values of n. (A) $n = 20$. (B) $n = 100$.

$= 40\%$]. The probability of 40% hits is becoming less likely with the larger sample, making the distribution more symmetric.

The Normal Approximation of the Binomial Distribution

Normal Distribution

A normal distribution is also known as a Gaussian distribution, and when standardized, a z distribution. The normal distribution has the property that about two-thirds of the distribution falls within one standard deviation (*SD*) of the mean, and most of the distribution falls within two standard deviations of the mean.

To *standardize* a score means to express the score in terms of the number of standard deviations from the mean. When a variable is normally distributed, use the following formula to compute z scores:

$$z = \frac{X - mean}{SD}.$$ (Eq. 1.11)

We shall use this equation repeatedly in this book.

One can compute the cumulative probability of different z scores using the EXCEL command: =NORMSDIST(z). So, for example, if a variable is normally distributed, the cumulative probability of obtaining a value of, say, 0.23, is 59%. With EXCEL syntax, =NORMSDIST(0.23) ➜ 59%.

It is frequently the case in statistics that we need to build a 95% confidence interval with 2.5% of the z distribution in each of the tails of the distribution. A 95% confidence interval around the mean can be built if we find the points that are about two standard deviations from the mean (1.96 SDs, to be precise). These would correspond to z scores of 1.96 and -1.96. Using EXCEL, one can verify that =NORMSDIST(-1.96) ➜ 0.025, and =NORMSDIST($+1.96$) ➜ 0.975.

EXPLORING THE CONCEPT

Some students wrongly expect that there should be a 95% chance of obtaining a z score of $+1.96$ because we are building a 95% confidence interval. Why is the probability of obtaining a z score of $+1.96$ or less 97.5% rather than 95%?

Suppose that X is normally distributed, has a mean of 3 and a standard deviation of 2. What is the probability of obtaining a raw score of 6? One could first compute the z score (1.5) and then consult EXCEL for the probability: =NORMSDIST(1.5) ➜ 93.3%. One could also use the unstandardized normal distribution and just enter the mean and standard deviation into EXCEL using the command =NORMDIST (this one does not have an "S" after "NORM"). Specifically, enter =NORMDIST(X, mean, SD, 1); the 1 requests a cumulative distribution. In the example, =NORMDIST(6, 3, 2, 1) ➜ 93.3%. Which command to use is one of personal preference.

The Normal Distribution Is Continuous, not Discrete

A discrete distribution is one where X cannot take on fractional values, whereas a continuous variable can take on any fractional value. A nominal (0, 1) variable is always discrete, whereas metric variables can be either continuous or discrete. For example, the probability distribution for the variable in Table 1.2 is discrete. Here,

Table 1.2 A Discrete Probability Distribution

X	Probability
1	15%
2	33%
3	7%
4	12%
5	33%
Total	100%

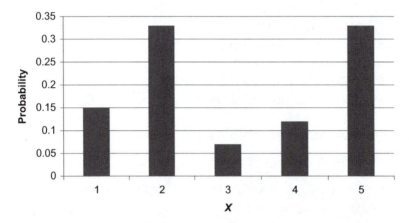

Figure 1.7 Example probability distribution for a discrete variable.

X only takes on the discrete values of 1, 2, 3, 4, and 5. On the other hand, if X could take on values such as 1.3, 2.67, or 4.3332, it would be continuous.

The probability distribution for the above discrete variable is displayed graphically in Figure 1.7. Ignoring the spaces between the bars, note that the area under the curve sums to 1.0. The area for the value $X = 2.0$ is given by the width of the bar (1 unit) times the height (which is 0.33), so the area is 0.33. The individual subareas sum to 1 because the probabilities sum to 1.

If X were continuous, the area for a value (such as 2.67, a number I arbitrarily chose) is given by the height (2.67) times the width, except that the width would be 0. This produces a paradox that the probability of any specific value of a continuous variable is 0. Obviously, this situation cannot occur, as the variable will take on specific values in particular cases. To resolve this paradox, we stipulate that for continuous variables, one can only meaningfully talk about the values of cumulative probabilities, for example, the probability of obtaining a value of 2.67 or less (or the probability of obtaining a value of 2.67 or more). The probability of obtaining exactly 2.67 is undefined.

The normal distribution is a continuous distribution. Z scores are therefore also continuous and the probability of obtaining a particular z score is cumulative. Therefore, while it is meaningful to write such expressions as $\text{Prob}(z \leq 0.23)$, it is not meaningful to write expressions such as $\text{Prob}(z = 0.23)$.

EXPLORING THE CONCEPT

In thinking about the normal curve and calculating the probability of z scores, why do we typically compute the area of the curve that is to the right (or left) of a certain point?

The Normal Approximation (the 100-Coin Example)

Suppose our problem is to compute the probability that if one tosses 100 fair coins, that 60 *or less* will come up heads. I present in this section two different methods for approaching this problem: (a) with the binomial formula, and (b) with the normal curve. The latter is more complicated with this particular problem, but the approach will prove to be useful on problems considered in other chapters.

The Binomial Formula

To use the binomial formula, simply use the EXCEL syntax:

=BINOMDIST(k, n, π, 1) (use 1 because we want cumulative probabilities).
=BINOMDIST(60, 100, 0.5, 1) ➔ 98.2%.

The Normal Approximation of the Binomial

It can be shown that as n approaches infinity, the binomial distribution approaches the normal distribution. When n is large, the differences between the distributions are negligible, so we speak about the normal *approximation* of the binomial. However, the binomial distribution is discrete, whereas the normal distribution is continuous, and this fact causes the approximation to be off a little, requiring what is known as a *continuity correction*. Let us set the continuity correction aside for a moment, as it will be the last step in our calculations.

We will first need to calculate a z score, and to do that we need to know the mean and standard deviation of the binomial distribution. Table 1.3 shows the mean and variances of both the Bernoulli and binomial distributions.

In our example, the mean is $n\pi$ (100 * 0.50 = 50). If we toss 100 coins in the air, the most likely outcome is 50 heads (see Figure 1.3). Although the likelihood of this occurring is only 8%, it is still the most likely outcome. It is certainly possible

Table 1.3 Means and Variances of the Bernoulli and Binomial Distributions

Parameter	Distribution	
	Bernoulli	Binomial
Mean	π	$n\pi$
Variance	$\pi(1 - \pi)$	$n\pi(1 - \pi)$

that we might get 49 or 53 coins coming up heads rather than exactly 50, but the number of heads is still likely to be around 50. The mean is also known as the *expected value.*

In our example, the variance is $n\pi(1 - \pi)$. This is sometimes written as *nPQ,* with P representing the probability of a hit and Q representing the probability of a miss. In statistics, by convention, Greek letters refer to population parameters and English letters to sample estimates of these parameters. Because we are not dealing with samples here, I use the Greek letters in writing the formulas.

To continue with the example, the variance is $n\pi(1 - \pi) = 100 * 50\% * 50\% = 25$. The standard deviation is the square root of the variance, or 5. The z score is then $z = \frac{X - mean}{SD} = \frac{60 - 50}{5} = 2.00$, $\text{Prob}(z \leq 2.00) = 97.7\%$. This result does not jive with the calculation using the binomial formula (98.2%) because we have not yet applied the continuity correction. When using z scores, it is assumed that the values are continuous. But we are here using discrete raw score values, for example, 62, 61, 60, 59, 58, and so forth. We want to find the probability of getting a value of $X \leq 60$. The area of the probability distribution corresponding to 60 is assumed with the normal-curve method to have a width of zero (because continuous variables are assumed), when in fact the bar has a width of 1. The area of the bar to the right of the midpoint corresponds to the probability excluded from the calculation with the normal-curve method. This method leaves out "half a bar" associated with the binomial formula method, which gives the exact value. The point of the continuity correction when using the normal approximation is basically to "add half a bar" to the probability estimate. We can do this if we use the value of $X = 60.5$ rather than 60 in the z-score calculation:

$$z = \frac{X - mean}{SD} = \frac{60.5 - 50}{5} = 2.10, \text{ Prob}(z \leq 2.10) = 98.2\%.$$

This result is equal to the result from using the binomial method.

The normal approximation is an important procedure for the material discussed in this book. It may seem unnecessary here to go through all the gyrations when we can calculate the exact probability with the binomial formula, but in many other situations we will not have recourse to the binomial formula or some other "exact" formula and therefore will need to use the normal approximation.

Continuity Corrections With X *or More Problems*

One last problem: What is the probability of obtaining 60 *or more* heads when toss-ing 100 fair coins? Once again, we should apply a continuity correction. However, there is a second adjustment we will need to make, namely, the one described a few pages back for "X or more" problems (specifically using $X - 1$ in the EXCEL formula). Therefore, instead of using $X = 60$ in the formula, we should use $X = 59$. But for the continuity correction, we also need to add a half a bar. Therefore, the correct X to use is 59.5:

$$z = \frac{X - mean}{SD} = \frac{59.5 - 50}{5} = 1.90, \ \text{Prob}(z \leq 1.90) = 97.1\%.$$

Subtracting from 100%, the probability of obtaining 60 or more heads is 2.9%.

EXPLORING THE CONCEPT

The probability of obtaining 60 or more heads is 2.9%. The probability of obtaining 60 or less heads is 98.2%. These probabilities sum to more than 100%; they sum to 101.1%. Can you explain why?

Problems

1. What is the scale of measurement for the following random variables?
    a) Someone's weight (in pounds).
    b) The state in which a person resides (Nevada, California, Wisconsin, etc.).
    c) Someone's IQ score.
2. What is the scale of measurement for the following random variables?
    a) A student's percentile ranking on a test.
    b) Self-report of the number of close friends an adult has.
    c) Political party affiliation.
    d) A child's categorization as normal, mildly emotionally disturbed, and severely emotionally disturbed.
3. One throws two dice. What is the probability of
    a) obtaining a 7 (combining the two numbers)? (Hint: Try to think of the different permutations that would produce a 7, and use the multiplication rule for joint events); and
    b) obtaining a 9?
4. In a single throw of two dice, what is the probability that
    a) two of the same kind will appear (a "doublet")?
    b) a doublet or a 6 will appear? (Hint: These are not mutually exclusive events.)
5. In three tosses of a fair coin, what is the probability of obtaining at least one head?

6. Three cards are drawn at random (*with replacement*) from a card deck with 52 cards. What is the probability that all three will be spades?

7. Three cards are drawn at random (*without replacement*) from a card deck with 52 cards. What is the probability that all three will be clubs? [Hint: When two events are not statistically independent, and A occurs first, Prob(A & B) = Prob(A) * Prob(B | A).]

8. Evaluate the following expression $\binom{4}{1}$. [Hint: Remember that $\binom{n}{k} = \frac{n!}{k!(n-k!)}$.]

9. Evaluate the following expression: $\binom{4}{3}$.

10. You toss four coins. Using the binomial coefficient formula, compute how many permutations are associated with a combination of two heads and two tails.

11. You toss four coins. Using the binomial formula, compute the probability of obtaining exactly two heads if the coin is fair. [Hint: The probability is given by $\binom{n}{k} * \pi^k (1-\pi)^{(n-k)}$.]

12. In a toss of four coins, use the binomial formula to compute the probability of obtaining exactly two heads if the coin is biased and the probability of a head is 75%.

13. Repeat the previous problem using EXCEL.

14. Using EXCEL, find the probability of obtaining two heads or *less* (again, with $n = 4$ and $\pi = 75\%$).

15. Using EXCEL, find the probability of obtaining two heads or *more* (again, with $n = 4$ and $\pi = 75\%$).

16. Chart the frequency distribution when $n = 4$ and $\pi = 75\%$.

17. You construct a science achievement test consisting of 100 multiple-choice questions. Each question has four alternatives, so the probability of obtaining a correct answer, based on guessing alone, is 25%. If a student randomly guesses on each item, what is the probability of obtaining a score of 30 or more correct?

    a) Use the binomial function in EXCEL. (Hint: Find the probability of 29 or less, and then subtract from 1.)

    b) Use the normal approximation method, without a continuity correction. You will need to calculate a z score, which means you will need to calculate the mean expected correct ($n\pi$) and standard deviation of this $\sqrt{n\pi(1-\pi)}$. (Hint: Find the probability of 29 or less, and then subtract from 1.)

    c) Use the normal approximation with a continuity correction. (Hint: Find the probability of 29.5 or less, then subtract from 1.)

18. If Y is a binomial random variable with parameters $n = 60$ and $\pi = 0.5$, estimate the probability that Y will equal or exceed 45, using the normal approximation with a continuity correction.

19. Let X be the number of people who respond to an online survey of attitudes toward social media. Assume the survey is sent to 500 people, and each has a probability of 0.40 of responding. You would like at least 250 people to respond to the survey. Estimate Prob($X \geq 250$), using the normal approximation with a continuity correction.

APPENDIX 1.1

LOGIC BEHIND COMBINATION FORMULA (EQ. 1.9)

The number of ways of combining n things so that there are k hits is $\binom{n}{k}$, for example, combining 100 coins so that there are 60 heads showing. The formula is $\binom{n}{k} = \frac{n!}{k!(n-k)!}$. This appendix explains the logic behind this formula.

First, consider a simpler scenario. Suppose three people (Harry, Dick, and Tom) go to the theatre. There are a number of possible seating arrangements:

HDT
DHT
HTD
DTH
THD
TDH

There are six permutations. Any one of the three people can fill the first seat. Once the first seat is filled, there are two people left, so either one could fill the second seat. Once the second seat is filled, there is just one person left, and so there is just one possibility for filling the third seat. The total number of possibilities is $3 * 2 * 1 = 6$. This is the factorial of 3 (3!). More generally, there are $n!$ distinct ways of arranging n objects.

Suppose now that Harry and Dick are identical twins. Some of the six permutations will no longer appear distinct. If H = D and substituting H for D, the six permutations become

HDT ➜ HHT
DHT ➜ HHT
HTD ➜ HTH
DTH ➜ HTH
THD ➜ THH
TDH ➜ THH

The number of distinct arrangements has been cut in half. We could just modify the formula by dividing 6 by 2, or more specifically, 2!. In general, when k of the objects (or people) are identical, the number of distinct permutations is given by $\frac{n!}{k!}$.

We need to divide by $k!$, and not just k, because there are $k!$ ways of arranging the k identical objects, and each of these correspond to a possibility that is duplicative of another possibility. There are therefore only three possible outcomes: HHT, HTH, and THH. If instead of people, we let H represent a coin toss coming up heads and T represent tails, we can see that there are three permutations corresponding to a combination of two heads and one tail, and that $\binom{3}{2} = \frac{3!}{2!(3-2)!} = 3$. More generally, $\binom{n}{k} = \frac{n!}{k!(n-k)!}$, where $(n-k)!$ is the number of misses. Although in the example, $(n-k)! = 1$, in more complex examples $(n-k)!$ may reflect duplicative possibilities, so we need to divide by this term as well as $k!$.

Reference

Nussbaum, E. M. (2011). Argumentation, dialogue theory, and probability modeling: Alternative frameworks for argumentation research in education. *Educational Psychologist, 46,* 84–106. doi:10.1080/00461520.2011.558816

CHAPTER 2

ESTIMATION AND HYPOTHESIS TESTING

In this chapter, we examine how to estimate a value of a population parameter such as a probability (π) or mean (μ). We will also examine the binomial test, which tests hypotheses about π.

Estimation

The Estimation Problem(s)

In teaching this topic in my course on categorical data analysis, I begin the lesson by asking students to answer on a piece of paper, "Have you ever been married?" This question asks about a nominal variable and therefore it is appropriate to calculate or estimate proportions. The goal of the exercise is to infer what percentage of students at the university as a whole have ever been married ($N = 30,000$). In other words, we need to estimate the population proportion. Doing so requires taking a sample and using sample statistics to estimate the population parameter. (A population parameter is some numerical feature of a population that we are trying to estimate.) The sample should be random, but in this exercise we pretend that the class represents a random sample. The population is all the students at the university.

By convention, population parameters are represented by Greek letters: in this case, pi (π) for a proportion. (For a metric variable, we would estimate μ, the population mean.) By convention, English letters (e.g., P or $\bar{X}$) represent sample statistics: Note that π, the population proportion, is also a probability. Therefore, if 60% of the students at the university have been married, then if one *randomly* draws an individual out of the population, there will be a 60% chance that she or he has been married. Suppose our sample consists of 20 students ($n = 20$). If X is the number of students in our sample who have been married, the expected value ($n\pi$) is the most likely value of X in the sample. This number also corresponds to an expected value of P of $\frac{12}{20} = 60\%$. This example shows why it is important to use random samples: The laws of probability make it likely that the characteristics of the sample will be representative (i.e., similar) to the population. In this case, if the sample contained 12 students who had been married, the P will be 60%, just like the population, where $\pi = 60\%$.

However, recall that this expected value is only the most likely value of our sample estimate P; it is possible that we could draw a sample with 11 or 14 married students. We should expect that we will obtain a sample of *about* 12 students who have been married but that our estimate will be off a little. The "off a little" reflects *sampling error,* the fact that even with a random sampling the sample might not represent the population perfectly. Because of sampling error, we always represent statistical estimates in the form of confidence intervals, for example .58 ± .02 (with 95% confidence).

Forming a Confidence Interval for a Proportion

In the above example, we stipulated that 60% of the population had been married. However, we typically will not know the population parameters, which is why we must estimate them. Suppose we take a random sample of 20 individuals, and 13 of them indicate that they have been married. (This category would include people who are presently married as well as those who have been divorced or widowed.) Then $P = 13/20 = 65\%$. P is our sample estimate of π. How do we form a confidence interval?

To address this question, we need to understand the concept of a *sampling distribution.* This concept is, next to probability, the most foundational one in all of inferential statistics. A sampling distribution for a sample statistic, such as P, given a particular sample size (e.g., $n = 20$), is the distribution of the statistic associated with all the different possible samples of size n that could be drawn from the population. For each such sample, there is a P. The sampling distribution of P is the distribution of all the possible Ps (one for each sample).

In the above example, we drew a sample with a P of 65%. If we were to draw another sample, we might obtain a P of 70%. If we were to draw a third sample, we might obtain a P of 55%. Theoretically, we could go on indefinitely, drawing samples and plotting the values of the sampling distribution. In practice, we typically draw only one sample, at most, two. We therefore only observe one or two points of the sampling distribution.

Although we never observe most of the points of the sampling distribution, we can still make some theoretical inferences about it. In order to construct a confidence interval, we need to make inferences about the sampling distribution's (a) mean, (b) standard deviation (standard error), and (c) shape.

It can be shown that the mean of the sampling distribution is π, the population proportion. (This claim will be demonstrated in the next chapter and in Appendix 2.1.)

EXPLORING THE CONCEPT

The mean of a sampling distribution is by definition the expected value. If the population mean π were 50%, why is the most likely value of P, from all the samples that could be drawn, 50%?

The population distribution is a Bernoulli distribution (see Figure 1.4 in the previous chapter), and so the graph of the population distribution will have two bars: one for $X = 1$ [with $\text{Prob}(X = 1) = \pi$] and one for $X = 0$ [with $\text{Prob}(X = 0) = 1 - \pi$]. Recall that the variance of a Bernoulli distribution is $\pi(1 - \pi)$. Because we are using P to estimate π, we can use PQ to estimate the population variance, where Q is the probability of a miss ($Q = 1 - P$).

However, our interest is in finding the standard deviation of the sampling distribution, not of the population distribution. (The former is known as the *standard error*, or *SE*). The applicable formula is

$$\textit{Estimated SE of a Proportion} = \sqrt{\frac{PQ}{n}}. \qquad \text{(Eq. 2.1)}$$

Appendix 2.1 presents the proof (see also Chapter 3).

EXPLORING THE CONCEPT

Conduct a thought experiment by answering the following questions: If your sample consisted of just one observation ($n = 1$) from a population of 30,000, (a) How many different samples could you draw? (b) Would the shape of the sampling distribution be just like the population distribution (i.e., Bernoulli)? (c) Would the standard deviations of the two distributions be the same?

Note from Eq. 2.1 that as n increases, the standard error decreases, meaning that the sampling distribution becomes "thinner." Eq. 2.1 implies that the standard error approaches zero as n approaches infinity. In fact, the standard error would approach zero as n approaches N, which is the size of the population. Technically, therefore, we should include this additional constraint in the formula by writing

$$\textit{Estimated SE of a Proportion} = \sqrt{\frac{PQ}{n} - \frac{PQ}{N}}. \qquad \text{(Eq. 2.2)}$$

When the population size is large, the second term is negligible, so it is typically left out of the formula for the standard error. One should use Eq. 2.2 when the population size is small.

EXPLORING THE CONCEPT

Suppose you sample the entire population where $n = N = 10$, and $P = 50\%$. To what value would Eq. 2.2 reduce? How many different samples could one draw out of the population? Does it make sense that the standard error would be zero?

In summary, the mean of the sampling distribution of P is π and the estimated standard error is $\sqrt{\frac{PQ}{n}}$. In regard to shape, whereas the shape of the *population* distribution is Bernoulli, the shape of the *sampling* distribution will be binomial when $n > 1$. This fact is illustrated in Figure 2.1. The sampling distribution for $n = 1$ is Bernoulli (same as the population distribution). For larger n ($n = 2$ and $n = 3$ are illustrated), the shape is binomial. Here is the rationale. The shape of the sampling distribution for each observation will be Bernoulli (identical to the population distribution), and a series of independent Bernoulli trials yields a binomial distribution. Thus in a sample of size $n = 2$, there are four possible outcomes, as shown in Table 2.1. Combining the middle two permutations where the number of hits is 1 yields the sampling distribution of P, shown in Table 2.2 and Figure 2.1. Just as the binomial distribution characterizes the probabilities resulting from flipping two coins, it also characterizes the distribution associated with selecting a sample of two.

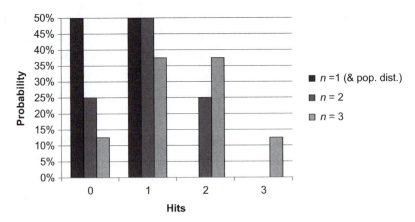

Figure 2.1 Bernoulli and binomial sampling distributions for varying sample sizes. For $n = 1$, the distribution is Bernoulli (and equal to the population distribution). For $n = 2$ and $n = 3$, the distributions are binomial.

Table 2.1 Sampling Distribution of P with $n = 2$, by Permutations

Possible outcomes

X_1	X_2	P	Prob(P)	Prob(P) if $\pi = 50\%$
1	1	100%	π^2	25%
1	0	50%	$\pi(1 - \pi)$	25%
0	1	50%	$(1 - \pi)\,\pi$	25%
0	0	0%	$(1 - \pi)^2$	25%

Table 2.2 Sampling Distribution of P with $n = 2$, by Combinations

X_{total}	P	Prob(P) if $\pi = 50\%$
2	100%	25%
1	50%	50%
0	0%	25%

In the previous chapter, we noted that the variance of a binomial distribution could be estimated with the expression nPQ. However, the astute reader may have noticed that I indicated above that the variance of the sampling distribution is $\frac{PQ}{n}$, where we are dividing by the sample size rather than multiplying by it. The difference is that in this chapter we were dealing with a proportion and not a straight number. In the previous chapter, we considered the example of tossing 100 coins and asking what is the probability that 60 coins will come up heads.

The mean chance expectation was $n\pi$, which, if $\pi = 50\%$, was 50 coins. Expressed as a proportion, the mean chance expectation is 50% (50 coins out of 100 coming up heads is 50%). The variance of the binomial distribution, when we are predicting the number of hits, is estimated by nPQ ($100 * 50\% * 50\% = 25$). The variance, when we are predicting a *proportion* of hits, is $\frac{PQ}{n}$ ($\frac{50\% * 50\%}{100} = 0.25\%$) . (The proof is in Chapter 3). In the former case, the standard error is $\sqrt{25} = 5$. In the latter case, it is $\sqrt{.25\%} = 5\%$, which is the same as in the former case but cast in percentage terms.

The Central Limit Theorem and the Normal Approximation

Although the shape of the sampling distribution of P is binomial, we saw in the last chapter that the sampling distribution can be approximated by the normal distribution when $n \geq 20$ and a continuity correction is applied. This is actually a special case of a more general, important theorem, the central limit theorem (CLT). The CLT holds that the sampling distribution of any random variable will approach normality as the sample size approaches infinity (see Technical Note 2.1). So if the sample size is large enough, the sampling distribution will be approximately normal. When we assume normality because the sampling distribution is so close to being normal, this is known as using a *normal approximation*. How large the sample needs to be for a normal approximation to be reasonable is a complex question that will be addressed later in this chapter.

For now, I wish to build the reader's conceptual intuitions about the CLT with another thought experiment. Suppose I take a sample of students at my university ($N = 30,000$), but my interest is now in a metric rather than a categorical variable, specifically, estimating the average height of males at the university. If we were to take a sample of $n = 1$ and drew an individual with a height of 6 feet, then for that sample, $\bar{X} = 6'$. If we drew someone of 5 feet, then $\bar{X} = 5'$. The distribution of possible $\bar{X}$'s would be identical to the population distribution. Because it is well known that the distribution of heights by gender in a large population is normal, it would follow that the sampling distribution is normal. This leads to the following important result, which holds even for larger samples ($n > 1$): *If the population distribution is normal, then the sampling distribution will be exactly normal.*

However, many if not most population distributions are not normal. Many are skewed—for example, the distribution of household income or the number of high school students in different schools who become professional actors. In addition, we usually have little information about the shape of the population distribution; although the sample can provide some information, we will nevertheless still be uncertain about the shape of the sampling distribution.

EXPLORING THE CONCEPT

Examine Figure 2.2 on the next page. Can you explain the difference between the empirical distribution (i.e., distribution of X in the data) and the sampling distribution? Which one provides us with a hint regarding the shape of the population distribution?

Fortunately, when we are uncertain about the shape of the population distribution, or when we know it is not normal, we can invoke the CLT if the sample is large enough. Suppose we want to estimate the average annual income of students at the university and (going to the other extreme) we sampled the entire population, so $n = 30,000$. There is only one possible sample we could draw, so the sampling distribution will have no variance, as indicated by Eq. 2.1. Graphically, it would be a spike, as shown in Figure 2.3 (light gray line in the middle). This spike is the central limit.

Now let us take the next step and give the sampling distribution a little variance. Suppose we sample all individuals in the population but one, so that $n = 29,999$. Because there are 30,000 people in the population, there are 30,000 different

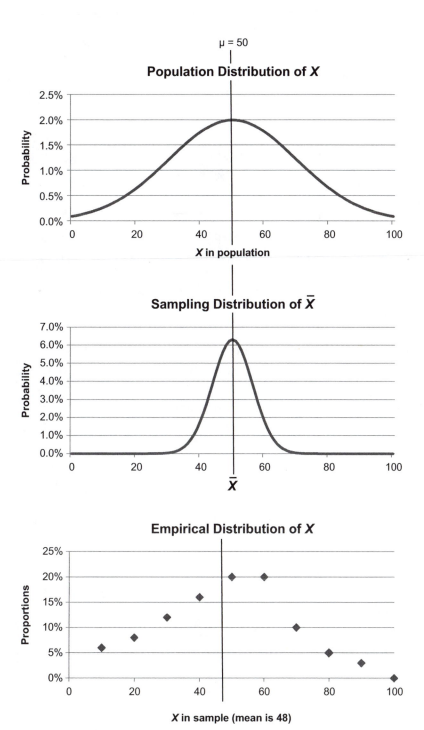

Figure 2.2 Empirical and population distributions of X and sampling distribution of $\bar{X}$. The population mean (μ) is 50 and the empirical mean in the sample ($\bar{X}$) is 48. The standard deviation of X in the population is 20 and in the sample is 19.51. The standard deviation of the sampling distribution (standard error) is 6.32.

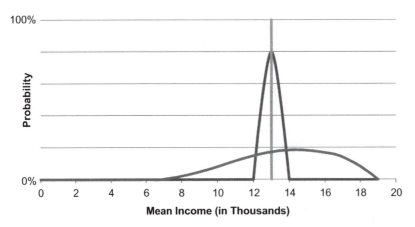

Figure 2.3 The sampling distribution becomes more symmetrical (and normal) as the sample sizes increase. Sample sizes are not shown. The line in the middle represents the distribution if the entire population is sampled ("the central limit").

samples that we can draw, each one corresponding to one person that we will omit from the sample. However, excluding one person from a sample of 30,000 is not going to have much effect; the variance (and *SD*) of the sampling distribution will be very small. The distribution will also be highly symmetrical, almost a spike.

The proof is too complicated to be included in the Appendices, but hopefully this conceptual demonstration will satisfy the reader. Graphically, Figure 2.3 shows the distribution becoming increasingly more symmetrical (and increasingly more normal) as the sample size increases. Suppose the skewed distribution in the figure represents the population distribution (and sampling distribution if $n = 1$). As the sample size increases, the sampling distribution will become more symmetrical (and normal) until it becomes a spike (i.e., the central limit).

The CLT holds that the sampling distribution will approach normality as n approaches infinity. However, in the above example, the central limit would be reached sooner than that, once the entire population is sampled (at $n = 30,000$). That difference will not matter, because 30,000 is still a very large number and—as shown below—the sampling distribution will approximate a normal distribution with much smaller sample sizes.

Requisite Sample Size

How big a sample is needed to be justified in assuming normality? This question is complex, because the answer depends on three factors:

- How close to normality is the population distribution? (Remember, if the population distribution is normal, so will be the sampling distribution.) The closer to normality the population distribution is, the smaller the sample needs to be. Conversely, the more skewed the population distribution is, the bigger the sample needs to be.

- What sort of statistical test is being conducted? With some tests, the sampling distribution approaches normality quicker than with others.
- How good an approximation is needed? If our estimates need to be accurate to, say, one-half percentage point, we can get by with a smaller sample than if we need precision to one-tenth of a percentage point.

There is an entire subfield of statistics, known as *asymptotic theory*, that is devoted to these questions. However, most researchers rarely have the time or expertise to obtain an exact answer to these questions and thereby fall back on rules of thumb. The rule of thumb that is used for estimating a population mean is that one needs a sample size of at least 20 or 30 for the CLT to apply (20 if the underlying population distribution is fairly normal, 30 if it is very skewed). However, some textbook authors put the lower bound at 15 rather than 20. One has to remember that these are not hard and fast rules but just rules of thumb (and the rules may vary somewhat depending on the type of test being considered). Using the normal approximation with a sample size of 15 is risky and makes the research open to criticism by potential journal editors and reviewers.

One can, however, use alternative methods with a small sample, such as binomial tests, that are based on an exact distribution and not on a normal approximation. These alternative tests are also often more statistically powerful in large samples when the sample is skewed (Tukey, 1960). Therefore, even when a researcher is justified in using a normal approximation (because the sample is large), that might not be the best choice on the grounds of statistical power (which is the ability to establish a result as significant). This point is a major one made in this book. Right now, however, we are just concerned with the issue of when the normal approximation may be justified.

Although the examples in this section have used metric variables, the same principles apply when categorical variables are used. There is, however, one major difference. The standard deviation of a sampling distribution is a function of the standard deviation of the population distribution (see Eq. 2.1). We do not generally know, however, what the population standard deviation is, so it must be estimated. With categorical procedures that involve estimating a sample proportion, the standard error $\left(\text{i.e., } \sqrt{\frac{\pi(1-\pi)}{n}} \right)$ is a direct function of π, which we estimate with P. With metric variables, when we are estimating a population mean $(\overline{X})$, we have to make a *separate* estimate of the population standard deviation. As a result, in large samples we assume the sampling distribution follows a t distribution rather than the standard normal distribution (z distribution). Hence the use of t tests with metric variables. The t distribution is explained more in the next chapter, but in categorical and nonparametric data analysis, z distributions are almost always used instead of t distributions. This is a good thing because the *SEs* associated with the *z* distribution are smaller, and so the statistical power is higher. (The higher standard errors associated with *t* statistics reflect the greater uncertainty stemming from estimating a second parameter.)

Example Confidence Interval (CI)

Now let us put this theory into practice. In the example with which this section started, we took a random sample of students and found that 65% of them had been previously married. Therefore, because *P* is our estimate of π, we estimated that 65% of the students in our population have been married. We also, though, want to form a CI around our estimate. We can conclude three things about the sampling distribution of *P*:

- The mean is π (and we are estimating this to be 65%).
- We estimate the standard error to be $\sqrt{\frac{PQ}{n}} = \sqrt{\frac{0.65 * 0.35}{20}} = 10.67\%$.
- The sampling distribution follows the *z* distribution, because $n \geq 20$.

For a 95% CI, the end points will be 1.96 *SE*s around the mean. [To see this for yourself, using EXCEL find =NORMSINV(.025) and =NORMSINV(.975); just type the expression into a cell.] Our CI will therefore be $P \pm 1.96 * SE$, or more specifically, $65\% \pm 1.96 * 10.67\%$, or $65\% \pm 20.9\%$. In conclusion, we can be 95% confident that between 44% and 86% of the students at this university have been previously married. This is because 95% of the possible CIs constructed through this procedure will contain the population mean.

An interval constructed in this manner is known as a Wald CI. Studies have shown that the Wald CI tends to be too inaccurate with extreme values of π (or *P*) and small *n*. Conventional wisdom is that $nP \leq 5$ or $n(1 - P) \leq 5$ makes the Wald CI inadequate. However, Brown, Cai, and DasGupta (2002) showed the actual coverage is unreliable (and can oscillate with *P*) even in large samples and even with non-extreme values of *P*. Chapter 14 covers alternative methods for constructing CIs.

If your sample size is less than 20 or does not otherwise meet the previously stated criterion, you can still make statistical inferences because you know the exact shape of the sampling distribution: It is binomial! Ideally, we would choose the ends of the CI so that the interval includes 95% of the distribution. This means that there will be 2.5% of the distribution in each tail. Constructing a 95% CI may be difficult to do exactly given the discrete nature of the variable, but we can shoot for including *at least* 95% of the distribution in our confidence interval.

In IBM SPSS, this interval can be calculated by going to "Nonparametric Tests" ➔ "One Sample," and on the settings tab, choosing customize and then the binomial test. Select the Clopper-Pearson and Jeffreys CIs. Click on the output and set the view tab, at the bottom of the page, to "Confidence Interval Summary View." The coverage of the Clopper-Pearson CI is at least 95% and conforms to the one described above. The Jeffreys CI contains a continuity correction, and its usage is preferable, as the interval is closer to being 95% in coverage. It also has some other desirable characteristics (see Technical Note 2.2). According to output from SPSS, the Jeffreys CI ranged from 43.5% to 91.7%. Both CIs are based on the binomial distribution and can be used in small samples. More explanation of these CIs is contained in Chapter 14 of this book.

A pattern that we will follow repeatedly in this book is to use exact tests in small samples and the normal approximation in large samples. Exact tests may be based on an exact distribution (not an approximation) or on exact probability calculations. For example, the probability of tossing two coins and obtaining one head can be determined exactly by using the binomial formula. However, if we toss a large number of coins, we can use the normal distribution to approximate the sampling distribution. In this case, statistical software like SPSS will label the solution as *asymptotic*, the asymptote being the central limit (such as μ or π) to which the distribution collapses as $n \rightarrow \infty$.

Hypothesis Testing

In statistical hypothesis testing, we start with a hypothesis about what the population parameter might be. For example, because students at my university tend to be older, I previously hypothesized that π was equal to two-thirds (66%). To test this hypothesis statistically, we formulate a null and alternative hypothesis, specifically,

H_0: π = 66%
H_a: π ≠ 66%

In this case, our research hypothesis is also the null hypothesis. By convention, the null hypothesis is always an equality, and the alternative hypothesis is always an inequality. The rationale is that we need to center a region of acceptance on a specific mean number under the assumption that the null hypothesis is true, so the null hypothesis has to be equal to a specific number.

To reiterate, one always starts by assuming the null hypothesis to be true. Under that assumption, the mean of the sampling distribution (which is π) can be set to 66%. This alleviates the need for having to estimate the mean, which is how hypothesis testing differs from estimation with confidence intervals. The second step is to set the critical significance level (α). It is typically set at 0.05, or 5%. The third step is to calculate critical values for the test. If our sample statistic is ≥ the upper critical value or ≤ the lower critical value, we then reject the null hypothesis and accept the alternative hypothesis. To find the critical values, construct a 95% interval around the hypothesized mean, in this case 66%.

To create the interval with a large sample ($n \geq 20$), one can use the normal approximation method. The mean of the sampling distribution is assumed to be 66%, so the standard error is $\sqrt{\frac{PQ}{n}} = \sqrt{\frac{0.66 * 0.34}{20}} = 10.6\%$, and the interval is

66% $\pm$ 1.96 $*$ *SE* (i.e., 66% $\pm$ 1.96 $*$ 10.6%). This expression reduces to 66% $\pm$ 20.8%. The upper critical value is 86.8%; the lower critical value is 45.2%. (Note that this is similar to the CI example in the previous section except that the interval is centered around H_0: $\pi = 66$ %.) If we were to observe a *P* of 45.2% or less, or 86.8% or more, we would reject the null hypothesis.

One-Tailed Tests With Critical Values: Large Sample

Suppose we want to conduct a one-tailed test, specifically,

H_0: $\pi = 66$%; H_a: $\pi > 66$%

This test assumes that the population parameter π cannot be less than 66%. One is only justified in making such an assumption if it is supported by strong evidence. For example, suppose a prior study found that the vast majority of students at the university are married. Based on this study, you firmly believe that π cannot be less than 66%. If *P* is less than 66% (say it is 60%), you would infer that this result is due to sampling error, not that the null hypothesis is wrong. In constructing the critical values, we want to keep the probability of a Type I error to 5% or less. (A Type I error is when we incorrectly reject the null hypothesis when it is true.) To do this, we specify a region of rejection corresponding to 5% of the sampling distribution. With a one-tailed test, we should place the entire 5% in the upper tail of the distribution. The question then becomes, what *z* value is associated with the 95th quantile of the *z* distribution? To compute this, one can use the inverse normal distribution function in EXCEL:

=NORMSINV(0.95) ➔ 1.65

EXPLORING THE CONCEPT

How does this procedure compare to the one for finding a CI? In finding a CI, which did we find first, the *z* value or the corresponding quantile value?

To find the upper critical value, calculate:

66% + 1.65 $*$ *SE* (= 66% + 1.65 $*$ 10.6%)
66% + 17.49% = 83.49%

Note that the upper critical value of 83.49% is less than the one for a two-tailed test, so a one-tailed test affords more statistical power; it is easier to reject the null hypothesis.

Probability Value Method

The p value (p) is the probability of obtaining your actual results if the null hypothesis is true. Note that it is not the same as P (the sample proportion).

To use this method, reject the null hypothesis if the p value is less than .05. This method is simpler than computing a critical value, so the p-value method is more widely used in practice. Most statistical packages compute p values, so one can simply examine the output to see which p values are above or below 0.05. However, keep in mind that if one is conducting a lot of statistical tests, some of these will likely be Type I errors (about 5% of them). If one is conducting 20 tests, at least one of them will likely be a Type I error if all the null hypotheses are true.

One-Tailed Calculation

Suppose we collect our sample and we obtain a P of 72%. To compute the p value, we first need the standard error (calculated above as 10.6%).

The next step is to calculate the z value, which is how far the observed value is from the mean in standard deviation (SE) terms:

$$z = \frac{P - mean}{SE} = \frac{P - 66\%}{10.6\%} = \frac{72\% - 66\%}{10.6\%} = \frac{6.0\%}{10.6\%} = 0.57.$$

The cumulative probability of obtaining a z value of 0.57 or less is 71.2%. (In EXCEL, =NORMSDIST(0.57) ➔ 71.57%.) The probability of obtaining a z value of 0.57 or more is simply $1 - 71.57\%$ or 28.43%. So in a journal article, you could report "$z = 0.57, p = 0.28$, one-tailed test," which means that the chance of obtaining your results or something more extreme if the null hypothesis is true is 28%. Since the p value is greater than 5%, the results are not significant. (Note: The previous calculations contain a rounding error. One can perform the calculations in EXCEL without accumulating a rounding error, yielding a $p = .285547 \approx .29$.)

Two-Tailed Calculation (Double It!)

To conduct a two-tailed test, one can simply double the p value from the one-tailed test if the sampling distribution is symmetric around the mean. This situation always holds when using the normal approximation method, because the normal distribution is symmetric. In this case, we could report "$z = 0.57, p = .57$." (Some statistical software packages such as SPSS only compute one-tailed p values for the binomial test, so if in doubt, check your computer results against manual calculations.)

EXPLORING THE CONCEPT

When using the critical values method, we split values in half (specifically, the 5% region of rejection is split into 2.5% per tail). When using the *p*-value method, why do we double the values rather than splitting them in half?

Small Samples

The *p*-value method can also be used in small samples. With $n < 20$, one cannot assume the sampling distribution is approximately normal, so it is inappropriate to calculate a *z* value. However, one does know the exact shape of the population distribution: It is binomial (and so will be the sampling distribution). One can then calculate a one-tailed *p* value.

For example, suppose we take a sample of 11 students and find that 8 of them had been previously married. *P* in this case is 72.7%. This value is higher than our hypothesized value ($\pi = 66\%$). Under the null hypothesis, the probability of obtaining a raw count of 8 or more (8, 9, 10, or 11) can be calculated in EXCEL as follows:

$$= 1 - \text{BINOMDIST}(7,11,0.66,1) \rightarrow 0.454, \text{ or } 45.4\%.$$

This *p* value is one-tailed and should only be reported if we are conducting a one-tailed test. We should only do this if we are very confident, on theoretical and empirical grounds, that a very low population parameter value is not possible and that, therefore, a very low estimated value could only be due to sampling error.

EXPLORING THE CONCEPT

Is the previously stated assumption reasonable in this example? If not, we should conduct a two-tailed test. Why are two-tailed tests more common than one-tailed tests?

For a two-tailed *p* value, one cannot simply double the one-tailed value because, unlike the normal distribution, the binomial distribution is not symmetrical unless $\pi = 50\%$. Here, the expected value is 7 (66% of 11). The sample value is 8, which reflects a deviation of one unit. To be as deviant in the opposite direction would be to obtain a count of 6 or less. The probability of this outcome is

$$= \text{BINOMDIST}(6,11,0.66,1) \rightarrow 30.59\%.$$

The overall, two-tailed *p* value is 76%. This value is far higher than 5%, so again we cannot reject the null hypothesis.

Table 2.3 Binomial Probabilities for $\pi = 66\%$

Number of hits	Probability
0	0.00
1	0.00
2	0.00
3	0.01
4	0.03
5	0.09
6	0.17
7	0.24
8	0.23
9	0.15
10	0.06
11	0.01

One could also find the critical values for rejecting the null hypothesis in EXCEL by computing the individual (noncumulative) probabilities for each possible outcome.

=BINOM.INV(11,0.66,0.025) ➜ 4
=BINOM.INV(11,0.66,0.975) ➜ 10

To reject the null hypothesis, one would need a count of 4 or less, or 10 or more. However, this proves to be too liberal, because if one consults the binomial probabilities in Table 2.3, the region of rejection sums to 11%, not 5%. A more conservative test would use 3 and 11 as critical values; here the region of rejection sums to 2%. This result is too conservative. Because of the discrete nature of the variable, we cannot specify critical values that sum to 5%, unless we consult the individual binomial probabilities, shown in Table 2.3.

Consulting these, we can define a 5% region of rejection with critical values of 4 and 11. A value equal or less than 4 or equal to 11 would be considered significant.

The Binomial Test

The example discussed in the previous section for small samples is an example of *the binomial test*. The binomial test examines the hypothesis that π is equal to a certain value, such as 66%. However, just because we failed to reject the null hypothesis does not provide very strong evidence that $\pi = 66\%$, because the evidence is consistent with many other hypotheses as well. Using the data presented in the previous section, if we then estimate what π is, our CI ranges from 39% to 94%. This interval is wide because we used such a small sample. With such little precision, what good is the binomial test?

Applications of the Binomial Test

Fortunately, statisticians have developed several applications of the binomial test that are more compelling than the previous example.

Is π Different From Another Population Mean?

Suppose that for high schools in a certain state, the statewide graduation rate is 70%. At the Excelier Academy, however, the graduation rate is 75% ($n = 100$). Is the graduation rate at Excelier significantly different from the statewide average?

To answer this question, we need to suppose that the students at Excelier represent a random sample from the general population. This assumption may not be reasonable, because Excelier is a charter school and the parents of the students specifically chose to send their children to Excelier. However, it is this very assumption that we are testing. The null hypothesis assumes that the students are a random sample; the alternative hypothesis reflects that the sample is not representative of the general population, either because of selection effects or because the teaching at Excelier makes the students special (i.e., less likely to drop out).

Because we are assuming that the students at Excelier represent a random sample, there is a sampling distribution with mean π. The null and alternative hypotheses are H_0: $\pi = 70\%$; H_a: $\pi \neq 70\%$. Let us set alpha to .05. P in this case is 75%. The one-tailed probability of obtaining P under the null, where $\pi = 70\%$ is

$$=1 - \text{BINOMDIST}(74, 100, 0.7, 1) \rightarrow .163.$$

However, we are interested in a two-tailed p value because our alternative hypothesis is two sided. Fortunately, because $n > 20$, the sampling distribution will be approximately normal and symmetrical, and therefore we can just double the p value. Specifically, $.163 * 2 = .326$. This is greater than .05, so we cannot reject the null hypothesis.

One could also use the normal approximation to calculate the p value. In performing a normal approximation, one has to apply a continuity correction, which involves adding or subtracting half a bar. So the question becomes, what is the probability of obtaining a count of 74.5 (and therefore a P of 74.5%)? The associated p value will be greater than that for $P = 75\%$, because it is easier to obtain 74.5% than 75% when $\pi = 70\%$. Continuity corrections always involve increasing the p value.

We next need to calculate a z value, but to do this we need to calculate the standard error under the null hypothesis. Because we are assuming that $\pi = 70\%$, the standard error is $\sqrt{\frac{\pi(1-\pi)}{n}} = \sqrt{\frac{0.70 * 0.30}{100}} = .046$. The z value is

$$z = \frac{P - \pi}{SE} = \frac{0.745 - 0.70}{0.046} = 0.98.$$

The Prob($z < 0.98$), calculated in EXCEL, is =NORMSDIST(0.98) ➔ .837. Remember that the p value is the probability of obtaining your results or something more extreme, in this case, a P of 75% or more. So we need to calculate the probability—under the null hypothesis—of drawing a z *greater than* (not *less than*) 0.98.

The Prob($z > 0.98$) = $1 - .837 = .163$.

Doubling this value to obtain a two-tailed p value yields .326, which is the same value produced by using the binomial formula.

In a Contingency Table, Is One Category Proportion Greater Than the Overall Proportion?

The following example relates to college graduation rates in five northern Midwestern states. Table 2.4 shows, by state, the number of students who graduated by summer 2009 from the cohort who entered college in 2003. The overall graduation rate is 45.4%. However, the graduation rate in Wyoming is 55.4%. We can use the binomial test to determine if the difference is statistically significant.

First, let us consider the meaning of this question. From a descriptive standpoint, Wyoming's rate for this cohort is different from the other states; the data show this. But from the standpoint of inferential statistics, we may want to draw a more general conclusion and to infer a probability from the proportion. In other words, we want

Table 2.4 Contingency Table: College Graduation Data in Five Northern Midwestern States for Fall 2003 Cohort

State	No. graduate[a]	No. not graduate	Total[b]	Graduation rate
Wyoming	769	619	1,388	55.4%
Idaho	2,718	3,687	6,405	42.4%
Montana	2,715	3,290	6,005	45.2%
North Dakota	3,289	3,721	7,010	46.9%
South Dakota	2,750	2,386	5,136	53.5%
Total	12,241	14,703	26,944	45.4%

[a]Graduated by summer 2009.
[b]Reflects cohort beginning fall 2003.
Source: The National Center for Higher Education Management Systems.

to ask if the probability of graduating from college in Wyoming is different from the overall average. The probability could be the same but, because of chance factors, more people graduated from college in Wyoming from this particular cohort.

Second, we need to formulate the null and alternative hypotheses. These can be written as follows: H_0: $\pi = 45.4\%$; H_a: $\pi \neq 45.4\%$, where π represents the probability of graduating in Wyoming. (Under the null hypothesis, we think of the actual data for Wyoming as representing one of many possible samples that could have been drawn.)

Because the cell sizes are large, we can use the normal approximation with a continuity correction. We want to find the probability of drawing a sample of $X = 55.4\%$ *or greater* ($X = 769$) if the null hypothesis is true (and $\pi = 45.4\%$). The number of trials is 1,388 (the size of the freshmen entering class in Wyoming); this quantity is analogous to the number of coins flipped in a coin-flipping example.

As explained in the previous chapter, for the continuity correction, we need to increase the probability by half a unit (continuity corrections always increase the p values), so we need to find the probability of obtaining $X = 768.5$ (55.4%).

EXPLORING THE CONCEPT

(a) Do you see why we subtract half a unit here rather than adding a half a unit? Which is more likely, $X = 768.5$ (55.3674%) or $X = 769$ (55.4035%) if $\pi = 45.4\%$? How does this result relate to the statement that a continuity correction always increases the p value? (b) Do you think the continuity correction would make much difference here? What about a problem where the numbers were smaller, say $X = 45$ and $n = 100$?

We next need to calculate a z value, and then, from that, the p value. But to calculate the z value, we need the SE, assuming the null hypothesis is true: $\sqrt{\frac{\pi(1-\pi)}{n}} = \sqrt{\frac{.454 * (1-.454)}{1,388}} = 1.34\%$. The z value is: $z = \frac{P-\pi}{SE} = \frac{55.4\%-45.4\%}{1.34\%} = 7.47$.

The Prob($z \geq 7.47$), calculated in EXCEL, is $= 1 - \text{NORMSDIST}(7.47) \rightarrow 4.0079110 * 10^{-14}$. We need to double the value for a two-tailed test, but because the p value is so incredibly small, doing so will make little difference. We conclude that Wyoming's graduation rate is significantly above average ($z = 7.47$, $p < .001$).

Although some textbook authors recommend conducting one-tailed binomial tests, I argue that a one-tailed test is inappropriate if we wish to keep the Type I error rate to 5%. Even though Wyoming's observed rate is above the regional average, we had no a priori reason to believe that Wyoming's graduation rate had to be above average. If the graduation rate were, say, 10%, we would be testing whether the state's graduation rate was significantly below the average, and we would be rejecting the null if the difference were large enough. So, conceptually we need to have rejection regions in both tails of the null distribution with the total area equal to 5% (or use two-tailed p values); however, the results would still be highly significant.

Table 2.5 Contingency Table: Members of a Club, by Ethnicity, that Give or Withhold Consent to Participate in a Psychology Experiment

Ethnicity	Consent	Withhold consent	Total	%
African American	4	1	5	80.0
Asian	5	2	7	71.4
Caucasian	9	3	12	75.0
Latino	9	17	26	34.6
Total	27	23	50	54.0

Source: Hypothetical data.

The results are so significant in part because the n is big. Let us now consider a case with a more modest n. Table 2.5 contains data on the number of individuals of a community club, by ethnicity, that give consent to participate in a psychology experiment. Note the percentage for Latinos is much lower than that for the other groups. To test significance, the null hypothesis is that the probability of a Latino member giving consent is $\pi = 54.0\%$, the overall table average. The observed proportion (P) is 34.6%. For the Latinos, $X = 9$ and $n = 26$. The probability of obtaining 9 or fewer hits out of 26 trials with $\pi = 54.0\%$ is given by the binomial function in EXCEL:

=BINOMDIST(9, 26, .54, 1) ➔ .037.

Because π is close to 50%, we can double this value for an approximate two-tailed p value of .07. (See Technical Note 2.3 for a more exact estimate.) We cannot therefore conclude that Latinos are less likely to give consent than other individuals.

Given that the sample size is greater than 20, one could also use the normal approximation method on this problem. First, correct for continuity by setting $X = 9.5$. (This correction adds a half unit and will make the p value larger.) Then $P = \frac{9.5}{26} = 36.5\%$. The expected value ($\pi$) is 54%. The $SE = \sqrt{\frac{54\%(1-54\%)}{26}} = 9.8\%$. $z = \frac{36.5\% - 54.0\%}{9.8\%} = -1.79$. Prob($z \leq -1.79$) = .037. Doubling this value yields the two-tailed p value of .07.

Conducting the Test in SPSS

The binomial test can also be implemented in SPSS. It is one of the options under Analysis ➔ Nonparametric. Before conducting the analysis, one has to enter the data. The variable "consent" is nominal—a person either gives consent or does not. It is customary to use dummy variables to represent binary nominal variables; a dummy variable is one that is coded 0 or 1. In our example, a 1 will represent cases where an individual has given consent. (Be sure to specify in the SPSS box, "Define Success for Categorical Fields," that a 1 indicates success. If you are not careful, the program might interpret a 0 as indicating success, which will yield an incorrect result. This is illustrated in the following *Exploring the Concept* box.) There are 50 data points, one for each individual in the club. Rather than entering 50 lines of data, one can define a second variable; in this illustration it is called FREQUENCY. One can then go to the DATA option and select WEIGHT CASES. The idea is to weigh the CONSENT variable by FREQUENCY. This is equivalent to having 17 zeroes in the database and 9 ones (for a total of 26 cases). Figure 2.4 presents a screen shot of this step.

In conducting the binomial test, set the hypothesized proportion to 0.54, because under the null hypothesis, 54.0% is the hypothesized value of π. This step is illustrated in Figure 2.5 using IBM SPSS (using the Settings tab and "Customize tests") under "Nonparametric Tests" ➜ "One Sample."

Please note that SPSS provides here a one-tailed p value. One has to double the value to obtain a two-tailed p value.

EXPLORING THE CONCEPT

Redo this example in SPSS coding 1 for those not giving consent, 0 for those giving consent. The p value is .167, which is incorrect (the p value should be .037). Using EXCEL, find the probability of obtaining 17 hits or more (rather than 9 hits or less). Can you give an explanation for why SPSS is providing a misleading value?

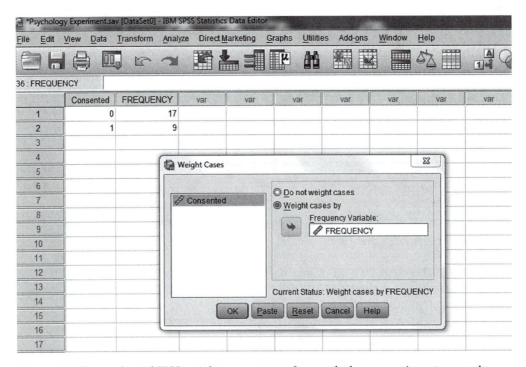

Figure 2.4 Screenshot of SPSS weight cases screen for psychology experiment example. The weight cases option can be found in the Data menu.

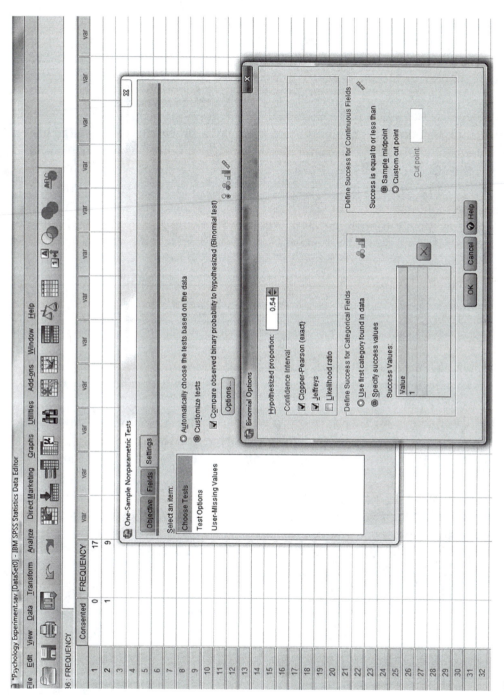

Figure 2.5 Screenshot of SPSS binomial test screen for psychology experiment example. The binomial test can be found under the nonparametric tests option in the Analyze menu.

Cox and Stuart Test of Trend

Trend tests apply to time-series data, to test whether the values of a variable are increasing (or decreasing) over time. The variable must be at least ordinal in nature. The Cox and Stuart test of trend is special case of a "sign test" (described in Chapter 7); a sign test is based on a binomial test. The idea is that given two points from the time series, the two points will either reflect an increase (coded with a plus sign) or a decrease (coded with a minus sign). If there is no change, the comparison is considered a "tie" and is not considered.

The null hypothesis is one of no trend, so that the probability of a "+" is equal to the probability of a "−." One is not more likely than another. The probability of a "+" is therefore equal to 50%. If we let π represent the probability of a plus, we have $H_0: \pi = 50\%$; $H_a: \pi \neq 50\%$. In the Cox and Stuart trend test, we split the time series in half and pair each point in the first half with one in the second half (in sequence). We then rate each pair as a + or a −, throwing out any ties. Assuming the null hypothesis to be true ($\pi = 50\%$), we use the binomial formula to find the cumulative probability of obtaining your results. Because the sampling distribution will be symmetrical with $\pi = 50\%$, double the probability to obtained a two-tailed p value.

Examples

In a certain city, the mortality rates per 100,000 citizens are given in Table 2.6 and also shown in Figure 2.6. Is there any basis for saying that the mortality rates display an increasing trend?

In the example, $n = 15$. Because this is an odd number, splitting the time series in half is more complicated. With an odd n, define the constant c as $(n + 1)/2$, which in this case is $(15 + 1)/2 = 8$. So in the first half of the time series, match each point at time t with the point at time $t + c$ (in this case, $t + 8$). Looking at the data, each of the seven pairs is rated as plus, because the second time point is greater than the first. What is the probability of obtaining 7 pluses? It is 1 minus the probability of obtaining 6 pluses or less: In EXCEL, $=1 - \text{BINOMDIST}(6,7,0.5,1)$ → 0.008. Doubling this value to obtain a two-tailed p value yields 0.016. This value is less than 0.05, so the results are significant. In a journal article we would write $T = 7$, $p < 0.05$. The letter T is for "test statistic." (Statisticians often use T generically to represent a test statistic if another symbol, such as z or χ^2, is not available or appropriate.) Here, T is the number of pluses.

Table 2.7 provides another example of the Cox and Stuart trend test (adapted from Walker & Buckley, 1968, but the example is fabricated). The data are for a single subject, Philip, an underachieving fourth grader with attention problems. The intervention was to provide Philip with positive reinforcement every time he was on task (the experimenter made a click, which earned Philip points that could be traded for a prize at the end of each session). The dependent variable was, for

Table 2.6 Cox and Stuart Test of Trend Example ($n = 15$)

Time (t)	X_t	t + 8	$X_{(t+8)}$	Change (+ or −)
1	17.3	9	17.7	+
2	17.9	10	20	+
3	18.4	11	19	+
4	18.1	12	18.8	+
5	18.3	13	19.3	+
6	19.6	14	20.2	+
7	18.6	15	19.9	+
8	19.2			

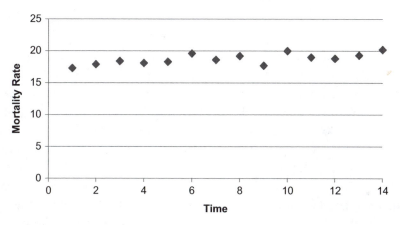

Figure 2.6 Mortality rates per 100,000 individuals over time (hypothetical data). For Cox and Stuart test of trend example.

each observation session, a percentage of attending behavior. Table 2.7 shows that Philip generally improved. However, there is one "minus" (decrease) in the data which caused the trend to be statistically nonsignificant.

Note that the Cox and Stuart test of trend can be used with small samples. There is no minimum sample size because the test does not rely on the normal approximation, just the binomial test. However, statistical power may be quite low, especially if one only has 5 or 6 points.

Other applications of this test demonstrate its flexibility. One can test for a correlation between two variables (X, Y) in a small sample by using the previous procedure, substituting X for time and ordering the data on X (from low to high). Then test whether there is an upward trend in Y. For example, suppose you collect

Table 2.7 Philip's Percentage of Attending Behavior by Session

Session (t)	X_t	t + 8	$X_{(t+8)}$	Change (+ or −)
1	70	9	91	+
2	68	10	91	+
3	71	11	70	−
4	73	12	95	+
5	80	13	96	+
6	84	14	100	+
7	80	15	98	+
8	85			

Notes: $T = 6$, $n = 7$, $p = .063$ (one-tailed), $p = .126$ (two-tailed). [In EXCEL, $=1 - \text{BINOMDIST}(5, 7, .5, 1) = > .063$.] Data adapted from graph (Figure 2) in Walker and Buckley (1968), for intervention phase only. Treatment was positive reinforcement. The score for Session 11 was changed from 89 to 70 for illustrative purposes.

information on the annual earnings of six married couples and want to see if there is a correlation between the earnings of the wife and those of the husband. The raw data are contained in the upper panel of Table 2.8. The table also shows the rank order of the wife's earnings. The variable will function just like "time" did in the previous examples. The lower panel of Table 2.8 puts the observations in order, based on the wife's earnings, and then forms matches. There are six pluses out of six; the one-tailed p value is .016 and the two-tailed p value is .032 (calculations are shown in the table).

One can also use the Cox and Stuart procedure to test for whether there is a cyclical pattern (e.g., whether there is a seasonal pattern in employment). For example, Table 2.9 shows levels of U.S. nonfarm employment averaged by quarter. It is hypothesized that employment during the fourth quarter (October–December) will be highest due to holiday spending and that employment during the third quarter (July–September) will be the second highest due to consumer spending on summer vacations and outdoor recreation. For these reasons, spending in the second quarter is expected to be third highest; consumers are expected to start spending more in spring than during the dreary winter months of January–March (First Quarter), where consumer spending is expected to be lowest. In summary, the farther a quarter is from the first quarter, the higher employment is expected to be.

To test for a cyclic trend, one merely rearranges the data so that the First Quarter data comes first, then Second Quarter, then Third Quarter, and then Fourth Quarter. Then one tests for a linear trend. To do that with the Cox and Stuart procedure, one splits the data in half (as shown in the bottom half of Table 2.9), and pairs the first

Table 2.8 Cox and Stuart Procedure: Hypothetical Data of 12 Couples' Annual Earnings (in Thousands)

Raw data

Couple	Wife's earnings (in $)	Wife's rank	Husband's earnings (in $)
1	47	6	50
2	24	1	48
3	38	3	51
4	28	2	41
5	39	4	39
6	50	8	90
7	74	10	84
8	77	11	95
9	48	7	56
10	40	5	86
11	60	9	74
12	100	12	120

Ranked and paired data

Wife's rank	Husband's earnings (in $)	Wife's rank	Husband's earnings (in $)	Sign
1	48	7	56	+
2	41	8	90	+
3	51	9	74	+
4	39	10	84	+
5	86	11	95	+
6	50	12	120	+

Notes: $T = 6$, $n = 6$, $p = .016$ (one-tailed), $p = .032$ (two-tailed). [In EXCEL, $=1-\text{BINOMDIST}(5, 6, .5, 1) .016$.]

half of the data with the second half. As shown in the table note, the p values were significant at the .05 level. (I recommend using the two-tailed values because one cannot rule out there being other types of cycles.) One can reject the null hypothesis and infer that there is a cyclical pattern in the data. The averages by quarter (not shown in table) are: Q1 = 131,442, Q2 = 133,438, Q3 = 132,757, and Q4 = 133,725. These averages are roughly consistent with our research hypothesis, in that employment is lowest in Quarter 1 and highest in Quarter 4. However, employment in Quarter 2 is higher than we originally hypothesized.

Table 2.9 Cox and Stuart Procedure for Cyclical Data: National Employment (in Thousands)

Raw data

Year	Jan.–Mar.	April–June	July–Sept.	Oct.–Dec.
2001	131,073	132,691	131,724	131,814
2002	129,114	130,895	130,187	131,168
2003	128,685	130,417	129,801	131,093
2004	129,120	131,894	131,584	133,142
2005	131,201	134,011	134,036	135,564
2006	133,905	136,482	136,265	137,693
2007	135,709	138,130	137,638	138,917
2008	136,380	137,919	136,691	136,172
2009	131,348	131,475	129,880	130,524
2010	127,880	130,469	129,761	131,168

Reformatted data

First half of data (Quarters 1 and 2)	Second half of data (Quarters 3 and 4)	Sign of difference
131,073	131,724	+
129,114	130,187	+
128,685	129,801	+
129,120	131,584	+
131,201	134,036	+
133,905	136,265	+
135,709	137,638	+
136,380	136,691	+
131,348	129,880	−
127,880	129,761	+
132,691	131,814	−
130,895	131,168	+
130,417	131,093	+
131,894	133,142	+
134,011	135,564	+
136,482	137,693	+
138,130	138,917	+
137,919	136,172	−
131,475	130,524	−
130,469	131,168	+

Notes: Data reflect nonfarm employment (Source: Bureau of Labor Statistics, retrieved from ftp://ftp.bls.gov/pub/suppl/empsit.compaeu.txt, revised 2010). Research hypothesis is that seasonal employment levels increase in each quarter (Quarter 4 > Quarter 3 > Quarter 2 > Quarter 1). Reformatted data compares Quarters 1 and 2 data with Quarters 3 and 4 data. $T = 16$ (number of pluses), $n = 20$, $p = .006$ (one-tailed), $p = .012$ (two-tailed). EXCEL Syntax for binomial test is $= 1 - \text{BINOMDIST}(15, 20, 0.5, 1)$.

Summary

In this chapter we first considered methods for estimating a population proportion as well as how to conduct binomial hypothesis tests. We then considered several applications of the binomial test, for example, whether a proportion is different from a population average or an overall average in a contingency table. Another application is the Cox and Stuart test of trend. This test can be used to test for a linear trend, although in Chapter 6 I will present a statistically more powerful test for evaluating linear trends (the Daniel's test). However, the Cox and Stuart test of trend is more flexible in that it can also be used to test for cyclical trends. The examples in this chapter also demonstrate, in simple terms, the logic and elegance of nonparametric tests and how these tests can be used with a small number of observations.

Problems

1. Twenty graduates of the University of Maryland's (UM) school psychology program took a certification exam, and 17 of them passed. Assuming the sample is representative of UM graduates, does this indicate that the probability of a UM graduate passing the certification exam is higher than the state average, which is 60%? Perform a one-tailed and two-tailed binomial test (do not use the normal approximation).

2. Senator Sly introduced into Congress a bill to subsidize health care costs for underinsured children. He claims that 80% of his constituents support the bill. As a policy analyst for Congressman Reyes (from the same state), you poll a random sample of 100 state voters and find that 63 support the bill. Conduct a binomial test to assess whether your evidence contradicts Senator Sly's claim. Use SPSS or an equivalent software package. Note whether the p value is one-tailed or two-tailed.

3. Using Table 2.4, test whether Idaho's graduation rate is significantly below the regional average. Conduct both a one-tailed and two-tailed test.

4. Using Table 2.5, if one averages the different consent rates together, the straight average is 65.25%. Is the Latino consent rate significantly below this unweighted average? (Perform a two-tailed test.) (Note: This problem reflects testing a different hypothesis.)

5. In Sun City, a survey identifies the following number of high school dropouts in different age ranges. Use an overall average dropout rate of 9.4%.
   a) Is there a significantly lower dropout rate in the Age 16–17 category than overall? (Conduct a two-tailed test.)
   b) Is there a significantly higher dropout rate in the Age 22–24 category than overall? (Conduct a two-tailed test.)

Age	Dropouts	Completers	Total	Dropout Rate
16–17	30	860	890	3.4%
18–19	66	694	760	8.7%
20–21	100	732	832	12.0%
22–24	148	1,049	1,197	12.4%
Total	344	3,335	3,679	9.4%

6. In Sun City, the high temperatures in July from 1990 to 2009 are contained in the EXCEL file *Temperatures.2.Problem_6*. Using the Cox and Stuart test, determine if there is a trend in the data.

7. The EXCEL file *Homicides.2.Problem_7* has data on the number of homicides per month in Mexico from drug wars, for the period 2007 to 2010. Use the Cox and Stuart trend test to determine if there is a significant positive upward trend. Because the initial number is so small, conduct only a one-tailed test.

8. As a class demonstration, the teacher in a large statistics class randomly chooses 12 students and asks them to estimate the average number of hours that they study per day. After the first exam, the teacher presents the class with the following data. Using the Cox and Stuart test adapted for correlation, determine if there is a significant association between exam scores and amount of studying. Conduct both a one-tailed and two-tailed test. The data are in the EXCEL file *Exams.2.Problem_8*.

9. The file *Turnout.2.Problem_9* has data on the percentage of the voting-age population who turned out to vote in federal elections. Use the Cox and Stuart procedure to test the hypothesis that turnout will be higher during presidential election years (PEY) than during off years. (In other words, test for a cyclical pattern.) Report the one-tailed and two-tailed p values.

Technical Notes

2.1 More precisely, the CLT states that the mean of an infinite number of random variables will have a normal distribution.

2.2 The Wald confidence interval can be biased because it is centered on P, not π. The latter is the mean of the sampling distribution, and 95% of the distribution of a normal curve is within two standard errors of π, not P. Although centering a CI around a population estimate is not a problem with continuous variables because the CI will still bracket the population mean 95% of the time, this situation does not hold with discrete data. Furthermore, the estimate of the standard error with the Wald method is based on P, rather than π, which creates further bias. The problem is that one does not typically know what the value of π is. Jeffreys (1946) addressed this problem by assuming a noninformative prior distribution for π reflecting a beta distribution with parameters $\alpha = 0.5$ and $\beta = 0.5$. (The distribution can be created and displayed in EXCEL using =BETA.DIST on X values from .01 to .99.) The distribution is U shaped because it asymptotes at 0 and 100% but otherwise is flat in the middle; prior to looking at the data, we believe that almost all values of π are about equally as likely. The Bayesian posterior distribution is $Beta(T + 0.5, N - T + 0.5)$, where T is the number of hits. The 95% Jeffreys CI is based on the 0.025 and 0.975 quantiles of this beta distribution, which can be looked up in EXCEL. Computer simulations show superior performance of the Jeffreys CI over the Wald (Brown et al., 2002). The Jeffreys CI is further discussed in Chapter 14.

2.3 In the consent example, for a more exact estimate of the two-tailed p value, compute the expected value $n\pi = 26 * 54\% = 14.04 \approx 14$. The deviation $X - nP = 9 - 14 = -5$. The other tail (as deviant in the opposite direction) is $\text{Prob}(X \geq 19) = 1 - \text{Prob}(X \leq 18) = .038$. The two-tailed p value is again .07.

APPENDIX 2.1

DERIVATION OF THE MEAN AND STANDARD ERROR OF THE SAMPLING DISTRIBUTION FOR P

Mean

It will be shown that the $E(P) = \pi$. Let the variable X be nominal (so $X = 1$ or $X = 0$). It is assumed that the sample is random, consisting of independent observations. Each observation has its own sampling distribution. We established previously that a sample of one observation will have a sampling distribution identical to the population distribution (with mean equal to π, the population proportion). Each of the sampling distributions will therefore also be identical to one another. We therefore have

(1) $E(x_1) = E(x_2) = E(x_3) = \cdots = E(x_n) = \pi$.

(2) $E(P) = E\left(\frac{\Sigma x_i}{n}\right) = \left(\frac{E(\Sigma x_i)}{n}\right) = \frac{(\Sigma(Ex_i))}{n}$ because $E(A + B) = E(A) + E(B)$ (see Chapter 3).

Because $E(x_i) = \pi$ from (1), (2) reduces to:

(3) $E(P) = \frac{(\Sigma E(x_i))}{n} = \frac{(\Sigma \pi)}{n} = \frac{n\pi}{n} = \pi.$ QED.

Standard Error

The standard deviation of the sampling distribution is known as the standard error. It will be shown that the *standard error* is equal to $= \frac{\sigma}{\sqrt{n}} \simeq \sqrt{\frac{PQ}{n}}$, where σ is the standard deviation of X in the population. The square (σ^2) is the population variance. Because the sampling distributions for each observation are identical, it follows that

(4) $Var(x_1) = Var(x_2) = Var(x_3) = \cdots = Var(x_n) = \sigma^2$.

Invoking the definition of a proportion $\left(P = \frac{\Sigma x_i}{n}\right)$, it also follows that

(5) $Var(P) = Var\left(\frac{\Sigma x_i}{n}\right) = \frac{1}{n^2}[Var(\Sigma x_i)]$.

The coefficient $\frac{1}{n}$ is squared because of the theorem $Var(aX) = a^2 Var(X)$ (see Chapter 3). The expression $Var(\Sigma x_i)$ can be reduced to

(6) $Var(\Sigma x_i) = [\Sigma Var(x_i)] = \Sigma \sigma^2 = n * \sigma^2$ (from adding σ^2 to itself n times).

Substituting (6) into (5) yields

(7) $Var(P) = \frac{1}{n^2}(n\sigma^2) = \frac{\sigma^2}{n}.$

Now, because the population variance σ^2 of a proportion is $\pi(1-\pi)$, it follows that

(8) $Var(P) = \dfrac{\pi(1-\pi)}{n}$, and

(9) $SE(P) = \sqrt{\dfrac{\pi(1-\pi)}{n}}$. The estimated standard error is therefore

(10) $\widehat{SE(P)} = \sqrt{\dfrac{PQ}{n}}$, where $Q = 1 - P$. QED.

References

Brown, L. D., Cai. T. T., & DasGupta, A. (2002). Confidence intervals for a binomial proportion and asymptotic expansions. *The Annals of Statistics, 30,* 160–201.

Jeffreys, H. (1946). An invariant form for the prior probability in estimation problems. *Proceedings of the Royal Society of London: Series A, Mathematical and Physical Sciences, 186* (1007), 453–461. doi:10.1098/rspa.1946.0056. JSTOR 97883

Tukey, J. W. (1960). A survey of sampling from contaminated normal distributions. In I. Olkin, S. Ghurye, W. Hoeffding, W. Madow, & H. Mann (Eds.), *Contributions to probability and statistics: Essays in honor of Harold Hotelling* (pp. 448–485). Stanford, CA: Stanford University Press.

Walker, H. M., & Buckley, N. K. (1968). The use of positive reinforcement in conditioning attending behavior. *Journal of Applied Behavior Analysis, 1*(3), 245–250.

CHAPTER 3

RANDOM VARIABLES AND PROBABILITY DISTRIBUTIONS

This chapter covers different types of probability distributions and some one-sample tests (e.g., for testing the shape of a distribution). It also introduces concepts useful for understanding statistical proofs. Except for the section on the chi-square distribution, readers with less interest in these topics could skip reading this more technical chapter without necessarily compromising their understanding of the remaining parts of this book.

With that said, I do encourage readers to peruse this chapter and acquire tools for understanding proofs. Understanding proofs is important for several reasons. First, proofs are the primary warrant for knowledge claims in statistical theory. Second, it is important to understand the initial steps in a proof that relate to a statistical test so that one can identify the assumptions underlying the test. (Failure to check if the assumptions are met sometimes has serious consequences, depending on the robustness of the test.) Third, proofs can sometimes enhance one's conceptual understanding of a test (i.e., why it works and under what conditions). Fourth, reviewing proofs helps build statistical literacy, enhancing readers' ability to review other statistical textbooks and articles. Finally, there is a beauty and elegance to proofs that help build critical thinking dispositions and condition the mind.

In this chapter, I review and extend the previous discussion of random variables (RVs) and probability distributions. First, however, the concept of a weighted average is reviewed, which will be needed to define the mean and variance of a random variable.

Random Variables

Weighted Averages

Consider the following set of numbers: 2, 3, 5, 5, 5, 7. The average is $\frac{\Sigma X_i}{n} = \frac{27}{6} = 4.5$. Instead of dividing the sum by 6, one could have divided each number by 6, producing $(\frac{1}{6}*2)+(\frac{1}{6}*3)+(\frac{1}{6}*5)+(\frac{1}{6}*5)+(\frac{1}{6}*5)+(\frac{1}{6}*7) = 4.5$. Note that the value of 5 occurs three times, so collecting those terms produces the following: $(\frac{1}{6} * 2) +$ $(\frac{1}{6} * 3) + (\frac{3}{6} * 5) + (\frac{1}{6} * 7) = 4.5$. One can also write the fractions as decimals, yielding: $(0.166 * 2) + (0.166 * 3) + (0.500 * 5) + (0.166 * 7) = 4.5$. This expression reflects the weighted average method of calculating a mean. The numbers 0.166 and 0.500 are weights attached to the different values of X; the weights reflect the percentage of times that a particular value occurs in the data set. The weights sum to 100%.

The weighted average method is used in many different situations. For example, when I teach a course, I use weighted averages to calculate course grades. Each test or assignment has a particular weight. Suppose a student, Marcus, received an 85 on his midterm exam and a 97 on the final. He was also required to complete a research paper, on which he received a 92. Suppose the midterm counts towards 35% of his grade, the final exam counts 40%, and the paper counts 25%. Using the weighted average method, his course grade would be: $35\%(85) + 40\%(97) + 25\%(92) = 91.55$. His overall course grade would be 91.55 (A-). The score 91.55 is his weighted average for the course. Next, I will show how the weighted average method is used to define the mean and variance of a random variable.

Random Variables Defined

A random variable (RV) is a variable with a probability distribution attached to it. Table 3.1 provides an example of a random variable.

The variable X takes on values at random. This does not mean that all values have an equal chance of occurring; the chance is given by the probability distribution. With the distribution shown in the table, X will take on the value of 3 about half the time and a value of 5 about 8% of the time. The values would have an equal chance of occurring only if the distribution was uniform, with each value having a probability of 20%. Figure 3.1 compares the shape of the probability distribution with a uniform distribution. The mean of a sample of data is given by the formula: $\frac{\sum_{i=1}^{n} X_i}{n}$.

Table 3.1 Example of a Random Variable (RV)

X	Prob(X)
1	5%
2	30%
3	50%
4	7%
5	8%
Total	100%

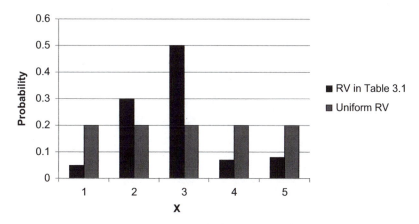

Figure 3.1 Distribution histograms of the random variable (RV) in Table 3.1 and a discrete, uniformly distributed random variable. In a uniform distribution, each value of the random variable has the same probability.

Table 3.2 Calculation of the Mean of a Random Variable

X	Prob(X)	X * Prob(X)
1	.05	0.05
2	.30	0.60
3	.50	1.50
4	.07	0.28
5	.08	0.40
Total	1.00	2.83 = E(X)

However, RVs relate to populations of infinite size, so n is infinite. This fact makes the formula meaningless. One can, however, compute the mean of a probability distribution with a weighted average:

$$E(X) = \Sigma[X_i * \text{Prob}(X_i)], \qquad \text{(Eq. 3.1)}$$

where $E(X_i)$ denotes the mean of the RV. In the above example, the mean would be calculated as shown in Table 3.2. The mean of this probability distribution is 2.83, that is, $E(X) = 2.83$.

The variance of a random variable is defined as follows:

$$Var(X) = \Sigma[(X_i - \bar{X})^2 * \text{Prob}(X_i)]. \qquad \text{(Eq. 3.2)}$$

In our example, the calculation of the variance is shown in Table 3.3. The variance is 0.86 and the standard deviation (square root) is 0.93, which is approximately

Table 3.3 Calculation of the Variance of a Random Variable

X	Prob(X)	$(X - \bar{X})$	$(X_i - \bar{X})^2$	$(X_i - \bar{X})^2 * Prob(X)$
1	.05	−1.83	3.35	0.17
2	.30	−0.83	0.69	0.21
3	.50	0.17	0.03	0.01
4	.07	1.17	1.37	0.10
5	.08	2.17	4.71	0.38
Total	1.00		2.83	0.86 = Var(X)[a]

[a]Table column sums to 0.87 due to rounding error.

one. The values two standard deviations plus or minus the mean creates an interval (with rounding) from 1 to 5 (3 ± 2). This encompasses the entire range of the distribution.

EXPLORING THE CONCEPT

The standard deviation is somewhat similar to the average deviation. (How does it differ mathematically?) If one divides the distribution in half at the mean, the point 1 average deviation above the mean gives a typical value for the above-average group. The interval between the points 1 average deviation above and below the mean would encompass 50% of the distribution. Why, then, in the case of the normal distribution does the interval between the points 1 standard deviation above and below the mean take in somewhat more of the distribution (68%)? Why would the interval 2 standard deviations take in somewhat less than the entire range (95.5% of the distribution)?

Linear Transformations

In performing statistical calculations, it is often necessary to add numbers to the values of random variables. If one were to add a constant k to each value of the RV, the mean would be increased by k. If one were to multiply each value of the distribution by k, the mean would be transformed into $k * \bar{X}$.

For example, suppose one were to add the value of 1.0 to each value of X, as shown in Table 3.4. All the values have been shifted up by one unit, and so has the mean, which increased from 2.83 to 3.83. The variance remains unchanged because the overall shape and spread of the distribution has not changed; only the location of the mean has changed. This fact is illustrated in Figure 3.2(A).

Table 3.4 Effect on Mean and Variance of Adding a Constant (1.0) to Each Value of a Random Variable ($Z = X + 1$)

Z	$Prob(Z)$	$Z * Prob(Z)$	$(Z - \bar{Z})^2$	$(Z_i - \bar{Z})^2 * Prob(Z)$
2	.05	0.10	3.35	0.17
3	.30	0.90	0.69	0.21
4	.50	2.00	0.03	0.01
5	.07	0.35	1.37	0.10
6	.08	0.48	4.71	0.38
Total	1.00	$3.83 = E(Z)$		$0.86 = Var(Z)$

(A)

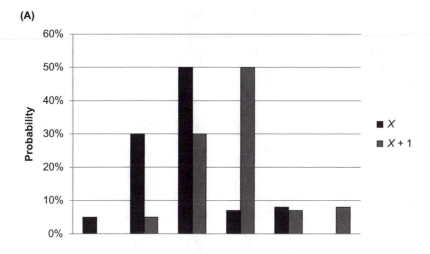

(B)

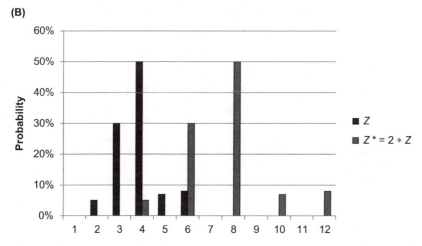

Figure 3.2 Linear transformations: Adding or multiplying a random variable by a constant. Panel A shows the effect of adding a constant ($k = 1$) to a random variable (X). The distribution is shifted to the right but the variance/standard deviation do not change $E(X) = 2.83$, $E(X + 1) = 3.83$, $SD = 0.93$). Panel B shows the effect of multiplying a random variable Z by a constant ($k = 2$), so that $Z^* = 2 * Z$. The standard deviation is doubled (from 0.93 to 1.86) and the variance is quadrupled.

On the other hand, if one were to *multiply* all the values by a constant rather than add them, the spread of the distribution would also change. For example, if one were to double all the values, the mean and standard deviation would double. Because the variance is the square of the standard deviation, the variance would become quadrupled, as indicated by the following formula:

$$Var(k * Z) = k^2 * Var(Z), \quad \text{(Eq. 3.3)}$$

which in this example is $Var(2 * Z) = 4 * Var(Z)$.

This formula would predict that the variance would be 3.44 if all the values were doubled, because $0.86 * 4 = 3.44$. One could also calculate the variance by hand, as shown in Table 3.5. The transformed variable is more spread out. The standard deviation (1.86), which is the square root of the variance, has doubled, the variance had quadrupled, and the values range from 4 to 12 rather than 2 to 6. This is illustrated in Figure 3.2(B).

The following equations summarized what has been concluded so far. The letters a and b refer to constants. I shall refer to $E(..)$ and $Var(..)$ as the expectation and variance operators, respectively.

$$E(a + bX) = a + bE(X). \quad \text{(Eq. 3.4)}$$

$$Var(a + bX) = b^2 * Var(X). \quad \text{(Eq. 3.5)}$$

In our examples, a is equal to 1 because we added 1 to all the values, and b is equal to 2 because we doubled all the values. Substituting these values for a and b yields $E[2 * (X + 1)] = 2 * E(X + 1) = 2 * E(X) + 2 * E(1)$. Since $E(1)$ is just 1, the expression reduces to $2 * 2.83 + 2 = 7.66$, which agrees with the mean calculated in Table 3.5. With the variance operator, any coefficient of X is squared and any constant is eliminated, so it follows: $Var[2 * (X + 1)] = 4 * Var(X) = 4 * 0.86 = 3.44$, which agrees with the variance result calculated in Table 3.5. (This is explained further in Appendix 3.1.)

Table 3.5 Effect on Mean and Variance of Multiplying a Constant (k) to Each Value of a Random Variable ($Z^* = 2 * Z$).

Z^*	Prob(Z^*)	$Z * Prob(Z^*)$	$(Z^* - \bar{Z}^*)^2$	$(Z^*_i - \bar{Z})^2 * Prob(Z^*)$
4	.05	0.20	13.40	0.67
6	.30	1.80	2.76	0.83
8	.50	4.00	0.12	0.06
10	.07	0.70	5.48	0.38
12	.08	0.96	18.84	1.52
Total	1.00	$7.66 = E(Z^*)$		$3.44 = Var(Z^*)$

Linear Combinations

The rules governing the linear combination of two random variables (rather than a linear transformation of just one) are

$$E(X + Y) = E(X) + E(Y). \tag{Eq. 3.6}$$

$$Var(X + Y) = Var(X) + Var(Y) + 2Cov(X,Y). \tag{Eq. 3.7}$$

When X and Y are assumed to be independent, the second equation reduces to

$$Var(X + Y) = Var(X) + Var(Y). \tag{Eq. 3.8}$$

Probability Distribution for Nominal Variables

Besides the mean and variance of an RV, one also needs to consider the shape of the probability distribution associated with an RV. Depending on the context, shapes may follow an established distribution, such as the Bernoulli or binomial (for nominal variables), or the normal, chi-square, t, or F distributions (for metric variables). Given the focus of this book, the distributions for nominal variables are discussed first. The other distributions are important for various large sample tests involving the central limit theorem (CLT), such as the normal approximation. The Bernoulli distribution is associated with a binary (0, 1) nominal variable.

For the Bernoulli distribution, the probability distribution, mean, and variance are shown in Table 3.6. The table demonstrates that the mean of the Bernoulli distribution is π and the variance $\pi(1 - \pi)$.

The binomial distribution is associated with an RV reflecting a linear combination (sum) of a series of independent Bernoulli trials. Each trial has an expected

Table 3.6 Derivation of Mean and Variance of a Bernoulli Random Variable

X_i	$Prob(X_i)$	$X_i * Prob(X_i)$	Mean deviation $(X_i - \pi)$	(Mean deviation)^2	(Mean deviation)$^2 *$ $Prob(X_i)$
1	π	π	$1 - \pi$	$(1 - \pi)^2$	$(1 - \pi)^2 * \pi$
0	$(1 - \pi)$	0	$0 - \pi$	$(-\pi)^2$	$\pi^2(1 - \pi)$
Total	$\pi = E(X)$[a]				$\pi(1 - \pi) = Var(X)$[b]

[a]Based on Equation 3.1.
[b]Based on Equation 3.2. Factoring $\pi(1 - \pi)$ out of the sum produces: $\pi(1 - \pi) * (1 - \pi) + \pi$; the last two π terms cancel one another out, producing $\pi(1 - \pi)$.

value of π. If T is the number of hits, the expected value of the binomial distribution can be derived as follows:

$$T = X_1 + X_2 + X_3 + \ldots + X_n, \text{ so}$$
$$E(T) = E(X_1) + E(X_2) + E(X_3) + \ldots + E(X_n),$$
$$= \pi + \pi + \pi + \ldots + \pi = n\pi.$$

Therefore,

$$E(T) = n\pi, \text{ when } T \sim \text{binomial.} \tag{Eq. 3.9}$$

The variance of each Bernoulli trial is $\pi(1 - \pi)$, so using similar reasoning,

$$Var(T) = n\pi(1 - \pi), \text{ when } T \sim \text{binomial.} \tag{Eq. 3.10}$$

Sometimes one needs to estimate the proportion of hits out of all trials. Let $P = \frac{T}{n}$. Then the variance is given by

$$Var(P) = Var\left(\frac{T}{n}\right) = \frac{Var(T)}{n^2} = \frac{n\pi(1-\pi)}{n^2} = \frac{\pi(1-\pi)}{n}. \tag{Eq. 3.11}$$

This proof demonstrates the validity of the formula used in Chapter 2 for estimating the standard error (specifically, Eq. 2.1).

The Normal Family of Random Variables

The normal distribution and distributions derived from it are used frequently in nonparametric and categorical data analysis, for example, in the context of the normal (asymptotic) approximation method described in Chapters 1 and 2.

Gaussian

The normal curve, also known as the Gaussian distribution, has certain important properties (e.g., about 95% of the distribution falls within 1.96 standard deviations from the mean). When standardized, it becomes the z distribution with a mean of 0 and an standard deviation of 1. Remember that the formula for z is

$$z = \frac{X - E(X)}{\sqrt{Var(X)}}. \tag{Eq. 1.11}$$

So taking the expectation (i.e., mean) of both sides, and separately the variance, yields

$$E(z) = E\left(\frac{X - E(X)}{\sqrt{Var(X)}}\right) = \frac{E(X) - E(X)}{\sqrt{Var(X)}} = 0, \text{ and}$$

$$Var(z) = Var\left(\frac{X - E(X)}{\sqrt{Var(X)}}\right) = \frac{Var(X) - 0}{Var(X)} = 1.$$

EXPLORING THE CONCEPT

(a) Conceptually, why is the mean of the standardized normal distribution zero? Do the positive values cancel out the negative ones? (b) In proving that the $Var(z) = 1$, the square root sign was dropped from the denominator. Why was this? Eq. 3.3 provides a hint [where $Var(X)$ is treated as the constant k].

Chi-Square Distribution

If one takes an RV with a standardized normal distribution and squares the values of the RV, the resulting probability distribution will, by definition, have a chi-square distribution (χ^2) with one degree of freedom (df). (Degrees of freedom are the number of chi-square random variables involved.) Now, if one squares all the values of a z distribution, all the negative values become positive. The mean of the squared distribution is the number of degrees of freedom (1 in this case), and the variance is $2 * df$ (2 in this case). Figure 3.3(A) displays this particular chi-square distribution.

In EXCEL, the quantiles of the chi-square distribution are given by the expression =CHIDIST(x, df). EXCEL has several statistical functions, some of which provide left-tail probabilities (including =CHISQ.DIST) and some of which provide right-tail probabilities (e.g., =CHISQ.DIST.RT). Earlier versions of EXCEL use the function =CHIDIST, which only provides a right-tailed probability. This function is used in this book because it is simpler. Some of the functions also provide noncumulative probabilities (see Technical Note 3.1 at end of this chapter).

What is the probability of obtaining a z value of -1.2 or less, and a chi-square value of 1.2^2 ($= 1.44$) or less? One finds that

= NORMSDIST(-1.2) ➔ 11.5%, and that
= 1—CHIDIST (1.44,1) ➔ 77%.

Note that CHIDIST(1.44,1) is a right-tailed probability and so it is subtracted from 1 to obtain the left-tailed probability.

EXPLORING THE CONCEPT

Will the probability of a z value less than -1.2, say -2.0, be included in the 77% (given that these values will be squared)? What range of z values would be included?

(A)

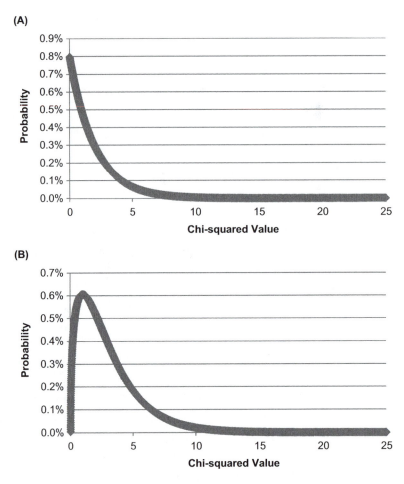

Figure 3.3 Graphs of representative chi-square distributions. Panel (A) is the distribution with one degree of freedom, Panel (B) for three degrees of freedom. The means are equal to the degree of freedom and the variances to 2*df*.

Now suppose one takes two independent, chi-squared random variables and adds them together. The resulting variable will also have a chi-square distribution, but with two degrees of freedom. Before, the probability of obtaining a chi-square value of 1.44 or *greater* was $1 - 77\% = 23\%$. The probability when combining two random variables is now 48.7% [according to EXCEL, =CHIDIST(1.44,2) ➔ 48.7%]. The probability has more than doubled. This result is because there is now more than one way to obtain a value of 1.44 from combining two random variables. For example, one variable might have a z score of 1.2 and the other a score of 0. (Note that the sum squared is 1.44.) Another possibility is that the first variable has a z score of 0 and the second of 1.2. A third possibility is that both variables had z scores of 0.85. Many other permutations are possible. The probabilities will increase with more degrees of freedom, and, for $\chi^2 = 1.44$, will reach a 99% asymptote with 8 degrees of freedom.

T *Distribution*

When one takes an RV with a z distribution and divides it by an RV with a chi-square distribution, the new variable will have a t distribution. This outcome is essentially what happens when one conducts a t test of the hypothesis that a mean (or a difference between two means) is equal to zero. The t value is

$$t = \frac{\bar{X} - (hypothesized\ mean)}{SE} = \frac{\bar{X} - 0}{SE}.$$ (Eq. 3.12)

The quantiles of the t distribution are given by the EXCEL syntax =TDIST(x, df, one- or two-tails).

F *Distribution*

If you take one chi-squared RV and divide it by another such RV, the resulting variable will be F distributed. Because each chi-square RV has degrees of freedom, for the F distribution one needs to specify both numerator and denominator degree of freedom to find the corresponding probability. The EXCEL command for the F distribution is =FDIST(X, $df1$, $df2$), where $df1$ is for the numerator, $df2$ is for the denominator.

So for example, the probability for a value of 1.44, with $df1 = 2$ and $df2 = 2$, would be =FDIST(1.44, 2, 2) ➜ 41%. The probability of obtaining an F value of 1.44 or greater is 41%. This value is somewhat less than 48%, which was the probability of 1.44 in the chi-square distribution with 2 df. If there were only 1 denominator df associated with the F value, the probability would be 50.8%, which is more comparable to the chi-square probability.

Summary

In summary, the chi-square, t, and F distributions are all derived from the standard normal (z) distribution. Because the z distribution is considered continuous, all these other variables are considered continuous as well.

A metric RV would be considered discrete if there are no fractional values, as in the example in the first section of this chapter. Although many authors use the terms synonymously, in my view, to be metric is not the same as being continuous. A metric (interval/ratio) variable can be continuous or discrete (e.g., the number of people in a room is a discrete, ratio variable). Continuous variables are a subset of metric variables. Nominal variables are always discrete.

Testing the Shape of a Distribution

The way that data are distributed in a sample is known as the *empirical distribution*. Although an empirical distribution will differ from the population distribution because of sampling error, it will give a hint as to the shape of the population distribution. We sometimes need to make an assumption about the shape of the population distribution, for example, that it is normal or follows some other standard type of shape. We may need to test, however, how closely the empirical distribution actually follows the hypothesized population distribution. In this section, two tests of distribution shape are examined: (a) the one-sample nonparametric Kolmogorov-Smirnov test, and (b) the one-sample chi-square test. Limitations and cautions associated with these tests are then discussed.

Kolmogorov-Smirnov Test

The Kolmogorov-Smirnov (KS) test is versatile in that it is a nonparametric test that can be used with small samples or with ordinal data (as well as with metric data). The test can be used with both continuous and discrete data (but is overly conservative in the latter case). In SPSS, the test can be found under "Nonparametric Test" → "One Sample." One can test whether the empirical distribution follows one of the following four shapes: normal, uniform (i.e., flat), Poisson, or exponential (aka, negative exponential). The Poisson distribution is typically associated with count data and is described in more detail in Chapter 11. The exponential distribution is often associated with times between events (e.g., times between when customers arrive at a bank teller station) and is used heavily in queuing theory. In psychology, learning curves for individual students typically have exponential distributions (Anderson & Tweney, 1997). The general shape of an exponential distribution is shown in Figure 3.4.

Arbesmana (2011) reported that the times between the rise and fall of empires could be described by an exponential distribution. Table 3.7 presents his data and Figure 3.5 displays the empirical distribution and the hypothesized shape.

To perform the KS test, let $S(X)$ be the cumulative sample distribution, and $F^*(X)$ be the cumulative hypothesized population distribution. One conceptually makes the following null and alternative hypotheses (see Technical Note 3.2 at end of this chapter for the exact hypotheses):

H_0: $F^*(X) =$ an exponential distribution
H_a: $F^*(X) \neq$ an exponential distribution

If the goal is to show that the population distribution is exponential, then one wants to fail to reject the null hypothesis. One can fail to reject the null hypothesis

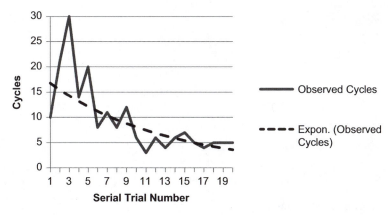

Figure 3.4 Learning curve graph.

Data from Ritter and Schooler (2001).

Table 3.7 Duration of Empires (in Centuries)

Empire	Duration	Empire	Duration
Yuen-Yuen (C. Asia)	0.3	Mitanni (Mesopotamia)	1.4
Assyrian	0.5	Late Period (Egypt)	1.9
Bactria (Indo-Iran)	0.6	Avar (Europe)	2.0
Lydia (Anatolia)	0.6	Kushan (Indo-Iran)	2.0
Phrygia (Anatolia)	0.6	Babylon-Hummurabi	2.0
Western Turk	0.7	Liu-Sung (China)	2.1
New Babylon (Mesopotamia)	0.7	Visigoth (Europe)	2.4
Hun (Europe)	0.8	Babylon (Mesopotamia)	2.5
New Assyrian	0.8	Ptolemaic (Africa)	2.9
Hykso (Syria)	0.8	Ch'in (China)	2.9
T'u Chueh Turk	0.9	Middle Empire (Egypt)	3.0
Gupta (India)	0.9	Achaemenid (Iran)	3.2
Maghada-Marurya (India)	0.9	Byzantine (Europe)	3.5
Urartu (Mesopotamia)	0.9	Andhra (India)	3.7
White Hun (Indo-Iran)	1.0	Rome (Europe)	4
Hsiung Nu Hun (C. Asia)	1.0	Hsia-Shang (China)	4
Old Assyrian (C. Asia)	1.0	New Empire (Egypt)	5
Akadia (Mesopotamia)	1.0	Old Empire (Egypt)	5
Saka (Indo-Iran)	1.2	Parthia (Iran)	7
Toba (China)	1.3	Elam (Mesopotamia)	10
Hittite (Anatolia	1.3		

Data retrieved from http://blogs.discovermagazine.com/gnxp/2011/08/the-fall-of-empires-as-an-exponential-distribution/.
Source: Arbesmana (2011).

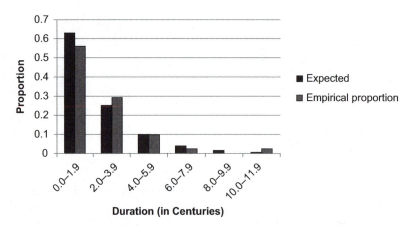

Figure 3.5 Comparison of the expected (exponential) distribution and the empirical distribution for Duration of Empires data (Arbesmana, 2011). Equation for expected values was $Y = e^{0.46 * X}$.

if the empirical distribution is fairly exponential. For each value of the RV, one can compute how much the empirical distribution departs from the hypothesized distribution. The KS test focuses on the largest of these differences, which is known as the *supremum.* The test statistic (*T*) is defined as

$$T = \sup_{n} | F^{*}(X) - S(X) |. \tag{Eq. 3.13}$$

One then finds the probability of obtaining *T* if the null hypothesis is true. The sampling distribution of *T* will be normal when $n > 40$. When $n \leq 40$, tables with exact *p* values can be consulted (e.g., those found in Conover, 1999); however, the information in these tables has been programmed into SPSS and other statistical packages, eliminating the need to consult tables. When the data in Table 3.7 was entered into SPSS and the KS test performed, the *p* value was .09. (The largest absolute difference, *T*, was 0.19.) The null hypothesis of an exponential shape is therefore not rejected because the test was not statistically significant, providing support that the distribution is exponential. Arbesmana (2011) therefore concluded that given the exponential shape, while a few empires last a long time, most do not (the mean duration is 200 years). His findings are of theoretical interest because the exponential distribution has been found to characterize survival times in other complex systems, such as in evolutionary science, how long species survive before they become extinct.

Returning to statistical theory, Smirnov extended the one-sample KS test to two samples. In the two-sample case, one tests whether the population distributions in two groups (e.g., experimental vs. control group) are the same. The group sizes should be equal. This test can be found in SPSS under "Nonparametrics" ➔ "Independent Samples." This test can be used with ordinal scales if the underlying ("latent") distribution is continuous. For example, one might rank the artistic quality of a group of students' paintings. The scale is ordinal, but it reflects an underlying continuous construct (degree of artistic talent).

Although the KS test can be used to test whether a distribution is normal, some other procedures have been developed for this purpose, for example, the Shapiro-Wilks test (available in SPSS by going to "Explore" ➔ "Plots" ➔ "Normality plots with tests"). In addition, visual inspection of distribution fit can be made with Q-Q plots, found in SPSS under the descriptive statistics menu, or by making a histogram and comparing the figure to the normal curve (this can be done under the FREQUENCY option). A crude but simple test for normality is to examine skew and kurtosis; a skew or kurtosis value twice its standard error in absolute value reflects significant nonnormality in samples with $n \geq 50$. (See Chapter 6 for an explanation of skew and kurtosis.)

One-Sample Chi-Square Test

Whereas the KS test requires at least ordinal data, the one-sample chi-square can be used with nominal data. The test can only be used with variables at higher levels of measurement if the variables are grouped and reduced to the nominal level.

The simplest application of the procedure is to test whether frequencies in all the categories are equal. Table 3.8 shows the distribution of births among the days of the week in 2009 from a U.S. sample ($n = 399$). The null hypothesis is that the probabilities of a birth are equal for each day of the week, so the expected frequencies are all the same ($\frac{n}{7} = 57$). The deviations of the observed from the expected frequencies are then computed; these sum to zero because the positive deviations cancel out

Table 3.8 One-Sample Chi-Square Test Data: Average Births by Day of Week ($n = 399$)

Day	Observed frequency (O)	Expected frequency (E)[a]	Deviation (O) – (E)	Deviation squared then weighted $\frac{(O-E)^2}{E}$
Sunday	37	57	−20	7.02
Monday	61	57	4	0.28
Tuesday	67	57	10	1.75
Wednesday	66	57	9	1.42
Thursday	64	57	7	0.86
Friday	62	57	5	0.44
Saturday	42	57	−15	3.95
Sum	399	399	0	$\chi^2 = 15.72$

[a] $E = \frac{399}{7} = 57$.

Notes: With $\chi^2 = 15.72$, $df = k - 1 = 6$, $p = .015$. Data based on population proportions, Table I-3, Martin et al. (2011).

the negative ones. To avoid a zero sum, all the deviations are squared (becoming positive) and then weighted by the expected frequency. This last step controls for the number of observations, except rather than dividing by actual cell sizes, the expected cell sizes are used. All the expected cell sizes must be at least five for the CLT to apply. Conover (1999) argued that the standard criterion of at least five expected cases in each cell is too stringent and instead recommended the following criteria: (a) all expected values > 0.5, and (b) at least one half of the expected values > 1.0.

The chi-square statistic is the overall sum, given by the following equation:

$$\chi^2 = \sum_{i=1}^{k} \frac{(O_i - E_i)^2}{E_i},$$

(Eq. 3.14)

where k is the number of categories. The χ^2 statistic is distributed chi-square with $k - 1$ df given a large enough sample ($n \geq 20$ and each expected cell value ≥ 5).

Table 3.8 shows that $\chi^2(6) = 15.72$. There are 7 days of the week, so $df = 6$. Consulting EXCEL, we find that =CHIDIST(15.72, 6) ➔ .015, so $p < .05$. We reject the null hypothesis and conclude there are statistically significant differences among the number of births by day. The data show that the number of births is much lower on weekends.

To conduct the test in SPSS, first create a categorical variable (on the variable tab, set "Type" to "String" and "Measure" to "Nominal"). In conducting the test, I called the categorical variable "Day" and then went to the Data menu and weighted the variable by the observed frequencies. I then went to "Nonparametric Tests," "One-Sample" to conduct the test. As the default, SPSS tests whether all categories have equal probability as the null hypothesis. One also has the option of testing other null hypotheses by inputting the expected probability for each category. So, for example, one might hypothesize that birth on weekends are half of those on weekdays. This particular model would involve the following expected values:

Saturday or Sunday: 33.25 births ($\pi_i = 8.3\%$),
Monday–Friday: 66.50 births ($\pi_i = 16.6\%$).

These values were derived as follows. Define "a unit" as the hypothesized number of births on one weekend day. Each weekday is therefore worth two units. There are 12 units in a week, under the null hypothesis. Dividing 399 (i.e., n) by 12, each unit is equal to 33.25 births. Inputting the values into SPSS yielded an insignificant p value of .73. This result implies that the data adequately fit the model (because we failed to reject the null hypothesis given the sample size).

When metric or ordinal variables are involved, one can use this procedure to test for a particular distributional shape. For example, one could test whether the empire duration data in Table 3.7 (and Figure 3.5) reflects an exponential distribution by reducing the data to nominal data (as was done in Figure 3.5) and deriving the expected value for the midpoint of each bin (i.e., category). Using the expected values shown in Figure 3.5 in a one-sample chi-square test yielded the following results: $\chi^2(5) = 3.92$, $p = .56$. As with the KS test, the null hypothesis of an exponential distribution is not rejected. The p value for a chi-square test is much higher, suggesting that the statistical power of the test is lower than the power of the KS test. (Statistical power is the ability to reject the null hypothesis.) Statistical power is often lost when data are reduced to the nominal level. This loss is not necessarily a bad thing when the goal is to not reject the null hypothesis. However, if one's sample is very small, there will be a general lack of statistical power and so a variety of null hypotheses could be accepted (or more precisely, not be rejected). This fact should be taken into account when interpreting the results.

A Note of Caution

Tests of distributional shape have been criticized, sometimes severely, on the grounds that they are very sensitive to sample size. In small samples, one may lack statistical power and tend to not reject many null hypotheses. In large samples, these tests can be sensitive to small departures from the hypothesized distribution. Some authors therefore recommend for this reason to use visual methods (e.g., to check for normality), but this introduces more human error into the process. Both numerical and visual methods have their place but must be used judiciously.

Tests of normality in particular are often used to decide whether or not to use a parametric or nonparametric test. This practice is unfortunate because with sufficiently large samples, the CLT shows that sampling distributions may be approximately normal even with nonnormal population distributions. Of more concern is how the power of a test may be affected; with skewed or kurtotic population distributions, nonnormal methods may be more statistically powerful, even if one is justified in using methods based on the normal distribution. Helping to convey this message is one reason that I decided to write this book.

Problems

1. Find the average household income for 2009 in the U.S. by calculating the weighted average for the following data. (To minimize rounding error, use weights to five decimal places.)

Ethnicity[a]	Income (in $)	No. Households[b]
White	51,861	95,489
Black	32,584	14,730
Asian/PI[c]	65,469	4,687
Hispanic	38,039	13,289
Total		128,195

[a]Race and Hispanic origin. Hispanics may be of any race.
[b]In thousands. Excludes those selecting more than one race.
[c]Includes Asians and Pacific Islanders.
Source: U.S. Census Bureau, Tables 691/693.

2. Using the following data, find the (weighted) average number of regional suicides in the U.S. for 1990–1994.

Region	Suicides	Population (in thousands)
Northeast	23,734	50,809
Midwest	34,492	59,669
South	57,509	85,446
West	38,709	52,789
Total	154,444	248,713

3. Rewrite each expression, moving the expectation or variance operator directly to the random variables, X or Y: (a) $E(3 + 7X)$, (b) $Var(3 + 7X)$, (c) $Var(3 - 7X)$, (d) $E(3 + 7X + 2Y)$, (e) $Var(3 + 7X + 2Y)$ if $Cov(X, Y) = 0$, (f) $Var(3 + 7X + 2Y)$ if $Cov(X, Y) = 1$.

4. Rewrite each expression, moving the expectation or variance operator directly to the random variables, X or Y: (a) $E(10 + 3X)$, (b) $Var(10 + 3X)$, (c) $Var(10 - 3X)$, (d) $E(10 + 3X + 3Y)$, (e) $Var(10 + 3X + 3Y)$ if $Cov(X, Y) = 0$, (f) $Var(10 + 3X + 3Y)$ if $Cov(X, Y) = 0.5$.

5. Consider the following random variable X:

X	Prob(X)
0	.19
1	.08
2	.23
3	.08
4	.33
5	.09
Total	1.00

a) Using the weighted average method, find the mean and variance of X. Use figures rounded to two decimal places.

b) Using Equations 3.4 and 3.5, find the mean and variance of $2X$.

c) Find the mean and variance of $Z = X + Y$, if X and Y are statistically independent, and the probability distribution of Y is as follows:

Y	Prob(Y)
0	.12
1	.17
2	.20
3	.25
4	.20
5	.06

6. Consider the following random variable X:

X	$Prob(X)$
0	.09
1	.14
2	.20
3	.47
4	.10
Total	1.00

a) Using the weighted average method, find the mean and variance of X. Use figures rounded to two decimal places.

b) Using Equations 3.4 and 3.5, find the mean and variance of $2X$.

c) Find the mean and variance of $Z = X + Y$, if X and Y are statistically independent, and the probability distribution of Y is as follows:

Y	$Prob(Y)$
0	.20
1	.30
2	.50

7. In the early 1930s, Karl Zener invented a test of extrasensory perception (ESP), which involved a deck of cards with five different symbols printed on each one (wave, square, circle, triangle, and cross). Behind a screen (blocking the participant's view), the experimenter would look at each card and the participant would guess which symbol the experimenter was viewing. Under the null hypothesis of no ESP, each participant's guesses were assumed to be random and independent.

a) Out of 80 trials, what is the mean proportion of hits that can be expected by chance alone?

b) What is the variance and standard error of the sampling distribution?

c) What is the mean and standard error if the random variable was the number, rather than the proportion, of hits?

d) At a .01 level of significance (two-tailed), how many hits would be required to reject the null hypothesis and conclude a participant has ESP? (To simplify the problem, do not include a continuity correction.) Round to the nearest whole number.

8. Suppose the Educational Testing Service (ETS) administered a standardized test consisting of 120 multiple-choice questions, each with four alternatives.

a) If a test taker answers the questions completely at random, what is the mean proportion of correct answers that can be expected by chance alone?

b) What is the variance and standard error of the sampling distribution?

c) What is the mean and standard error if the random variable was the number of correct answers, rather than a proportion?

9. The difference in proportions test evaluates whether the proportion of hits in one group is significantly different from that in another. (For example, is the incidence of a certain diagnosis such as depression significantly different among males than females?)

   a) A test statistic T is constructed by dividing the difference in sample proportions by the standard error, which is derived from estimates of the population variance, pooling the variances of the two groups. In a large sample, T would have what sort of distribution: (a) z, (b) t, (c) χ^2, or (d) F?

   b) What sort of distribution would T^2 have?

10. True or false? If a random variable with a chi-square distribution is divided by another RV with a normal distribution, the resulting variable will have an F distribution.

11. Learning-curve data for individual participants usually follow an exponential function (Anderson & Tweney, 1997). Ritter and Schooler (2001) presented data on a participant's time to perform a troubleshooting task in seconds. A computer also performed the tasks, with "decision cycles" used as a proxy for time. The computer model fit the participant's data extremely well.

    Below are data from the computer model.

Trial no.	Number of decision cycles	Trial no.	Number of decision cycles
1	10	11	3
2	21	12	6
3	30	13	4
4	14	14	6
5	20	15	7
6	8	16	5
7	11	17	4
8	8	18	5
9	12	19	5
10	6	20	5

    a) Perform a one-sample KS test (with statistical software) to evaluate whether an exponential function fits these data. Support your answer by reporting the KS test statistic (z) and p value.

    b) Plot the data. Do the data seem to follow an exponential function? If using EXCEL, you may wish to add an exponential trend line (under "Chart Tool" ➜ "Layout").

12. Using the data in the previous problem, perform a one-sample KS test (with statistical software) to evaluate whether the data would fit a uniform distribution. Report the KS z and p values.

13. Below are the heights and frequencies of 60 females from the College of Micronesia-FSM.

a) Perform a one-sample KS test (with statistical software) to determine if the sample follows a normal distribution. Report the KS z, p value, and your conclusion. Hint: Be sure to weight the data.

b) Optional: Create a histogram (by hand or using EXCEL, SPSS, or other computer software).

Height	Frequency
59.6	6
61.2	16
62.8	18
64.4	16
66.0	4

Source: Retrieved from http://www.comfsm.fm/~dleeling/statistics/notes06.html.

14. Below are the frequencies of 100 random numbers from 1 to 10 generated in EXCEL and then made discrete by being placed in "bins" numbered from 1 to 10.

a) Perform a one-sample KS test (with statistical software) to determine if the sample follows a uniform distribution. Report the KS z, p value, and your conclusion. Hint: Be sure to weight the data.

b) Create a histogram (by hand, or using EXCEL, SPSS, or other software).

c) EXCEL uses a uniform distribution to generate continuous random numbers. The numbers without binning followed a uniform distribution (KW $z = 0.98$, $p = .16$). What effect might binning have on the uniformity of the distribution, and why? Data: (Number: Frequency) (1: 11) (2: 8) (3: 13) (4: 17) (5: 6) (6: 10) (7: 13) (8: 8) (9: 4) (10: 10).

15. For the data in the previous problem, perform a one-sample chi-square test with statistical software, testing for equal probabilities among categories. Be sure to code the variable as nominal (e.g., in the SPSS measures list).

a) State the chi-square, degree of freedom, p value, and your conclusion.

b) Use EXCEL to find the p value for the chi-square test (show your work).

c) Which test is more statistically powerful, the one-sample chi-square or KS test? What evidence from these data support your conclusion?

16. When the raw data in the two previous problems was binned into four equal groups, with cut point at 25, 50, and 75, the following data resulted: (Bin, Frequency): 1: 27, 2: 28, 3: 28, 4: 17. Answer the same questions as in the previous problem.

17. Optional: Use EXCEL to generate at least 100 random numbers [e.g., using =RAND()]. Copy and paste "values" so the random numbers won't change. Copy and paste the data into an SPSS data file. Go to "Transform" → "Visual Binning" and select "Make Cutpoints" and "Equal Width Intervals." Divide your data into four bins (use three cutpoints) and then test for a uniform distribution using the chi-square and KS tests.

18. Optional: Using the 100 random numbers from the previous problem, calculate or generate the following using EXCEL:

    a) A normal distribution. Calculate the mean (=AVERAGE(array)), standard deviation (=STDEVP(array)), and z values (using =STANDARDIZE, using F4 to make absolute cell references for the mean and standard deviation so you can copy the cell formula to other cells). It should look something like =STANDARDIZE(B3,$B$104,$B$105). Then, for these 100 random z values, calculate the noncumulative probabilities, using =NORM.DIST(cell reference, 0). Then select both the z values and probabilities, and insert a scatterplot. (Under chart tool design, click "Select Data" to make sure your selections are correct.)

    b) A distribution of your choice. EXCEL can generate the following distributions: beta, binomial, chi-square, exponential, F, gamma, hypergeometric, lognormal, negative binomial, normal, Poisson, t, and Weibull (search for "statistical functions" in HELP). Look up the distribution on the Internet and in EXCEL HELP and report whether the distribution is applicable to (a) nominal, ordinal, and/or metric variables, (b) discrete or continuous, (c) the number and type of parameters that must be specified, and (d) the type of phenomena the distribution tends to characterize.

Technical Notes

3.1 The probability mass function measures the height of the probability density function at a particular point. EXCEL uses these values to provide noncumulative probabilities, although technically these are not probabilities when continuous variables are involved, because the probability is given by the area under the curve at a particular point, which is zero (because of zero width).

3.2 For the KS test, the null hypothesis is that the actual population distribution, $F(X)$, is equal to a particular, fully specified distribution, $F^*(X)$. To be fully specified means that it has a particular shape (e.g., normal), with certain parameters specified (e.g., mean and variance). The null and alternative hypotheses are therefore H_0: $F(X) = F^*(X)$ for all X; H_a: $F(X) = F^*(X)$ for at least one value of X. In SPSS, the parameters may be estimated from the data, because what is being tested is distributional shape. (Technically, this entails using modified KS tests known as the Lilliefors tests, in the case of normal and exponential distributions.) Other modified tests include the popular Shapiro-Wilk test for normality and the Anderson-Darling test, which places more weight on the tails of a distribution.

APPENDIX 3.1

LINEAR COMBINATION OF TWO RANDOM VARIABLES

I demonstrate, through a simulation (tossing two coins), that

$E(X + Y) = E(X) + E(Y)$.
$Var(X + Y) = Var(X) + Var(Y) + 2Cov(X, Y)$.

The example involves nominal variables, but the same principles apply to metric variables. Assume first that the tosses are independent. Let X be the outcome of the toss of the first coin and Y the second. Let $Z = X + Y$ (Z will be the total number of heads). The permutations and probabilities are

X	Y	Z = X + Y	Prob(Z) = Prob(X) * Prob(Y)
0	0	0	.25
1	0	1	.25
0	1	1	.25
1	1	2	.25

Using the weighted average method, the mean is $(0 * .25) + (1 * .50) + (2 * .25) = 1.0$. This result also follows from Equation 3.4: $E(Z) = E(X) + E(Y) = .5 + .5 = 1$. The probability function for Z is more spread out than for X and Y. The range is greater (0 to 2) and the variance is $Var(Z) = Var(X) + Var(Y) = .25 + .25 = 0.50$.

Now assume there is complete statistical dependence between X and Y, so that $Prob(Y \mid X) = 1.0$. The permutations and probabilities are now

X	Y	Z* = X + Y	Prob(Z*) = Prob(X) * Prob(Y | X) = Prob(X) * 1
0	0	0	.50
1	1	2	.50

According to EXCEL, $Cov(X,Y) = 0.25$ [=COVAR(array1, array2) ➜ 0.25]. Using Equation 3.7, the variance is $Var(X) + Var(Y) + 2Cov(X,Y) = 0.25 + 0.25 + 2(.25) = 1.0$. This result is double the variance under independence. Conceptually, the variance is greater because when $Cov(X,Y)$ is positive, extreme outcomes are more likely.

References

Anderson, R. B., & Tweney, R. D. (1997). Artifactual power curves in forgetting. *Memory & Cognition, 25,* 724–730.

Arbesmana, S. (2011). The life-spans of empires. *Historical Methods: A Journal of Quantitative and Interdisciplinary History, 44,* 127–129.

Conover, W. J. (1999). *Practical nonparametric statistics* (3rd ed.). New York, NY: John Wiley & Sons.

Martin, J. A., Hamilton, B. E., Ventura, S. J., Osterman, M. J., Sirmeyer, S., Mathews, T. J., & Wilson, E. (2011). Births: Final data for 2009. *National Vital Statistics Report, 60*(1). Center for Disease Control. Washington, DC: U.S. Government Printing Office.

Ritter, F. E., & Schooler, L. J. (2001). The learning curve. In *International encyclopedia of the social and behavior sciences* (pp. 8602–8605). Amsterdam, The Netherlands: Pergamon. Retrieved from http://www.iesbs.com/

CHAPTER 4

CONTINGENCY TABLES

The Chi-Square Test and Associated Effect Sizes

This chapter examines contingency tables and how to conduct a chi-square test for statistical independence between two nominal variables in such tables. We also examine associated effect size measures.

The Contingency Table

The prior chapters dealt with "one sample" cases that involved analysis of only one random variable (RV) (e.g., testing hypotheses about the mean, π). This chapter discusses how to examine the relationship between two nominal RVs. Data on the relationship are typically presented in the form of a *contingency table*, which conveys the number of cases falling into different combinations of the two RVs.

Introduction of the Hamilton High Example

As an example, consider a school I shall call Hamilton High. (The data are hypothetical but closely reflect national census data on dropout rates.) Hamilton High had 500 students entering as freshmen in 2000. Suppose one randomly selects 100 of these freshmen for the sample and subsequently measures how many of the participants completed high school four years later. The data are shown in Table 4.1.

The totals 49, 51, 81, and 19 are called *marginal* totals, because they appear on the edge of the contingency table. The marginal row totals sum to the overall sample size, as do the marginal column totals. If the row totals and column totals do not sum to the same number, this indicates a mathematical error.

The contingency table can also be expressed in proportions. All one has to do is divide each number by the grand total (100), producing the values shown in Table 4.2(a).

Table 4.1 Contingency Table: Gender × High School Completion Status at Hamilton High

	Male	Female	Total
Completers	35	46	81
Noncompleters	14	5	19
Total	49	51	100

Table 4.2 Unconditional and Conditional Estimated Probabilities: Hamilton High Data

	(a) Unconditional			(b) Conditional (on gender)		
Gender	M	F	Total	M	F	Total
Completers	35.0%	46.0%	81.0%	71.4%	90.2%	81%
Noncompleters	14.0%	5.0%	19.0%	28.6%	9.8%	19%
Total	49.0%	51.0%	100.0%	100.0%	100.0%	100.0%

These percentages are estimates of unconditional probabilities. For example, based on our sample (of 100), we estimate that there is a 35% chance that a student in our cohort (of 500 students) will be male and will complete high school in 4 years.

Conditional Probabilities

Now, let us take the data in Table 4.1 and instead of dividing by the grand total ($n = 100$), divide by the column totals, 49 and 51 (i.e., number of males and females, respectively). This produces the *conditional* probability estimates shown in Table 4.2(b).

The table shows that if a student is male, then the chance of completing high school is 71.4%. If a student is female, then the chance is 90.2%. The conditional probabilities could also be derived from the proportions in Table 4.2(a). For example, if one takes the first cell of Table 4.2(a) (35%) and divides it by the respective column total (49%), the result is 71.4%, which is the same value as in Table 4.2(b).

EXPLORING THE CONCEPT

Why does it not matter whether the conditional probabilities are derived from a table of frequencies or proportions? (a) Verify that this is the case using one of the female cells. (b) Is multiplying or dividing the numerator and denominator of a fraction by the same number the same as multiplying by one?

Given that conditional probabilities can be defined in terms of other probabilities, we formally define conditional probability as follows:

$$\text{Prob}(B\,|\,A)=\frac{\text{Prob}(B\,\&\,A)}{\text{Prob}(A)},$$ (Eq. 4.1)

where $\text{Prob}(B\,|\,A)$ is read "The probability of B, given A."

In this example, $\text{Prob}(B\,|\,A)$ is the conditional probability that a student will complete high school (B) given that they are male (A). The $\text{Prob}(B\,\&\,A)$ is the unconditional joint probability (35%). The conditional probability is $\frac{35\%}{49\%}=71.4\%$, giving the same result as before.

EXPLORING THE CONCEPT

Suppose Events A and B are not statistically independent. Then the joint probability, as given by Eq. 1.4, is Prob(A & B) = Prob(A) * Prob(B | A). Derive Eq. 4.1 by dividing both sides of Eq. 1.4 by Prob(A).

Chi-Square Test of Statistical Independence

In the previous example, you may have noticed that the conditional probability that a student would complete high school, given that they are male, was 71.4%, while for females, the probability was 90.2%. Is this difference statistically significant? One way of thinking about this problem is to ask if the two variables (Completing High School and Gender) are statistically independent or not. If the two variables are not statistically independent, then gender may somehow influence, or at least predict, completion status. A convenient definition of statistical independence is

If $\text{Prob}(A)$ and $\text{Prob}(B)$ are independent,
then $\text{Prob}(A\,\&\,B) = \text{Prob}(A) * \text{Prob}(B)$. (Eq. 4.2)

Recall that Eq. 4.2 is also the multiplication rule for joint probabilities under independence.

EXPLORING THE CONCEPT

Statistical independence can also be defined as Prob(B | A) = Prob(B), showing that Event A has no statistical relationship with B (other than being independent). Derive this equation from Eq. 4.2, using Eq. 4.1.

We will use Eq. 4.2 in the chi-square test of statistical independence. Chi-square tests have many applications in statistics, for example, in model fitting (see Chapter 3). The idea behind a chi-square test is that one calculates what the distribution of the data should be under the null hypothesis; this distribution represents the *expected* values. One then calculates the *observed* values and calculates how much they deviate from the expected values. If the deviations are large, one will likely reject the null hypothesis. The deviations, however, are first squared so that the negative ones do not cancel out the positive ones. One divides each squared deviation by its expected value to control for the number of observations; the sum is the chi-square test statistic (χ^2). Assuming a large enough sample, the sampling distribution for the statistic will be chi-squared (see Figure 3.3 in the previous chapter, and Eq. 3.14 for the statistical formula).

To conduct a chi-square test for a contingency table, one would follow these steps. (The test applies not only to 2 × 2 tables, where both nominal variables are dichotomous, but also to larger tables where one or both variables are multinomial.)

1. *Define the null and alternative hypotheses.* The null hypothesis is that the two variables are statistically independent, that is:

   H_0: Prob(A & B) = Prob(A) * Prob(B).
   H_a: Prob(A & B) ≠ Prob(A) * Prob(B).

2. *Assume the null hypothesis is true, and calculate the expected values using the observed marginal probabilities.* Consulting Table 4.2(a), one finds that Prob(A = Male) = 49% and Prob(B = Completer) = 81%. These are reproduced in Table 4.3(a). Table 4.3(a) also shows the expected proportions for each cell using the definition of statistical independence (Eq. 4.2). (So, for example, the expected probability for male completers is 81% * 49% = 39.7%.) The expected probabilities are then multiplied by n (100) to find the expected frequencies [shown in Table 4.3(b)]. Note that we first divided by n to derive proportions and then multiplied by n to derive frequencies. Because these two operations cancel one another out, one can simplify the calculations by deriving the expected frequencies directly from the observed marginal frequencies, using the following equation:

$$E_{ij} = \frac{O_{i\bullet} * O_{\bullet j}}{n},$$ (Eq. 4.3)

   where $O_{i\bullet}$ and $O_{\bullet j}$ are the column and row totals for *cell$_{ij}$*.

EXPLORING THE CONCEPT

Why does Eq. 4.3 still involve dividing by n? In the more complicated method described above using proportions, did we divide by n once or twice?

Table 4.3 Expected Probabilities and Frequencies Under Independence: Hamilton High Data

Gender	(a) Expected probabilities[a]			(b) Expected frequencies[d]		
	M	F	Total[c]	M	F	Total[c]
Completers	39.7%[b]	41.3%	81.0%	39.7	41.3	81
Noncompleters	9.3%	9.7%	19.0%	9.3	9.7	19
Total[c]	49.0%	51.0%	100.0%	49.0	51.0	100

[a]Derived from multiplying marginal values together.
[b]For example, 81% $*$ 49% = 39.7%.
[c]Observed values from Table 4.2.
[d]Derived from multiplying expected probabilities by n.

Table 4.4 Chi-Square Calculations: Hamilton High School Completion Data

	(a) $O_{ij} - E_{ij}$		(b) $(O_{ij} - E_{ij})^2$		(c) $\dfrac{(O_{ij} - E_{ij})^2}{E_{ij}}$	
	M	F	M	F	M	F
Completers	−4.69	4.69	22.0	22.0	0.55	0.53
Noncompleters	4.69	−4.69	22.0	22.0	2.36	2.27

(d) $\chi^2 = \sum \dfrac{(O_{ij} - E_{ij})^2}{E_{ij}} = 5.72.$

Note: The subscript i indexes the row number; j indexes the column number.

3. *Calculate the deviations, square them, and divide by the expected frequencies.* For the Hamilton High example, the calculations are shown in Table 4.4.
4. *Sum the values to find the chi-square statistic.* Table 4.4 shows that for the Hamilton High example, $\chi^2 = 5.72$.
5. *Calculate the degrees of freedom.* The degrees of freedom are the number of values that could vary and still yield the marginal probabilities given in the tables. The formula for calculating the degrees of freedom is

$$df = (r - 1)(c - 1), \qquad\qquad\qquad\qquad \text{(Eq. 4.4)}$$

where r is the number of rows (2) and c the number of columns (2), so $df = 1$. For example, if one only knew the number of male completers (35) and the marginal values, one could calculate all the remaining cell values.

EXPLORING THE CONCEPT

For the following contingency table (Table 4.5), calculate all the remaining cell values.

Table 4.5 Incomplete Contingency Table

	Male	Female	Total
Completers	35	?	81
Noncompleters	?	?	19
Total	49	51	100

Note that the number of male completers would likely vary in other samples, thus the label "degree of freedom" (*df*).

Although a 2 × 2 table has one degree of freedom, a larger table, say a 3 × 3 table, would have more (4 *df* in the 3 × 3 case). In essence, four of the cells could vary and still produce the marginal totals. Conceptually, one would have four random variables, each with a chi-square distribution, that are then combined.

6. *Calculate the p value.* The *p* value is given by the amount of area in the chi-square distribution to the right of the test value. Recall our result is $\chi^2(1) = 5.72$, where the number in the parentheses is the number of degrees of freedom. According to EXCEL, the *p* value is =CHIDIST(χ^2 value, *df*), and more specifically, =CHIDIST(5.72, 1). ➔ .017.

Because this result is less than .05 (the α level), we are justified in rejecting the null hypothesis and concluding that the two variables (gender and graduating from high school) are statistically dependent, at least at Hamilton High.

Using SPSS to Perform the Chi-Square Test

To perform the test in SPSS, one needs to create three variables: gender, completion, and frequency. The first two can be "string" variables that use text, such as "Male" and "Female" instead of numeric values. Alternatively, one can use numeric values such as "1" for females and "0" for males. In the database, there will also be a third column for the number of students that fall into each cell, as shown in Table 4.6.

Once the data are entered, go to "Data" ➔ "Weight Cases" and weight the data by the frequencies. This weighting creates (internally) in the database 35 cases that have "Completer" for completion status and "Male" for gender, 46 cases that have "Completer" for completion status and "Female" for gender, and so on. These duplicates will not show up in the data set that you see, so it is easy to forget this step, but don't—it is vital.

The next step is to go to "Analyze" ➔ "Descriptive Statistics" ➔ "Crosstabs," and to place "Gender" in the column box and Completion in the row box. Then

Table 4.6 SPSS Data Format: Hamilton High Data

Completion	Gender	Count
Completer	Male	35
Completer	Female	46
Noncompleter	Male	14
Noncompleter	Female	5

Completion ∗ Gender Crosstabulation

Count

		Gender		Total
		Female	Male	
Completion	Completer	46	35	81
	Noncompleter	5	14	19
Total		51	49	100

Chi-Square Tests

	Value	df	Asymp. Sig. (2-sided)	Exact Sig. (2-sided)	Exact Sig. (1-sided)
Pearson Chi-Square	5.719[a]	1	.017		
Continuity Correction[b]	4.565	1	.033		
Likelihood Ratio	5.897	1	.015		
Fisher's Exact Test				.022	.016
N of Valid Cases	100				

a. 0 cells (.0%) have expected count less than 5. The minimum expected count is 9.31.
b. Computed only for a 2 × 2 table.

Figure 4.1 SPSS output of crosstab chi-square test using Hamilton High data.

click on "statistics" and check the box for the chi-square test. Run the analysis and the following output results (shown in Figure 4.1)

We are interested in the first row of the test table. The Pearson chi-square statistic is 5.72 and the p value is .017, which conforms to our previous calculations.

The next row in the table reports the results with a continuity correction. The chi-square distribution is a continuous distribution, but our variables are discrete, so a continuity correction is needed. One should therefore write in a journal article: $\chi^2(1) = 4.57$, $p = .033$. The continuity correction is calculated by reducing one of the cell frequencies by half a unit (e.g., reducing female completers from 46 to 45.5) before the chi-square value is calculated.

The chi-square test should only be performed with $n \geq 20$ and with at least five expected observations in each cell of the table. (More exact criteria are discussed in Technical Note 4.1.) Footnote (a) in the SPSS output indicates that these conditions have been met. If the conditions are not met, a warning message will result and results for Fisher's exact test (discussed in the next chapter) should be reported instead.

Effect Sizes

In reporting results, one should always include an effect-size statistic. Whereas p values indicate whether an effect is *statistically* significant, effect sizes convey information as to whether or not an effect is *practically* significant. One can have a very small effect, but if one has a lot of data, the results could be highly statistically significant. With a very large sample, one has a lot of statistical power. This situation is analogous to having a very powerful telescope that can see small things very far away.

This section considers two different effect size measures: (a) the difference between two proportions and (b) the odds ratio.

Difference Between Two Proportions

The difference between two proportions involves computing the difference between two conditional probabilities. For example, in Table 4.2(b), 71.4% of the male students completed high school, whereas 90.2% of the female students did. The difference between the proportions (ΔP) is 18.8%. This difference seems big. Although whether or not something is practically significant is ultimately a value judgment. Researchers need to report effect sizes so that readers can make these judgments.

Fixed and Random Variables

One limitation of the ΔP effect size measure is that it assumes that the independent variable (IV) is *fixed* and the dependent variable (DV) is random. A variable is fixed if it has a causal effect on or is used to predict the value of a random variable. The fixed variable is the IV and the random variable is the DV. The DV should have no causal effect on the IV; otherwise, the IV will have a random component and will be neither fixed or truly independent.

With a fixed variable, in principle one could know what the values are before looking at the rest of the data. If one is conducting an experiment or quasiexperiment, whether an individual is in the experimental or control condition is fixed before the experiment begins. Certainly if one is assigning participants randomly to conditions, in that context the variable is random. But once the assignments have been made, the variable becomes fixed. An analogous situation is that while the probability that a certain coin will come up heads may be 50% of the time, once the coin is flipped (and say it comes up heads), the probability of a head is now 100%. The outcome is no longer random and the variable has been fixed at a certain level.

In the Hamilton High example, "Gender" is a fixed variable, as it was fixed at birth. One might not know if someone is a male or female until one looks at the data, but one could conceivably know ahead of time. Considering a conditional probability [e.g., Prob($Y \mid X$)] makes most sense when X is fixed. In this case, it is possible that X has a causal influence on Y or that at least X can be used to predict Y. When using the effect size statistic $\Delta P\, (= P_2 - P_1)$, remember that the proportions (Ps) are conditional probabilities, specifically $\text{Prob}(Y|X) = \frac{\text{Prob}(X\,\&\,Y)}{\text{Prob}(X)}$. This is why X needs to be fixed.

Suppose X were not fixed because Y exerts a causal influence on X, or the causal relationship is reciprocal. For example, variables such as *anxiety* and *self-efficacy* are negatively correlated, but from this fact one does not know if anxiety affects self-efficacy (i.e., being less anxious makes one feel more capable at a task) or whether self-efficacy affects anxiety (if one feels more competent, one will be less anxious). It is possible, even likely, that anxiety and self-efficacy exert a reciprocal causal effect on one another. If Y exerts a causal effect on X, then X has a random component driven by Y and could not be considered fixed. The statistic, ΔP, would be a misleading measure of effect size. Let us consider the extreme case where X is completely determined by Y and has no causal effect at all on Y. Then changing the value of X from 0 to 1 or from 1 to 0 will have no effect on Y. But the proportions P_1 and P_2, conditional on X, might still be substantially different. To use to ΔP (conditional on X) as an effect size of the impact of X on Y would be misleading because X has no influence on Y at all. This point is why the difference of proportion statistic should only be used when the IV (X) is fixed.

EXPLORING THE CONCEPT

Examine Table 4.1. Find the conditional probability of being a female if one completes high school. Find ΔP, conditional on completing high school. Could completing high school affect gender status? Why is ΔP a misleading effect size measure here?

Odds Ratio

The odds ratio (OR), being symmetric, is a more appropriate effect size measure when X is random. For a 2×2 contingency table, it will be the same regardless of on which variable one conditionalizes and is thus insensitive to whether the variables are fixed or random.

To understand the OR, one first needs to understand the difference between odds and probability. The definition of a probability, Prob(X), is the proportion that X appears in the population, that is, $\lim_{n \to \infty} \frac{frequency(X)}{n}$. The odds of X are

$$Odds(X) = \lim_{n \to \infty} \frac{frequency(X)}{frequency(\sim X)}, \qquad \text{(Eq. 4.5)}$$

where $\sim X$ is read, "Not X." The odds can also be expressed as a ratio of two probabilities:

$$Odds(X) = \frac{P(X)}{P(\sim X)}. \qquad \text{(Eq. 4.6)}$$

For example, suppose one were to toss a coin 1,000 times, and the coin came up heads 600 times and tails 400 times. The coin is likely biased and one would estimate the odds of a head to be $\frac{600}{400}$ (or $\frac{60\%}{40\%}$) $= 1.5$. (The value can also be expressed as 3:2 odds.) The odds of a head are 1.5, whereas the probability of a heads is 60%.

Probabilities can always be converted into odds and vice versa. As an example, the probability 70% is equivalent to odds of $\frac{70\%}{30\%} = 2.33$. Conversely, one can convert odds to probabilities using the following formula:

$$\text{Prob}(X) = \frac{Odds(X)}{1 + Odds(X)}. \qquad \text{(Eq. 4.7)}$$

In the example, $\frac{Odds(X)}{1 + Odds(X)} = \frac{2.33}{3.33} = 70\%$. There is little difference between probability and odds when both are small; for example, for a probability of .01, the odds are $\frac{.01}{.99} = .01$. When a probability is high, the two can diverge substantially; probabilities asymptote at 1.0 whereas odds can increase to infinity.

Why use odds if one could also use probabilities? Professional gamblers use odds because they are easier to calculate. Statisticians sometimes use odds because a certain statistical model may yield output in terms of odds rather than probabilities.

Having now defined the concept of *odds*, I next define the concept of an *odds ratio*. The odds ratio is the odds of a hit on Y when $X = 1$ divided by the odds when $X = 0$. In other words,

$$\text{Odd Ratio (OR)} = \frac{Odds(Y|X)}{Odds(Y|\sim X)}. \qquad \text{(Eq. 4.8)}$$

Table 4.7 Contingency Table With Variable for Frequency Counts

	X	~X	Total
Y	a	c	a + c
~Y	b	d	b + d
Total	a + b	c + d	a + b + c + d

For the Hamilton High example, and consulting the data in Table 4.1, the odds that a student is a completer, given that he is male, are estimated to be $\frac{35}{14} = 2.5$. The odds that a student is a completer, given that she is female, are $\frac{46}{5} = 9.2$. The OR is $\frac{9.2}{2.5} = 3.68$. If there is no effect, the OR would be 1.0, so in predicting whether a student will complete high school, being female substantially increases the odds.

Let us review again how odds differ from probabilities. Table 4.7 is a contingency table with variables (a, b, c, and d) used for the cell counts.

The $\text{Odds}(Y|X) = \frac{a}{b}$, while the $\text{Prob}(Y|X) = \frac{a}{a+b}$. This formulation shows that calculating odds is simpler. (Research has shown that people are better at processing counts and odds than probabilities; Gigerenzer & Hoffrage, 1995.) The OR would be

$$OR = \frac{\frac{a}{b}}{\frac{c}{d}} = \frac{ad}{bc}.$$

(Eq. 4.9)

Furthermore, unlike ΔP, one does not have to assume that X is fixed. The OR is symmetric, and one could have just as well conditioned on Y rather than X.

EXPLORING THE CONCEPT

For Hamilton High, find the odds that a student is female, given that she is (a) a completer, and (b) a noncompleter. Calculate the OR. It should again be 3.68 (may differ slightly due to rounding error). How does the symmetry of the OR follow from Eq. 4.9?

One could also calculate the OR by dividing the larger odds into the smaller (i.e., $\frac{2.5}{9.2} = 0.27$). I do not recommend this practice because it is easier to work with and interpret ORs greater than 1.0. Note that 0.27 is the inverse of 3.68 ($\frac{1}{0.27} \approx 3.68$).

How should an OR such as 3.68 be interpreted? One could say that the odds of completing high school are 368% greater if one is female than if one is male. Another useful interpretation is to take the OR, subtract 1.0 from it, and to interpret the results as the percentage increase in the odds. Therefore, an OR of 3.68 represents a 268% increase in the odds of completion when moving from a male to a female case. The rationale for subtracting 1.00 comes from the definition of a percentage increase: $\% \text{ Increase in } Y = \frac{(Y_2 - Y_1)}{Y_1}$. So with the example, $\frac{9.5 - 2.5}{2.5} = 2.68 = 268\%$. This follows from the definition because

$$\% \text{ Increase in } Y = \frac{(Y_2 - Y_1)}{Y_1} = \frac{Y_2}{Y_1} - 1 = OR - 1. \qquad \text{(Eq. 4.10)}$$

This result will be important later in the book. For now, remember that to interpret ORs, subtract 1 and interpret them as percentage increases (or decreases). Thus, for example, the OR 0.27 can be interpreted as $0.27 - 1 = -0.73$, that is, as a 73% decrease in the odds.

Measures of Association

A measure of association is a measure of how strongly two variables are related on a scale from 0 (*no association*) to 1 (*perfect association*). Sometimes the scale ranges from -1 to 1. The canonical example of a measure of association is the Pearson product-moment correlation for two metric variables:

$$\rho_{x,y} = \frac{E[(X - \mu_x)(Y - \mu_y)]}{\sigma_x \sigma_y}. \qquad \text{(Eq. 4.11)}$$

The numerator is the covariance; the denominator is the product of the two standard deviations. The denominator standardize the measure onto a -1 to 1 scale. There is a corresponding correlation coefficient, known as the *phi* coefficient, when both X and Y are nominal:

$$\phi = \frac{ad - bc}{\sqrt{(a+b)(c+d)(a+c)(b+d)}}. \qquad \text{(Eq. 4.12)}$$

The letters refer to the frequency count in Table 4.7. If most of the frequencies are concentrated in cells a and d, there will be a large positive correlation. If most of the frequencies are concentrated in cells b and c, there will be a large negative correlation. If the frequencies are evenly distributed, there will be a zero correlation. It can be shown that the phi coefficient is also equal to

$$\phi = \sqrt{\frac{\chi^2}{n}}. \qquad \text{(Eq. 4.13)}$$

Therefore, in the Hamilton High example, we have: $\phi = \sqrt{\frac{\chi^2}{n}} = \sqrt{\frac{5.72}{100}} = .239$. The correlation between X and Y is therefore 23.9%.

One limitation of the phi coefficient is that it is only meaningful for a 2 × 2 contingency table. Sometimes one needs to analyze larger tables (3 × 2, 4 × 4, etc.). These are referred to generically as $r \ x \ c$ tables. A measure of association that can be used for these larger tables is Cramer's V. Cramer's V reduces to the phi coefficient in the 2 × 2 case. Cramer's V is calculated by taking the ratio $\frac{Actual\ \chi^2}{Maximum\ possible\ \chi^2}$. This measure of association will be 0 if the actual χ^2 is zero (no association), and it will be 1.0 if the actual χ^2 is at its maximum value. The maximum possible χ^2 is given by the formula: $n * (k - 1)$, where k is the number of rows or columns in the contingency table, whichever is smaller. Therefore, the formula for Cramer's statistic is

$$V = \sqrt{\frac{\chi^2}{n(k-1)}}.$$ (Eq. 4.14)

Phi and Cramer's V are measures of association, not effect sizes. Some textbook authors do classify measures of associations as types of effect sizes, but I do not. I consider an effect size an estimate of how much we predict one variable would change given a change in the other variable, whereas measures of association are symmetric measures of how closely two variables covary.

Using SPSS

To obtain the statistics discussed in this chapter, click on "statistics" under the crosstab options, and then check "chi-square," "phi and Cramer's V," and "risk" (for the OR). Also go to "cells" on the crosstab option and check "column"; this option will give you the column percentages for calculating ΔP. There is no necessary reason that these percentages should sum to 100% and, if they do, you may have checked the wrong box.

For the Hamilton High example, the resulting output should look like that shown in Figure 4.2.

The numbers are slightly different from what was calculated above from rounding error.

I have highlighted the relevant statistics in bold. The values are close to, or identical with, the values that we calculated in this chapter (differences being due to rounding error). Depending on how the contingency table is constructed, your software package might instead provide an OR value of 0.272; this is simply the reciprocal of 3.680.

Completion ∗ Gender Crosstabulation

			Gender		Total
			Female	Male	
Completion	Completer	Count	46	35	81
		% within Gender	90.2%	71.4%	81.0%
	Noncompleter	Count	5	14	19
		% within Gender	9.8%	28.6%	19.0%
Total		Count	51	49	100
		% within Gender	100.0%	100.0%	100.0%

Chi-Square Tests

	Value	df	Asymp. Sig. (2-sided)	Exact Sig. (2-sided)	Exact Sig. (1-sided)
Pearson Chi-Square	5.719[a]	1	.017		
Continuity Correction[b]	4.565	1	.033		
Likelihood Ratio	5.897	1	.015		
Fisher's Exact Test				.022	.016
N of Valid Cases	100				

a. 0 cells (.0%) have expected count less than 5. The minimum expected count is 9.31.
b. Computed only for a 2 × 2 table.

Symmetric Measures

		Value	Approx. Sig.
Nominal by Nominal	**Phi**	**.239**	**.017**
	Cramer's V	.239	.017
N of Valid Cases		100	

Risk Estimate

	Value	95% Confidence Interval	
		Lower	Upper
Odds Ratio for Completion (Completer/Noncompleter)	**3.680**	1.211	11.186
For cohort Gender = Female	2.158	.993	4.689
For cohort Gender = Male	.586	.406	.846
N of Valid Cases	100		

Figure 4.2 SPSS output of crosstab chi-square test with column percents, ϕ and Cramer's V, and odds ratio. Key statistics are in bold.

Summary

In summary, all these analyses show that being a female increases one's chance of completing high school at Hamilton High (with $\chi^2(1) = 4.57$, $p < .05$):

- the difference in proportions (ΔP) is 18.8%;
- the OR is 3.68, indicating a 268% increase in odds; and
- the correlation (ϕ) is 23.9%.

Of these, it may seem like the magnitude of the OR is the most impressive of the three. One has to keep in mind that a percent increase on a small base can sound more impressive than it really is. Also, odds increase faster than probabilities at the high end of a spectrum. So it is easy to exaggerate the importance of one's findings when using odds ratios; ΔP should be preferred. However, when both variables are random, do not use ΔP. In that case, reporting ϕ, in addition to the OR, would provide a more balanced picture than just reporting the OR alone.

Problems

1. Fernandez and Cannon (2005) surveyed public school teachers from the United States and Japan on teachers' goal when preparing lesson plans. The following table indicates whether a teacher had a goal of understanding students' thinking/feelings.

	Japanese	U.S.
Yes	15	5
No	10	31

    a) What is the estimated conditional probability that a Japanese teacher would answer "yes"? That a U.S. teacher would?
    b) Perform a chi-square test of statistical independence, *without a continuity correction*. Perform the calculations step by step, using EXCEL.
    c) Using EXCEL, calculate the OR and ϕ.
    d) Calculate ΔP. Would it be appropriate to report ΔP for these data? Why or why not?
    e) Using EXCEL and showing your work step by step, perform a chi-square test of statistical independence, *with a continuity correction*. The following numbers reflect a continuity correction.

	Japanese	U.S.
Yes	14.5	5.5
No	10.5	30.5

    f) Perform a chi-square test using statistical software such as SPSS. Attach the output. What is the value of Pearson's chi-square, the p value, ϕ, and the OR? How do these compare with your previous answers?

2. The following contingency table from Blalock (cited in Kritzer, 1978) shows whether having high versus low classroom grades is associated with high versus low levels of intelligence.

Grades	Intelligence	
	High	Low
High	158	72
Low	98	122

a) What is the estimated conditional probability that someone with high intelligence would receive high grades? That someone with low intelligence would?

b–f) Answer the same questions as in problem 1.

3. Altham and Ferrie (2007) reported data on occupation categories in 1910 for 17,733 males, cross classified by their fathers' occupational category (as of 1880). Their interest was how much occupational change was occurring and what this implied for social mobility. The data were:

Son's 1910 occupation	Father's 1880 occupation			
	White collar	Farmer	Skilled/semiskilled	Unskilled
White collar	1,538	1,622	1,203	529
Farmer	550	3,371	363	409
Skilled/semiskilled	907	1,486	1,736	858
Unskilled	500	1,428	622	611

a) Perform a chi-square test using statistical software such as SPSS. Attach the output. What is the value of Pearson's chi-square? What is the significance level?

b) Calculate Cramer's V. What does the square of V represent?

c) Calculate the estimated unconditional probabilities for each cell. What cell has the highest (off-diagonal) unconditional probability? Which column has the highest unconditional probabilities? What are the implications in terms of social mobility?

d) What are the estimated probabilities, conditionalized on Father's 1880 occupation, that the son will have the same occupation? Conduct a chi-square test to determine if the percentage of unskilled workers is significantly different from the other categories combined. (Hint: construct a 2×2 table.) Report the test statistic, significance level, ΔP, OR, and your conclusion.

e) Why can Father's 1880 occupation be considered a fixed variable in the previous problem? Is it appropriate to report ΔP?

    f)   Interpret the OR in part d). By what percentage did the odds that a son has the same occupation increase (or decrease) if the father is skilled rather than unskilled?

4.   For this problem, use the data on Sun City dropout rates (Problem 5, Chapter 2) and statistical software such as SPSS.

    a)   Perform a chi-square test. Attach the output. What is the value of Pearson's chi-square? What is the significance level?

    b)   Calculate Cramer's V. What does the square of V represent?

    c)   Conduct a chi-square test to determine if the dropout rate is significantly different for the 16–17 age group than the 18–19 age group. Report the test statistic, significance level, ΔP, OR, and your conclusion.

    d)   Is age a fixed or random variable? Is it appropriate to report ΔP?

    e)   Interpret the OR in part (c). By what percentage do the odds of dropping out appear to increase from being in the 18–19 age group as compared to the 16–17 age group? Is it appropriate to draw a causal conclusion?

Technical Note

4.1   For the chi-square test of independence, Cochran (cited in Agresti, 2002) showed that for a 2 × 2 table ($df = 1$) all expected values should be 5 or greater if $N < 40$; otherwise, a chi-square test can be used. If $df > 1$, expected values of approximately 1 are permissible as long as no more than 20% of the cells have expected values less than 5.

References

Agresti, A. (2002). *Categorical data analysis* (2nd ed.). Hoboken, NJ: Wiley-Interscience.

Altham, P. M., & Ferrie, J. P. (2007). Comparing contingency tables: Tools for analyzing data from two groups cross-classified by two characteristics. *Historical Methods, 40,* 3–16.

Fernandez, C., & Cannon, J. (2005). What Japanese and U.S. teachers think about hen constructing mathematics lessons: A preliminary investigation. *The Elementary School Journal, 105,* 481–498.

Gigerenzer, G., & Hoffrage, U. (1995). How to improve Bayesian reasoning without instruction: Frequency formats. *Psychological Review, 102,* 684–704.

Kritzer, H. (1978). An introduction to multivariate contingency table analysis. *American Journal of Political Science, 22,* 187–226.

CHAPTER 5

CONTINGENCY TABLES
Special Situations

This chapter examines tests for 2×2 contingency tables where it is inappropriate to use the chi-square test for statistical independence described in Chapter 4. The situations examined reflect important considerations in choosing the best statistical techniques, not only for nominal variables but also for ordinal and metric ones. These situations are

- small sample size;
- lack of statistical independence among observations; and
- the need to statistically control for confounding variables.

We specifically consider Fisher's exact test (for small samples), the McNemar test (for related samples), and the Mantel-Haenszel test (for controlling a third variable).

The Small Sample Case: Fisher's Exact Test

Because the chi-square test is asymptotic, the sampling distribution of the χ^2 statistic can be assumed to be chi-squared (i.e., square of a normal distribution) only if the sample is sufficiently large. For 2×2 tables, the minimum sample size is about $n = 20$, but more specifically, according to asymptotic theory (which is the branch of statistics that studies how the central limit theorem applies to various distributions), there should not be any empty or sparse cells. The usual criterion, and the one used by SPSS, is that the expected values in each cell should be at least five. (For larger $r \times c$ tables, at least 80% of the cells should meet this criterion, but see Chapter 4 for alternative criteria.)

Consider Table 5.1 (from Le, 1998). The research question is, "Which approach is more effective at controlling cancer, radiation or surgery?" If one were to conduct a chi-square test in SPSS, the output in Table 5.2 would result.

Table 5.1 Contingency Table of the Frequency of Controlling Cancer by Treatment Type

Disease status	Radiation	Surgery	Total
Cancer not controlled	3	2	5
Cancer controlled	15	21	36
Total	18	23	41

Table 5.2 SPSS Output for Cancer Example

	Chi-Square Tests				
	Value	df	Asymp. Sig. (2-sided)	Exact Sig. (2-sided)	Exact Sig. (1-sided)
Pearson chi-square	.599[a]	1	.439		
Continuity correction[b]	.086	1	.769		
Likelihood ratio	.595	1	.441		
Fisher's exact test				.638	.381
N of valid cases	41				

[a]Two cells (50.0%) have expected count less than 5. The minimum expected count is 2.20.
[b]Computed only for a 2×2 table.

Note the error message in footnote a: two cells have a minimum expected count less than 5, making the chi-square test invalid according to the usual criteria. With a larger, $r \times c$, table, one could combine cells to address this problem or use log-linear analysis (described in Chapter 12) with an equality constraint (this practice is preferable because combining cells can change results). With a 2×2 table, however, one can use Fisher's exact test; SPSS will automatically provide the p values when a chi-square test is run. However, to facilitate understanding of Fisher's exact test, in this section I will illustrate Fisher's exact test step by step using EXCEL. It is unnecessary to perform step-by-step calculations in practice, although we shall see that there is an adjustment to Fisher's exact test, related to the method of mid-p values, which should be calculated in EXCEL.

Basic Considerations of Fisher's Exact Test

Before beginning, consider a few preliminary points. First, why bother to conduct tests with small samples? Although larger samples provide more statistical power, it is sometimes desirable to first conduct a small pilot study. Also, even with large studies, attrition can reduce the sample size, or one might decide to conduct an ancillary analysis on a subset of the data. Finally, resource limitations may necessitate a small n.

The second point is that Fisher's exact test assumes that the marginal totals in the contingency table are fixed (i.e., determined or known in advance). This assumption is typically not a reasonable one to make. Typically, at least one of the variables will be random. In the cancer example, cancer status is a random variable that may depend on treatment type. Violation of the fixed variables assumption makes the test very *conservative*. In a conservative test, the estimated p value is too big, making it harder to establish statistical significance. So if the results are in fact significant according to Fisher's exact test, there likely is a real effect there. If the test were too liberal, it would be difficult to justify using it; but because the results are conservative, and scientists are usually more concerned about avoiding Type I rather than Type II errors, Fisher's exact test is widely used in categorical data analysis.

EXPLORING THE CONCEPT

Why does violation of the assumption of fixed marginal totals make the test conservative? The test works by calculating the number of possible tables, and the probability of each table, given the fixed marginal totals. Are there another set of possible tables associated with other possible marginal totals? If the number of possible outcomes is greater than that assumed by the test, would the true p value be greater or less than the one calculated by the test? See Eq. 1.1 for a hint.

One last preliminary point: The test is called *exact* in contrast to asymptotic tests. I used the expression *normal approximation* in previous chapters when the sampling distribution was approximately normal in a large enough sample. In Chapter 2, it was pointed out that if one were unable to use the normal approximation to estimate a proportion because of sample size, one could still use the (exact) binomial distribution.

The same is the case with 2×2 contingency tables, except the exact shape of the sampling distribution is *hypergeometric* rather than binomial. In a binomial distribution, all the trials are independent. When there is sampling without replacement, however, the probabilities of a hit for different trials are not statistically independent. The probability that a card in a 52-card deck will be spades is 13/52 (or 25%), but the probability that the second card will also be spades, if the first card is not put back, is $\frac{(13-1)}{(52-1)} = \frac{12}{51} = 23.5\%$. The probabilities for sampling without replacement follow a hypergeometric distribution, which does not reflect independent trials.

The Sampling Frame Underlying Fisher's Exact Test

To be more specific about the theoretical underpinning of the test, consider Table 5.3, a generic 2 × 2 table where the frequency counts are represented by letters as follows:

Table 5.3 Cell Variables for a 2 × 2 Contingency Table

	Column 1	Column 2	Row totals
Row 1	x	$r - x$	r
Row 2	$c - x$	$(n - r - c + x)$	$n - r$
Column totals	c	$n - c$	n

Here c and $(n - c)$ are the column totals, r and $n - r$ are the row totals. Remember, all four of these values are fixed. The only random variable (RV) in this table is x, which is one of the cell counts. Because the table only has one degree of freedom (df), once x is known, all other cell values can be derived.

EXPLORING THE CONCEPT

As an example, try to complete all the cell values of Table 5.4.

Table 5.4 Incomplete Contingency Table

	Column 1	Column 2	Total
Row 1	3		5
Row 2			
Total	18		41

With just four pieces of information (x, c, r, and n), you should be able to complete the table. It should be an exact duplicate of Table 5.1.

Fisher's genius was in defining n as the size of the population, not the sample. In Table 5.1, the population is defined as the 41 cancer patients in the study. Normally, one thinks of the 41 patients as constituting some sort of sample from a larger population, but because we are assuming n, r, and c to be fixed, we can for purposes of the test define the population as the 41 patients. One has some latitude on how a population is defined.

Now, if n is the size of the population, we can conceive of c (in Table 5.3) as the size of a sample randomly drawn from the population. Note that this is

Table 5.5 Variable References for Fisher's Exact Test

	Column 1	Column 2	Row totals
Row 1	x = number of hits in sample	$r - x$	r = number of hits in population
Row 2	$c - x$	$(n - r - c + x)$	$n - r$
Column totals	c = size of sample	$n - c$	n = size of population

Note: A "hit" is defined as having the characteristic for inclusion in Row 1 (i.e., having uncontrollable cancer).

sampling without replacement. As summarized in Table 5.5, let x be the number of hits, and $\frac{r}{n}$ be the probability of a hit (analogous to π). In our example, there are 5 patients in the population with uncontrollable cancer, so on the first trial the probability of cancer being uncontrollable is $\frac{5}{41} = 12.2\%$. From the population of 41 patients, one draws a sample of 18 patients who received radiation treatment (denoted by c). Make the null hypothesis that there is no relationship between type of treatment and whether one's cancer is controllable. One then asks, given these 18 patients (trials), where $\pi = 12.2\%$, what is the probability of $x = 3$, where a hit is defined as having uncontrollable cancer? The binomial formula cannot be used to calculate this probability because sampling without replacement is involved. One can, however, use the function in EXCEL for the hypergeometric distribution: =HYPGEOMDIST(x, c, r, n), which here is =HYPGEOMDIST(3, 18, 5, 41) ➜ 27.55%.

The probability (27.55%) turns out to be higher than if the binomial formula is used (21.1%). The calculation in EXCEL is based on a formula explained in Appendix 5.1.

Calculating the *P* Value

For this test, 27.55% is not the p value; the p value is the probability, under the null hypothesis, that you will obtain your results *or something more extreme.* In our sample, $x = 3$, but one could have also obtained an x of 4 or 5.

To calculate the p value, one needs to calculate the probability of each possible table. This would include tables for $x = 0, 1, 2, 3, 4,$ or 5. In the example, x cannot be > 5 or c would be negative. The value of x cannot be greater than c or r, whichever is less. Table 5.6 uses EXCEL to calculate the probability of each possible table.

Figure 5.1 is a graph of this hypergeometric distribution. One can see from the figure that the hypergeometric distribution is not perfectly symmetrical.

Table 5.6 Probability of Possible Contingency Tables

x	Prob(x)	EXCEL formula
0	0.045	=HYPGEOMDIST(0,18,5,41)
1	0.213	=HYPGEOMDIST(1,18,5,41)
2	0.362	=HYPGEOMDIST(2,18,5,41)
3	0.275	**=HYPGEOMDIST(3,18,5,41) observed value**
4	0.094	=HYPGEOMDIST(4,18,5,41)
5	0.011	=HYPGEOMDIST(5,18,5,41)
TOTAL	1.000	

Note: Prob(x) values do not sum to 100% in table due to rounding error.

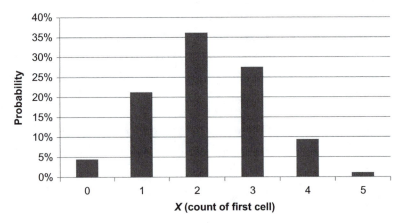

Figure 5.1 Graph of the hypergeometric sampling distribution in Table 5.6.

One-Tailed Test

One can couch the null and alternative hypotheses as, for example,

H_0: Prob(hit in sample) = Prob(hit in population), i.e., $\frac{x}{c} = \frac{r}{n}$.
H_a: Prob(hit in sample) > Prob(hit in population), or
Prob(hit in sample) < Prob(hit in population).

The one-sided p value is the sum of the probabilities for the x values ranging from 0 to 3, or from 3 to 5, depending on the direction of the alternative hypothesis. The sum of the probabilities for x ranging from 3 to 5 is 38.1%, which is a nonsignificant result.

In conducting a test of statistical significance using p values, one asks whether the p value is greater or less than the α level (5%). An alternative is to use the method of critical values, where one asks whether the RV x is greater or less than some critical value, where the critical value has, in the one-sided case, a p value of 5%. (For purposes of the present analysis, if the observed value is equal to the critical value, then one rejects the null hypothesis.) These procedures may result in a fractional value, making it difficult to implement when using discrete values. For example, in Table 5.6, the $Prob(x \geq 4) = 10.5\%$ (the sum of 9.4% + 1.1%), and so setting the critical value equal to 4 would be too liberal. On the other hand, $Prob(x = 5) = 1.1\%$, so making 5 the critical value would be too conservative. If one made 4.5 the critical value, then one could have a p value close to 5%, but because x is discrete, it's not possible to have fractional values.

One solution to the problem is to use the method of mid-p values. In this method, one reduces the p value obtained from Fisher's exact test by half a unit. This adjustment is essentially conducting a continuity correction in reverse. Recall that the p value for the one-sided test was 38.1%. If the test were conducted in the other direction, the p value would be 89.5%. Note that the two p values sum to more than 100% (127%) because both include the possibility that $x = 3$ (the observed value), which has a probability of 27%. Some have argued that to make one-sided hypothesis tests more comparable with those performed with continuous variables, the two p values should sum to 100%. This situation is the case with continuous variables because the probability of the observed value (any point along the continuum) is zero. (Recall this fact is why only cumulative probabilities are meaningful with continuous variables.) To put discrete variables on the same par, those advocating the method of "mid-p values" would only include half of the bar in the probability distribution corresponding to the observed value ($x = 3$). The probability is 27%, so half of that probability is 13.5%. When that is used in the calculations of the one-sided p value, the value would be 24.3% (or 75.7% in the other direction); the two p values now sum to 100%.

Note that the method of mid-p values improves the statistical power of the test. Agresti (2002) recommends the adjustment as a reasonable compromise for the fact that the Fisher exact test is too conservative. We have seen that the method of using unadjusted p values is too conservative by about a "half a bar" and also because of the assumption of fixed marginal totals. However, SPSS does not currently implement the correction.

If one accepts Agresti's recommendation, one can easily implement the method of mid-p values without performing the entire calculation in EXCEL. One can use the p value from SPSS and then just reduce the value by half the probability of x using EXCEL. In our example, $x = 3$, and =HYPGEOMDIST(3, 5, 18, 41) ➔ 27.5%; half of this probability is 13.75%. Subtracting this value from the p value given by SPSS (38.1%) yields 24.4%, which is almost exactly what was calculated in the step-by-step method (24.3%), the difference being due to rounding error.

Two-Tailed Test

Performing one-tailed tests is usually not justified. In our example, one would need strong arguments that radiation could not be less effective than surgery, so that a small x could only be attributed to sampling error. Without a strong rationale, use a two-tailed test but increase statistical power by using mid-p values.

The computational details are as follows:

1. Compute the probability of x (27.5% in the cancer example).
2. Add to it the probability of all other tables with a probability less than that of the observed one (in our example, this is all the possible tables except the one corresponding to $x = 2$). This calculation yields a two-sided p value of 63.8%.
3. To implement the method of mid-p values, subtract one half the probability in Step 1 (13.75%), yielding a p value of 50.1%.

McNemar Test (for Related Samples)

Two samples are related if every individual (or object) in one sample is matched with an individual (or object) in the second. These are also referred to as paired or matched samples. Because the two samples are not statistically independent, special methods must be used to analyze related samples.

Types of Dependence

Matching

Matching involves selecting a second sample so that it is similar to the first. For example, McKinney et al. (2005) evaluated the Family Writing Project (FWP), in which family members were involved in teaching students to write and in helping them practice at home. The FWP was implemented in eight different schools in a large urban school district. As part of the evaluation, a comparison group of eight schools that were not implementing the program was chosen. For each school in the experimental group, the evaluation team looked for a school with similar characteristics to include in the comparison group. Schools were matched on three characteristics: grade levels at school, average reading scores, and average language scores. Some matching was also performed at the student level. Through these matching procedures, the comparison group was made similar to the experimental group. All participants then wrote essays, and the quality of the essays between the two groups was compared.

The research design here was quasiexperimental because it was not possible to assign schools or students randomly to condition; inclusion in the experimental group was based on whether the school and students were participating in the FWP. In an experimental design, randomization is used to make the experimental and control groups similar on all variables except the independent variable (IV) and dependent variable (DV); that is, similar on all "third variables". Because randomization was not possible, the matching procedure was used to make the experimental and comparison groups similar on at least some extraneous variables.

Pre–Post Designs

A second situation in which pairing occurs is in pre–post research designs, where each individual serves as her own control. This design is simpler than the matching design discussed above, and so in the book we will only consider pre–post or other longitudinal designs when related samples are discussed. The reader must keep in mind that the same statistical tests that apply here also apply to matched designs.

As an example of a pre–post design, consider a hypothetical school, Basic Middle School. (The example is loosely based on an actual study by Nussbaum and Edwards [2011].) In the experimental classes, students were surveyed as to whether they thought it was a good idea for the federal government to raise taxes to help feed people who are hungry or homeless. The students then discussed the issue in several whole-class discussions and then were resurveyed as to their attitudes. I will refer to the two surveys as the *presurvey* and *postsurvey*. The research question was whether participating in oral discussions caused student attitudes to change. The outcome of this study is shown in Table 5.7.

At pre, the percentage of students who supported the proposed policy (raising taxes to provide more services to the hungry or homeless) was $\frac{84}{100} = 84\%$. At post, the percentage was $\frac{67}{100} = 67\%$. The drop in support was 17 percentage points. We want to ask if this drop was statistically significant. The two percentages (P_{pre} and P_{post}) are estimates of the population parameters π_{pre} and π_{post}, which are the values that one would obtain if the experiment was conducted with everyone in the population (e.g., all seventh graders). The population parameters represent the "truth" of whether the intervention had any effect. The parameters are estimated

Table 5.7 Support for Raising Taxes to Feed Homeless (in 7th-Grade Classroom)

	Support at post	Oppose at post	Total
Support at pre	63	21	84
Oppose at pre	4	12	16
Total	67	33	100

Table 5.8 Contingency Table With Nonmatching Design

	Support	Oppose	Total
Pre	84	16	100
Post	67	33	100
Total	151	49	200

by the sample proportions; however, small differences in the proportions could be due to sampling error. Thus, one needs to ask whether the decline in proportions is statistically significant.

It so happens that the DV in this example was nominal (students either support the policy or they do not). In this example, the survey instrument did not give them the opportunity to be undecided. The IV was also nominal. If a nonmatching design were implemented, one could use Table 5.8.

Note the sample size is 200, double that for Table 5.7, because there are now different people in the experimental group than in the control group. A chi-square test would show that $\chi^2(1) = 6.92$, $p = .009$, $\Delta P = -17.0\%$. A chi-square test assumes, however, that all the cases are statistically independent from one another, for example, that no student copied responses off another student's survey.

As noted previously, pre–post and other paired designs involve nonindependent observations. Student A's attitude at pre surely affects her attitude at post. The post attitude could conceivably just be a "copy" of her attitude at pre. The post attitude might also be influenced by the treatment and other factors, but one needs to calculate statistical significance by taking the pairing into account, as shown in Table 5.7. To analyze the table, one needs to use the McNemar test.

McNemar Test

The McNemar test only assumes that the individual cases are mutually independent but that for each individual there are two dependent observations (pre and post). One classifies each individual into one of the four cells in Table 5.9. Let us designate these four cells with the letters *a, b, c,* and *d,* as shown in the table.

Recall from before that the percentage who supported the proposed policy decreased from 84% to 67%, a drop of 17%. One wants to find out if this drop is significant. Note that (*b*) – (*c*) is also 17. The cells (*b*) and (*c*) represent individuals whose attitudes changed. Under the null hypothesis of no treatment effect, attitudes may change for various reasons, but the attitudes are just as likely to increase as decrease. Note that if there is no attitude change, (*b*) and (*c*) will both be zero and so will (*b*) – (*c*). The statistic (*b*) – (*c*) can therefore be thought of as an effect size measure, analogous to ΔP.

Table 5.9 Support for Raising Taxes to Feed Homeless (Cells Labeled With Letters)

	Support at post	Oppose at post	Total
Support at pre	63 (a)	21 (b)	84
Oppose at pre	4 (c)	12 (d)	16
Total	67	33	100

Table 5.10 Observed and Expected Counts in McNemar Test

X	Observed count	Expected count	Deviations: observed − expected
1	c	$\dfrac{(b+c)}{2}$	$c - \dfrac{(b+c)}{2}$
0	$a + d$		
−1	b	$\dfrac{(b+c)}{2}$	$b - \dfrac{(b+c)}{2}$

The McNemar test just focuses on cells (b) and (c). The null and alternative hypotheses are expressed as: H_0: $(b) = (c)$; H_a: $(b) \neq (c)$. Because the table was constructed so that the individual cell values are independent, one can perform a chi-square test on the contingency table. Let X be a random variable with the coding −1, 0, 1 (representing, respectively, lose support, no change, gain support). Then one can compute the following expected values (see Table 5.10; from McNemar, 1947).

In our example, $b = 21$ and $c = 4$, so $b + c = 25$. There are a total of 25 students whose attitudes changed. Under the null hypothesis, b should equal c, so there should be just as many students whose attitudes changed in one direction as the other. This hypothesis is why the expected count in Table 5.10 is $\frac{(b+c)}{2} = 12.5$. Now the test statistic T is

$$T = \frac{(b-c)^2}{(b+c)} \tag{Eq. 5.1}$$

In the example, $T = \frac{(21-4)^2}{21+4} = \frac{289}{25} = 11.6$. Because T follows the chi-square distribution (in a large sample), one can write: $\chi^2(1) = 11.6$. The probability of obtaining a chi-square statistic of 11.6 or greater is, from EXCEL, =CHIDIST(11.6,1) ➔ .0007, so, $p = .0007$. One should also report $b - c = 17\%$ as an effect size measure. Readers may not know what b and c represent, so the result should be described as the difference between those who changed to "support" and those who changed to "oppose." Seventy-five percent did not change their attitudes. Overall, it appears that the treatment (discussion of the issue) had a negative impact on students' attitudes. Of course, one also needs to consider other possible explanations for the change; for example, perhaps there was a TV program on the issue that some of the students watched at home. One can still say, however, that the change in attitudes does not appear to be caused by chance.

It may seem that one is not using the information from the a and d cells in this test, but actually it is being done implicitly. Remember that for every student who falls into the a and d cells, that is one less student that will fall into the b and c cells, so the information is still being taken into consideration by the test.

EXPLORING THE CONCEPT

How is the McNemar test a form of a chi-square test? In Table 5.10, the deviation for $X = -1$ is $b - \frac{(b + c)}{2}$. Multiplying the first term by $\frac{2}{2}$ to create common denominators yields $\frac{2}{2}b - \frac{(b + c)}{2} = \frac{b - c}{2}$. Squaring the deviation produces $\frac{(b - c)^2}{4}$. (a) Why is the next step to multiply by the reciprocal of the expected value $\left(\frac{2}{(b + c)}\right)$? (b) Go through similar calculations for when $X = 1$. Is the result identical, given that $(c - b)^2 = (b - c)^2$? (c) Does summing the two results $\left[\text{each } \frac{1}{2} * \frac{(b - c)^2}{(b + c)}\right]$ produce Eq. 5.1?

Small Sample Binomial Test

According to asymptotic theory, the McNemar test requires $(b + c) \geq 25$. If this condition does not apply, one can instead assume the sampling distribution is binomial, using b as the test statistic. Perform a binomial test, with b representing positive cases (hits) and c representing negative cases (misses). The total number of trials is $b + c$.

Consider Table 5.11. The table is the same as Table 5.7 except that b has been reduced to 15. Because $b + c = 19$, one needs to conduct the binomial test, not the chi-square test. The number of hits (b) is 15, and the number of trials ($b + c$) is 19. Under the null hypothesis, $b = c$, or equivalently, the probability of a hit (π) is 50%.

To perform the test, first calculate a one-tailed p value and then double it for the two-tailed value. The probability of obtaining 15 or more hits is 1 minus the probability of obtaining 14 or fewer hits. Recall that in EXCEL, the cumulative probability can be calculated by =BINOMDIST(k, n, π, 1), where k is the number of hits. In our example: =2 * (1 – BINOMDIST(14, 19, 0.5, 1)) ➔ .019. The p value is therefore .019. For a one-tailed test (in small or large samples), also perform the binomial test (but do not double the result).

Table 5.11 Support for Raising Taxes With Smaller Cell Sizes

	Support at post	*Oppose at post*	*Total*
Support at pre	63 (*a*)	15 (*b*)	78
Oppose at pre	4 (*c*)	12 (*d*)	16
Total	67	27	94

Calculations in SPSS

To perform the McNemar test in SPSS, be careful about how the contingency table is set up. As a check, it is a good idea to also perform the test in EXCEL. Figure 5.2 shows how the data should be formatted to perform the McNemar test. Be sure to then weight the data set by the frequency counts. Then go to "crosstab" and "statistics" and check the McNemar test.

In setting up the data, note that the two variables are "Pre" and "Post" and the values of these variables are "support" and "oppose." It would be incorrect to set up a variable called "Time" and to make the values "Pre" and "Post" (and then another variable called support, coded Yes or No). That would only work for unpaired data.

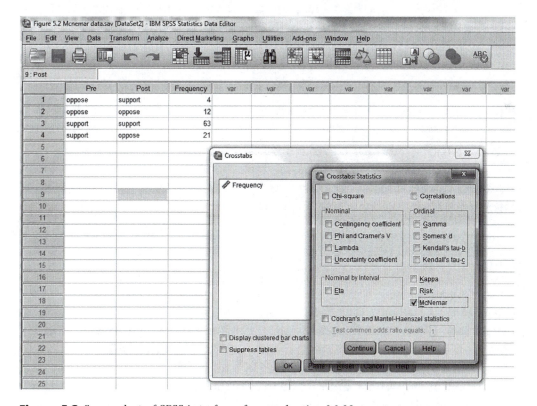

Figure 5.2 Screenshot of SPSS interfaces for conducting McNemar test.

After pressing "continue" and then "OK," SPSS returns a *p* value of .001, which is consistent with our previous calculation. There is a footnote that the binomial distribution was used. SPSS calculations could have also used the chi-square distribution; it would have returned the same results, however. If one wishes to report a χ^2 value in the text of one's manuscript, it can be calculated in EXCEL. Reporting a chi-square statistic may be more familiar to many readers than just reporting *b*.

Another way of conducting the test is to go to the nonparametric statistics option ➔ 2 related samples. Then check the box for the McNemar test on the "custom" tab.

Cochran Test

One other test that needs to be noted is the Cochran test. The Cochran test is an extension of the McNemar test for three or more related samples. For example, students might complete a presurvey, postsurvey, and delayed postsurvey. Another scenario is that students complete three treatments, and a survey is administered after each treatment. The Cochran test is the nominal equivalent to a repeated-measures ANOVA involving one group. To implement the Cochran test in SPSS, go to "nonparametric statistics," then to "related samples." (The Cochran test is for a $2 \times k$ table; for a $k \times 2$ table, perform the marginal homogeneity test.) The Cochran test can also be performed with the crosstab option; there is a box under "statistics" called "Cochran's and Mantel-Haenszel statistics." The Mantel-Haenszel test is considered next.

Controlling for a Third Variable: The Mantel-Haenszel Test

The Mantel-Haenszel (MH) test is used to statistically control for a third variable, which is useful if one needs to understand the causal structure among a set of variables. One can also use log-linear analysis for this purpose (see Chapter 12), but the MH test is simpler to conduct and explain to readers.

Causal Structure

The chi-square, Fisher exact, or McNemar tests are used to establish that two variables are statistically associated. It is well known, however, that association (e.g., correlation) does not establish that one variable exerts a causal influence on the other. If *X* is associated with *Y*, it is possible that *X* causes *Y* (*X* ➔ *Y*), but it is also possible that *Y* ➔ *X*, or that both are caused by a third variable, *Z*. In the latter cause, the relationship between *X* and *Y* is said to be *spurious*. Figure 5.3 is a causal diagram of a third variable causing a spurious relationship between two other variables.

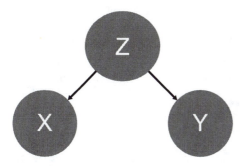

Figure 5.3 Illustration of a spurious relationship between X and Y. The arrows connote causal relationships. Any association between X and Y is caused by a third variable, Z.

In this situation, if one intervenes to increase the X variable through some sort of intervention (e.g., an educational program), it will have no effect on Y because Z is not increased. The time and energy spent on the intervention would be wasted. One can test for a spurious relationship by performing a randomized experiment where X is intentionally varied. It is not always possible, however, to use random assignments in field settings. In addition, conducting such experiments can be expensive.

Another option is to statistically control for Z. If one holds Z constant statistically, then the association between X and Y should disappear. What causes the spurious relationship in the first place is that, in the absence of intervention, when Z is high, so will be both X and Y (and whenever Z is low, so will be both X and Y). Under this scenario, if one keeps the value of Z constant, the association will disappear. (Of course, if there is a fourth variable, Q, that also has a causal influence on X and Y; the association may not disappear entirely but it likely will be reduced.)

All these scenarios assume that the causal links between the variables are positive. If one of the links is negative, then controlling for Z may increase the association between X and Y. In this case, it is still helpful to statistically control for Z so as to tease out the true pattern of relationships among the variables. Finally, one might control for Z and find that the association between X and Y is unchanged. This outcome would eliminate Z as a possible confounder. Such a procedure does not establish a causal relationship between X and Y because there still could be a fourth variable that may cause a spurious relationship between X and Y, but it does make a causal relationship between X and Y more plausible by eliminating one possible explanation for the association.

For these reasons, it is often useful to control statistically for a third variable. This can be done with the MH test when all the variables are nominal. X and Y must be dichotomous; Z can have multiple categories. The test is presented here for nonmatched, independent samples; however, the test can also be used for repeated measures if "person" is used as a control variable (this approach is equivalent to the Cochran test).

Statistical Analysis

Preliminary Analysis

Use of the procedure first requires analyzing the 2×2 contingency table using the chi-square or Fisher exact tests.

Consider the following hypothetical contingency (Table 5.12) where X represents whether or not a child regularly watches the TV show *Power Girls* and Y is whether the child shows aggressive tendencies on the playground (pushing, shoving, etc.).

A chi-square analysis shows that $\chi^2 (1) = 1.63$, $p = .67$, $\Delta P = 1.5\%$, and odds ratio (OR) $= 1.06$. There is little evidence here that watching *Power Girls* has any adverse effects.

Partial Contingency Tables

Use of the MH procedure next requires that one specifies two or more partial contingency tables for the different values of Z. In the example, we create the following two partial tables (see Table 5.13), one for each gender.

The data suggest that among both males and females, watching *Power Girls* is associated with a higher percentage of aggressive behavior. That was not apparent when examining the overall table, where there were no differences in the percentages. However, a third variable, gender, was masking the association. When one constructs partial tables, one statistically control for the effect of gender and are thus able to see the relationship.

Table 5.12 Number of Aggressive Children Who Watch *Power Girls*

Aggressive	Do not watch	Watch	Total
Yes	170	450	620
No	200	500	700
Total	370	950	1,320

Table 5.13 Number of Aggressive Children Who Watch *Power Girls* by Gender

	Males		Females	
	Not watch	Watch	Not watch	Watch
Aggressive	150	300	20	150
Not aggressive	100	100	100	400
Total	250	400	120	550
% Aggressive	60%	75%	17%	27%

EXPLORING THE CONCEPT

What data in the table support the claim that males are more aggressive than females? That males watch *Power Girls* less? Would these factors, taken together, tend to reduce the association between watching *Power Girls* and aggression?

It is important to keep in mind that in each partial table, the third variable (Z = gender) does not vary. This is why making partial tables is so important; it holds the third variable, Z, constant. In contrast to the partial tables, the overall table (Table 5.12) is known as the *marginal table*, because it combines the cells in Table 5.13.

Statistical Significance

One next needs to ask whether the differences are statistically significant. Table 5.13 shows the percentage differences, but one can also calculate the odds ratios. The ORs for the partial tables are: $OR_{males} = 2.00$, $OR_{females} = 1.88$. The partial odds ratios are close to one another: they average 1.94. This value is called the *common OR* (see Appendix 5.2 for the exact formula). The OR for the overall table is 1.06. A relationship may look different when one controls for a third variable.

The MH test examines whether X and Y are statistically independent when controlling for Z. The test statistic is

$$M = \frac{\left(\left|\sum x_i - \sum \frac{r_i c_i}{n_i}\right| - 0.5\right)^2}{\sum \frac{r_i c_i (n_i - r_i)(n_i - c_i)}{n_i^2 (n-1)}}, \text{ with } M \sim \chi^2. \tag{Eq. 5.2}$$

The test statistic is explained in greater detail in Appendix 5.2.

Implementation in SPSS

One can perform the Mantel-Haenszel test in SPSS using the crosstabs procedure. Figure 5.4 illustrates the appropriate steps. In the figure, the variable gender was entered in the crosstabs box under "Layer 1 of 1." In the statistics box, "Cochran's and Mantel-Haenszel statistics" was checked.

A portion of the resulting output is shown below in Table 5.14.

The table shows that $\chi^2(1) = 21.18$, $p < .001$, so one can accept the alternative hypothesis of conditional dependence. Note again that the estimated common OR

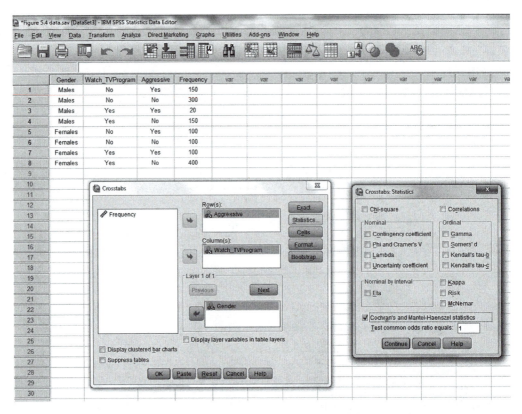

Figure 5.4 Screenshot of SPSS interfaces for conducting Mantel-Haenszel test.

Table 5.14 SPSS Output for Mantel-Haenszel Test

Tests of Conditional Independence

	Chi-Squared	df	Asymp. Sig. (2-sided)
Cochran's	21.855	1	.000
Mantel-Haenszel	**21.176**	**1**	**.000**

Mantel-Haenszel Common Odds Ratio Estimate

Estimate			**.511**
ln(Estimate)			−.672
Std. Error of ln(Estimate)			.145
Asymp. Sig. (2-sided)			.000
Asymp. 95% Confidence Interval	Common Odds Ratio	Lower Bound	.384
		Upper Bound	.679
	ln(Common Odds Ratio)	Lower Bound	−.956
		Upper Bound	−.387

Note: The Mantel-Haenszel common odds ratio estimate is asymptotically normally distributed under the common odds ratio of 1.000 assumption. So is the natural log of the estimate.

is 0.511, or if one takes the reciprocal, 1.96. These values are substantially different than the marginal (unconditional) OR of 1.06.

Preliminary Test

The MH test assumes that there is no interaction effect. An interaction effect is different from a confound. The presence of a third variable such as Z that is correlated with both X and Y is a confound. The definition of an interaction effect is that the association of X with Y differs for the different levels of Z. In this case, the different levels of Z are male and female. The OR_{male} and OR_{female} did not appear to differ much, so it did not appear that there was an interaction effect. This assumption can be formally tested with the Breslow-Day test, which SPSS will automatically conduct when one runs the Mantel-Haenszel procedure. (In the output, the result of the Breslow-Day statistic is reported as a test of the homogeneity of the odds ratio.) The null hypothesis here is that, in the population, the $OR_{male} = OR_{female}$. Because the Breslow-Day chi-square is 0.042 ($p = .84$), we fail to reject the null hypothesis of no interaction effect. The assumption of the MH test has been satisfied.

The Tarone statistic is an adjustment to the Breslow-Day statistic to make it more powerful when the common OR estimator is used. In practice, the results of the two tests usually do not differ, and that is the case here.

Because there is no interaction effect, it is meaningful to compute the common OR. SPSS estimates it to be 0.511, or if one takes the reciprocal, 1.96, which is very close to what was calculated previously (1.94). The value is substantially greater than the marginal (unconditional) OR of 1.06. The analysis shows that there is a strong *conditional association* (that is, an association when one conditions on gender) even though there is weak *marginal* association.

One application of the MH test is when an experiment with a nominal outcome is repeated on an independent sample. For example, a medical treatment (X) might be conducted on a second sample to see if it prevents the occurrence of a disease (Y). Z would be whether the cases are from the first or second sample. By performing the MH test and controlling for Z, one can calculate the common OR, which would be an effect size that employs data from both the samples. The Breslow-Day test can be used to determine if the treatment was equally effective in both samples. This scenario was the original context for which the MH test was developed (Mantel & Haenszel, 1959). For more background on the MH test, see Technical Notes 5.1 and 5.2.

Log-Linear Mode

The MH test is only appropriate when analyzing a 2×2 contingency table (technically a $2 \times 2 \times k$ table). With a larger $r \times c \times k$ table, such as a $3 \times 2 \times k$ or a $3 \times 3 \times k$ table, one should perform a log-linear analysis. Log-linear analyses will be discussed in Chapter 12. Log-linear models can also handle interaction effects, so if the Breslow-Day test is significant, one might then also consider performing a log-linear analysis. The alternative is to analyze the two partial tables separately, performing a chi-square test of statistical independence on each. To do that in SPSS, just check the chi-square box in crosstabs (be sure to have your Z variable, gender, included in the "Layer 1 of 1" box).

Summary

All the tests discussed in this chapter are only appropriate for 2×2 tables. Each section described a special situation: small samples, related samples, or the need to control for potentially confounding variables. (The one exception was the Cochran test, which involves 2×3 or larger tables, because three or more time periods are involved.) For larger tables, one can still perform a chi-square test of statistical independence, and then the Fisher exact test (or Mantel-Haenszel test) to compare two cells. One can also use log-linear models to analyze $r \times c$ tables. Keep the three special situations in mind when deciding what statistical tests to perform.

Problems

1. A researcher takes a poll of 30 voters to determine whether they support allocating city funds for a new recycling center.
   a) With the data below and statistical software, use Fisher's exact test to determine if there is a statistically significant association between party affiliation and support for reform at the $\alpha = .05$ level.
   b) Use EXCEL to adjust the results using the method of mid-p values.

	Democrats	Republicans	Total
Support	14	3	17
Oppose	6	7	13
Total	20	10	30

2. A researcher takes a poll of 67 voters to determine whether they support eliminating the capital gains tax.
   a) With the data below and statistical software, use Fisher's exact test to determine if there is a statistically significant association between party affiliation and support for eliminating the capital gains tax at the $\alpha = .05$ level.

b) Use EXCEL to adjust the results using the method of mid-p values. What is the size of the adjustment (to 4 decimal places)?

	Democrats	Republicans	Total
Support	1	20	21
Oppose	19	47	66
Total	20	67	87

3. Following are data on whether students passed a mathematics proficiency test after taking an online course designed to improve proficiency skills.

		After course	
		Unproficient	Proficient
Before course	Unproficient	25	22
	Proficient	9	21

a) Is the sample sufficiently large to use the (asymptotic) McNemar chi-square test rather than the binomial test?
b) Conduct a McNemar test by hand. Compute T (the chi-square statistic for this test), and the two-sided p value. For this problem, do not perform a continuity correction.
c) Does the course appear to be effective? What is $b - c$ (where b and c are expressed as percentages of the total)?
d) Conduct the test with statistical software. Report the p value.

4. During the campaign for votes in a Republican primary, 50 Republicans were selected at random and asked whether they would vote for Mitt Romney. Thirty-five supported Romney and 15 did not. After several weeks of campaigning, they were again asked their opinion. Of the 35 who were initially supportive, 15 continued to support Romney but 20 were now opposed. Of the 15 who were initially opposed, 7 remained opposed but 8 changed to supporting Romney. Hint: You may wish to use dummy variable (0, 1) coding.
a) Perform the McNemar chi-square test by hand or using EXCEL.
b) Perform the small-sample binomial test by hand or using EXCEL. Are the results more valid than those for the chi-square test?
c) Perform the McNemar test using statistical software. In SPSS, the figures may differ from those done by hand due to application of a continuity correction.

5. Below are data on predictors of reading disabilities, specifically whether a child's father has less than (or equal to) 12 years of education. (Variables based on St. Sauver, Katusic, Barbaresi, Colligan, & Jacobsen, 2001.)

Gender	Paternal education ≤ 12 years (EdLow)	Reading disability (RD)	
		Yes	No
Boys	Yes	111	718
	No	104	1,499
Girls	Yes	46	736
	No	42	1,571

Input the data into SPSS (or a similar computer program) and attach to your work a printed copy of the data set. You will need three variables (Gender, EdLow, and RD), plus a column for the frequencies. The first three variables may be either string variables or numeric (0 vs. 1) variables. There needs to be a separate row for each cell in the above table. I recommend

1) carefully inputting the first row corresponding to the first cell (Gender = "Boys," EdLow = "Yes," RD = "Yes") leaving off the frequency;
2) then copying or retyping the first row into the second and changing RD to "No";
3) then copying both these rows into new rows and changing Gender to "Girls";
4) then inputting all the frequencies into a new column called "Freq");
5) weighting the data by the frequencies.
   a) Perform a Mantel-Haenszel (MH) test using a statistics program such as SPSS. In making the contingency tables, make RD the columns, LowEd the rows, and "layer" by Gender. (SPSS will generate partial tables as well as perform the MH test if you check that box.) Report the results of the Breslow-Day (BD) test, MH test, and the estimate of the common OR. Note: If the BD test is significant, also report chi-square tests for statistical independence for each partial table, for example by checking "chi-square" in SPSS.
   b) For each of the partial tables, find the OR (using EXCEL or SPSS).
   c) Interpret your results. What story do they tell? Be sure to first interpret, when appropriate, the statistics for the BD and MH tests and the common OR.

6. Answer the questions in the previous problem for the following data:

Gender	Maternal education ≤ 12 years	Reading disability	
		Yes	No
Boys	Yes	103	836
	No	112	1,381
Girls	Yes	50	752
	No	38	1,555

7. Answer the questions in Problem 5 for the following data:

Gender	Low birth weight (< 2,500 grams)	Reading disability	
		Yes	No
Boys	Yes	86	990
	No	129	1,299
Girls	Yes	53	692
	No	35	1,615

8. Answer the questions in Problem 5 for the following data. The Apgar test measures the health of a newborn quickly after birth.

Gender	Apgar score (1 minute) < 8	Reading disability	
		Yes	No
Boys	Yes	97	987
	No	118	1,230
Girls	Yes	44	1,130
	No	44	1,177

9. A research team scored a persuasive essay for elements of a reasoned argument. For each sentence, a decision was made as to whether the sentence contained a claim, counterclaim, rebuttal (i.e., refutation of the counterclaim), or none of the above (null). The essay was scored independently by the professor and his graduate assistant (GA) to assess interrater reliability, with the following results:

GA codes	Professor codes				
	Claims	Counterclaims	Rebuttals	Null	Total
Claims	39	8	12	1	60
Counterclaims	3	22	0	0	25
Rebuttals	20	0	13	1	34
Null	8	5	0	18	31
Total	70	35	25	20	150

a) Enter the data into a statistical program such as SPSS. One variable should be PROF (can take values of claims, counterclaims, rebuttals, null). Be sure to set "variable type" to "string." The other variable should be GA. A third (numeric) variable should be the frequency (FREQ). Remember to weight by FREQ.

b) Create a contingency table. (In SPSS, go to "crosstabs." Also, click on "cells" and check "percentages" for total.) From the output, compute the total rate of agreement (sum the diagonals). Is the agreement rate satisfactory? In what cell is the disagreement rate the highest?

c)   Run *crosstabs* again, but this time check "kappa" (under "statistics"). Kappa (κ) controls for agreement that occurs just by chance. What is the value of κ? Do you think the value is satisfactory? (See Technical Note 5.3 for how kappa is calculated.)

Technical Notes

5.1   The Mantel-Haenszel test is an extension of one developed by Cochran and so is sometimes referred to as the Cochran-Mantel-Haenszel (CMH) test. Mantel and Haenszel included a continuity correction and a "variance inflation factor." The MH estimator is a consistent estimator, meaning the amount of bias decreases as the sample size increases. So technically, there should either be a large number of partial tables or the marginal frequencies should be large. However, in smaller samples, one can perform Fisher's exact test for each table to verify the significance of the results. Some versions of the MH test are also based on Fisher's exact test rather than the asymptotic test.

5.2   Some argue that the Mantel-Haenszel test is appropriate even in the presence of an interaction if one is only interested in knowing that X and Y are statistically dependent and not in estimating the ORs. However, one of the partial ORs might be insignificant, so this argument is weak.

5.3   Here is how kappa is calculated (Landis & Koch, 1977). Let θ_1 refer to the total percent agreement ("overall concordance"). Let θ_2 refer to probability of agreement just by chance ("chance concordance"). Chance concordance is found by multiplying the row and column marginal probabilities for each cell and summing them up. Let $\theta_1/100\%$ be the ratio of percent agreement over the total possible (100%). Then if one subtracts out of the numerator and denominator θ_2 (agreement due to chance), we get $\kappa = \frac{(\theta_1 - \theta_2)}{(100\% - \theta_2)}$.

APPENDIX 5.1

UNDERSTANDING THE HYPERGEOMETRIC DISTRIBUTION

The exact sampling distribution for 2×2 contingency tables with fixed marginal totals is hypergeometric. Given a particular n and set of marginal totals, one can find the probability of each possible table. Let x be the frequency count in the upper left cell. Given that there is 1 df, the question can be reframed as, "What is the probability of each possible value of x?" It is best to arrange the rows and columns so that the row and column totals are displayed in ascending order. In calculating the probabilities, the following formula is used:

$$\frac{\binom{r}{x}\binom{n-r}{c-x}}{\binom{n}{c}}, \tag{Eq. 5.3}$$

where r and $(n - r)$ are the row totals, n is the grand total, and c is the first column total.

Table 5.3 shows how this notation maps onto a 2×2 contingency table. Suppose you have 7 blue cars and 3 green cars. Define the population as these 10 cars. The probability of randomly selecting a blue car is therefore 70%. If you randomly select a sample of 5 cars, what is the probability of selecting 2 blue cars [i.e., Prob($x = 2$)]? These values are shown in Table 5.15.

Given the marginal totals, the probability of obtaining $x = 2$ is, using Eq. 5.3,

$$\frac{\binom{7}{2}\binom{3}{3}}{\binom{10}{5}} = 8.3\%.$$ To better understand this equation, remember one is drawing a sample of five. Think of five parking spaces, where at least two are reserved for blue cars.

| B | B | _____ | _____ | _____ |

Table 5.15 Car Example

Column 1	Column 2	Row total
$x = 2$	5	$r = 7$ blue cars in population.
$c - x = 3$	0	$n - r = 3$ green cars in population
Total: $c = 5$ in sample	$n - c = 5$	$n = 10$ cars in population

Because $\binom{7}{2} = \frac{n!}{k!(n-k)!} = 21$, there are 21 different ways that 2 blue cars (out of 7) could be drawn to fill those two parking spaces (e.g., you could select blue car 1 and 2, 1 and 3, 1 and 4 and so on, up to 6 and 7). In this example, $r = 7$ and $x = 2$ in the first term of the numerator of Eq. 5.3. Now, there is only one way that 3 green cars (out of 3) could be drawn, so the second term in the numerator is equal to 1. There are therefore 21 different ways that 2 blue cars and 3 green cars could be drawn from the population. These represent favorable possibilities (favorable in the sense of yielding the observed cells). The number of total possibilities is represented by the denominator in Eq. 5.3; it represents the number of different ways that the 5 parking spaces can be filled by the 10 cars. Because each permutation is equally likely, the probability is given by the number of favorable possibilities divided by the number of total possibilities, as reflected in Eq. 5.3.

APPENDIX 5.2

UNDERSTANDING THE MANTEL-HAENSZEL FORMULA

This appendix examines the computational details behind the *Power Girls* example.

Chi-Square Test

First, calculating the chi-square statistic for the overall table (Table 5.12) with SPSS yields a value of 0.163.

Mantel-Haenszel Test

When there are partial 2×2 tables, and no interaction effects, the Mantel-Haenszel test should be used. The formula for the test is

$$M = \frac{\left(\left| \sum x_i - \sum \frac{r_i c_i}{n_i} \right| - 0.5 \right)^2}{\sum \frac{r_i c_i (n_i - r_i)(n_i - c_i)}{n_i^2 (n - 1)}}, \text{ with } M \sim \chi^2. \tag{Eq. 5.4}$$

For this test, the null hypothesis is that the OR for each partial table is equal to one, and so the common OR is equal to one (H_0: $OR_{common} = OR_i = 1.0$).

With this formula, as with Fisher's exact test, x_i (the value in the first cell) is the random variable. It will differ for each partial table, so if there are two partial tables, there will be an x_1 and x_2. In the *Power Girls* example, x_1 (males) is 150; x_2 (females) is 20. The sum $\sum x_i = 170$. To perform a chi-square test, subtract the expected values of x_i from the observed values, where $E(x) = \sum \frac{r_i c_i}{n_i}$. Applied to *the Power Girls* example, and deriving the marginal totals from Table 5.12, we have $E(males) = \frac{250 * 450}{650} = 173.08$ and $E(females) = \frac{120 * 170}{670} = 30.45$. The overall expected value is 203.52. Because $\sum x_i = 170$, the deviation is -33.52. One half unit is then subtracted from the absolute value ($33.52 - 0.5 = 33.02$) for a continuity correction. The next step in a chi-square test is to square the value: $32.97^2 = 1,090.32$, producing the numerator of the test statistic, M.

The denominator is the *SE*, $\sum \frac{r_i c_i (n_i - r_i)(n_i - c_i)}{n^2 (n - 1)}$, as derived by Mantel and Haenszel (1959). For the *Power Girls* example, the calculations are, for males,

Table 5.16 EXCEL Calculations of Common Odds Ratio

Cell	Males	Females	
a	150	20	
b	300	150	
c	100	100	
d	100	400	
n	650	670	
			Below are the sums of the two columns:
ad/n	23.08	11.94	35.01
cb/n	46.15	22.39	68.54
			Divide the second value above into the first: Common OR = 0.51 or 1.96 (taking inverse).

$\frac{450 * 250 * (650 - 450) * (650 - 250)}{650^2 * (650 - 1)} = 32.82$, and for females $\frac{170 * 120 * (670 - 170) * (670 - 120)}{670^2 * (670 - 1)} =$ 18.68. The sum for the two tables is 51.50. M is therefore $\frac{1090.32}{51.50} = 21.17$. (Because of rounding error above, the value in the SPSS output is 21.18). Because M has a chi-square distribution, the p value, calculated in EXCEL, is =CHIDIST(21.01,1) → 0.000005. These results can be reported as $\chi^2_{MH} = 21.17, p < .001$.

Estimating the Common Odds Ratio

The formula for estimating the common odds ratio is (using the a, b, c, d) notation introduced in Table 5.9 is

$$OR_{common} = \frac{\sum(a_k d_k / n_k)}{\sum(c_k b_k / n_k)}. \tag{Eq. 5.5}$$

One can perform the calculation in EXCEL as shown in Table 5.16.

The result is basically the average of the individual ORs, weighted by sample size.

References

Agresti, A. (2002). *Categorical data analysis* (2nd ed.). Hoboken, NJ: Wiley-Interscience.

Landis, J. R., & Koch, G. G. (1977). The measurement of observer agreement for categorical data. *Biometrics, 33,* 159–174.

Le, C. T. (1998). *Applied categorical data analysis.* New York, NY: John Wiley & Sons.

Mantel, N., & Haenszel, W. (1959). Statistical aspects of the analysis of data from the retro-spective studies of disease. *Journal of the National Cancer Institute, 22,* 719–748.

McKinney, M., Lasley, S., Nussbaum, M., Holmes-Gull, R., Kelly, S., Sicurella, K., . . . King-horn, M. (2005). *Through the lens of the Family Writing Project: The Southern Nevada Writing Project's impact on student writing and teacher practices* (Submitted to sponsor). Berkeley, CA: National Writing Project.

McNemar, Q. (1947). Note on the sampling error of the difference between correlated propor-tions or percentages. *Psychometrika, 12,* 153–157.

Nussbaum, E. M., & Edwards, O. V. (2011). Argumentation, critical questions, and integrative stratagems: Enhancing young adolescents' reasoning about current events. *Journal of the Learning Sciences, 20,* 433–488. doi:10.1080/10508406.2011.564567

St. Sauver, J. L., Katusic, S. K., Barbaresi, W. J., Colligan, R. C., & Jacobsen, S. J. (2001). Boy/girl differences in risk for reading disability: Potential cues. *American Journal of Epidemiol-ogy, 154,* 787–794.

CHAPTER 6

BASIC NONPARAMETRIC TESTS FOR ORDINAL DATA

Ordinal data come in several forms. One form consists of *ranking* the cases in a sample from high to low on some attribute, such as creativity, motivation, and so forth. Another form is *ordered categories*, where cases are classified into three or more groups that have some natural ordering, such as *high, medium,* and *low*. A third form consists of scores from a *rubric*, such as the one shown in Table 6.1. In this rubric, the quality of the arguments in students' essays was rated on a 7-point scale. While it is tempting to analyze these scores as metric, that assumes that the distances between the adjacent points on the scale are roughly equal on some underlying continuous variable. That assumption might be reasonable in some circumstances, but when it is not, treating the scale as metric can introduce additional measurement error into the scores, thereby reducing statistical power. If in doubt, it is safest to treat rubric scores as ordinal.

This chapter addresses how to conduct nonparametric statistical tests when one or more of the variables are ordinal. The meaning of the term *nonparametric* is discussed first. I then consider the case when both variables (X and Y) are ordinal in the context of measures of association. The subsequent section examines the situation when the dependent variable (DV) is ordinal and the independent variable (IV) is nominal, focusing on the median and Wilcoxon-Mann-Whitney tests. Tests for multiple or related samples are discussed in the next chapter.

The statistical procedures examined in this and the following two chapters are summarized in Table 6.2. Table 6.2 also shows the tests appropriate for contingency tables (both variables nominal). The table is a matrix that can *help* determine what statistical procedure to use in various situations. The possible levels of measurement for the X variable are shown along the vertical axis and those for the Y variable along the horizontal. (Unless the procedure is nondirectional, X is the IV, and Y is the DV.) Only procedures for nominal and ordinal IVs are shown; those for metric IVs will be introduced in later chapters. For the complete matrix, see Chapter 15 (Table 15.1).

Table 6.1 Holistic Essay Scoring Rubric Used

Coder _____   Subject No. _____

0   **Response to topic.** Paper responds to the topic in some way but does not provide an opinion on the issue.
1   **Undeveloped opinion.** Paper states an opinion, but no reasons are given to support the opinion, or the reasons given are unrelated to or inconsistent with the opinion, or they are incoherent.
2   **Minimally developed.** Paper states a clear opinion and gives one or two reasons to support the opinion, but the reasons are not explained or supported in any coherent way. The reasons may be of limited plausibility, and inconsistencies may be present.
3   Between the standards for 2 and 4. (If only 1 reason developed and 1 reason undeveloped.)
4   **Partially developed.** Paper states an opinion and gives reasons to support the opinion, plus some explanation or elaboration of the reasons. The reasons are generally plausible though not enough information is provided to convince a reader. There may be some inconsistencies, irrelevant information, or problems with organization and clarity (more poorly organized than 5).
5   Between the standards for 4 and 6. (A "It depends on what watching" would go here due to strength.) Could have counterclaims but are not developed.
6   **Well developed.** Paper states a clear opinion and gives reasons to support the opinion. The reasons are explained clearly and elaborated using information that could be convincing. Should mention opposing opinion. The essay is generally well organized and may include a concluding statement.
7   **Elaborates and addresses opposition.** Meets the criteria for previous level. In addition, the paper deals with the opposing opinions either with refutation or alterative solutions, or explaining why one side is stronger/more convincing than the other. Overall, the essay is positive. The paper is free of inconsistencies and irrelevancies that would weaken the argument. (Counterargument is strong, elaborated, or there are multiple counterarguments.)

Scoring rubric from Nussbaum and Schraw (2007).

Introduction to Nonparametric Statistics

Definitions of "Nonparametric"

The term *nonparametric test* seems to imply a procedure where no population parameters are estimated. This impression is incorrect. Many nonparametric tests estimate parameters such as the population median or mean rank. Most authors define a nonparametric test as one that is distribution free. This definition is also a bit misleading, because statistics that estimate population parameters always have a sampling distribution. The key idea is this: If one reduces a metric variable to the ordinal level through ranking, the transformed variable will have a uniform distribution regardless of the population distribution of the original metric variable

Table 6.2 Matrix of Nonparametric and Categorical Techniques Involving Nominal or Ordinal IVs

		X variable(s)	
		Nominal	*Ordinal*

	Metric	(A)
		Two matched samples:
		    Wilcoxon signed ranks
		Interactions: ART
		ANOVA/Regression with dummy
		    variables; Poisson regression for
		    count data (Repeated measures or
		    GEE with matched samples)
			(D)
			Reduce *X* to nominal or *Y* to
			    ordinal

Y variable	**Ordinal**	(B)
		Two samples:
		    Wilcoxon-Mann-Whitney
		Multiple samples:
		    Kruskal-Wallis, Median
		Two matched samples:
		    Sign test
		Multiple matched samples:
		    Friedman test
		Interactions/confounding variables:
		    ANOVA-type statistics (ATS)
		Moses extreme reactions
		Ordered categories on Y:
		    Ordinal regression
		    (GEE with matched samples)
			(E)
			Spearman's $\hat{\rho}$
			Many ties/ordered categories:
			    Kendall's $\hat{\tau}$, Somer's *d*
			    (directional)
			Ranks (*Y*) and ordered categories
			    (*X*): Jonckheere-Terpstra test
			    (Page test with matched
			    samples)

	Nominal	(C)
		Chi-square test
		Small sample: Fisher's exact test
		Matched samples: McNemar (2×2);
		    Cochran ($r \times c$)
		Confounding variables: Mantel-
		    Haenszel ($2 \times 2 \times k$) or
		    Log-linear (GEE with matched
		    samples)
			(F)
			Kendall's $\hat{\tau}$
			    ($2 \times c$)

Note: Procedures in italics are discussed in Chapters 9–14.

(Randles & Wolfe, 1979, p. 37). The sampling distribution of a nonparametric statistic will therefore be a function of the uniform distribution, although it will become normal in large samples. The sampling distribution is therefore "freed" from the distributional shape of the original metric variable. One might also define a nonparametric test as one that involves the analysis of ranks. (This definition is consistent with the above definition but simpler to comprehend.) The limitation with this definition, however, is that the analysis of contingency tables (using only nominal data) is also often considered nonparametric. Here I need to make recourse to the distinction between categorical and nonparametric data analysis. These are two separate but overlapping subfields of statistics, each with their own textbooks and journals. Both subfields have laid claim to the analysis of contingency tables but, other than that similarity, have evolved in different directions. Categorical data analysis evolved to incorporate linear and log-linear regression techniques where the DV is nominal and the IV(s) are nominal (as in log-linear models) or metric (as in logistic regression and generalized linear models). Nonparametric statistics evolved to focus more on the analysis of rank data.

When Should One Use Nonparametric Statistical Procedures?

There are four conditions under which one should consider using a nonparametric test. First, when a variable involved in the analysis is ordinal. Second, when all the variables are metric but the sample is small and so one cannot assume that the sampling distribution is approximately normal. Third, when some other assumption of parametric analysis is violated, such as the homogeneity of variances assumption and when a nonparametric test does not make a similar assumption. Fourth, if the DV is notably skewed or has high kurtosis (meaning that the distribution is unusually peaked), one should consider using nonparametric or categorical analysis techniques.

The fourth condition requires some explanation. High levels of skew or positive kurtosis can significantly reduce the statistical power of parametric tests (Hodges & Lehmann, 1956; Randles & Wolf, 1979; Tukey, 1960). These conditions increase the probability of a Type II error, where one fails to establish a real effect as statistically significant. Given that conducting a study involves considerable time and expense, it can be argued that researchers should be as concerned with avoiding Type II errors as they are with avoiding Type I errors. Another condition that can reduce statistical power is the presence of outliers (extreme observations). (The presence of skew is often equivalent to having a number of semiextreme outliers, so the following explanation for why outliers can reduce statistical power also applies to skew.)

Figure 6.1(a) involves two variables that are perfectly correlated (the values of one are just doubles of the other). Figure 6.1(b) charts the same data but where one of the data points has been multiplied by 5, making it an outlier. As shown in Figure 6.1(b), the outlier has created a point off the best-fitting line. The outlier biases the slope of the line upward, throwing the remaining points off the line as well, thereby reducing the magnitude of the correlation. Specifically, r has dropped from 1.0 to .56 (and the Y variable has become more skewed). The lower correlation makes it more difficult to establish the correlation (and all associated statistics) as statistically significant, thereby reducing statistical power. (There is a loss of power also because the standard errors are increased.) Skew in a distribution can have the same effect, as can high positive kurtosis.

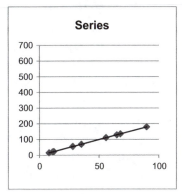

(A) Perfect correlation

Variable X	Variable Y
8	16
11	22
12	24
28	56
35	70
56	112
65	130
68	136
90	180

Skew	0.35	0.35
Kurtosis	−1.28	−1.28
Correlation	1.00	

Series

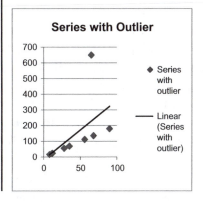

(B) Correlation with outlier

Variable X	Variable Y	
8	16	
11	22	
12	24	
28	56	
35	70	
56	112	
65	650	**Outlier**
68	136	
90	180	

	0.35	2.56
	−1.28	7.01
	0.56	

Series with Outlier

Figure 6.1 Illustration of how outliers can often reduce correlations.

EXPLORING THE CONCEPT

Why would kurtosis also reduce statistical power? Peakedness in a distribution is also associated with thicker tails. Would this create more semiextreme observations compared to the case of a normally distributed variable?

As a rule of thumb, some researchers use nonparametrics when the skew or kurtosis of a DV is greater than 1.0. This particular rule of thumb is based on professional judgment. Osborne (2013) noted that traditional parametric tests require remediation with skew levels as high as |0.8|. (The remediation could take the form of a data transformation or the use of nonparametrics.) Some researchers only take corrective action when skew or kurtosis significantly differ from zero (i.e., about twice the standard error); but this criterion is not based on statistical power considerations and so is not recommended.

Whichever criterion is used, a researcher should consistently use the same one from study to study. If the researcher is inconsistent, it creates the appearance and possibly the reality that one is being opportunistic, meaning that one is picking the statistical test that will give one the best results. It is highly inappropriate to analyze data with both parametric and nonparametric techniques and then to report the test that has the lowest p value. If one conducts both a parametric and nonparametric test and they both return the same result (i.e., both indicate significance or both indicate nonsignificance), the situation is not problematic: One can briefly report that one performed the analysis a different way but found the same general result. But if the two tests return different substantive results, then what can one conclude? It is tempting to use the test giving the lowest p value, but this practice inflates the Type I error rate.

Remember that a Type I error is accepting the alternative hypothesis when the null hypothesis is in fact true (tantamount to reporting a "false" finding). Statistical tests are designed with a particular α level, usually 5%, meaning that about 5% of the time one will make Type I errors when the null hypothesis is true and will correctly not reject the null in 95% of the cases. Now if in the latter type of situation one also uses a nonparametric test and, when that test is significant, erroneously rejects the null, then one will be making Type I errors more than 5% of the time (when the null hypothesis is true).

If one runs both a parametric and nonparametric test on the same data, it is imperative that one states a decision rule in advance indicating which test will be binding. There should also be a rationale for preferring one type of test over the other. Having some consistent decision rule can serve as one sort of rationale, but a criterion such as skew or kurtosis values greater than 1.0 is still somewhat arbitrary. Zimmerman (2011) proposed a decision rule based on a simulation study: In comparing two groups with the similar variances, if a t statistic based on metric data and one based on ranks differ by more than 0.40, then use the nonparametric test (see Technical Note 6.1).

For even more precision, one should consult each test's *asymptotic relative efficiency* (ARE). The ARE is a rough measure of how powerful one's statistical test is compared with another in large samples. It is also known as *Pitman efficiency*. Suppose we desire a Type I error rate of $\alpha = .05$ and a Type II error rate of $\beta = .10$. The goal is to decide which one of two different tests (say the t test and the median test) is more efficient, that is, has more statistical power. Let n_1 be the sample size of Test 1 to achieve these error rates, and n_2 be the sample size of Test 2. Pitman (1948) showed that the ratio n_1/n_2 tended toward a limit (the ARE) as n_1 approached infinity, and that this limit was the same regardless of our choice of α and β. Although technically only applicable to large samples, the ARE is still useful when using small samples, because simulation studies indicate that small-sample relative efficiency is usually close to the ARE (Randles & Wolfe, 1979; Sawilowsky, 1990; Sprent, 1989).

AREs are computed empirically from mathematical derivations and computer simulations. Table 6.3 presents some ARE results summarized from Conover (1999) and Randles and Wolf (1979). An ARE greater than 1.0 indicates that Test A is more

Table 6.3 Asymptotic Relative Efficiency (ARE) for Selected Statistics

Test 1	Test 2	ARE	Comment
Cox and Stuart Test of Trend	Regression	0.78	Normality required.
	Spearman $\hat{\rho}$, Kendall's $\hat{\tau}$	0.79	
Spearman $\hat{\rho}$, Kendall's $\hat{\tau}$	Regression	0.98	Normality required.
Friedman	t test	0.64	Two samples.
Friedman	F test	$0.864k/(k+1)$	or greater
		$0.955k/(k+1)$	if normal.
		$k/(k+1)$	if uniform[a]
		$3k/[2(k+1)]$	if double expon.[b]
Kruskal-Wallis Test	Median Test	1.50	if normal
		3.00	if uniform[a]
		0.75	if double expon.[b]
	F test	0.86	to infinity if i.i.d.[c]
		0.96	if normal
		1.00	if uniform
		1.50	if double expon.[b]
Wilcoxon-Mann-Whitney	t test	0.86	to infinity if i.i.d.[c]
		0.95	if normal and X, Y = shapes.
		1.00	if uniform[a]
		1.50	if double expon.[b]
	Median Test	1.50	if normal
		3.00	if uniform[a]
		0.75	if double expon.[b] and large sample.
Wilcoxon Sign Rank Test	Paired t	0.86	to infinity, if random sample and i.i.d.[c]
		0.96	D is normal
		1.00	D is uniform[a][d]
		1.50	D is double expon.[b]
Sign Test	Paired t	0.64	if normal
		0.33	if uniform[a]
		2.00	if double expon.[b]

[a]High negative kurtosis (flat distribution).
[b]Double exponential: high positive kurtosis.
[c]Groups should be independently and identically distributed, except for the locations of the means.
[d]Simulation study by Randles and Wolfe (1979) found t to be more powerful when $n = 15$ or 20.

powerful than Test B in large samples. The table mainly demonstrates the effect of high kurtosis on the power of nonparametric tests relative to parametric ones. For example, the presence of a double exponential distribution—a distribution with high positive kurtosis (see Chapter 8, Figure 8.3)—makes nonparametric tests more powerful than parametric ones.

The table does not demonstrate, however, the detrimental effect of high skew on the power of parametric tests. Although parametric tests may be valid in large samples because of the central limit theorem, *that does not mean that those tests are the most statistically powerful ones.* If the assumptions underlying a test are violated (because, for example, of small sample sizes or the use of variables at the wrong level of measurement), then that should override statistical power as the decisive consideration in choosing a test. However, when it is valid to conduct either type of test, statistical power should be a major consideration. Unfortunately, an underappreciation of the effect of skew and kurtosis on statistical power, and the underdevelopment of methods for testing statistical interactions (see Chapter 8), led to a waning in the popularity of using nonparametric statistics in the 1970s and beyond, after a peak of popularity in the 1960s (Sawilowsky, 1990). During the last couple decades, however, there has been a resurgence of interest in using alternatives to traditional parametric methods.

Measuring Skew and Kurtosis

One can easily measure the skew and kurtosis of a variable using either EXCEL or statistical software. In EXCEL, use the =SKEW and =KURT functions. In SPSS, there are a variety of options that will provide skew and kurtosis statistics. I am partial to the "Frequency" option under "Descriptive Statistics" (then, under "Frequency," go to "Statistics"). By using the "Frequency" option, one can also obtain a histogram showing the shape of the distribution and a list of the frequencies of different categories. (A frequency listing can be used to spot errors, such as when one entry is misspelled; the "Frequency" procedure will create a category just for that mistyped entry.) One can also obtain skew and kurtosis values from the "Descriptive" or "Explore" options.

The formulas for skew and kurtosis reference the *moments of the distribution.* The moments of a RV are $\frac{\sum [X_i - \bar{X}]^m}{n}$ where m is the number of the moment. More specifically, $\frac{\sum |X_i - \bar{X}|^1}{n}$ is the first moment and $\frac{\sum |X_i - \bar{X}|^2}{n}$ is the second moment (variance). The third moment, $\frac{\sum |X_i - \bar{X}|^3}{n}$, when standardized, gives the skew

$$Skew = \frac{\sum [X_i - \bar{X}]^3}{(n-1) * SD(X_i)^3}. \tag{Eq. 6.1}$$

The term $(n - 1)$ is used in the denominator because one degree of freedom (df) has been lost in estimating the mean, $E(X_i)$. Unlike the second moment, which is positive because of squaring, the third moment can be negative because cubing a negative deviation results in a negative.

EXPLORING THE CONCEPT

Does the magnitude of any outliers or semiextreme observations become magnified when the deviation from the mean is skewed? How would this affect the magnitude of the skew statistic?

The formula for kurtosis is a function of the fourth moment, $\dfrac{\sum \left| X_i - \bar{X} \right|^4}{n}$. The numerator of the expression relates to the deviations from the mean, $(X_i - \bar{X})$. When these deviations are more varied than in a normal distribution, some will be smaller (with X_i values nearer the center of the distribution) and some will be larger (in the tails of the distribution), thus producing both more peakedness and heavier tails. (The actual formula for kurtosis: $\dfrac{\sum [X_i - \bar{X}]^4}{(n-1) * SD(X_i)^4}$, but because under this formula, the standard normal distribution has a kurtosis of 3; to make the standard normal distribution have a kurtosis of zero, the usual formula for kurtosis also involves subtracting 3:

$$Kurtosis = \frac{\sum [X_i - \bar{X}]^4}{(n-1) * SD(X_i)^4} - 3.0. \tag{Eq. 6.2}$$

Some statistical packages use the unadjusted formula, but SPSS and EXCEL use Eq. 6.2 (and it is the one used in this book). Distributions that have high positive kurtosis are known as *leptokurtic* and those with high negative kurtosis are known as *platykurtic*. The former are more peaked than the normal distribution and the latter more flat.

In summary, high levels of skew or kurtosis may cause a parametric test to lose statistical power compared to a nonparametric test. In other situations, parametric tests may be more powerful. Nonparametric tests are designed to be used with ordinal or nominal data, so conducting a nonparametric test on metric data essentially reduces the level of measurement from metric to ordinal, thereby throwing away information. This can reduce statistical power in large samples when the population distribution is fairly normal, but sometimes not by much (see Table 6.3). However, when the DV is in fact ordinal or nominal, or the sample size is small, nonparametric tests should be used.

Analysis of Association or Trend

When both variables are ordinal, it is appropriate to use a nonparametric measure of association. This section discusses Spearman's $\hat{\rho}$ and Kendall's $\hat{\tau}$. Before introducing these tests, however, I first clarify the notion of a tie in rank.

Ties in Rank

Suppose students are asked to complete an art project that displays their creativity, and a panel of judges ranks the projects in terms of creativity. To keep the number of cases manageable for purposes of illustration, I use an *n* of 6. Suppose two of the art projects (3 and 4) are judged to be equally creative. In this case both projects are given the same rank, as shown in Table 6.4. The table shows three methods for assigning ranks in the presence of ties. Method A assumes that both cases could be assigned either a 3 or a 4, and then assigns the mean of 3 and 4 (3.5). Method B just assigns the smaller value (3) to both cases, and Method C is similar to Method B except all numbers are kept in sequential order. Method A is the one that is almost always used in statistical analysis. It is known as the method of *midranks*.

In SPSS, if one needs to reduce metric data to ranks, the "Rank" option (under "Transform") can be used; midranks is the default. Most of the nonparametric options automatically rank data, typically making it unnecessary to use the "Rank" option, but there are exceptions. As a precaution, the "Rank" option should be used with rubric-based scores with many ties (e.g., ordered categories), where the raw numeric scores are not in rank sequential order. For example, 10, 20, 30, or 1, 1.5, 2, should both be transformed to 1, 2, 3.

A few of the nonparametric tests considered in this book are not available in SPSS but can be calculated step by step in EXCEL. Doing so will require one to use the ranking function in EXCEL, =RANK.AVG. Only this function produces midranks but is not available in earlier versions of EXCEL. (One can still use the older rank function, =RANK, and then calculate the midranks by hand.) The syntax for the RANK.AVG function is: =RANK.AVG(number, **ref**, [order]). "Number" is typically a cell address of the number being ranked, **ref** is a reference to the array of data among which the number is being ranked, and [order] refers to whether the data are to be ranked in ascending or descending order. A "1" (or any nonzero entry) for [order] yields a descending order of values where high numbers are given high ranks; this option should almost always be the choice when conducting nonparametric

Table 6.4 Illustration of Methods for Assigning Ranks in the Presence of Ties (Projects 3 and 4 Tied)

Project No.	Method A	Method B	Method C
1	1	1	1
2	2	2	2
3	3.5	3	3
4	3.5	3	3
5	5	5	4
6	6	6	5

tests. (It may or may not make a major statistical difference but makes the results easier to interpret.) An example of the syntax is =RANK.AVG(B2,$B$2:$C$21,1), where in this example, B2 is the cell address of the number being ranked and B2 through C21 are the cell addresses of the data. The dollar signs are inserted using the F4 key and these create absolute cell addresses, so that when the syntax is copied to other cells, the cell addresses of the data array will not change. Sometimes the data array will be a single column, but in this example the data are in two columns (one each for an experimental and control group).

Spearman's $\hat{\rho}$

Spearman's rho ($\hat{\rho}$) is a measure of association between two rank variables based on the following formula:

$$\hat{\rho} = 1 - \frac{6\sum d^2}{n(n^2 - 1)},$$ (Eq. 6.3)

where n is the number of paired ranks and d is the difference between the ranks for a specific pair. The formula is not intuitive, but fortunately when there are not ties it reduces to Pearson's correlation coefficient (r) applied to the ranks. If there are more than a couple ties in rank on one or both variables, Kendall's $\hat{\tau}$ is a bit more suitable. (Kendall's $\hat{\tau}$ is described in the following section.)

Robustness

Spearman's $\hat{\rho}$ can also be used with metric variables, rather than Pearson's r, when there are outliers. This is because Spearman's $\hat{\rho}$ is more robust to outliers than (r). Table 6.5 shows the same data as in Figure 6.1(b), where there was one substantial outlier. Recall that there was a perfect correlation between X and Y until the outlier (650) was introduced, after which the r dropped to .56. The note to the table shows that $\hat{\rho}$ is more robust, dropping only to .95. The outlier was created by multiplying the original value by 10 but the *rank* of the observation only increased from 7 to 9.

Because $\hat{\rho}$ is robust to outliers, it is sometimes a good idea, with metric data to generate $\hat{\rho}$ in addition to r for each cell of a correlation matrix, and to compare the two. Substantial differences can be used as one screening device for outliers.

Using Spearman's $\hat{\rho}$ should also be considered when one or both variables are skewed, because skew is like having a lot of semiextreme points, which can distort the values of Pearson's r. It is best to use r when both variables are fairly symmetrical. One can also sometimes use a data transformation (e.g., logarithms, square roots) to remove skew from a metric variable. A data transformation has the advantage of not reducing the data to the ordinal level but makes the analysis more difficult to explain. Reducing a variable to the ordinal level does throw away

Table 6.5 Pearson's r and Spearman's $\hat{\rho}$ Compared in the Presence of an Outlier

Variable X	Variable Y	Rank$_X$	Rank$_Y$
8	16	1	1
11	22	2	2
12	24	3	3
28	56	4	4
35	70	5	5
56	112	6	6
65	**650**	7	9
68	136	8	7
90	180	9	8

Notes: Pearson's $r_{xy} = .56$. Spearman's $\hat{\rho}_{xy} = .95$. Variable $Y = 2X$ with the exception of the outlier (650), which is in bold.

some information, which can reduce statistical power, but this may be more than compensated by the advantage of reducing skew, which can increase power.

When there are ties in the rank, one should use the midrank. A statistical correction for ties can be applied to $\hat{\rho}$ if desired (see Daniel, 1990, for the specific formulas).

Statistical Tests and Required Sample Sizes

One can test whether an estimate of ρ is statistically different from zero. The null hypothesis is that, in the population, ρ is equal to zero. SPSS will automatically conduct the test for you (go to "Correlate," then to "Bivariate," and then check "Spearman"). To conduct the test manually in small samples, use the table of quantiles from Glasser and Winter (1961), a portion of which is reproduced in Table 6.6. (Use of a quantile table assumes that there are no ties in the data; if there are ties, the test will likely be conservative.) One can also conduct a t test, which provides a reasonably good approximation when $n > 10$, using the following formula (Sheskin, 1997):

$$t = \frac{\hat{\rho}\sqrt{n-1}}{\sqrt{1-\hat{\rho}^2}}(df = n-2). \tag{Eq. 6.4}$$

With larger samples ($n > 30$), use the normal approximation. In the normal approximation, it can be shown that:

$$z = \hat{\rho}\sqrt{(n-1)}. \tag{Eq. 6.5}$$

Table 6.6 Exact Upper Quantiles of Spearmen's $\hat{\rho}$ for $n \leq 10$

n	$p = .900$	.950	.975	.990	.995	.999
4	.80					
5	.70	.80	.90	.90		
6	.60	.77	.83	.89	.94	
7	.54	.68	.75	.86	.89	.96
8	.50	.62	.71	.81	.86	.93
9	.47	.58	.68	.77	.82	.90
10	.44	.55	.64	.73	.78	.87

Notes: Take the negative of each value to find the lower quantile. (For example, $\hat{\rho} = .80$ corresponds to the 90th quantile, $\hat{\rho} = -.80$ corresponds to the 10th quantile.) Critical regions correspond to areas greater than but not including the appropriate upper quantile, and less than (but not including) the appropriate lower quantile. Table from Glasser and Winter (1961), Critical values of the coefficient of rank correlation for testing the hypothesis of independence, *Biometrika, 48,* 444–448 (Appendix). Reprinted by permission of the publisher Oxford University Press.

For the data in Table 6.5, $n = 9$, so a quantile table should be used to test the significance of $\hat{\rho}$. Now $n = 9$, and $\hat{\rho} = .95$, which corresponds to at least the .999 quantile (see Table 6.6), therefore $p < .001$. Quantile tables are exact tables, because the tables give the exact probabilities, based on calculations of the number of permutations of different ranks that yield particular values of $\hat{\rho}$ or $\hat{\tau}$ over all possible permutations.

Daniel's Trend Test

An interesting application of Spearman's $\hat{\rho}$ is Daniel's test of trend. The test examines whether there is a linear trend in longitudinal data by correlating X with Time. This test has been found to be somewhat more powerful than the Cox and Stuart test of trend, but is also less flexible, as the Cox and Stuart procedure can also test for cyclical patterns. If there are ties in the data, one can use Kendall's $\hat{\tau}$ instead of Spearman's $\hat{\rho}$ in the test of trend.

Kendall's $\hat{\tau}$

Like Spearman's $\hat{\rho}$, Kendall's tau ($\hat{\tau}$) is a measure of association between two ordinal variables and can range from -1 to 1. Kendall's $\hat{\tau}$ accommodates ties better than Spearman's $\hat{\rho}$; also its sampling distribution approaches normality more quickly, specifically with $n \geq 8$. One additional advantage of $\hat{\tau}$ is that the sample estimate is unbiased in small samples whereas the one for $\hat{\rho}$ is not.

The idea behind Kendall's $\hat{\tau}$ is that if two variables, X and Y, are associated, then as X increases, so should Y. If the two variables are inversely associated, then as X increases, Y should decrease. If two variables are unassociated, Y should increase or decrease independently from X. These same ideas also hold for Pearson's r, except the latter also quantifies the covariance. Kendall's $\hat{\tau}$ is not based on the covariance but rather on the number of concordant and discordant pairs. A pair of points is concordant if Y increases when X does.

To calculate $\hat{\tau}$, each case will have a rank on X and a rank on Y. So for each case there will be a pair of numbers representing the coordinates of that observation on the X-rank and Y-rank axes. For example, if a case ranks first on X but fifth on Y, then the coordinates for that case will be (1, 5). Next one pairs each case with every other case in the sample. If (1, 5) is matched with (3, 6), the pair of cases would be considered *concordant,* because as X increases, Y also increases (from 5 to 6). If (1, 5) is matched with (3, 2), the pair of cases would be considered *discordant,* because as X increases, Y decreases. Pairs can also reflect ties on either X or Y. For example, the following pair of cases would be considered tied on Y: (1, 5), (3, 5).

When There Are No Ties

When there are no ties, $\hat{\tau}$ is defined as follows:

$$\hat{\tau} = \frac{N_c - N_d}{N},$$
(Eq. 6.6)

where N_c is the number of concordant pairs, N_d is the number of discordant pairs, and the denominator is the total number of pairs N.

EXPLORING THE CONCEPT

Examine Eq. 6.6. If all the pairs are concordant, why will $\hat{\tau} = 1.0$? If all the pairs are discordant, why will $\hat{\tau} = -1.0$? If there is no association, why will $\hat{\tau} = 0.0$? (Hint: In the last case, one is just as likely to have concordant as discordant pairs; that is, $N_d = N_c$).

Consider a simple example with the four data points shown in Table 6.7.

One first makes a matrix of the two variables to find all the possible pairs of points, as shown in Table 6.8.

The table shows that there are six total pairs of points: five concordant and one discordant. The number of possible pairs is given by:

$$\text{Total possible pairs} = \frac{n(n-1)}{2},$$
(Eq. 6.7)

Table 6.7 Data for Kendall $\hat{\tau}$ Example (No Ties)

Ranks on Variable (X)	Ranks on Variable (Y)
1	1
2	2
3	4
4	3

Table 6.8 Matrix of Values for Finding All Possible Pairs of X and Y for $\hat{\tau}$ (No Ties) Example

Pair	1, 1	2, 2	3, 4	4, 3
Pair				
1, 1	—	Concordant	Concordant	Concordant
2, 2	—	—	Concordant	Concordant
3, 4	—	—	—	Discordant
4, 3	—	—	—	—

or $\frac{4*3}{2} = 6$ in this example. In calculating $\hat{\tau}$, the numerator is $N_c - N_d = 5 - 1 = 4$, so $\hat{\tau} = \frac{4}{6} = .67$.

You may be wondering why a hat is placed over the letter tau. This notation is because $\hat{\tau}$ is an estimate of the population parameter, τ. The parameter is the difference between the probability of a pair being concordant and the probability of being discordant. However, the hat symbol is not consistently used in the literature.

When There Are Ties

The presence of ties in the rankings, for example (1, 2.5) and (4, 2.5), complicates matters. Let us redo Tables 6.7 and 6.8 so that there is one tie in the Y ranking on the first two observations, yielding Tables 6.9 and 6.10.

The tables show that there are four concordant pairs, one discordant pair, and one tie (as this case is neither concordant nor discordant). What is done with the tie? One option is to not count the ties in the numerator. The resulting statistic, known as $\hat{\tau}_a$, is defined as

$$\hat{\tau}_a = \frac{N_c - N_d}{N},$$

(Eq. 6.8)

Table 6.9 Kendall $\hat{\tau}$ Example (One Tie)

Ranks on Variable (X)	Ranks on Variable (Y)
1	1.5[a]
2	1.5[a]
3	4
4	3

[a]Reflects the mean of 1 and 2, as the first two cases are tied for first place.

Table 6.10 Matrix of Values for Finding All Possible Pairs of X and Y for $\hat{\tau}$ Ties Example

Pair	1, 1.5	2, 1.5	3, 4	4, 3
Pair				
1, 1.5	—	Tie	Concordant	Concordant
2, 1.5	—	—	Concordant	Concordant
3, 4	—	—	—	Discordant
4, 3	—	—	—	—

which, for the example, is $\frac{4-1}{6} = .50$. In contrast, the *gamma* statistic (Goodman & Kruskal, 1963) also excludes the tie from the denominator. Gamma is defined as:

$$G = \frac{N_c - N_d}{N_c + N_d},$$

(Eq. 6.9)

which, in the example, is $\frac{4-1}{4+1} = .60$. It is a better use of information, however, if the tie is somehow counted instead of thrown away. A better alternative is to count the pair as half concordant and half discordant. However, one would need to distinguish between ties on X values versus Y values, as these do not have the same status. Formally, the following rules define whether a case is concordant or discordant:

If $\dfrac{Y_2 - Y_1}{X_2 - X_1} > 0$, then concordant,

If $\dfrac{Y_2 - Y_1}{X_2 - X_1} < 0$, then discordant,

If $\dfrac{Y_2 - Y_1}{X_2 - X_1} = 0$ (i.e., tie on Y), then half concordant, half discordant,

If $X_2 = X_1$, then $\dfrac{Y_2 - Y_1}{X_2 - X_1}$ is undefined, and no comparison is made.

No comparison is made in the last case because $\frac{Y_2 - Y_1}{X_2 - X_1}$ has a zero denominator. Therefore, pairs involving a tie on X would be thrown out of the analysis. In the example, however, the tie is on the Y value. If one can consider X to be a fixed IV that is used to predict Y, Somers (1962) proposed using the following directional measure:

$$Somers'\ d = \frac{N_c - N_d}{N_c + N_d + T} \tag{Eq. 6.10}$$

where T is the number of ties on the DV. In the example, Somers' $d = \frac{4-1}{4+1+1} = .50$.

EXPLORING THE CONCEPT

Note that Somers' d is mathematically equivalent to counting the tie as half concordant and half discordant (e.g., $\frac{4.5 - 1.5}{4 + 1 + 1}$ also equals .50). Increasing N_c and N_d by one-half unit each has what effect on the numerator and denominator?

Note that Somers' d is asymmetrical. If we redo the previous example, switching X and Y, the tie would now relate to X and so would not be included in the calculations. Specifically $N_c = 4$, $N_d = 1$, Somers' $d = \frac{4-1}{4+1} = .60$. Now in many situations, what is considered the X versus the Y variable is arbitrary. If one truly wants a symmetric measure, it should not matter which variables one considers X vs. Y. Yet in this procedure, it does make a difference. One way of rectifying this problem is to average the two estimates, using the geometric mean for the denominator (see Appendix 6.1). This statistic is known as $\hat{\tau}_b$ and is equal to 54.8% in this example.

Statistical Significance

The null hypothesis is that X and Y are independent. As was the case with Spearman's $\hat{\rho}$, the p values in small samples may be extracted from an exact quantile table, but the table only applies when there are no-ties. If there are ties, the normal approximation must be used; the approximation is only valid if $n \geq 8$. (See Technical Note 6.2 for the normal approximation procedure.) Fortunately, the exact quantile table goes up to $n = 60$.

Table 6.11 is the exact table for $\hat{\tau}$. Now for the no-ties example, $\hat{\tau} = .67$, $n = 4$. The corresponding quantile is 90%; the one-tailed p value is 10% (the two-sided p value is 20%). To perform the analysis in SPSS, choose "Bivariate" under "Correlate," check the box for Kendall's $\hat{\tau}_b$. SPSS uses the normal approximation (even though the sample size is too small) and gives a two-sided p value of .17. With very small sample sizes ($n \leq 7$), do not use the SPSS results but use exact tables.

Table 6.11 Exact Upper Quantiles of Kendall's $\hat{\tau}$

n	.900	.950	.975	.990	.995	n	.900	.950	.975	.990	.995
4	.67	.67	1.00	1.00	1.00	40	.14	.18	.22	.25	.28
5	.60	.60	.80	.80	1.00	41	.14	.18	.21	.25	.28
6	.47	.60	.73	.73	.87	42	.14	.18	.21	.25	.27
7	.43	.52	.62	.71	.81	43	.14	.17	.21	.24	.27
8	.36	.50	.57	.64	.71	44	.14	.17	.21	.24	.27
9	.33	.44	.50	.61	.67	45	.13	.17	.20	.24	.26
10	.33	.42	.47	.56	.60	46	.13	.17	.20	.24	.26
11	.31	.38	.45	.53	.56	47	.13	.17	.20	.23	.26
12	.27	.36	.42	.52	.55	48	.13	.16	.20	.23	.26
13	.28	.33	.41	.49	.53	49	.13	.16	.19	.23	.25
14	.25	.34	.38	.45	.49	50	.12	.16	.19	.23	.25
15	.25	.31	.37	.45	.49	51	.12	.16	.19	.22	.25
16	.23	.30	.37	.42	.47	52	.12	.16	.19	.22	.24
17	.24	.29	.35	.41	.46	53	.12	.16	.19	.22	.24
18	.23	.28	.33	.40	.44	54	.12	.15	.18	.22	.24
19	.22	.27	.32	.37	.43	55	.12	.15	.18	.21	.24
20	.21	.26	.32	.37	.41	56	.12	.15	.18	.21	.24
21	.20	.26	.30	.36	.40	57	.12	.15	.18	.21	.23
22	.19	.26	.30	.35	.39	58	.12	.15	.18	.21	.23
23	.19	.25	.29	.34	.38	59	.12	.15	.17	.21	.23
24	.19	.24	.28	.33	.37	60	.11	.15	.17	.21	.23
25	.19	.23	.28	.33	.36						
26	.18	.23	.27	.32	.35						
27	.17	.23	.27	.32	.35						
28	.17	.22	.26	.31	.34						
29	.17	.22	.26	.31	.34						
30	.17	.21	.25	.30	.33						
31	.16	.21	.25	.29	.32						
32	.16	.21	.24	.29	.32						
33	.16	.21	.24	.28	.31						
34	.16	.20	.23	.28	.31						
35	.15	.19	.23	.27	.30						
36	.15	.19	.23	.27	.30						
37	.15	.19	.23	.26	.29						
38	.15	.19	.22	.26	.29						
39	.14	.18	.22	.26	.29						

Notes: To find the lower quantile, take the negative of each value. (For example, $\hat{\tau} = .67$ corresponds to the 90th quantile, $\hat{\tau} = -.67$ corresponds to the 10th quantile.) Critical regions correspond to areas greater than but not including the appropriate upper quantile, and less than (but not including) the appropriate lower quantile. Data from Best (1974).

As an example of a study using Kendall's $\hat{\tau}$, Johnson and Nussbaum (2012) examined the effect of mastery orientation on the coping skills of college students. (Mastery orientation indicates that a student's achievement goal is to understand and master the material.) Because of skew, mastery-orientation scores were reduced to the ordinal level. Mastery-approach scores were found to be correlated with task-oriented coping ($\hat{\tau} = .32$, $p < .001$), whereas mastery-avoidance goal-orientation scores were correlated with emotion-oriented coping scores ($\hat{\tau} = .27$, $p < .001$).

Ordered Categories

Because Kendall's $\hat{\tau}$ can accommodate ties, one can use Kendall's $\hat{\tau}$ when both X and Y consists of ordered categories. As mentioned in the previous section, Kendall's $\hat{\tau}_b$ can be used with a contingency table with ordered categories, such as high, medium, and low. The categories here could be ranked 3, 2, 1, where 1 represents the lowest-ranking category. Within any category, the ranks of the cases are tied (e.g., all the cases in the medium category would be tied at 2). The categories cannot be coded with string variables (e.g., *high*), as most statistical packages will not recognize the ranking. Here, the high category should be coded as "3."

To calculate $\hat{\tau}$ in SPSS, go to "Descriptive Statistics," "Crosstab," and then to "Statistics." One is given a choice of four statistics to choose from: Somers' *d*, gamma, $\hat{\tau}_b$, and $\hat{\tau}_c$. Although the data are not in categories, the crosstab procedure will still work. If one checks the $\hat{\tau}_b$ option, the output gives a value of .548. The final statistic that is reported in SPSS is $\hat{\tau}_c$ (.56). The statistic $\hat{\tau}_c$ is very close to $\hat{\tau}_b$ but includes a factor that adjusts for table size. One should use $\hat{\tau}_c$ for ordered categories where the contingency table is rectangular but not square.

As an example using ordered categories, consider the data in Table 6.12. The table is square, so one should use $\hat{\tau}_b$ rather than $\hat{\tau}_c$. Ranks must be used to code the categories (i.e., *high* = 3, *medium* = 2, *low* = 1). The results are $\hat{\tau} = .21$, $p = .03$. One can reject the null hypothesis and conclude that X and Y are dependent; furthermore, because $\hat{\tau}$ is positive, there is a tendency for Y to increase as X increases, and vice versa.

2 × c Ordered Categories

In a situation where one variable consists of ordered categories and the other variable is nominal, one can use Kendall's $\hat{\tau}$ because the nominal variable can be coded 1 and 2. Which category to assign to 1 and which to 2 is arbitrary, but that will only affect the sign of the $\hat{\tau}$ statistic, not the magnitude.

Table 6.13 displays a 2 × c contingency table where X represents whether students were in the high, medium, or low academic tracks in high school and Y is whether or not the student completes high school. In this case Y is dependent on X, so one can use the directional measure, Somers' *d*. The results are Somers' $d = .31$,

Table 6.12 Contingency Table of Ordered Categories

	High (Y)	Medium (Y)	Low (Y)	Total (Y)
High (X)	21	4	14	39
Medium (X)	15	17	6	38
Low (X)	5	3	15	23
Total (X)	41	24	35	100

Table 6.13 2 × c Contingency Table With Ordered Categories

	High (X)	Medium (X)	Low (X)	Total (X)
Completers	21	18	2	41
Noncompleters	4	17	3	24
Total (Y)	25	35	5	65

$p = .002$. Further calculations show that 84% of the high-track students graduated, but only 51% of the medium-track students and 40% of the low-track students did so.

Jonckheere-Terpstra Test for r × c Tables

The scenarios considered so far assume that X and Y are both either ranks or ordered categories. The Jonckheere-Terpstra test is appropriate if one variable consists of ranks and the other consists of ordered categories. For example, Y might consist of ranks for the variable creativity and X might consist of the categories high, medium, and low motivation. The test is based on a comparison of all possible pairs of groups. The test statistic (J) is the sum of the tie-corrected Wilcoxon-Mann-Whitney test statistics for comparing two groups, to be described later in this chapter. It turns out, however, that J is equivalent to $\hat{\tau}$, so a $\hat{\tau}$ test may be employed instead, given that both variables are ordinal.

Partial $\hat{\tau}$

The partial $\hat{\tau}$ is a measure of association that controls for a third variable. A partial $\hat{\tau}$ can be computed by some statistical software, such as SAS. In SPSS, one can calculate $\hat{\tau}_{xy}$, $\hat{\tau}_{xz}$, and $\hat{\tau}_{yz}$, and then apply the following formula:

$$\hat{\tau}_{xy.z} = \frac{\hat{\tau}_{xy} - \hat{\tau}_{xz} * \hat{\tau}_{yz}}{\sqrt{(1 - \hat{\tau}_{xz}^2)(1 - \hat{\tau}_{yz}^2)}}. \qquad \text{(Eq. 6.11)}$$

$\hat{\tau}$ and $\hat{\rho}$ Compared

Although $\hat{\tau}$ and $\hat{\rho}$ are both measures of association, they are based on different methodologies. The $\hat{\tau}$ has the advantage of approaching asymptotic normality more quickly and has also been extended to contingency tables. Other than that, the two statistics almost always produce the same statistical decisions. They also have the same ARE against Pearson's r (0.91) when the underlying population distribution is normal and potentially larger if skewed. The choice of statistic may often largely depend on which one the researcher finds easiest to explain.

Comparing Two Groups

In reference to Table 6.2, previous chapters discussed tests appropriate when both X and Y variables are nominal (Cell C), and this chapter has so far discussed tests when both variables are ordinal (Cell E). The one exception is the use of Kendall's $\hat{\tau}$ for a $2 \times c$ contingency table (shown in Cell F), where the X variable is ordinal and the Y variable is nominal (and dichotomous). Although Kendall's $\hat{\tau}$ could also be used when the X variable is nominal and the Y variable ordinal, it is more customary to use a median or Wilcoxon-Mann-Whitney test in that situation, especially when X is a fixed variable. This section discusses these and other tests that fall into Cell B of the matrix, specifically those involved in comparing two groups.

There is one important note for users of IBM SPSS. To compute the test statistics discussed in this section (and the following chapter as well), the ordinal variables must be set to "scale" on the variable view tab. Technically, these tests assume that the DV reflects an underlying (i.e., latent) continuous variable that cannot be measured exactly, and so the measurement scale is only ordinal (for example, self-reports of the amount of stress one is under). If the existence of a continuous latent variable is not plausible and the variable is purely ordinal, then inferences using nonparametric tests can still be made but should be confined to those relating to distributional tendencies.

Median Test

The median test can be used when X is dichotomous or has multiple categories; however, the dichotomous case is easier to understand. The test is loosely analogous to the difference of means test (t test) where the Y variable is metric. In the median test, Y consists of ranks and one tests whether the medians of the two groups are significantly different. (If Y consists of metric data, Y is reduced to ranks.) If there are n ordered observations, the overall median is the $\left(\frac{n}{2}\right)$th observation, rounding up if the number of observations is odd $\left(\frac{n}{2} + 0.5\right)$, or averaging the two middle observations if n is even. As an example, if $n = 7$, the overall median would be given by the fourth ordered observation. In addition to the overall median, there are also medians for each group. For example, consider the data in Table 6.14. (The ranks are of the performance of seven surfers in a contest—7 being the best—with those in Group A having used a specially designed surfboard.) The overall median is 4, whereas the median for group A is 6 and for group B is 3.5. Is the difference between 6 and 3.5 significant? The null and alternative hypotheses are

H_0: Mdn(A) = Mdn(B) = Mdn(Overall) in the population.
H_a: Mdn(A) ≠ Mdn(B) ≠ Mdn(Overall) in the population.

Table 6.14 Example Data for Median Test

Data ranks and medians

Group	Rank (Y)	Median
A	7	
A	6	Mdn(A) = 6
B	5	
B	4	Mdn(Overall) = 4
B	3	
B	2	Mdn(B) = 3.5
A	1	

Contingency table: Is observation above the overall median?

	Group A	Group B	Total
Frequency: >Mdn(Overall)	2	1	3
Frequency: ≤Mdn(Overall)	1	3	4
Total	3	4	7

Probability of different tables

Upper left frequency	Probability (hypergeometric distribution)
0	11.4%
1	51.4
2	34.3
3	2.9
Total	100.0%

Sum of probabilities equal to or less than the one for the observed value of 2: 48.6%.
P value of median test: 48.6%, accept null hypothesis.

Table 6.14 also classifies each case into a 2 × 2 contingency table regarding whether each case is above the overall sample median. The table can then be analyzed using Fisher's exact test or a chi-square test if the sample is large enough. However, for purposes of the step-by-step illustration in Table 6.14, the random variable is the value contained in the upper left cell of the contingency table (it here takes on the value of 2). The possible values of the RV range from 0 to 3. (The value cannot be greater than the smallest marginal total, or otherwise one of the cell frequencies would be less than zero.) The results of a Fisher exact test indicate that the probability of obtaining an observed value of 2 or something more extreme is 48.6%. This probability is then taken to be the p value of the median test. The results are nonsignificant; with only seven observations, statistical power is very low. It should be recognized that Fisher's exact test is quite conservative. One can improve on the power of the test by using the method of mid-p values when conducting this test (see Chapter 5), which reduces the p value to 31.4%.

To conduct the median test in IBM SPSS, go to "Nonparametric Tests," "Independent Samples." In older versions of SPSS, go to "Nonparametrics," "K Independent Samples."

Wilcoxon-Mann-Whitney Test

Whereas the median test examines whether the medians of two (or more) groups significantly differ, the Wilcoxon-Mann-Whitney (WMW) test examines whether the average ranks are different. The computational formulas are actually based on the sum of the ranks in each group. It makes no mathematical difference whether one uses total or average ranks. Use of the total ranks simplifies the formulas, but use of average ranks may make more intuitive sense. In Table 6.14, the ranks in group A are 7, 6, and 1, which average to 4.67. The ranks in group B are 5, 4, 3, and 2, which average 3.50.

The test used to be based on a complicated statistic known as the Mann-Whitney U (see Technical Note 6.3). Wilcoxon independently developed a simpler statistic, W, which returns the same p values. Therefore the test is now sometimes referred to as the Wilcoxon-Mann-Whitney test. So what is W? W is the sum of ranks for one of the two groups. It is more common for W to be based on whichever sum is *smallest*, and that is the convention adopted in this book. (One could also use the other sum of ranks for W—newer versions of SPSS do this—it should not affect the results.) In Table 6.14, the ranks in both groups sum to 14, so one does not have to choose a group; $W = 14$. The p value is the probability of obtaining a W of 14 or something more extreme.

As is the case with many of the tests considered in this book, there are two methods for testing significance: one for small samples based on exact tables or tests and one for larger samples based on a sampling distribution approximating the normal distribution or some derivative of it (e.g., chi-square or F). The latter are known as *asymptotic tests*. For Table 6.14, because the sample is small, one can use an exact table. Table 6.15 presents an exact table for the WMW test. (Because exact tables have been programmed into most statistical software, there are only a few exact tables presented in this book.) Note that the table only provides critical values (for $\alpha = .05$), not p values. If we let n represent the size of the group on which W is based, $n = 3$ and $m = 4$; because W could be based on either group, I assigned n to the smaller group. (We will also use a capital letter N in this context to refer to the total number of observations, so that $n + m = N$). According to the exact table, the critical value (leaving 2.5% in the lower tail of a two-tailed test) is 6. The upper critical value is obtained from the following equation:

Upper critical value $= n(n + m + 1) - $ *Lower critical value.* (Eq. 6.12)

In the example, we have:

Upper critical value $= 3(3 + 4 + 1) - 6 = 18.$

The region of rejection is a $W < 6$ *or* $W > 18$ (obtaining a W of 6 or 18 would not be considered significant). Because the actual value of W is 14, the result is not statistically significant.

Null Hypotheses

What is the null hypothesis for the WMW test? There are actually two different null hypotheses in common use, and it is important that one be clear as to which hypothesis one is choosing to test.

1. *Identical distributions hypothesis.* This hypothesis is that the two samples are drawn from the same population distribution. Therefore, the population distribution for sample A is the same as sample B. Formally, this can be stated as $F(A) = F(B)$, where F refers to the cumulative probability density function. This null hypothesis is the one originally used by Mann and Whitney (1947).
2. *Identical medians hypothesis.* This hypothesis states that population medians of the two groups are the same ($Mdn_A = Mdn_B$).

Rejection of the identical distributions hypothesis only implies that some of the moments of the two distributions are not the same. For example, the population means or medians could be the same but the variances or skews could differ. The population means and medians are measures of a distribution's *location* on

Table 6.15 Exact Critical Values of the Wilcoxon-Mann-Whitney W Statistic Corresponding to the .025[a] Quantile

n	m = 2	3	4	5	6	7	8	9	10	11	12	13	14	15	16	17	18	19	20
2	3	3	3	3	3	3	4	4	4	5	5	5	5	5	5	6	6	6	6
3	6	6	6	7	8	8	9	9	10	10	11	11	12	12	13	13	14	14	15
4	10	10	11	12	13	14	15	15	16	17	18	19	20	21	22	22	23	24	25
5	15	16	17	18	19	21	22	23	24	25	27	28	29	30	31	33	34	35	36
6	21	23	24	25	27	28	30	32	33	35	36	38	39	41	43	44	46	47	49
7	28	30	32	34	35	37	39	41	43	45	47	49	51	53	55	57	59	61	63
8	37	39	41	43	45	47	50	52	54	56	59	61	63	66	68	71	73	75	78
9	46	48	50	53	56	58	61	63	66	69	72	74	77	80	83	85	88	91	94
10	56	59	61	64	67	70	73	76	79	82	85	89	92	95	98	101	104	108	111
11	67	70	73	76	80	83	86	90	93	97	100	104	107	111	114	118	122	125	129
12	80	83	86	90	93	97	101	105	108	112	116	120	123	128	132	136	140	144	148
13	93	96	100	104	108	112	116	120	125	129	133	137	142	146	151	155	159	164	168
14	107	111	115	119	123	128	132	137	142	146	151	156	161	165	170	175	180	184	189
15	122	126	131	135	140	145	150	155	160	165	170	175	180	185	191	196	201	206	211
16	138	143	148	152	158	163	168	174	179	184	190	196	201	207	212	218	223	229	235
17	156	160	165	171	176	182	188	193	199	205	211	217	223	229	235	241	247	253	259
18	174	178	184	190	196	202	208	214	220	227	233	239	246	252	258	265	271	278	284
19	193	198	204	210	216	223	229	236	243	249	256	263	269	276	283	290	297	304	310
20	213	219	225	231	238	245	251	259	266	273	280	287	294	301	309	316	323	330	338

Notes: n is the sample size of the group on which W is based. Values of W that are less than the critical values are significant; values equal to the critical values are nonsignificant.

[a]Upper critical values are $w_{.975} = n(n + m + 1) - w_{.025}$, where $w_{.025}$ is the lower critical value. Exact values for other α levels can be found in Conover, 1999.

an underlying latent scale. Differences in location usually are of interest (e.g., does a particular intervention tend to cause scores to increase). One might argue that the second null hypothesis (identical medians) is more informative, but use of that hypothesis assumes that the two population distributions are identical in all respects except for location (i.e., they have the same shape, including the same variances). This assumption is necessary because a significant WMW test can be caused by other distributional differences than differences in the medians. (Remember that the test statistic is related to mean ranks, and means can be affected by things such as skew.) The assumption is very restrictive and often not met in practice. My recommendation is to use the first null hypothesis but to supplement it with some measure of effect size (to be discussed later in this chapter).

The WMW test is sometimes used with metric data when the assumption of homogenous variances for parametric tests is violated. However, unless the data display high skew or kurtosis, reducing the data to ranks may decrease statistical power (see Technical Notes 6.1 and 6.4). Some authors therefore recommend using a parametric Welch t test in this situation (or always when conducting a t test), as the Welch test does not require equal variances (Hayes & Cai, 2007; Rasch, Kubinger, & Moder, 2011).

Performing the Wilcoxon-Mann-Whitney Test in SPSS

The WMW test may be performed in SPSS by going to "Nonparametric Tests," and then to "Independent Samples" (or "2 Independent Samples" in older versions). SPSS will provide information on the mean ranks by group, the Mann-Whitney U statistic (see Technical Note 6.3), and Wilcoxon W.

Theory Behind the WMW Test

Although not absolutely essential, it is useful to understand some of the theory (which follows) on which the WMW test is based.

Definition of Expected Value

The underlying theory makes reference to the expected values of the ranks. There is a theorem that states

$$Sum\ of\ Consecutive\ Intergers\ (\textstyle\sum_{i=1}^{N} i) = \frac{N(N+1)}{2}. \tag{Eq. 6.13}$$

For example, the sum of 1, 2, 3, 4 is $\frac{4*5}{2} = 10$. (The proof of the theorem is in Appendix 6.2.) Let S be the sum of ranks on a particular variable; S is just a sum of consecutive integers. To find the mean rank, just divide S by N. Under either version of the null hypothesis, the group distributions are assumed identical, so the means of any group are equal to the overall mean. In other words, when considering average ranks

$$E(R_{\bullet A}) = E(R_{\bullet B}) = \frac{(N+1)}{2}.$$ (Eq. 6.14)

As an example, the expected values for the data in Table 6.14, where $N = 7$, is $\frac{(7+1)}{2} = 4$.

Derivation of the Exact Tables

It is important to note that exact tables are not necessarily based on any one formula. The entries in an exact table are based on logical, mathematical arguments for each entry, and the arguments may not necessarily be exactly the same for each entry. Deriving an exact table is labor intensive; however, once an exact table is derived, the author can publish it for posterity. Due to the effort involved, exact tables are confined to small sample sizes.

As an example of the type of logical reasoning involved, Table 6.16 gives the derivation for one entry in the WMW exact table, specifically for $n = m = 2$ (and N, the total sample size, is therefore 4). The table shows that W could range from 3 to 5, and that the probability of each permutation is one-sixth. Because each permutation is listed twice, the probability of each combination is one-third.

Table 6.15 shows that the lower critical value is $W = 3$. Table 6.16 implies that $W = 3$ has a one-sided p value of .33. The value of W would therefore have to be below 3 to be significant, but the minimum value of W is 3; therefore it would not be possible to establish statistical significance with such a small sample.

Table 6.16 Illustration of Exact Table Derivations for Groups of Size Two Each ($n = m = 2$)

Possible sum of ranks for one group	Sum of ranks other group	W (smaller of the two)	Probability
$1 + 2 = 3$	$3 + 4 = 7$	3	1/6
$1 + 3 = 4$	$2 + 4 = 6$	4	1/6
$1 + 4 = 5$	$2 + 3 = 5$	5	1/6
$2 + 3 = 5$	$1 + 4 = 5$	5	1/6
$2 + 4 = 6$	$1 + 3 = 4$	4	1/6
$3 + 4 = 7$	$1 + 2 = 3$	3	1/6

Normal Approximation

One can use the normal approximation if either *n* or *m* is ≥ 10.

EXPLORING THE CONCEPT

Why do you think the sampling distribution for *W* approaches normality more quickly than many of the statistics discussed in previous chapters? (Hint: The population distribution of ranks is uniform and therefore not skewed.)

To use the normal approximation method, one must first compute a *z* statistic:

$$z = \frac{W - E(W)}{SE_w}.$$
(Eq. 1.11)

The formula for $E(W)$ is:

$$E(W) = \frac{n(N+1)}{2}.$$
(Eq. 6.15)

The rationale for Eq. 6.15 is as follows: If from Eq. 6.14 the expected value for each case in the applicable group is $\frac{(N+1)}{2}$, then the expected value of the sum of these is $\frac{n(N+1)}{2}$ because there are *n* cases in the group. (Recall that *n* is the size of the group on which *W* is based, and *m* is the size of the other group.) The formula for the standard error of *W* is

$$SE_w = \sqrt{\frac{nm(N+1)}{12}}.$$
(Eq. 6.16)

Compared to expected value, it is harder to develop a conceptual understanding of formulas for standard errors, so I do not attempt to do so here.

We can, however, apply the formulas to a new example. In this example, suppose the sum of ranks for Group A is 128 ($m = 10$), and for Group B is 62 ($n = 9$). $W = 62$ (the smaller of these). Because the size of one of the groups is at least 10, one can use the normal approximation.

Plugging these numbers into the formulas, one finds that

$$E(W) = \frac{9 * (19+1)}{2} = 90.$$

$$SE_w = \sqrt{\frac{9 * 10 * (19+1)}{12}} = 12.25.$$

$$z = \frac{62 - 90}{12.25} = -2.29.$$

In EXCEL, =NORMSDIST(-2.29) = .011.

Double this for a two-sided *p* value (.011 $*$ 2 = .022).

One is allowed to double the one-sided p value into a two-sided one because the normal distribution is symmetrical. One could report the results in a journal article as $W = 62$, $p = .022$. This result would be considered significant at the 5% level.

If there are more than a couple ties in the data, the following formula for z should be used:

$$z = \frac{\left(W - n\frac{N+1}{2}\right)}{\sqrt{\frac{nm}{N(N-1)}\sum_i^N R_i^2 - \frac{nm(N+1)^2}{4(N-1)}}}.$$ (Eq. 6.17)

Without these adjustments, the WMW test will be too conservative.

Effect Sizes

It is good practice to report an effect size, especially if one is using Version 1 of the null hypothesis. Two effect size measures are considered here: (a) The Hodges-Lehmann estimator, and (b) the probability of superiority.

Hodges-Lehmann Estimator

The Hodges-Lehmann (HL) estimator (ΔHL, or just $\hat{\Delta}$) quantifies the treatment effect when comparing two groups when the DV is metric (Hodges & Lehmann, 1956; Lehmann, 2006). The HL estimator can therefore be used as an effect size measure with the WMW test. The statistic is computed by newer versions of SPSS (i.e., IBM SPSS) or can be calculated by hand by pairing every metric observation in Group 1 with each metric observation in Group 2, calculating the difference ($d_{ij} = y_j - x_i$), and finding the median difference. (There will be mn possible pairings.) Let us assume that the data in Table 6.14 are not rankings but are instead metric measurements. As shown in Table 6.17, for these data $\Delta HL = \text{median}(d_{ij}) = 2$. Note that

Table 6.17 Calculation of the Hodges-Lehmann Estimator

	Group B values			
	2	3	4	5
			d_{ij}	
Group A values				
1	−1	−2	−3	−4
6	4	3	2	1
7	5	4	3	2

Notes: $\Delta HL = \text{median}(d_{ij}) = 2$. Confidence interval is −4 to 5 (see Technical Note 6.5). It is assumed that the scores are metric.

this is not the same as the difference in the sample medians (which is 2.5), but ΔHL is an unbiased "median" estimator of the difference in the population medians.

EXPLORING THE CONCEPT

Note that the observations in Group A contain an "outlier" (the score of 1 is substantially different from the score of 6 and 7). Why is ΔHL more robust, that is, less affected by the outlier, than say comparing the means (4.7 for Group A and 3.5 for Group B)? For similar reasons, would ΔHL be more robust (and less variable) than just taking the difference between the sample medians?

Because the d_{ij}s constitute a distribution, a confidence interval (CI) can be computed for ΔHL. When $mn \geq 100$, the normal approximation method can be used (see Technical Note 6.5 for methods to use in smaller samples). With metric data, ΔHL is a more efficient estimator than the difference between the means when the distributions are substantially nonnormal (Helsel & Hirsch, 1991). Greater efficiency means that the CI will be smaller. The test assumes, however, that the variances of the two groups are roughly equal. The test is robust to violations of this assumption if the inequality is only caused by the presence of a small number of outliers.

Probability of Superiority

Another useful and more robust effect size measure is the probability of superiority (PS). The PS is a measure that "a randomly sampled score from one population is greater than a randomly sampled score from a second population" (Erceg-Hurn & Mirosevich, 2008, p. 599). When comparing two independent groups, the PS can be calculated directly from the Mann-Whitney U statistic (which in SPSS is also reported along with W; see Technical Note 6.3). The formula is

$$PS_{est} = \frac{U}{mn},$$

(Eq. 6.18)

where m and n are the sample sizes of the two groups. For the data in Table 6.14, SPSS provides a U value of 4, so $PS_{est} = \frac{4}{4*3} = 33\%$. Therefore 33% is the chance that a randomly chosen value in Group B will be greater than one in Group A. The probability that a value in Group A will be greater than one in Group B is $1 - 33\% = 67\%$. Not too much should be made of this value because the results are not statistically significant due to the small samples sizes, but they are presented here for illustration purposes only. Specifically, one can see from Table 6.14 that two of the three observations in Group A have higher ranks than all observations in Group B (and the third ranks lowest), so the PS_A of 67% makes sense.

If there is a practically significant effect, the PS should be above 50%; 50% is associated with no effect. Grissom (1994) and Grissom and Kim (2005), based on finding equivalences of PS to Cohen's d (a popular parametric effect size measure), suggests that a PS of 56% is a small effect, 65% is medium, and 71% is large.

Moses Test of Extreme Reactions

This test is useful when an intervention may cause values to become more extreme in both the positive and negative directions (but not affect central tendency). Sprent (1989) gave the example of how receiving a positive screening for cervical cancer may cause some women to want more information on the topic and others—who go into denial—to want less (compared to those with negative screenings). Moses (1952) gave the example of how, with some "defensive" personality traits, individuals may act in extreme ways (e.g., very loud or very quiet) compared to normal individuals.

For this test, the null and alternative hypotheses are

H_0: Extreme values are equally likely in both groups.

H_a: Extreme values are more likely to occur in the group with the larger range.

The test assumes that the population medians are the same.

For the data in Table 6.14, SPSS results rejected the null hypothesis using an exact test ($p < .001$), indicating that Group A had a larger range (6 as compared to 3 in Group B). Group B is the control group (defined as the group with the smaller range). The experimental treatment is hypothesized to create extreme reactions in Group A, the group with the larger range. In SPSS, the control group must be coded with a 1 and the experimental group with a 2, with the level of measurement set to nominal.

The test statistic is based on the "span," which is the range of the control group plus 1. (Ranks must be used to compute the span.) The span for Group B is $5 - 2 + 1 = 4$. If one deletes the top and bottom value in the control group, the "trimmed" span is $4 - 3 + 1 = 2$. (The test statistic is the trimmed span.) Trimming is used because the span is highly affected by outliers. The span is more significant the smaller it is. The p value provides the probability, under the null hypothesis, of obtaining the trimmed span or something more extreme (i.e., smaller) and is one tailed.

EXPLORING THE CONCEPT

Why would a smaller span be more significant? What if there were no extreme reactions and ranks in Group B ranged from 1 to 7? What would be the span?

Robust Methods for Detecting Outliers

It is important to screen for outliers regardless of whether one uses parametric or nonparametric statistics. Outliers can be univariate or multivariate. With a univariate outlier, the value is extreme relative to the other values on that one variable. With a multivariate outlier, the value is extreme relative to what would be predicted by a regression line, such as the one shown in Figure 6.1(b). (This outlier is bivariate because only two variables are involved; for this discussion, I consider bivariate outliers a special case of multivariate ones.) Univariate outliers are likely also to be multivariate outliers, and the latter can bias the results and reduce statistical power. One can screen for univariate outliers by computing standardized scores (e.g., using "Descriptives" in SPSS) and examining those values in excess of 3.29 in absolute value, a threshold recommended by Tabachnick and Fidell (2001). However, the criterion is only useful when the data (excluding the suspected outliers) are approximately normal, not skewed. (How to screen for multivariate outliers is described at the end of this section.)

According to Hawkins (1980), outliers can be generated by one of two processes: (a) a causal mechanism that is different from one generating the other observations, and (b) sampling from a heavy-tailed distribution. Only in the first case should the outlier be deleted; in the second case, it is more appropriate to downgrade the influence of the outlier rather than eliminate it. In either case, the nature of the outlier should be investigated because frequently this produces substantively important information. Zimmerman (1995) reported that nonparametric statistics are more robust to Type I errors caused by outliers than parametric statistics but are not necessarily more robust to Type II errors. So even when nonparametric statistics are used, it is important to check for outliers. One approach recommended by Sprent (1989) that is fairly robust is the median absolute deviation (MAD) method. One finds the deviation of each observation from the median in absolute value and then finds the median of these (this value is the MAD). Divide each deviation by the MAD; any ratios greater than 5 indicate an outlier. Table 6.18 shows that and example where the median is 9.5, the MAD is 6.5, and—except for the outlier—the ratios range from 0.08 to 1.46, well under the threshold of 5. (Technical Note 6.6 gives the rationale for this threshold.) Once identified, the outlier can be downgraded or deleted. When comparing two groups, I recommend performing the procedure separately for each group, using either metric data or ranks from the combined sample, as this will help identify multivariate outliers, not just univariate ones.

Assessing Slippage

Outliers can cause the shape of the distribution in one sample to be different from that in the other. It is important to establish that the empirical distributions of the two samples have the same shape if using the second version of the null hypothesis, and addressing outliers can help.

Table 6.18 Median Absolute Deviation Method (MAD) for Detecting Outliers

X	|Deviation|	Ratio $\frac{\|Deviation\|}{MAD}$
1,000	990.5	152.38
19	9.5	1.46
18	8.5	1.31
17	7.5	1.15
16	6.5	1.00
16	6.5	1.00
16	6.5	1.00
14	4.5	0.69
11	1.5	0.23
10	0.5	0.08
9	0.5	0.08
6	3.5	0.54
6	3.5	0.54
4	5.5	0.85
3	6.5	1.00
3	6.5	1.00
3	6.5	1.00
2	7.5	1.15
2	7.5	1.15
1	8.5	1.31
Mdn = 9.5	Mdn = 6.5 (MAD)	

But one can also more generally assess a concept known as *slippage*. With the WMW test, if the null hypothesis is false, it could be that the mean or median of one group may have "slipped" away from the other. The mean and median are measures of *location,* so we say there has been slippage of location.

Scale Slippage

However, there instead could be *slippage of scale. Scale* is a more general term refer-ring to the dispersion of a random variable; the range and variance are measures of scale. A slippage of scale would occur when the population variance of one group is different from that of another.

A popular (but not very powerful) test for scale slippage is the Siegel-Tukey test. If two samples come from populations with the same means but different variances, the one with greater variance will have more extreme observations. The Siegel-Tukey test assigns Rank 1 to the smallest observation, Ranks 2 and 3 to the largest and next largest, Ranks 4 and 5 to the next two smallest observations, and so forth, and then computes the sum of ranks. One next applies the WMW test to these sums. The sum of ranks should be equal under the null hypothesis; if they are not, the group with greater variability will have the larger sum of ranks. If the assumption of location equality is violated (e.g., the medians are different), one can align the observations by adding the median difference to all the values in the group with a lower median. This alignment does not affect the variances and will increase the power of the test. However, the sample sizes must be large in order to obtain a reliable estimate of the location differences in the population (Sprent, 1989). The Siegel-Tukey test is often used to test for homogenous variances when using parametric tests when normality assumptions are violated (Sheskin, 1997).

There are a number of other nonparametric tests for testing whether the variances of two independent groups are the same. These include the Klotz test, Mood test, Ansari-Bradley test, and the Moses test (of equal variances, not extreme reactions). The Moses test can only be used for metric data and does not assume equal medians between the two groups (Sheskin, 1997). The two-sample Kolmogorov-Smirnoff (KS) test can also be used (see Chapter 3 for the one-sample description); however, this test is sensitive to any distributional differences (including mean differences), not just variance differences. The reader is referred to Sheskin (1997) for more information on tests of dispersion.

Slippage in Both Scale and Location

A problematic situation occurs when there is slippage in both scale and location. For example, relative to Group A, suppose the location for Group B has slipped upwards, and its scale has also increased. The higher variance will produce some low rankings and may reduce Group B's mean rank, reducing statistical power (Hawkins, 1980). The scale shift will not necessarily produce higher ranking over and beyond the increase in scale because there is an upper limit on the value of the ranks (e.g., the highest rank when $N = 10$ cannot be greater than 10).

Now consider the following scenario. If there is only slippage in scale and Group B has the largest variance, Group B will have the most extreme ranks. If the location of Group B also slips upwards, it will no longer have as many low ranks. This will reduce the magnitude of test statistics like Siegel-Tukey, again reflecting a loss of statistical power. (This point is why aligning the ranks increases the power of the test.) If the location of Group B were to slip downwards instead of upwards, it would also reduce the power of scale tests by reducing the number of high ranks.

Summary

This analysis of slippage shows how unequal variances can reduce the power of locational tests like WMW, as can outliers. The WMW test, used as a locational test, has also been found to have unsatisfactory Type I error rates (well in excess of .05, see Stonehouse & Forrester, 1998), but the higher error rates occurs because the WMW is sensitive to other distributional differences beside location. So unless there is good evidence for distributional equality, it is best practice to only use the WMW test as a test of distributional differences and to supplement it with a robust measure of effect size, particularly PS. However, slippage in both scale and location might still result in some reduction of statistical power. Checking and addressing outliers may cause some improvement.

If one is using metric data, another option is to consider using one of the generalized linear models presented in Chapters 9 through 13 if there is substantial skew or kurtosis. This solution is more complex but also more statistically powerful. Even here, though, one still needs to check for outliers and to make certain distributional assumptions. A final option is to use more robust nonparametric techniques (see Chapter 8); however, these techniques may require specialized software.

Problems

1. Find the skew and kurtosis of the following data using statistical software. Create a histogram of the data. (In SPSS, go to "Analyze Frequencies," with "Statistics" [skew and kurtosis] and "Chart" [histogram].)

   1 1 1 1 2 2 3 3 3 3 3 3 3 4 4 5 5 5 5

2. Find the skew and kurtosis of the following data using EXCEL (=SKEW, and =KURT) and create a graph. (To do the latter, you need to count the number of cases that fall into each category [1,2,3,4,5,6,7,8,9,10].) Do this by visual inspection or by using =COUNT for each category. Then to create the graph, highlight the frequency column, "insert chart," click on column bar chart as the subtype, then click on finish.

   1 1 1 1 1 1 2 2 2 2 2 3 3 3 3 3 3 4 4 4 4 5 5 7 7 10

3. A psychologist interviewed 10 adolescent girls referred for counseling to assess the extent they may have eating disorders. Based on questions from the Eating Disorders Examination (Fairburn & Cooper, 1993) regarding such things as dietary restrictions and frequency of self-induced vomiting, the psychologist rank ordered the girls based on the extent they had an eating disorder (12 = high). The psychologist also obtained each girl's score on the Beck Depression Inventory (BDI). Using the data below and statistical software, calculate (a) Spearman's $\hat{\rho}$, and (b) Kendall's $\hat{\tau}_b$ along with the p values. Is there a significant association?

Eating Disorder (EDE) rank	Depression (BDI) score
10	65
9	56
8	25
7	57
6	38
5	47
4	33
3	25
2	30
1	3

Based conceptually on Stiles-Shields, Goldschmidt, Boepple, Glunze, and Le Grange (2011).

4. Bromet et al. (2011) collected data on rates of depression in 10 high-income countries. Using the data on "positive lifetime screens" (in file *Depression.6.Problem_4*), determine if there is a significant correlation with the suicide rate (per 100,000/year), using Spearman's $\hat{\rho}$ (due to the small sample size). You may use statistical software.

5. A researcher collected data from 34 college students on the extent that (a) their psychology professor involves students in class discussions and (b) students attended the class. The theory being tested was that when professors stimulate more interest and participation, students will be more intrinsically motivated to learn course content (and come to class). The cell frequencies are shown in the table below. Both variables were measured by self-report surveys involving ordered categories.

	Class attendance		
	Attend a little	Attend somewhat	Attend a lot
Class participation			
Not at all	0	1	0
A little	0	1	1
Somewhat	1	4	4
A lot	0	5	17

Problem based conceptually on Chandler (2008).

a) Why is it better here to calculate $\hat{\tau}$ rather than Spearman's $\hat{\rho}$?
b) Why is it better here to calculate $\hat{\tau}_c$ rather than $\hat{\tau}_b$?
c) Calculate $\hat{\tau}_c$ and the p value using statistical software.
d) Calculate Somers' d and the p value using statistical software.

6. Olmedo and Padilla (1978) attempted to validate an instrument measuring acculturation of Mexican Americans. Below are data on education and acculturation levels from a sample of 71 adults in Southern California from various ethnic groups. Test whether there is an association between the two variables by using statistical software to compute Kendall's $\hat{\tau}_b$ and the p value.

Education	Acculturation level (quartile)			
	1	2	3	4
Less than 8 years	7	5	1	4
9–11 years	4	5	3	3
12 years	5	5	9	3
College	1	2	4	10

7. A researcher (attempting to replicate the findings in Sendağ, 2010) sends a survey to all preservice teachers at her university on attitudes towards electronic democracy (e-democracy), which holds that the Internet has a democratizing effect on society. There is also an item on the survey asking respondents to rate their Internet use skills (from *poor* to *excellent*). The hypothesis is that there is a positive relationship between Internet-use skills and attitudes towards e-democracy. Because the attitude scores are skewed, the researcher uses the nonparametric Jonckheere-Terpstra test for ranks by ordered categories. (The $\hat{\tau}$ test can also be used.) The data are in the file *EDemocracy.6.Problem_7*. Using statistical software, find the z score (or $\hat{\tau}_b$) and the two-sided p value for the following data. (In IBM SPSS, go to nonparametric tests for multiple independent samples. Code the Internet Use categories 1, 2, 3, 4, as an ordinal variable. In older versions of SPSS, code Internet Use as a scale variable and go to "Correlate," "Tau-b.")

8. Using statistical software (e.g., SPSS crosstabs), compute Kendall's $\hat{\tau}_b$ and the p value for the following $2 \times k$ ordered contingency table (from Le, 1998, p. 64):

Admission status	Cups of coffee per day		
	Low (0)	Medium (1 to 4)	High (>4)
Acute	340	457	183
Chronic	2,440	2,527	868

9. An educational experiment was performed in which students were taught to write using either outlines (as planning devices) or concept maps. The data collected were:

Condition	Writing scores
Outline	80 70 86 79 81 63 75
Concept map	67 72 73 70 66 59

a) Rank the data using EXCEL or statistical software (in SPSS, go to "Transform," "Rank Cases"). Be sure the lowest score has a rank of 1. Move the values for the second group into a second column (use "Paste Special" to copy on the values). Find the sum of ranks and mean ranks, by group.

b) Find W. (It is equal to the sum of ranks for one of the groups, whichever is smallest.) Note: Some versions of SPSS report the larger value.

c) Find n and m.

d) Use Table 6.15 to find the lower critical value for a two-tail test, using $\alpha = .05$ (so $\alpha/2 = .025$).

e) Find the upper critical value (see formula in Table 6.15 note).

f) Is W significant?

g) Use statistical software to find the exact p value. Why should the normal approximation not be used?

h) Using statistical software, find the effect size ΔHL and the associated CI.

i) Using the metric data, find the medians of each group and compare ΔHL with the difference between the medians.

j) Using Eq. 6.18, find the PS for Group A. What is the practical significance?

10. Attempting to replicate the findings of Nau, Halfens, Needham, and Dassen (2010), a researcher provided a 1-week course for nursing students on how to deal with patient aggression. Before and after the course, students responded to two hypothetical scenarios (Scenarios A and B) with simulated patients. Videotapes of their behavior were rated by an expert judge, who completed a 7-item Likert-scale from the Deescalating Aggression Behavior Scale (DABS). Half the students received Scenario A at pretest and Scenario B at posttest, the other half received the scenarios in reverse order. Below are data on the "reducing fear" item for Scenario A, comparing scores of the group who received the scenario before training with those receiving it after. Answer the same questions as in Problem 9.

Condition	"Reducing fear" scores
Before training	4 3 4 4 3 5 4 4
After training	2 1 2 3 2 3 4 2 2

11. Over a 1-year period, individuals suffering from schizophrenia receive either routine care or routine care combined with cognitive therapy (as in

Palma-Sevillano et al., 2011). After 6 months, the participants are assessed using a global functioning scale. The scores obtained are as follows:

Condition	Scores
Cognitive Therapy (CT)	17 13 26 29 33 39 14 31 42 35 37
Routine Care (RC)	10 12 15 16 18 19 20 21 22 23 25

a) For each group, calculate the sum of ranks and mean ranks. What is W?
b) Calculate the expected value of W.
c) Calculate the variance of W (not corrected for ties).
d) Calculate z and the two-tailed p value.
e) Using statistical software, perform a WMW test. (In SPSS, code condition as a NOMINAL dummy variable, 1 for CT, 0 for RT). What is the asymptotic p value?
f) For each group (and using ranks), find the mean, median, variance, range, skew, and kurtosis. (This can be done in EXCEL or using "Compare Means" and then "Mean" in SPSS.) Do the distributions look similar? Does this undermine the results or is there still evidence of significant differences in central tendency?
g) As an effect size measure, is it better to report ΔHL or PS (probability of superiority) for these data, and why?
h) Compute and interpret PS.
i) Perform a median test using statistical software. (In SPSS, go to "nonparametric tests, independent samples, objective: compare medians across groups" or "legacy dialog, k independent samples").

12. A researcher administers an anger scale to 117 university students and selects 32 with the highest scores to participate in a study on emotional intelligence (as in Yilmaz, 2009). The students are randomly assigned to an experimental or control group; the former receive instruction in emotional regulation over a 6-week period (12 sessions). Before the course, there was no significant difference between the mean anger scores of the two groups. After the course, the anger scale was readministered (results are shown below). Higher scores reflect better anger management. Because of the small sample sizes, use the WMW test to analyze the data, answering the same questions as in Problem 11. Because there are more than a few ties, use the following formula to compute z:

$$z = \frac{\left(W - n\dfrac{N+1}{2}\right)}{\sqrt{\dfrac{nm}{N(N-1)}\sum_i^N R_i^2 - \dfrac{nm(N+1)^2}{4(N-1)}}},$$

where $\sum_i^N R_i^2$ is the square of each rank (or mid-rank) in the data.

Condition	Anger Scores
Experimental	6 9 10 12 13 15 16 17 19 20 21 22 24 24 25
Control	1 3 4 6 7 9 10 11 13 14 15 16 19 20 22

13. Normally, plants have roots of variable length. Following Sinkkonen, Penttinen, and Strömmer (2009), a researcher hypothesizes that exposing plants, while in vitro, to a small dose of a toxic chemical (copper sulfate) will inhibit growth and cause these plants to develop less variable root lengths. The researcher wants to use the Moses test of extreme reactions to test her hypothesis. Using the data below, find the range of the experimental and control group root lengths. The data are in the files *Plants.6.Problem_13.xlxs* (EXCEL) and *Plants.6.Problem_13.sav* (SPSS).
    a) Compute the ranks. Find the range of the ranks in each group.
    b) Find the untrimmed span.
    c) Find the span trimming the top and bottom value of the control group.
    d) Use statistical software to conduct the Moses test. Report the *p* values for both the untrimmed and trimmed span.

14. Answer the same questions as in problem 13 using the following hypothetical data (from Moses, 1952):
    Experimental: 1.3, 1.5, 1.6, 3.4, 3.5, 3.8, 4.5, 6.7.
    Control: 1.7, 1.8, 1.9, 2.2, 2.7, 2.8, 3.1, 3.6, 3.7.

15. A program evaluator is hired to assess the effectiveness of a program to assess public speaking. Twenty high school students are randomly assigned to an experimental or control group (the experimental group receives the program; the control group works on improving their writing). Afterwards, all 20 students give speeches and a panel of three judges rates them on a 20-point scale (1 = *highest quality*); the average score was used in the analysis. The data are presented below.
    Experimental: 1, 3, 4, 5, 6, 7, 8, 9, 11, 12.
    Control: 2, 10, 13, 14, 15, 16, 17, 18, 19, 20.
    a) Use the MAD procedure to identify any outliers. For any outliers, report the value of the outlier, the median, and the MAD for the relevant group, and the ratio of the absolute deviation to the MAD.
    b) Delete any outliers. On the revised data, perform the WMW test, using statistical software.
    c) Using the "Means" and "Compare Means" option in SPSS (or similar software), examine the means, standard deviations, skews, and kurtosis values of the two groups. Do the two distributions look similar? Why or why not?
    d) Find the PS for each group.

16. Using a 20-point scale, a political scientist rates 31 randomly chosen Congressional representatives on the perceived degree their opinions on issues depart from the party platform. The data are presented below, broken down by party affiliation. Answer the same questions as in the previous problem.
Republicans: 9, 4, 5, 6, 5, 19, 20, 5, 2, 7, 6, 1, 0, 5, 9.
Democrats: 10, 7, 8, 9, 8, 10, 7, 8, 5, 10, 9, 4, 4, 4, 8, 12.

Technical Notes

6.1   Zimmerman (2011) proposed that when comparing two groups with similar variances using metric data, one should conduct a t test of the initial data and a t test on ranked transform data (t_r), and use a WMW test only if the two t-statistics differ by more than 0.40. (Conover & Iman, 1981, found that a t test on ranks produces the same results as WMW.) Zimmerman argues that it can be misleading to examine characteristics of the sample distribution, because in small samples these can differ substantially from the population distribution. Furthermore, preliminary tests of normality are themselves subject to Type I and Type II errors, which can alter the overall Type I and Type II error rates.

Zimmerman argued that if the population distributions are not normal, one would expect the difference between parametric and nonparametric tests to differ substantially. Based on various simulations, Zimmerman found the value of .40 to be "a good compromise for all distributions; that is, it resulted in a conditional test that performed like t in cases where t was superior to t_R and like t_R when the reverse was true" (Zimmerman, 2011, p. 400). His proposed decision rule produced very satisfactory Type I and Type II error rates (i.e., power). Zimmerman clarifies that the decision rule is not equivalent to choosing the test with the most favorable result (i.e., lowest p value) and sometimes favors the opposite.

6.2   To calculate p values of $\hat{\tau}$ using the normal approximation, follow these steps:

a)   find $T = N_c - N_d$;

b)   find $p_{(lower-tailed)} = \text{Prob}\left(z \leq \dfrac{(T + 1)\sqrt{18}}{\sqrt{n(n - 1)(2n + 5)}}\right)$;

c)   find $p_{(upper-tailed)} = \text{Prob}\left(z \geq \dfrac{(T - 1)\sqrt{18}}{\sqrt{n(n - 1)(2n + 5)}}\right)$;

d)   take the smaller of the two one-tailed p values and double it, to obtain the two-tailed p value (e.g., if $n = 10$, $N_c = 35$, and $N_d = 10$, then $T = 25$); $p_{(upper-tailed)} = \text{Prob}\left(z \geq \dfrac{(25 - 1)\sqrt{18}}{\sqrt{10(10 - 1)(20 + 5)}}\right) = \text{Prob}(z \geq 2.15) = .016$. [In EXCEL, $= 1 - \text{NORMSDIST}(2.15) \rightarrow .016$]. $p_{(lower-tailed)} = \text{Prob}\left(z \leq \dfrac{(25 + 1)\sqrt{18}}{\sqrt{10(10 - 1)(20 + 5)}}\right)$ $= \text{Prob}(z \leq 2.33) = 0.99$. The upper-tailed p value is smaller. $p_{(two-tailed)} = 2 * .016 = .032$.

6.3 Wilcoxon and Mann-Whitney independently developed two different tests which are statistically equivalent: Wilcoxon's W and Mann-Whitney's U. To calculate U, take each observation in Group A and find the number of observations in Group B with smaller ranks (counting ties at one-half). Sum up these frequencies. Do the same procedure with Group B relative to A. U is the smaller of these two values. W (as the WMW statistic) is often reported rather than U because the logic and computations behind W are easier to explain.

6.4 Randles and Wolfe (1979) conducted a Monte Carlo simulation study comparing the power of the WMW test and the t test when $m + n = 20$ ($\alpha = .05$) for various types of population distributions. With some symmetrical, kurtotic distributions (double exponential and Cauchy), the WMW was more statistically powerful (but there was no clear winner with a logistic distribution). Two skewed distributions were also tested. The WMW was more powerful with an exponential distribution; the performance of the two tests was similar with a Weibull distribution. The t test was slightly more powerful when the distributions were uniform or normal. The results provide empirical support of how and when statistical power can be increased by using WMW when the data are skewed or kurtotic. (Although not reliable in small samples, the fit to some of these distributional shapes can be assessed in SPSS using the Q-Q plot function under "Descriptive Statistics.") For a related study, see Skovlund and Fenstad (2001).

6.5 Lehmann (2006) provided a detailed discussion of calculating confidence intervals for ΔHL, along with a completed table of the quantiles of W. Briefly, rank order all the d_{ij}; let $D_{(1)}$ represent the lowest ranked d_{ij}, D_2 the second lowest ranked, etc., where D is subscripted by r (r ranges from 1 to mn). For a 95% CI, choose the D closest to the .025 quantile of the null distribution of W. To find the exact probability level, Lehman showed that Prob$(\Delta$HL $\leq D_r) =$ Prob$[W \leq (r - 1)]$, and this formula, plus an exact table, can be used to find the probability of each D_r. For the example in Table 6.17, $D_1 = -4$, $r = 1$, $n = 3$ and $m = 4$. According to the exact tables in Lehmann (2006) for W, the cumulative probability of $W = r - 1 = 0$ is .0286 $\approx$ .025. So one should use the D_1 ($= -4$) as the lower point of the CI. This value is the lowest-ranking D_r (or d_{ij}). Because the CI is symmetric around the median r, use the highest-ranked D_r ($D_{12} = 5$) as the upper end of the CI.

6.6 In the MAD method for detecting outliers, the rationale for a threshold of five is as follows. If one multiplies the deviations (in absolute value) by 1.4626, one obtains scores on a normalized scale, S. The value 1.4626 was chosen so that the expected value of S will equal the standard deviation of the initial scores when the population distribution is normal (Pearson, 2001). Multiplying 3.2 (the threshold for a univariate outlier on the normal scale) by 1.4626 yields 4.68, which rounds to 5.0, the recommended threshold (Sprent, 1989). S is also known as the Hampel identifier and is particularly resistant to multiple outliers that skew the mean and standard deviation of the original data. It is even more robust than Winsorization, which also requires an arbitrary decision as to the percent to Winsorize. In Winsorization, the top 10% or so of the ordered observations are all set equal to the next lowest observation, and the bottom 10% or so of the bottom observations are all set equal to the next highest observation.

APPENDIX 6.1

COMPUTATIONAL FORMULA FOR $\hat{\tau}_b$

The Kendall $\hat{\tau}_b$ coefficient is defined as:

$$\hat{\tau}_b = \frac{P-Q}{\sqrt{D_r D_c}},$$

(Eq. 6.19)

where $P = 2 * N_c$, $Q = 2 * N_d$, $X_0 = 2 *$ number of tied pairs on X, $Y_0 = 2 *$ number of tied pairs on Y, $D_r = P + Q + X_0$ (row denominator), $D_c = P + Q + Y_0$ (column denominator). P and Q represent the number of concordant and discordant pairs doubled so as to count each cell of the matrix (e.g., Table 6.10), excluding the diagonal. Ties could be added to the numerator in Eq. 6.19 (one-half for concordant, one-half for discordant) but these operations will have no effect because the concordant and discordant adjustments will cancel one another out. The denominator reflects the total number of pairs including ties and is a geometric mean of D_{row} and $D_{columns}$. Eq. 6.19 is written for ordered categories but can also be applied to ranked data if each observation is considered to have its own set of categories (one for X, one for Y).

For Tables 6.9 and 6.10, the formula for $\hat{\tau}_b$ takes on the following values: $P = 2 * 4 = 8$; $Q = 2 * 1 = 2$, $D_r = 8 + 2 + 0 = 10$; $D_c = 8 + 2 + 2 = 12$, $\hat{\tau} = \frac{8-2}{\sqrt{10 * 12}} = \frac{6}{10.95} = .548$.

APPENDIX 6.2

SUM OF CONSECUTIVE INTEGERS

The theorem states:

$$Sum\ of\ Consecutive\ Intergers\ (\textstyle\sum_{i=1}^{N} i) = \frac{N(N+1)}{2}.$$ (Eq. 6.13)

Following is a proof of the theorem with N arbitrarily set to 3. S is the sum of the consecutive integers starting with 1.

a) $S = 1 + 2 + 3$.
b) $S = 1 + (1 + 1) + (1 + 1 + 1)$.
c) $S = (1 + 1 + 1) + (1 + 1) + 1$.
d) $S = N + (N - 1) + (N - 2)$, where $N = 3 = (1 + 1 + 1)$.

Add lines (b) and (d).

a) $2S = (N + 1) + (N - 1 + 1 + 1) + (N - 2 + 1 + 1 + 1)$,
b) $2S = (N + 1) + (N + 1) + (N + 1)$.
c) $2S = 3(N + 1)$. Because $N = 3$, it follows that:
d) $2S = N(N + 1)$.
e) $S = \frac{N(N+1)}{2}$. QED.

References

Best, D. J. (1974). *Tables for Kendall's $\hat{\tau}$ and an examination of the normal approximation.* Division of Mathematical Statistics Technical Paper No. 39, Commonwealth Scientific and Industrial Research Organization, Australia (5.4, Appendix).

Bromet, E., Andrade, L. H., Hwang, I., Sampson, N. A., Alonso, J., de Girolamo, G., . . . Kessler, R. C. (2011). Cross-national epidemiology of DSM-IV major depressive episode. *BMC Medicine, 9*, 90. Retrieved from http://www.biomedcentral.com/1741-7015/9/90

Chandler, D. S. (2008). *The relational-behavior model: The relationship between intrinsic motivational instruction and extrinsic motivation in psychologically based instruction.* Retrieved from http://www.eric.ed.gov/PDFS/ED500815.pdf

Conover, W. J. (1999). *Practical nonparametric statistics* (3rd ed.). New York, NY: John Wiley & Sons.

Conover, W. J., & Iman, R. L. (1981). Rank transformations as a bridge between parametric and nonparametric statistics. *American Statistician, 35*, 124–134.

Daniel, W. W. (1990). *Applied nonparametric statistics* (2nd ed.). Boston, MA: PWS-Kent.

Erceg-Hurn, D., & Mirosevich, V. M. (2008). Modern robust statistical methods. *American Psychologist, 63*, 591–601.

Fairburn, C. G., & Cooper, Z. (1993). The eating disorder examination (12th ed.). In C. G. Fairburn, & G. T. Wilson (Eds.), *Binge eating: Nature, assessment, and treatment* (pp. 317–360). New York, NY: Guilford Press.

Glasser, G. J., & Winter, R. F. (1961). Critical values of the coefficient of rank correlation for testing the hypothesis of independence. *Biometrika, 48*, 444–448.

Goodman, L. A., & Kruskal, W. H. (1963). Measures of association for cross-classifications, III: Approximate sample theory. *Journal of the American Statistical Association, 58*, 310–364.

Grissom, R. J. (1994). Probability of the superior outcome of one treatment over another. *Journal of Applied Psychology, 79*, 314–316.

Grissom, R. J., & Kim, J. J. (2005). *Effect sizes for research: A broad practical approach.* Mahwah, NJ: Erlbaum.

Hawkins, D. M. (1980). *Identification of outliers.* New York, NY: Chapman & Hall.

Hayes, A. F., & Cai, L. (2007). Further evaluating the conditional decision rule for comparing two independent means. *Journal of Mathematical and Statistical Psychology, 60*, 217–244.

Helsel, D. R., & Hirsch, R. M. (1991). *Statistical methods in water resources.* New York, NY: Elsevier.

Hodges, J., & Lehmann, E. (1956). The efficacy of some nonparametric competitors of the *t*-test. *Annals of Mathematical Statistics, 27*, 324–335.

Johnson, M. L., & Nussbaum, E. M. (2012). Achievement goals, coping strategies, and traditional/nontraditional student status. *Journal of College Student Development, 53*(1), 41–54.

Le, C. T. (1998). *Applied categorical data analysis.* New York, NY: John Wiley & Sons.

Lehmann, E. L. (2006). *Nonparametrics: Statistical methods based on ranks* (Rev. ed.). New York, NY: Springer.

Mann, H. B., & Whitney, D. R. (1947). On a test of whether one of 2 random variables is stochastically larger than the other. *Annals of Mathematical Statistics, 18*, 50–60.

Moses, L. E. (1952). A two-sample test. *Psychometrika, 17*, 239–247.

Nau, J., Halfens, R., Needham, I., & Dassen, T. (2010). Student nurses' de-escalation of patient aggression: A pretest-posttest intervention study. *International Journal of Nursing Studies, 47*, 699–708.

Nussbaum, E. M., & Schraw, G. (2007). Promoting argument-counterargument integration in students' writing. *The Journal of Experimental Education, 76*, 59–92.

Olmedo, E. L., & Padilla, A. M. (1978). Empirical and construct validation of a measure of acculturation for Mexican Americans. *The Journal of Social Psychology, 105*, 179–187.

Osborne, J. W. (2013). *Best practices in data cleaning: A complete guide to everything you need to do before and after collecting your data.* Thousand Oaks, CA: Sage.

Palma-Sevillano, C., Cañete-Crespillo, J., Farriols-Hernando, N., Cebrià-Andreu, J., Michal, M., Alonso-Fernández, I., . . . Segarra-Gutiérrez, G. (2011). Randomised controlled trial of cognitive-motivational therapy program for the initial phase of schizophrenia: A 6-month assessment. *The European Journal of Psychiatry, 25*, 68–80.

Pearson, R. K. (2001). Exploring process data. *Journal of Process Control, 11*, 179–194.

Pitman, E.J.G. (1948). *Lecture notes on nonparametric statistical inference: Lectures given for the University of North Carolina.* Amsterdam, The Netherlands: Mathematisch Centrum.

Randles, R. H., & Wolfe, D. A. (1979). *Introduction to the theory of nonparametric statistics.* New York, NY: John Wiley & Sons.

Rasch, D., Kubinger, K. D., & Moder, K. (2011). The two-sample *t* test: Pretesting its assumptions does not pay off. *Statistical Papers, 52*, 219–231.

Sawilowsky, S. S. (1990). Nonparametric tests on interaction in experimental design. *Review of Educational Research, 60*, 91–126.

Sendağ, S. (2010). Pre-service teachers' perceptions about e-democracy: A case in Turkey. *Computers & Education, 55*, 1684–1693.

Sheskin, D. J. (1997). *Handbook of parametric and nonparametric statistical procedures.* New York, NY: CRC Press.

Sinkkonen, A., Penttinen, O.-P., & Strömmer, R. (2009). Testing the homogenizing effect of low copper sulfate concentrations on the size distribution of *Portulaca oleracea* seedlings in vitro. *Science of the Total Environment, 407*, 4461–4464.

Skovlund, E., & Fenstad, G. U. (2001). Should we always choose a nonparametric test when comparing two apparently nonnormal distributions? *Journal of Clinical Epidemiology, 54,* 86–92.

Somers, R. H. (1962). A new asymmetric measure of association for ordinal variables. *American Sociological Review, 27,* 799–811.

Sprent, P. (1989). *Applied nonparametric statistical methods.* New York, NY: Chapman & Hall.

Stiles-Shields, E. C., Goldschmidt, A. B., Boepple, L., Glunz, C., & Le Grange, D. (2011). *Eating Behaviors, 12,* 328–331.

Stonehouse, J. M., & Forrester, G. J. (1998). Robustness of the *t* and *U* tests under combined assumption violations. *Journal of Applied Statistics, 25,* 63–74.

Tabachnick, B. G., & Fidell, L. S. (2001). *Using multivariate statistics* (4th ed.). Boston, MA: Allyn & Bacon.

Tukey, J. W. (1960). A survey of sampling from contaminated normal distributions. In I. Olkin, S. Ghurye, W. Hoeffding, W. Madow, & H. Mann (Eds.), *Contributions to probability and statistics: Essays in honor of Harold Hotelling* (pp. 448–485). Stanford, CA: Stanford University Press.

Yilmaz, M. (2009). The effects of an emotional intelligence skills training program on the consistent anger levels of Turkish university students. *Social Behavior and Personality, 37,* 565–576.

Zimmerman, D. W. (1995). Increasing the power of nonparametric tests by detecting and downweighting outliers. *Journal of Experimental Education, 64,* 71–78.

Zimmerman, D. W. (2011). A simple and effective decision rule for choosing a significance test to protect against non-normality. *British Journal of Mathematical and Statistical Psychology, 64,* 386–409.

CHAPTER 7

NONPARAMETRIC TESTS FOR MULTIPLE OR RELATED SAMPLES

The previous chapter presented nonparametric tests for comparing two groups. One important point made was that there are two forms of each test: one for small samples using an exact table or test and one for large samples using an asymptotic approximation. This point also holds for the tests considered in this chapter. We consider the situations of comparing (a) more than two groups (Kruskal-Wallis test), (b) related groups (sign test or Wilcoxon matched signed-ranks test), or (c) multiple, related groups (Friedman or Page tests).

Because of the large number of statistical tests to be discussed, Table 7.1 presents a summary. The table can be used as a reference guide for quickly determining which test is most appropriate in specific situations. For this reason, I also included in the table some of the tests (median and Wilcoxon-Mann-Whitney) discussed in the previous chapter and some discussed in the next.

Kruskal-Wallis Test for Multiple Independent Groups

The Kruskal-Wallis (KW) test is used when there are three or more independent groups. It is an extension of the Wilcoxon-Mann-Whitney (WMW) test for two groups. The KW test is the nonparametric equivalent of the omnibus F test in a one-way ANOVA (which is used with metric dependent variables). The KW test is used when the dependent variable (DV) consists of ranks. It tests the null hypothesis that the location of each group is the same in the population. If the null hypothesis is rejected, then at least one of the locations is different from the others. When the KW test is significant, one then performs follow-up pairwise tests comparing two groups using the WMW test or using Eq. 7.3.

Not all researchers believe it is necessary to conduct an omnibus test if one has specific research hypotheses, which is more often the case than not. Karpinsky (2007) argued that in an ANOVA, if one is conducting planned comparisons, the

Table 7.1 Features of Nonparametric Statistical Tests for Comparing Groups

Test	Median	WMW	Kruskal-Wallis	Sign	Wilcoxon signed ranks	Friedman	Page
Compares	2+ groups	2 groups	3+ groups	pre–post	pre–post (requires metric data)	multiple measures— ordered or unordered treatments	repeated measured for ordered categories
Assumes . . .		groups identically distributed	groups identically distributed		symmetry of difference scores		
Exact	hypergeom.	table	table	binomial	table	table	table
Asymptotic	chi-square	z distribution	chi-square	z distribution	z distribution	chi-square or F	z or chi-square
df	$c - 1$	not applicable	$k - 1$	not applicable	not applicable	chi: $(k - 1)$; F: $(k - 1)$, $(b - 1)(k - 1)$.	1
Minimum sample	$n = 20$	n or $m \geq 10$	all $n_i > 5$	$n \geq 20$	10	$b \geq 12$	$n > 12$ $(k \geq 4)$, $n \geq 20$, or $k \geq 9$
Mean	df	$\dfrac{n(N + 1)}{2}$	$\dfrac{n_i(N + 1)}{2}$	$n\pi$	0	$\dfrac{b(k + 1)}{2}$	—
SE	$2\,df$	$\sqrt{\dfrac{nm(N + 1)}{12}}$	$\sqrt{\dfrac{n_i(N + 1)(N - n_i)}{12}}$	$\sqrt{n\pi(1 - \pi)}$	$\sqrt{\sum (R_i^s)^2}$	N/A	N/A
Test statistic	chi-square or Fisher	W (smaller sum of ranks)	H	# plusses	T or $T+$	χ^2 or F	L

omnibus F test is optional. (It is not optional if one is conducting post hoc tests or examining all pairwise comparisons because of the possibility of capitalizing on chance.) Suppose one is comparing three groups (A, B, and C), and your research hypothesis is that the mean of Group A will be greater than the average mean of B and C combined. Then one should just test that hypothesis by comparing those two means. Conducting the omnibus F test also examines whether the mean of Group A (or any group) is different from the mean of all three groups combined. However, if one's research hypotheses do not address this particular comparison, there is no good reason for conducting the omnibus test. One should minimize the number of statistical tests conducted to reduce the number of possible Type I errors and/or conflicting results from conducting multiple tests.

These same arguments apply to the KW test: If one is testing specific research hypotheses, one could dispense with the omnibus KW test and just conduct WMW tests.

EXPLORING THE CONCEPT

What are the advantages and disadvantages of conducting the omnibus KW test just "for completeness"? It is customary to conduct a KW test; is that a sufficient reason for doing so?

Formulas

An asymptotic KW test can be conducted if the size of each group is greater than five. Let the rank of each observation be denoted by R_{ij}, where i is the number of the observation ($i = 1 \ldots n$) and j is the index for the group ($j = 1 \ldots k$). One computes a test statistic, H, which has a chi-square distribution with $k - 1$ degrees of freedom, where k is the number of groups. The formula for H (assuming no ties) is

$$H = \frac{12}{N(N+1)}\left[\sum_{j=1}^{k}\frac{\left(R_{\bullet j}\right)^2}{n_j}\right] - 3(N+1), \tag{Eq. 7.1}$$

where $R_{\bullet j}^2$ is the sum of ranks for the jth group. (The formula is a computational shortcut derived from the asymptotic approximation method, as explained in Appendix 7.1.) Because the Kruskal-Wallis test is a chi-square test, it is always one sided.

The *exact* computational formula for the test statistic, which should be used in lieu of Eq. 7.1 when there are ties, is:

$$H = (N-1)\frac{\left[\sum_{j=1}^{k}\frac{\left(R_{\bullet j}\right)^2}{n_j} - \frac{N(N+1)^2}{4}\right]}{\left(\sum_{j=1}^{k}\sum_{i=1}^{n_j}R_{ij}^2\right) - \frac{N(N+1)^2}{4}}. \tag{Eq. 7.2}$$

Readers familiar with ANOVA may recognize that Eq. 7.2 conceptually reflects the ratio of the treatment sum of squares to the total sum of squares. The denominator involves the term R_{ij}^2, which is the square of each rank, and $\left[\sum_{j=1}^k \sum_{i=1}^{n_j} R_{ij}^2\right]$ is therefore just the sum of all the squared ranks regardless of condition. The numerator involves the term $(R_{\bullet j})^2$, which is the square of the sum of ranks for each condition. The term $\frac{N(N+1)^2}{4}$ is the square of the expected value regardless of condition (see Appendix 7.1).

EXPLORING THE CONCEPT

If the treatment has no effect, is the term $\sum_{j=1}^k \frac{(R_{\bullet j})^2}{n_j}$ likely to be greater than or equal to the expected value?

After conducting the KW test, multiple comparisons are conducted between pairs of groups, using the WMW test or the following equation:

$$z = \frac{|R_{\bullet a} - R_{\bullet b}|}{\sqrt{\frac{N(N+1)}{12}\left(\frac{1}{n_a} + \frac{1}{n_b}\right)}},$$ (Eq. 7.3)

where a and b are the two groups being compared and $R_{\bullet a}$ and $R_{\bullet b}$ are the sum of ranks for the respective groups. A variation of Eq. 7.3 for tie adjustments is used in IBM SPSS (see Dunn, 1964).

A statistically more powerful method is to use the *stepwise, stepdown* procedure. The procedure is designed to reduce the number of comparisons made, thereby reducing the frequency of Type I and II errors. There are several versions of the procedure; the one by Campbell and Skillings (1985), which is referenced in the SPSS documentation, is presented here. In this method, one sorts the samples from smallest to largest based on the mean ranks and then searches for *homogenous subsets* of the whole sample, which refers to groups with average ranks that are not significantly different.

An example using five groups is presented in Table 7.2. Samples 1 and 5 are compared first, using one of the procedures referenced above (e.g., WMW). These samples have the largest difference in mean rank, so if the mean ranks of the two samples are not significantly different, then none of the other comparisons could be significant either, and the process stops (there is one homogenous group). All the other comparisons are a subset of the more global, Samples 1 and 5 comparison, so one is implicitly comparing five groups (i.e., H_0: $Mdn_1 = Mdn_2 = Mdn_3 = Mdn_4 = Mdn_5$), even though one is only explicitly testing H_0: $Mdn_1 = Mdn_5$. The "size" of the comparison (p) is 5. If Samples 1 and 5 are significantly

Table 7.2 Example of Stepwise, Stepdown Multiple Comparison Procedure With $k = 5$ Samples

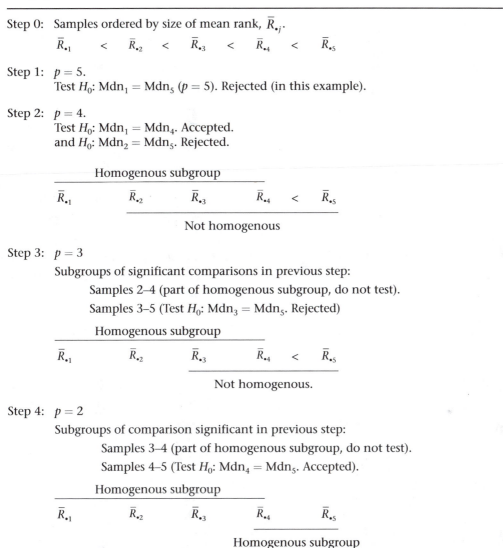

Step 0: Samples ordered by size of mean rank, $\bar{R}_{\bullet j}$.

$$\bar{R}_{\bullet 1} \quad < \quad \bar{R}_{\bullet 2} \quad < \quad \bar{R}_{\bullet 3} \quad < \quad \bar{R}_{\bullet 4} \quad < \quad \bar{R}_{\bullet 5}$$

Step 1: $p = 5$.
Test H_0: $\text{Mdn}_1 = \text{Mdn}_5$ ($p = 5$). Rejected (in this example).

Step 2: $p = 4$.
Test H_0: $\text{Mdn}_1 = \text{Mdn}_4$. Accepted.
and H_0: $\text{Mdn}_2 = \text{Mdn}_5$. Rejected.

Homogenous subgroup

$\bar{R}_{\bullet 1}$ $\bar{R}_{\bullet 2}$ $\bar{R}_{\bullet 3}$ $\bar{R}_{\bullet 4}$ < $\bar{R}_{\bullet 5}$

Not homogenous

Step 3: $p = 3$
Subgroups of significant comparisons in previous step:
  Samples 2–4 (part of homogenous subgroup, do not test).
  Samples 3–5 (Test H_0: $\text{Mdn}_3 = \text{Mdn}_5$. Rejected)

Homogenous subgroup

$\bar{R}_{\bullet 1}$ $\bar{R}_{\bullet 2}$ $\bar{R}_{\bullet 3}$ $\bar{R}_{\bullet 4}$ < $\bar{R}_{\bullet 5}$

Not homogenous.

Step 4: $p = 2$
Subgroups of comparison significant in previous step:
  Samples 3–4 (part of homogenous subgroup, do not test).
  Samples 4–5 (Test H_0: $\text{Mdn}_4 = \text{Mdn}_5$. Accepted).

Homogenous subgroup

$\bar{R}_{\bullet 1}$ $\bar{R}_{\bullet 2}$ $\bar{R}_{\bullet 3}$ $\bar{R}_{\bullet 4}$ $\bar{R}_{\bullet 5}$

Homogenous subgroup

Notes: The letter p is the size of the comparison. The table uses the identical medians version of the null hypothesis but the identical distributions version can also be used (see Chapter 6). The null hypotheses are tested by comparing the sample mean ranks ($\bar{R}_{\bullet j}$). Assuming the population distributions have the same shape among groups, groups with different mean ranks will also have different medians. Subgroups are formed based on which $\bar{R}_{\bullet j}s$ are significantly different from one another based on the procedure shown in the table.

different, one then stepdowns p to 4 and compares Samples 1 and 4 and Samples 2 and 5. Suppose that the former is not significant but the latter is. Samples 1 through 4 therefore form a homogenous subset. (By definition, $p = 4$ because four groups are implicitly involved.) In the next step of the procedure, one investigates the subsets of the significant comparison, stepping down p to 3. As shown in the table, the subsets are Samples 2 versus 4 and Samples 3 versus 5. However, the former is already part of a homogenous subset (Samples 1 through 4), so one only needs to test Samples 3 versus 5. Because that comparison is significant, one then stepdowns p to 2 and considers comparing Samples 3 to 4 and Samples 4 to 5. The former is part of a homogenous subgroup (Samples 1 through 4), so one only needs to compare Samples 4 and 5; the comparison is nonsignificant in the example.

One can therefore conclude that Samples 1 through 4 and Samples 4 through 5 comprise homogenous subsets. The implication is that Sample 5 is significantly different from those of Samples 1, 2, and 3 but not 4. (The fact that Sample 4 is not significantly different from all other samples does not mean that all the samples are the same. It only means that the sample sizes are sufficient to identify a difference between the first homogenous subset and Sample 5 but are not sufficient to identify any differences between Samples 4 and 5.)

The important point is that one can come to these conclusions by only performing five statistical tests, whereas making all pairwise comparisons would involve 10 tests. One is therefore less likely to make a Type I error or, if one controls the Type I error rate by dividing the α value based on the number of comparisons made (explained below), the adjustments will be less with fewer comparisons. In either case, statistical power is enhanced.

In IBM SPSS, stepwise, stepdown comparisons can be obtained by setting the multiple comparison menu to "Stepwise, Stepdown," and then setting the view menu to "Homogenous Subsets" on the model viewer results screen.

Examples

Let us consider several examples of both the KW omnibus test and the multiple comparison procedures. First, consider the data in Table 7.3. (Suppose these are the ranked performances of 18 singers trained by three different coaches.) There are three groups of six observations each, the minimum number needed to use the asymptotic test. Inspection of the average ranks ($R_{\bullet j}$) in the first part of the tables suggests that the average ranks are not equal. The bottom portion of the table shows that $H = 15.16$, $p = .0005$, indicating a significant result. The standard deviation, skew, and kurtosis statistics for the two groups are the same, indicating that the assumption of identical distributional shapes among groups (in the population) is likely met.

Table 7.3 Example 1 of Kruskal-Wallis Test

Data (ranks of purely ordinal data) and descriptive statistics

	Group A	Group B	Group C
	1	7	13
	2	8	14
	3	9	15
	4	10	16
	5	11	17
	6	12	18
Sum of Ranks ($R_{\bullet j}$)	21.0	57.0	93.0
Average Rank ($\bar{R}_{\bullet j}$)	3.5	9.5	15.5
Standard Dev.	1.7	1.7	1.7
Skew	0.0	0.0	0.0
Kurtosis	−1.2	−1.2	−1.2

Calculation of H and p value

$(R_{\bullet j})^2$	441	3,249	8,649
n_j	6	6	6
$\dfrac{(R_{\bullet j})^2}{n_j}$	73.5	541.5	1,441.5

(A) $\sum_j^k \dfrac{(R_{\bullet j})^2}{n_j} = 2{,}056.5$

(B) $\dfrac{12}{N(N+1)} = \dfrac{12}{18 * 19} = 0.035$

(C) $3(N+1) = 3(19) = 57$

(D) $H = (A) * (B) - (C) = 72.16 - 57 = 15.16$

(E) Degrees of Freedom $= k - 1 = 3 - 1 = 2$.

(F) p value for $\chi^2 (2) = 15.16$ is 0.0005.

Technically the assumption also applies to the underlying continuous popula-tion distributions, as we are making inferences about those distributions. If there are metric (i.e., equal interval) measurements of the latent continuous variable, then the assumption should be tested on the metric data. (However, it is still good practice to examine the distribution of the ranks to see how the rank transforma-tion is affecting the data.)

An example with metric data is shown in Table 7.4, along with descriptive sta-tistics. The shapes of the distributions do not appear similar, especially with regard to skew and kurtosis. Sawilowsky (1990) notes that with the KW test, when there are unequal variances, Type I error rates rise slightly when samples sizes are equal, and drastically when they are not. Fortunately, the variances (as well as sample

Table 7.4 Example 2 of Kruskal-Wallis Test

Data and descriptive statistics

	Metric data			Ranks		
	Group A	Group B	Group C	Group A	Group B	Group C
	15	3	12	15	3	11.5
	9	11	17	6	9.5	17
	11	12	18	9.5	11.5	18
	10	1	13	7.5	1.5	13
	8	10	16	4.5	7.5	16
	1	8	14	1.5	4.5	14
Sum	54	45	90	44	37.5	89.5
$\frac{(R_{\bullet j})^2}{n_j}$	—	—	—	322.7	234.4	1,335.0
Mean	9.0	7.5	5.0	7.3	6.3	14.9
SD	4.2	4.1	2.2	4.2	3.6	2.3
Skew	–0.9	–0.7	–0.0	0.7	0.2	–0.2
Kurtosis	2.2	–1.5	–1.9	0.9	–1.7	–1.6

Calculation of H and p value

$$\sum_j^k \frac{(R_{\bullet j})^2}{n_j} = 44^2 + 37.5^2 + 89.5^2 = 1,892.08.$$

$$\sum_{j=1}^k \sum_{i=1}^{n_j} R_{ij}^2 = 2,106.5 \quad \text{(Sum all the squared ranks.)}$$

$$\frac{N(N+1)^2}{4} = \frac{18 * 19^2}{4} = 1,624.5$$

$$H = (N-1) \frac{\left(\sum_{j=1}^k \frac{(R_{\bullet j})^2}{n_j} - \frac{N(N+1)^2}{4}\right)}{\left(\sum_{j=1}^k \sum_{i=1}^{n_j} R_{ij}^2\right) - \frac{N(N+1)^2}{4}} = 17 * \frac{1,892.08 - 1,624.5}{2,106.5 - 1,624.5} = 9.44$$

$df = k - 1 = 2$
$H \sim \chi^2(2) = 9.44, p = .009.$

sizes) are equal in this example, so inflating the Type I error rate is not a concern; however, inflating Type II error rates could be. (We shall later see that distributional differences can affect statistical power.) Nonetheless, the KW test is still significant ($H = 9.44, p = .009$). This result means that at least one of the groups is significantly different from the others.

One next performs multiple comparisons. The stepwise, stepdown procedure in SPSS indicated that Groups A and B (which had the lowest mean ranks) comprise a homogenous subgroup ($p = .10$), but Group C is statistically different from the other two. Compare this result with the methods making all pairwise comparisons. WMW follow-up tests also indicate significant differences between Groups A and C and between Groups B and C, but not between Groups A and B. For both significant comparisons, $W = 24$, $p = .015$. (Given the small sample size, the exact significance levels are reported.)

Many authors (e.g., Pallant, 2010) recommend that the alpha level ($\alpha = .05$) be divided by three because one is conducting three tests (for the three pairwise comparisons), so that the overall "familywise" Type I error rate can remain at 5% (this adjustment is known as a Bonferroni correction). Alternatively, one can leave the per test alpha level at .05 but multiply the p values by three. Even with a Bonferroni adjustment, the two significant comparisons remain so. (See Technical Note 7.1 for a more refined procedure.) With the other popular method for making all pairwise comparisons using Eq. 7.3 (and SPSS uses this method), the Bonferroni-corrected p values (which involves multiplying the unadjusted values by three) are .04 for Groups A and C, and .01 for Groups B and C, so both of these comparisons are again significant.

It is generally recommended that the stepwise, stepdown procedure be used over making all pairwise comparisons. It makes little difference with only three groups, but with more groups fewer comparisons can be performed and therefore the Bonferroni corrections can be smaller. Sheshkin (1997) also notes that the Bonferroni corrections are not necessary if only one or two planned comparisons are made. Suppose Group C is the control group, and one had previously hypothesized that the central tendency of both experimental treatments would be less than the control (but made no hypothesis about the effectiveness of Treatment A against B). One could then use the unadjusted p values from the SPSS "all pairwise" comparisons for just the two planned comparisons, which are .01 and .02, respectively. (One could also just perform two WMW tests, yielding p values of .015.) It is best practice to have specific research hypotheses and to use those to conduct planned comparisons. Otherwise, use the stepwise, stepdown procedure.

I now return to the issue of checking the assumption of identical distributional shapes, because this task becomes more manageable once specific comparisons have been identified as significant (or not). For ease of exposition, I will dispense with a comparison of Groups A and B and will concentrate on the significant ones. Inspection of Table 7.4 suggests the presence of other distributional differences than just in the means or mean ranks. These other distributional differences can potentially bias the estimates of the mean difference. So for example, there is more skew in Group A than C. The difference in average ranks between Groups C and A is $14.9 - 7.3 = 7.6$. The mean rank for Group A is skewed upwards by the 15. The overall difference in mean ranks (and significance level) would be even greater without the skew, so this diagnostic check does not suggest the violation would cause a Type I error, but it could cause a Type II error. A Type II error did not occur here given that the comparison was significant, but this example again illustrates that nonparametric tests are not necessarily robust against Type II errors. The recommendations in the previous chapter regarding use of the WMW test (e.g., checking for outliers and distributional differences) apply to the KW test as well.

If there are distributional differences, a significant KW result would indicate rejection of the null hypothesis that the distributions are all the same. As noted in the previous chapter, this conclusion can be supplemented with the calculation of effect sizes for each significant comparison, such as the probabilities of superiority, which are 92% for each significant comparison in the example. The result provides very strong evidence that the values in Group C tend to be higher than in the other groups. With multiple groups, it would also be informative to compute relative treatment effects, as explained in the next chapter.

EXPLORING THE CONCEPT

In the previous example, there was a difference in kurtosis between Groups B and C. Why would these differences in kurtosis be less likely to bias estimates of differences in population location parameters than would skew?

Test Considerations

The KW test can be used with samples of different sizes, as well as with a large number of ties in the ranks. It can therefore be used when the DV consists of ordered categories. If there is a natural ordering to the IV as well, then both the IV and DV will be ordinal and one can use the $\hat{\tau}$ test described in Chapter 6, which will be more statistically powerful than using the KW test.

With the KW test for ranks or ordered categories, if the samples are too small to use the asymptotic test, exact tables are available (see, for example, Table 7.5). There must be at least two observations in each group. SPSS will also provide exact values along with a message, "not corrected for ties." Ties make the exact values too conservative, so the asymptotic value is also reported. For small samples, the true significance levels would fall between the two p values.

Hawkins (1990) reviews a number of slippage tests when there are multiple samples. Slippage tests of scale can be useful for checking for outliers, as can use of the median absolute deviation (MAD) technique described in the previous chapter.

Table 7.5 Kruskal-Wallis Test: Exact 95th Quantile Table
(for Three Groups, Small Samples)

n_1	n_2	n_3	Statistic for 95th quantile ($w_{.99}$)
2	2	2	4.5714
3	2	1	4.2857
3	2	2	4.5000
3	3	1	4.5714
3	3	2	5.1389
3	3	3	5.0667
4	2	1	4.8214
4	2	2	5.1250
4	3	1	5.0000
4	3	2	5.4000
4	3	3	5.7273
4	4	1	4.8667
4	4	2	5.2364
4	4	3	5.5758
4	4	4	5.6538
5	2	1	4.4500
5	2	2	5.0400
5	3	1	4.8711
5	3	2	5.1055
5	3	3	5.5152
5	4	1	4.8600
5	4	2	5.2682
5	4	3	5.6308
5	4	4	5.6176
5	5	1	4.9091
5	5	2	5.2462
5	5	3	5.6264
5	5	4	5.6429
5	5	5	5.6600

Note: Values from Iman, Quade, and Alexander (1975) and Conover (1999).

Sign Test for Pre- and Posttest Data

The remaining tests considered in this chapter use related samples that are not statistically independent, due either to matching or because within-subject variables are used. The simplest situation is when there is a pretreatment and posttreatment score for each individual, such as a pretest and posttest, and each score reflects ranks or metric data reduced to ranks. In this case one can use a sign test. With metric data, one would normally use a paired t test, but if the sample is very small or the data are skewed, one might find a sign test preferable. It was the very first nonparametric test invented (Arbuthnott, 1710) and is based directly on the binomial test.

The idea behind the sign test is that each pair of observations is assigned either a plus or minus, depending on whether the rank for each person is greater at post-test than pretest (a plus) or is less (a minus). If there is no change in the rank, the case is thrown out. Let the test statistic T represent the number of plus cases. Under the null hypothesis, an increase is just as likely as a decrease, so the probability of a plus (π) is 50%. In other words: H_0: $\pi = 50\%$, H_a: $\pi \neq 50\%$.

To compute the p value, use the binomial formula to find the probability of T hits given $\pi = 50\%$ and n as the number of nontied cases. If $n \geq 20$, one can also use the normal approximation method and compute a z value. To do that, one can use SPSS or the following computational formula, which is derived in Appendix 7.2. (It will result in the same decision, but some researchers prefer reporting z rather than T because readers may be more familiar with z scores.)

$$z = \frac{2T - n - 1}{\sqrt{n}}. \tag{Eq. 7.4}$$

Remember that the probability of z, Prob(z), is the cumulative probability of obtaining a value of z or less. To derive a two-sided p value, double the value if z is negative; if z is positive, subtract the value from 1.0 and then double the resulting value. One can also double the one-sided p values to obtain two-sided values if the binomial distribution is used, as the distribution is symmetric under the null hypothesis of $\pi = 50\%$.

As an example, suppose one is evaluating the effectiveness of a creative writing course using a pretest–posttest design. Each student wrote a short story on the first day of the course (pretest) and one on the last day (posttest). There were 25 students in the class and therefore 50 stories in all. Two scorers read all the stories and then rank ordered them on the basis of quality based on some predefined criteria (creativity, suspense, organization, etc.). The two scorers worked together to resolve the differences between their scores, so there is just one set of rankings.

The data are shown in Table 7.6. There are several tied ranks. The ranks of Students 11 and 24 are tied on the pretest and an average rank assigned. The two scores for Student 17 are also tied; because she exhibited no change, her values are discarded. The effective n is therefore 24. In summary, there are 17 pluses out of 24 trials. If under the null hypothesis, the probability of a plus is 50%, then the probability of obtaining $T \geq 17$ is $1 - \text{Prob}(T \leq 16)$. Using EXCEL, we have $=1 - \text{BINOMDIST}(16, 24, 0.5, 1) \rightarrow .032$; doubling yields a two-sided p value of .064.

Table 7.6 Data for Sign Test Example on Creative Writing Course

Student	Pre	Post	Change
1	9	24	+
2	1	34	+
3	27	43	+
4	36	33	−
5	14	44	+
6	15	45	+
7	37	32	−
8	39	50	+
9	38	49	+
10	22	46	+
11	2.5	25	+
12	28	26	−
13	20	10	−
14	5	19	+
15	21	47	+
16	31	48	+
17	16.5	16.5	thrown out
18	35	18	−
19	23	41	+
20	40	30	−
21	7	11	+
22	8	12	+
23	6	13	+
24	2.5	4	+
25	42	29	−
Average	21.02	29.98	

Note: There are a total of 17 plusses out of 24. One tie was thrown out.

One therefore does not reject the null hypothesis at a .05 level of significance. The results could be reported as $T = 17$, $p = .06$.

If one used the normal approximation and Eq. 7.4, we would have

$$z = \frac{2T - n - 1}{\sqrt{n}} = \frac{(2*17) - 24 - 1}{\sqrt{24}} = 1.84,$$

Prob($z < 1.84$) = 96.7%. Subtracting from 100% and doubling the value yields almost the same p value of .066. The results could be reported as $z = 1.84$, $p = .07$.

When using metric scores, it is not necessary to rank the scores; just use the metric scores and note where the scores increase (for a plus) or decrease (for a minus). One could perform a sign test in small samples on metric data, but even here t tests are sometimes performed with 15–20 cases because the t test has been found to be robust to violations of normality (Scheffé, 1959). However, when the data are skewed (even in large samples) or kurtotic, the sign test may be more statistically powerful.

EXPLORING THE CONCEPT

Why is it necessary to throw out ties when conducting a sign test? (If computing the probability of 4 out of 10 coin tosses landing on heads, do we allow for the possibility that some coins will not land on either heads or tails?) If there are many ties, is it wise to use a sign test?

Wilcoxon Signed-Ranks Test

With metric data, one can also perform a Wilcoxon signed-ranks test. The idea behind this test is to compute the difference score for each individual (i.e., posttest–pretest score), then to rank the difference scores based on their absolute values. The signed ranks are then computed by reinserting the sign of the difference score into each rank. These should sum to zero under the null hypothesis, because under the null, an increase is just as likely as a decrease. More formally, the null hypothesis states that (a) the difference scores are distributed symmetrically around the median score, and (b) the median difference score is zero.

Because it uses more of the information in the data than the sign test (specifically the magnitude of each increase or decrease), the test is more statistically powerful than a sign test. However, the test can only be used with a metric DV.

The test statistic is the sum of the signed ranks, T.

$$T = \sum R_i^s,$$
(Eq. 7.5)

where R_i^s are the signed ranks. (Table 7.7 shows an example, where $T = -17$.)

The asymptotic version (normal approximation) of the test uses the following formula:

$$z = \frac{T}{\sqrt{\sum \left(R_i^s \right)^2}}, n \geq 10.$$
(Eq. 7.6)

Table 7.7 Data for Wilcoxon Signed Ranks Test Example

Person	Pretest	Posttest	D_i	$|D_i|$	Rank	Signed rank (R_i^s)	$(R_i^s)^2$
1	86	88	2	2	3	3	9.00
2	71	77	6	6	7	7	49.00
3	77	76	−1	1	1.5[a]	−1.5	2.25
4	68	64	−4	4	4	−4	16.00
5	91	96	5	5	5.5[a]	5.5[a]	30.25
6	72	72	0	0	—[b]	—	—
7	77	65	−12	12	10	−10	100.00
8	91	90	−1	1	1.5[a]	−1.5	2.25
9	70	65	−5	5	5.5[a]	−5.5	30.25
10	71	80	9	9	9	9	81.00
11	88	81	−7	7	8	−8	64.00
12	87	72	−15	15	11	−11	121.00

Sum						−17 (= T)	505.00
Sum of positive values						24.5 (= T^+)	

Summary of Steps
1. A difference score for each person is calculated (D_i).
2. Any cases where $D_i = 0$ are removed from the analysis.
3. The absolute value of each D_i is calculated.
4. The scores in Step 3 are ranked.
5. For any ties in the ranks, use the average ranks ("midranks").
6. Place the sign of the difference scores back into the scores in Step 5.
7. To perform the normal approximation procedure ($n \geq 10$):
   a) Calculate the sum of the signed ranks ($\sum R_i^s$).
   b) Calculate the square of the scores in Step 6, $(R_i^s)^2$, and the sum of these, $\sum (R_i^s)^2$.
   c) Calculate the z score using $\dfrac{\sum R_i^s}{\sqrt{\sum (R_i^s)^2}}$.
8. In smaller samples, if there are no ties, calculate the sum of the positive values in Step 6; this sum is the test statistic T^+. Then consult Table 7.7 to determine statistical significance. If the sum of the negative values (T^-) is $< T^+$ in absolute value, use T^- instead.

Note: Data from Conover (1999, p. 355).
[a]The average rank is assigned in the case of tied ranks.
[b]Case removed ($D_6 = 0$).

EXPLORING THE CONCEPT

A z score represents the number of standard deviation units that a statistic differs from the mean. Why under the null hypothesis is the mean of the sampling distribution (for T) zero? Is an increase just as likely as a decrease? The null hypothesis states that the *median* difference is zero, but will the mean and median differ if the distribution of the difference scores is symmetrical?

In Eq. 7.6, the denominator represents the standard error (standard deviation of the sampling distribution). In calculating the standard error, one does not need to explicitly calculate the difference of each value from the expected value because the expected value is zero. Because the test is relatively complex, Appendix 7.3 gives a more detailed explanation of the underlying theory.

In our example, the z score is $z = \dfrac{\sum R_i^s}{\sqrt{\sum (R_i^s)^2}} = \dfrac{-17}{\sqrt{505}} = \dfrac{-17}{22.47} = -0.76$, Prob$(z \geq 0.76) =$ 0.224. Double this value for the two-sided p value of 0.45. If your statistical software gives a different value, it may be because it read the data backwards, switching the negative and positive ranks.

The normal approximation requires a minimum of 10 cases. For smaller samples, one computes the test statistic:

$$T^+ = \sum_{i=1}^{n} R_i^s \text{ (where } D_i \text{ is positive)}, \tag{Eq. 7.7}$$

where n is the number of individuals, D_i is the difference score, and R_i^s is the signed rank. The formula assumes that T^+ is smaller in absolute value than T^- (sum of the negative ranks), otherwise use T^- instead.

There are tables reflecting the exact distribution of T^+; Table 7.8 shows the exact quantiles for $n \leq 12$. The negative difference scores are still taken into

Table 7.8 Wilcoxon Signed Ranks Test: Exact Quantiles of T^+ for Small Samples

Quantile	Sample size								
	4	5	6	7	8	9	10	11	12
0.05%	0	0	0	0	1	2	4	6	8
1.0%	0	0	0	1	2	4	6	8	10
2.5%	0	0	1	3	4	6	9	11	14
5.0%	0	1	3	4	6	9	11	14	18
10.0%	1	3	4	6	9	11	15	18	22
20.0%	3	4	6	9	12	15	19	23	28
30.0%	3	5	8	11	14	18	22	27	32
40.0%	4	6	9	12	16	20	25	30	36
50.0%	5	7.5	10.5	14	18	22.5	27.5	33	39
60.0%	6	9	12	16	20	25	30	36	42
70.0%	7	10	13	17	22	27	33	39	46
80.0%	7	11	15	19	24	30	36	43	50
90.0%	9	12	17	22	27	34	40	48	56
95.0%	10	14	18	24	30	36	44	52	60
97.5%	10	15	20	25	32	39	46	55	64
99.0%	10	15	21	27	34	41	49	58	68
99.5%	10	15	21	28	35	43	51	60	70

Note: Values from Conover (1999). Values for larger n (up to 50) can be found in Harter and Owens (1970).

consideration because n is a column entry in this table. The tables should be used cautiously if there are ties in the ranking (the results will be too conservative). For this reason, and because using tables sometimes requires interpolation, I would recommend using the normal approximation if one can, even though the exact tables have been worked out for up to 50 cases (see Harter & Owens, 1970).

If there are a lot of ties in the ranks, use the following formula to calculate z:

$$\frac{\sum R_i^s - \dfrac{n(n+1)}{4}}{\sqrt{\dfrac{n(n+1)(2n+1)}{24} - \dfrac{\sum t^3 - \sum t}{48}}}, \qquad \text{(Eq. 7.8)}$$

where t is the number of subjects with tied ranks (Sheshkin, 1997). If there are only a few ties, the adjustment will make little difference.

Test Considerations

The test assumes that the difference scores are symmetric. Violation of this assumption should not deter one from using the test, but if significant results are found, the mean, median, and skew of the difference scores should be examined. In rejecting the null hypothesis, it could be because Part (a) of the null hypothesis is false and not Part (b). A violation will not invalidate the test, but care should be taken in how the results are interpreted. If the difference scores are perfectly symmetric, then the mean will equal the median; if the difference scores are substantially skewed, then the median should be quite different from the mean. Only one of these might be significantly different from zero. When the test is significant, examination of the difference scores can indicate whether it is the mean or the median that is driving the effect. One can infer that there is an effect on the mean change, but not the median, or vice versa. In either case, there is still a locational effect on the population distribution, but the effect is not as clear cut as if there was an effect on both the mean and median.

Conover (1999) reported that the asymptotic relative efficiency (ARE) of the Wilcoxon signed-ranks test compared to the paired t test is never less than 0.84 and is sometimes substantially greater when the difference scores are not normally distributed. Also, compared to the t test, the Wilcoxon signed-ranks test can be used in much smaller samples. A Monte Carlo simulation study reported in Randles and Wolfe (1979)—which involved drawing 5,000 small random samples ($n \leq 20$) from various types of distributions—found that for various heavy-tailed, nonuniform distributions, the Wilcoxon test was more statistically powerful than both the t test and sign test, with one exception: The sign test was more powerful than both with a Cauchy distribution (see Table 4.1.7 in Randles & Wolfe, 1979). See Technical Note 7.2 at the end of this chapter for more details.

Friedman Test

The Friedman test is an extension of the sign test when there are multiple, related samples (e.g., repeated measures), just like the KW test is an extension of the WMW test from two to three or more samples. As with the sign test, the level of measurement must be at least ordinal, specifically, ranks. The variance in each group should be approximately the same (Harwell & Serline, 1989b, cited in Sawilowsky, 1990).

Unlike the other tests we have considered, however, the ranking is performed within subjects rather than among subjects. Consider the example again of students in a creative writing course. Suppose each student writes three stories that are formally assessed. One story is the pretest, the second is a posttest, and the third is a delayed posttest (written 6 months after the course has ended as part of an end-of-year proficiency exam). We hypothesize an increase in scores from pretest to posttest and that—because students' skills may show some deterioration from a lack of practice—the delayed posttest scores will be in between. For scoring purposes, the raters (blind to when each story was written) rank the quality of each student's essays against the other essays that the student has written, deciding which of the three stories is strongest, weakest, and in between. This scenario differs from the example for the sign test in that for the latter, the stories for each student were ranked against those written by other students (although the sign test would also work if scored in the same manner as with the Friedman test). The KW and WMW test also involve such between-subject rankings.

Another application of the Friedman test is when subjects rank order their preferences, for example, which of three brands of cereal they like best. Such preference ranking is commonly used in economics and market research. In fact, the Friedman test was developed by the Nobel Prize–winning economist Milton Friedman. In my cereal example, one can think of a "treatment" as consuming a particular brand of cereal. After being exposed to the treatments, participants in the study rank the cereals based on the one they like best, worst, and in between. More generally, the Friedman test can be used when subjects are exposed to multiple treatments, although it is assumed that the effect of one treatment does not influence the effect of another. If the Oak Odyssey cereal tastes better after eating Cherry Puffs than before (because there is a sweet aftertaste from Cherry Puffs), then this will bias the results. To prevent any consistent biases, the three treatments could be administered in a random order for each subject.

Table 7.9 shows data for this example, with the subject number and the rankings for each cereal as the column headings. If a subject likes two or three cereals equally well, then there is a tie and the average ranking is assigned (see Subjects 3, 5, and 10). The sum for each column is given at the bottom of the matrix and is denoted by R_j, where j runs from 1 to 3. (More generally, if there are k treatments,

Table 7.9 Data for Friedman Test Example on Cereal Preferences

Subject	Ranking		
	Cherry Puffs	Oak Odyssey	Sunbursts
1	3	2	1
2	2	3	1
3	3	1.5	1.5
4	3	1	2
5	2	2	2
6	3	1	2
7	3	1	2
8	3	2	1
9	3	2	1
10	2.5	2.5	1
11	3	2	1
12	1	3	2
Total (R_j)	31.5	23	17.5
Expected[a]:	24	24	24
Deviation:	7.5	−1.0	−6.5
Deviation squared:	56.25	1.00	42.25
Sum of deviations squared:	99.5		

Coefficient: 0.0833[b]
$T_1 = 99.5 * 0.0833 = 8.29$[c] ($\sim \chi^2$ with $k - 1$ $df.$)

Note: The highest ranking (3) indicates the cereal that the subject liked best.
[a]Given by $\frac{b(k + 1)}{2}$, where b is the number of rows and k the number of treatments.
[b]Given by $\frac{12}{bk(k + 1)}$. The coefficient standardizes the variances.
[c]As explained in the text, T_1 should not be used if ties are present.

then j runs from 1 to k.) The total is highest for Cherry Puffs, suggesting that Cherry Puffs is liked best and Sunbursts is liked least.

Significance Test

One needs to perform the Friedman test to determine if these differences are statistically significant. The asymptotic version of the Friedman test requires a minimum sample size of 12 ($b \geq 12$). The null and alternative hypotheses for the test are

H_0: Cereals are liked equally well (each ranking is equally likely).

H_a: Cereals are not liked equally well (at least one of the treatments tends to yield higher rankings than the others).

Inspection of the table will show that each row sums to 6. This is because each row contains three ranks (1, 2, and 3). The average rank for each row is $6/3 = 2$. The more general formula for the expected value of each row is

$$E(R_{i\bullet}) = \frac{(k+1)}{2}, \tag{Eq. 7.9}$$

where k is the number of treatments. The formula reflects the mean of the sum of consecutive integers (see Appendix 6.2). The expected value for the rows is the same under both the null and alternative hypotheses. However, under the null hypothesis, one expects that each ranking is equally likely, so the expected value of each observation is also 2.

What is the expected value of the column totals under the null hypothesis? It is a function of the expected values of each observation. In the example, the expected value of each observation is 2. Because there are 12 rows, the expected value of each column is $12 * 2 = 24$. Remember that under the null hypothesis b represents the number of subjects (also known as *blocks*), so that $b = 12$. More generally, the expected value of each column total is

$$E(R_{\bullet j}) = \frac{b(k+1)}{2}. \tag{Eq. 7.10}$$

EXPLORING THE CONCEPT

In the cereal example, under the null hypothesis the expected value of each observation is 2. So is the expected value (i.e., average) of each row. Does it therefore follow that $E(R_{i\bullet}) = E(R_{ij})$? Use that fact to derive Eq. 7.10.

The next step is to perform a chi-square analysis where one computes how the actual column totals differ from the expected values. One then squares the deviances and sums the results. The sum is then multiplied by the coefficient:

$$\text{coefficient} = \frac{12}{bk(k+1)}. \tag{Eq. 7.11}$$

The result is the test statistic T_1, which has a chi-square distribution with $k - 1$ degrees of freedom (2 df in this example). The table shows that $T_1 = 8.29$.

However, while this demonstrates the logic behind the Friedman test, 8.29 is not the correct value of T_1 because when there are ties in the data, a modified formula must be used. The modification is based on the square of the rankings, which are shown in Table 7.10; the formulas and calculations are shown in the table footnote (see formulas for A_1, C_1, & T_1). According to the calculations (and SPSS), $T_1 = 9.48$, $p = .009$. [In EXCEL, =CHISQ.DIST.RT(9.48,2) → .0087.] This result could be reported in a journal article as $\chi^2(2) = 9.48$, $p < .01$.

Table 7.10 Cereal Example Calculations for Tie Adjustment (Friedman Test)

Subject	Rankings Squared		
	Cherry Puffs	Oak Odyssey	Sunbursts
1	9	4	1
2	4	9	1
3	9	2.25	2.25
4	9	1	4
5	4	4	4
6	9	1	4
7	9	1	4
9	9	4	1
9	9	4	1
10	6.25	6.25	1
11	9	4	1
12	1	9	4

Note: Sum of all values $(A_1) = \sum_{i=1}^{b}\sum_{j=1}^{k} R_{ij}^2 = 165$. $C_1 = \frac{bk(k+1)^2}{4} = \frac{12*3*4^2}{4} = 144$. Sum of deviations squared (from previous table) $= 99.5$. $T_1 = \frac{(k-1)*(Sum\,of\,dev.squared)}{A_1 - C_1} = \frac{2*99.5}{165-144} = 9.48$.

Whether or not there are ties in the data, the value given by SPSS should be further adjusted to obtain a more accurate approximation of the exact sampling distribution using the F distribution. The adjustment must be performed by hand or with EXCEL. Specifically,

$$T_2 = \frac{(b-1)T_1}{b(k-1)-T_1}, \sim F[k-1,(b-1)(k-1)]. \tag{Eq. 7.12}$$

So in this example, $T_2 = \frac{(12-1)*9.48}{12(3-1)-9.48} = \frac{11*9.48}{24-9.48} = 7.18$. The sampling distribution of T_2 has $k-1$ df in the numerator and $(b-1)(k-1)$ degrees of freedom in the denominator of the F statistic. In our example, there are two df in the numerator. In the denominator, there are $(12-1)*2$ or 22 df. Consulting EXCEL, one finds that =FDIST(7.18, 2, 22) ➔ .004. This value is essentially half the p value without the adjustment. The adjustment should typically be performed to maximize statistical power. One could report the results in a journal article as $F(2, 22) = 7.18$, $p < .01$.

In the above example, the number of subjects (i.e., blocks) was 12. The Friedman test can also be performed on smaller samples. Although some exact tables have been worked out (e.g., see Neave, 2011, p. 49), Conover (1999) reported that the asymptotic tests provide a reasonable approximation of the sampling distribution even in small samples. In addition, the information from the exact tables has been programmed into SPSS and other statistical packages, alleviating the need to consult such tables.

The Friedman test assumes the variances of the rankings for each treatment are the same. If this assumption is drastically violated, more robust tests should be used (see Chapter 8).

Multiple Comparisons

The Friedman test is an omnibus test that determines if one of the total ranks (R_j) is different from the others, but it does not specify which one. The Friedman test therefore needs to be followed up with multiple comparisons, using Eq. 7.13:

$$z = \frac{\left| \bar{R}_{\bullet 1} - \bar{R}_{\bullet 2} \right|}{\sqrt{k(k+1)/6b}}.$$

(Eq. 7.13)

In our example, the standard error (denominator of Eq. 7.13) is $\sqrt{\frac{3 * 4}{6 * 12}} = .408$. The sum of ranks for Cherry Puffs, Oak Odyssey, and Sunbursts are, respectively, 31.5, 23, 17.5, so the average ranks are 2.63, 1.92, 1.46. Comparing Cherry Puffs and Oak Odyssey, $\left| \bar{R}_{\bullet 1} - \bar{R}_{\bullet 2} \right| = 0.71$, so $z = \frac{0.71}{.408} = 1.74, p = .08$. Likewise, the other p values are .004 (Cherry Puffs/Sunbursts) and .26 (Oak Odyssey/Sunbursts). Only the Cherry Puffs/Sunburst comparison is significant at a Bonferroni-adjusted $\alpha = .017$.

Measure of Association: Kendall's *W*

In reporting the results of the Friedman test, one should also report as an effect size a measure of association, specifically Kendall's W (the coefficient of concordance). It is a measure of how much the rankings agree. In our example, SPSS computes W as 39.5%. (One needs to check Kendall's W under "Nonparametric Tests" in addition to the Friedman test; however, the p values will be the same.) The specific formula for W is

$$W = \frac{T_1}{b(k-1)}.$$

(Eq. 7.14)

In the example, T_1 is 9.48. We therefore have: $W = \frac{9.48}{12 * 2} = 39.5\%$. This result indicates a substantial effect.

For pairwise comparisons, when the variances of both groups are similar, the Hodges-Lehmann estimate (ΔHL) for related samples can also be reported (see Chapter 6 for an explanation of ΔHL). Comparing Cherry Puffs and Sunbursts, ΔHL = 1.25. If one subtracts each Sunburst rating from the matching Cherry Puffs rating, the median difference is 1.25.

Durbin Test

In a balanced incomplete block design, where each subject is only administered some of the treatments, the Durbin test can be used in lieu of the Friedman test. The formula for the Durbin test is

$$T = \frac{12(t-1)}{rt(k-1)(k+1)}\sum_{j=1}^{t}(R_{\bullet j})^2 - \frac{3r(t-1)(k+1)}{k-1},$$ (Eq. 7.15)

where t is the total number of treatments, r is the number of times each treatment occurs, k the number of subjects per block, and $R_{\bullet j}$ is the sum of ranks for the jth treatment. If there are ties within a block, the calculations must be adjusted for ties (see Conover, 1999, for further details).

Page Test for Ordered Alternatives

In the Friedman test, the DV consists of ranks and is therefore ordinal. The IV is considered nominal. However, if there is some natural ordering to the treatments, then the Page test could be used instead of the Friedman test. The assumptions are the same. This situation is analogous to that for the Jonckheere-Terpstra ($\hat{\tau}$) test for independent samples. The Page test is potentially more powerful than the Friedman test, as it uses more information from the data. Suppose one ranks the three cereals in terms of the amount of sugar each contains, with Cherry Puffs the most, Oak Odyssey the second, and Sunbursts the least. The test statistic is L_1 and is a function of the column totals $R_{\bullet j}$s. To compute L_1, within-block rankings are performed and then the column totals are weighted by the hypothesized order that the treatments may have on the DV. So if one hypothesizes that Cherry Puffs, the cereal with the most sugar, will be preferred the most, Oak Odyssey second, and Sunbursts third, then one would multiply the column total for Cherry Puffs by 3, Oak Odyssey by 2, and Sunbursts by 1. Therefore,

$$L_1 = \sum_{j=1}^{k} jR_{\bullet j} = 3*31.5 + 2*23 + 1*17.5 = 158.$$ (Eq. 7.16)

Table 7.11 is an exact table for L_1 when there are no ties. The presence of ties makes the test conservative, but it is not recommended that any adjustments be made for ties (Sheshkin, 1997). Use a normal approximation for larger samples than those shown in the table using Eq. 7.17:

$$L_2 = \frac{L_1 - bk(k+1)^2/4}{\sqrt{b(k^3-k)^2/144(k-1)}}.$$ (Eq. 7.17)

In our example, $L_2 = \frac{158 - 12*3(3+1)^2/4}{\sqrt{12(3^3-3)^2/144(3-1)}} = \frac{14}{4.90} = 2.86.$

The sampling distribution of L_2 follows the normal (z) distribution. Consulting EXCEL, one finds that $= 1 - \text{NORMSDIST}(2.86) \rightarrow .002$.

Table 7.11 Page Test: Exact Table[a] of Critical Values of L_1

No. of participants (b)	α (one sided)	Number of treatments (k)					
		3	4	5	6	7	8
2	.001	—	—	109	178	269	388
	.01	—	60	106	173	261	376
	.05	28	58	103	166	252	362
3	.001	—	89	160	260	394	567
	.01	42	87	155	252	382	549
	.05	41	84	150	244	370	532
4	.001	56	117	210	341	516	743
	.01	55	114	204	331	501	722
	.05	54	111	197	321	487	701
5	.001	70	145	259	420	637	917
	.01	68	141	251	409	620	893
	.05	66	137	244	397	603	869
6	.001	83	172	307	499	757	1090
	.01	81	167	299	486	737	1063
	.05	79	163	291	474	719	1037
7	.001	96	198	355	577	876	1262
	.01	93	193	346	563	855	1232
	.05	91	189	338	550	835	1204
8	.001	109	225	403	655	994	1433
	.01	106	220	393	640	972	1401
	.05	104	214	384	625	950	1371
9	.001	121	252	451	733	1113	1603
	.01	119	246	441	717	1088	1569
	.05	116	240	431	701	1065	1537
10	.001	134	278	499	811	1230	1773
	.01	131	272	487	793	1205	1736
	.05	128	266	477	777	1180	1703
11	.001	147	305	546	888	1348	1943
	.01	144	298	534	869	1321	1905
	.05	141	292	523	852	1295	1868
12	.001	160	331	593	965	1465	2112
	.01	156	324	581	946	1437	2072
	.05	153	317	570	928	1410	2035

Notes: L should equal or exceed critical value to be significant. When $k = 3$, the exact critical values for α of .001, .01, and .05 respectively are: $b = 13$: 172, 169, 165; $b = 14$: 185, 181, 178; $b = 15$, 197, 194, 190; $b = 16$: 210, 206, 202; $b = 17$: 223, 218, 215; $b = 18$: 235, 231, 227; $b = 19$: 248, 243, 239; $b = 20$: 260, 256, 251. Critical values for larger b or k may be obtained through asymptotic methods, or see table in Page (1963). Table adapted from Page (1963). Ordered hypotheses for multiple treatments: A significance test for linear ranks. *American Statistical Association Journal, 58,* 216–230. Reprinted by permission of the publisher Taylor & Francis, Ltd, http://www.tandf.co.uk/journals.

[a]Table assumes no tied ranks.

The asymptotic test is only valid when (a) $b \geq 20$ for any number of conditions, (b) $b > 12$ when there are four or more conditions, or (c) for any b when there are nine or more conditions. In the above example, there are only 12 subjects and three treatments, so it is therefore *not* valid to test L_2. Instead one needs to use the exact tables, although the results will be conservative in the presence of ties. Table 7.11 indicates one-tailed critical values of 153 at the .05 level, 156 at the .01 level, and 160 at the .001 level. Because $L_1 = 158$, the result is significant at the .05 and .01 levels. One can see that the one-tailed p value is between .01 at .001 (because 158 is between 156 and 160); interpolating, the one-sided p value is about .006 (averaging .01 and .001).

The Page test cannot be performed in IBM SPSS, but it is fairly easy to perform in EXCEL using basic arithmetic (and =RANK.AVG). Note that the Page test is inherently a one-sided test. It involves essentially correlating the ranked observations with the hypothesized ordering of the treatment, as in a Spearman's ρ test. The null hypothesis is that the means (or mean ranks) of each group are the same (H_0: $\mu_1 = \mu_2 = \mu_3$), against the alternative reflecting the hypothesized ordering (H_a: $\mu_1 \leq \mu_2 \leq \mu_3$). Although other sorts of orderings may be possible, the test is confined to testing the hypothesized ordering; there is no interest in other types of orderings, and finding alternative orderings in the data should not be used to reject the null hypothesis. These conditions provide the rationale for conducting a one-sided test.

The Page test can be performed with repeated measures data, such as in the short story example. Recall that in that example, there were three sets of outcome scores: pretest, posttest, and delayed posttest. It was hypothesized the scores on the pretest would be lowest, those on the posttest highest, and those on the delayed posttest in between. Therefore, the posttest scores would receive the highest rank (3) on the DV, the delayed posttest the midrank (2), and the pretest scores the lowest rank (1). The actual data are shown in Table 7.12, with the within-block rankings shown in Section (D) of the table (these were calculated in EXCEL using =RANK. AVG). The column totals are 16, 22, and 34.

To check the original hypothesis (that T3 would be between T1 and T2), one should give T1 a weight of 1, T2 a weight of 3 and T3 a weight of 2. In this case, L_1 is 150. One can use the exact table because there are no ties in the within-block rankings. The table shows that the one-tailed critical value is 153 at the .05 level. The research hypothesis is therefore not supported.

EXPLORING THE CONCEPT

What sort of pattern is suggested by the column totals (16, 22, and 34)? If we were to test a research hypothesis of an increasing trend, what should the weights be? Calculate L_1 (it should equal 162). Are the one-tailed results statistically significant at the .001 level? Is it a problem that we ran a second (post hoc) test on the data, given the level of significance?

Table 7.12 Page Test Data on Creative Writing Scores

(A) Pretest (T_1)	(B) Posttest (T_2)	(C) Delayed Posttest (T_3)	(D) Rank Within Block		
			T_1	T_2	T_3
80	90	100	1	2	3
77	91	98	1	2	3
65	86	94	1	2	3
89.3	88	92	2	1	3
94.5	89.5	87.5	3	2	1
84	84.5	99	1	2	3
76	83	95	1	2	3
73	79	93	1	2	3
82	78	89.8	2	1	3
81	92.5	97	1	2	3
75	89	99.5	1	2	3
72	87	93.5	1	2	3

(A) Sums for Page test: $\sum = 16$ 22   34
(B) Weights[a] for Page test: 1   3   2
(C) = (A) * (B)   16   66   68
L_1 = Sum of Row C = 150

[a]Weights reflect expectation that scores on the delayed posttest will be in between those of the other tests. A different expected ordering would require different weights.

Summary

The logic behind comparing two samples nonparametrically (presented in Chapter 6) was extended in this chapter to multiple samples (Kruskal-Wallis test) and related samples (sign test or Wilcoxon signed-ranks test for metric data). We also considered tests applicable when there are multiple related samples, specifically, the Friedman test and the Page test (when the sample treatments can be ordered in some fashion). Table 7.1 summarizes the test statistic and minimum sample size for the asymptotic version of each test, and for smaller samples, the exact distribution (or whether an exact table should be used). The next chapter considers nonparametric tests for even more complex designs, such as those involving interactions between two factors.

Problems

1. As in Alock (1974), a researcher took 80 adults from two countries (Canada and India) and divided them into pairs ("dyads"); members of a dyad had the same gender and nationality. The members of each dyad had to bargain with

one another for "points" by passing back and forth slips of paper with offers and acceptances. An individual could, at any time, initiate a 2-minute time limit. If agreement could not be reached within the time limit, neither party received any payoff. The number of time-limit initiations was measured over 10 sessions as an index of competitiveness and coercion. The data are presented below:

(1) Canadian Males: 10 10 8 2 6 9 7 4 8 3
(2) Canadian Females: 4 6 7 0 3 1 7 0 2 0
(3) Indian Males: 8 6 4 3 0 2 3 1 0 0
(4) Indian Females: 0 0 0 0 1 0 0 0 0 0

Given the small sample sizes and the skew, you decide to use a nonparametric test. Answer the following using statistical software such as SPSS:

a) Perform a Kruskal-Wallis test: report H, df, and the p value. (Code the groups as 1, 2, 3, and 4 and label the variable "groups.")

b) Using the stepwise, stepdown procedure, find which samples form homogenous subgroups. (If this procedure is not available in your statistical software, then perform all pairwise comparisons using the WMW test with a Bonferroni-corrected alpha level of .008.)

c) According to the analysis in Part (b) of this problem, the distribution of Group 4 differs significantly from the other groups, as does the distribution of Group 1. Use the "Compare Means" function in SPSS or similar software to compare the means, standard deviations, skews, and kurtosis values of the metric data and also of the ranked data (you can generate ranks using the "Transform," "Rank Cases" option). Do the results for the significant comparisons appear to be driven by differences in location, scale, or both?

d) For each significant comparison with Group 4, use the WMW test to find U and then the probability of superiority.

2. Lewis-Beck and Tien (1996) compared three types of models for predicting outcomes for incumbents in presidential elections. One model (RETRO) used factors that occurred in the past (economic performance, presidential popularity). The second model (PROS) used predictors that were presumed to influence voters' perceptions of how an incumbent would perform in the future (e.g., leading economic indicators). The third model (COMBO) used both types of predictors. The performance of each model in predicting the incumbent party's share of the vote were ranked for the period 1952–1992 (with a low rank associated with a good prediction); the results are in the files *Elections.7.Problem_2.xlsx* (EXCEL) and *Elections.7.Problem_2. sav* (SPSS).

a) Find the sum of ranks for each model. Which one appears to predict best?

b) Using statistical software, perform a Kruskal-Wallis test. Report H and the p value. Are the results significant at the .05 level?

3. A program evaluator assesses the effectiveness of three different job training programs by examining the number of program graduates who are employed 9 months after graduating from the program. The results are shown below.

   a) In this problem, one variable is nominal and the other is ordinal (ordered categories). Why are tau-based tests like Jonckheere-Terpstra inappropriate for these data?

   b) Perform a Kruskal-Wallis test using statistical software. Code "Employed full-time" as 3, "Part-time" as 2, and "Unemployed" as 1. Enter the data as a contingency table (e.g., in SPSS, weight the data by the frequencies). Report H, df, and p value.

   c) Also use EXCEL to find the p value for H (remember it follows a chi-square distribution).

   d) Conduct stepwise, stepdown comparisons; what conclusions can be drawn. (If this procedure is unavailable, conduct WMW tests.)

   e) For any significant comparisons, calculate the probability of superiority.

Employment status	Program A	Program B	Program C
Full-time	14	3	5
Part-time	9	7	6
Unemployed	79	98	58

4. A random sample of 36 CEOs from three different cities were surveyed as to whether they perceived the city's business climate as favorable, neutral, or unfavorable.

   a) Using statistical software, conduct a Kruskal-Wallis test (report H, df, and the p value).

   b) Conduct stepwise, stepdown comparisons (or WMW comparisons if that procedure is not available). Which comparison(s) are significant?

   c) Calculate the probability of superiority for any significant comparisons.

Climate perceptions	City A	City B	City C
Favorable	7	2	2
Neutral	3	7	1
Unfavorable	2	3	7

5. Richard Delgado is running for governor of California. His campaign polls voters to ask how many would vote for Delgado, both immediately before and 2 days after a new campaign TV commercial airs in six key cities. Below are the data, using city as the unit of analysis.

a) Using the binomial formula, conduct a sign test by hand (or with EXCEL) to determine if there is a significant increase in support after the commercial is aired. State the null and alternative hypotheses, and report T, the one-sided and two-sided p values.

b) Conduct the test in statistical software and compare your results. (In IBM SPSS, go to "Nonparametric Tests," "Related Samples," or in older versions, "2 Related Samples.")

Percent support	City A	City B	City C	City D	City E	City F
Before	65	86	43	40	11	41
After	74	90	49	37	16	48

6. Delgado's opponent ran a TV ad accusing Delgado of racism. The opponent's campaign manager ran a poll in 10 media markets along with pre- and post-polls. Answer the same questions as in the previous question, using the following data:

Percent support	Media market									
	A	B	C	D	E	F	G	H	I	J
Before	73	80	45	37	18	45	46	35	53	62
After	60	50	52	9	16	40	35	29	43	40

7. A science teacher gives her students a pretest, has them read Chapter 10 of the chemistry textbook, and then a posttest (with the same items). Of the 30 students, 20 showed an increase in scores, 7 showed a decrease, and 3 showed no difference. Using the normal approximation, conduct a sign test by hand or in EXCEL to determine if these changes are significant. Report T, the z score, and two-sided p value.

8. The performance on a mathematics proficiency test of 50 special education students was assessed before and then while using a certain assistive technology that helps with reading. Of these, 30 improved their proficiency percentile, 15 declined, and 5 showed no difference.

a) Percentiles are considered ordinal measures. Why would a sign test be more appropriate in this situation than a paired t test?

b) Using the normal approximation, conduct a sign test by hand or in EXCEL to determine if the changes are statistically significant. Report T, the z-core, and two-sided p value.

9. a) Perform a Wilcoxon Signed Ranks Test in EXCEL using the Richard Delgado data from problem 5. The column headings you should use are given below, along with the first two subjects. Compute T^+. What are the

one- and two-tailed p values? Is T^+ significant? How do the p values compare with those for the sign test in the previous problem?

City	Before	After	D	ABS(D)	Remove 0s	Rank	Signed Rank (R_i)	R_i^2
1	65	74	9	9	9	6	6	36
2	86	90	4	4	4	2	2	4

Example EXCEL Syntax

A	B	C	D	E	F[a]	G[b]	H[c]	I[d]
			= C1 − B1	= ABS(D1)				

[a] Remove by hand.
[b] =RANK.AVG(F1,$F$1:$F$6,1).
[c] =IF(D1>0,G1,–G1).
[d] =H1².

b) Although the sample is not large enough, for practice, find the asymptotic solution. Specifically, find T, z, and the one- and two-tailed p values.

c) Test the underlying assumption that the D scores are symmetric by finding the mean, median, and skew. Is the assumption confirmed? What are the implications?

d) Perform the test with statistical software.

e) Using statistical software, perform a two-samples Hodges-Lehmann (ΔHL) test. Does the confidence interval bracket 0?

10. Answer the same questions as in the previous problem with the data from Problem 6. (For these data, the asymptotic p values are valid because $n \geq 10$.)

11. Following Marascuilo and Dagenais (1974), a sociologist has 10 high school students rank order four different definitions of the word *integration* (e.g., integration as free association, as forced mixing). A rank of 1 indicates that the definition is closest to the student's own. The data are in the EXCEL file *Careers.7.Problem_11*.

a) Using the equations for the Friedman test, find the expected value for each column. (The answer will be the same for each column.)

b) How much does the sum of each column (R_j) deviate from the expected value for each column?

c) Using either EXCEL or statistical software, find T_1? What is the p value (to three decimal places)?

d) Using Eq. 7.12, find T_2. Using F tables or EXCEL, find the p value.

e) Using statistical software and the stepwise, stepdown procedure, find which comparisons are statistically significant? (If the procedure is not available, conduct all pairwise comparisons using Eq. 7.13 and report the Bonferroni-adjusted p value(s) of the significant comparisons.)

f) Calculate Kendall's W using Eq. 7.14 or with statistical software. Interpret W.

12. Similar to a larger study by Syed et al. (2008), a researcher surveys seven prospective medical students on their attitudes toward various medical specialties as a career. On one question, the students rate their expectations for job satisfaction on a 5-point Likert scale, with 1 indicating *high expected satisfaction*. The data for four specialties (other than internal medicine) are given below. Using EXCEL, perform a Friedman test. In EXCEL, rank across the rows and calculate any midranks for ties (using =RANK.AVG). Calculate (a) the sum of each column (R_j); (b) the deviations from the expected value for each column; (c) T_1; and (d) T_2, df, and the associated p value. Are the results significant?

Student	Surgery (S)	Psychiatry (Ps)	Ob-Gyn (OG)	Pediatrics (Pe)
A	3	4	2	1
B	2	4	1	3
C	1	2	3	4
D	1	3	2	4
E	3	5	4	1
F	2	5	3	1
G	2	5	4	1

13. As in Haffajee, Thompson, Torresyap, Guerrero, and Socransky (2001), a researcher measures how regular use of a powered toothbrush reduces plaque over a 6-month period, measured at baseline, 3 months, and 6 months ($n = 19$). The data are in the files *Plaque.7.Problem_13.xlsx* (EXCEL) and *Plaque.7.Problem_13.sav* (SPSS).
    a) Using EXCEL, find the within-block rankings and conduct a Page test to assess the effectiveness of the intervention. Because a decline in plaque buildup was hypothesized, use a weight of 3 for baseline, 2 for 3 months, and 1 for 6 months. Report the test statistic and whether it is significant (and at what one-sided α level).
    b) Conduct a Friedman test (you may use statistical software) and report T_1 and T_2 and the respective df and p values.
    c) To obtain a feel of their relative power, compare the p values of the three tests (Page's L_1 and Friedman's T_1 and T_2).
    d) For the Friedman test, perform multiple comparisons.
14. Thirteen students enrolled in Dr. Nussbaum's doctoral research class. At three different points in the semester, they prepared short research proposals. At the end of the semester, Dr. Nussbaum rank ordered the quality of the three research proposals for each student, but did so blind to time. (In other words, he did not know which one was written first, second, etc.) The data are in the files *Research.7.Problem_14.xlsx* (EXCEL) and *Research.7.Problem_14.sav* (SPSS).
    a) Conduct a Friedman test to determine if the quality of the research proposals increased over time. Report T_2, the p value, and the mean rank for each time period. Are the research reports becoming better on average?

b) Perform multiple comparisons.

c) Using EXCEL and the exact tables, conduct a Page test of the hypothesis. Does it yield a lower *p* value than the Friedman test?

Technical Notes

7.1 In conducting pairwise comparisons following a Kruskal-Wallis omnibus test, Daniel (1990) recommends selecting a familywise error rate (α) of 0.15, 0.20, or 0.25, depending on the number of groups (k). With larger groups, there will be more comparisons made and so the researcher may choose to be more liberal, but this choice will depend on one's methodological preferences and standards. Next, find the value of the z distribution corresponding to the $1 - \frac{\alpha}{k(k-1)}$ quantile. As an example, if $\alpha = .10$ and there are three groups, $1 - \frac{.10}{3*2} = .983$, which corresponds to a z-value of 2.12 (in EXCEL, =NORM.S.INV(.983) ➔ 2.12).

To be significant at the α level, the mean ranks of two groups must differ in absolute value by more than $z_{(1 - |\alpha/(k(k-1)|)} * \sqrt{\frac{N(N+1)}{12}\left(\frac{1}{n_a} + \frac{1}{n_b}\right)}$, where the first term is the z value (e.g., 2.12) and n_a and n_b are the sample sizes of the two groups being compared. If the sample size of group 1 is 10, group 2 is 12, and group 3 is 14, then the critical difference in comparing Groups 1 and 2 would be $2.12 * \sqrt{\frac{36(37)}{12}\left(\frac{1}{10} + \frac{1}{12}\right)} = 9.56$. For the more conservative familywise $\alpha = .05$, the critical difference would be 10.80.

If the only comparisons to be made are to a control group, then divide α by just $2(k-1)$. In the above example, suppose Group 1 is the control group. Therefore $1 - \frac{\alpha}{2(k-1)} = 1 - \frac{.10}{4} = .975$, which corresponds to a z value of 1.96. The critical difference comparing groups 1 and 2 would then be just 8.84. The only other comparison to be made would be Groups 1 and 3 (with a critical difference of 8.55). Because fewer comparisons are being made, one can afford to be more liberal to maintain an α of .10.

7.2 Randles and Wolfe's (1979) simulation study compared the power of the Wilcoxon signed-ranks test, sign test, and paired *t* test in small samples ($N \le 20$; $\alpha = .05$). The Wilcoxon test was more powerful than the other tests with a logistic or double exponential distribution (the *t* test was second); the sign test was more powerful with a Cauchy distribution (followed by the Wilcoxon test), and the *t* test was more powerful with a uniform or normal distribution (although a *t* test, being asymptotic, should not be used without an adequate sample size). The one exception was that the sign test was more powerful than the Wilcoxon text with a double exponential (i.e., Laplace) distribution with a small sample ($n = 10$) and small effect (0.2 standard deviation); power was .239 (sign test) and .226 (Wilcoxon test). Although not highly reliable in small samples, fit to some of these distributional shapes can be tested in SPSS using the Q-Q plot function under "Descriptive Statistics."

APPENDIX 7.1

LOGIC BEHIND KRUSKAL-WALLIS TEST (EQ. 7.1)

To show $H = \frac{12}{N(N+1)}\left(\sum_{j=1}^{k} \frac{(R_{\bullet j})^2}{n_j}\right) - 3(N+1)$.

Let $R_{\bullet j}$ be the sum of ranks for the jth group.

Step 1: $z_j \sim \frac{R_{\bullet j} - E(R_{\bullet j})}{SD(R_{\bullet j})}$.

Step 2: $\chi_j^2 \sim \left[\frac{R_{\bullet j} - E(R_{\bullet j})}{SD(R_{\bullet j})}\right]^2$.

Step 3: Similar to the WMW test, $E(R_{\bullet j}) = \frac{n_j(N+1)}{2}$ and standard deviation $(R_{\bullet j}) = \sqrt{\frac{n_j(N+1)(N-n_j)}{12}}$.

Step 4: Because the sum of the $R_{\bullet j}$'s is $\frac{N(N+1)}{2}$ (i.e., the sum of consecutive integers), if you know all the other $R_{\bullet j}$'s you can derive the last one. Therefore, the $R_{\bullet j}$'s are not statistically independent. Kruskal showed that the following term corrected for this problem (it shows what percent of N is not in the jth group): $\frac{N-n_j}{N}$.

Step 5: For each group, the term in Step 2, with those for Steps 3 substituted in, is multiplied by the final term in Step 4. Summing terms and simplifying the resulting expression yields H.

Note: One can transform the ranks to have a normal distribution by using van der Waerden's method: compute rank divided by "$n + 1$," and then find the appropriate quantile from the standard normal distribution. (This transformation can be performed in SPSS using the RANK option.) Denote these scores by a_{ij}. The transformation simplifies the calculation of the test statistic to $T = (N-1)\frac{A_t^2}{A_r^2}$, where A_t^2 is the treatment sum of squares $\left(\sum_j \frac{a_{ij}^2}{n_j}\right)$ and A_r^2 is the total sum of squares $\left(\sum_{ij} a_{ij}^2\right)$ (each van der Waerden score is squared regardless of group). This approach may increase power slightly (Sprent, 1989), but is presented here mainly to illustrate the logic of the approach.

APPENDIX 7.2

DERIVATION OF EQ. 7.4 FOR THE ASYMPTOTIC SIGN TEST

Let T equal the number of plus pairs (e.g., showing an increase over time). The sampling distribution for T will be binomial, with H_0: $\pi = 50\%$ and H_a: $\pi \neq 50\%$. If $n \geq 20$, one can use the normal approximation method for a binomial test, where

$$z = \frac{T - E(T)}{SD_T}.$$ (Eq. 7.A1)

For a binomial distribution, the $E(T) = n\pi$ and the $SD(T) = \sqrt{n\pi(1-\pi)}$, therefore $z = \frac{T - n\pi}{\sqrt{n\pi(1-\pi)}}$. Because under the null hypothesis, $\pi = 50\%$, $z = \frac{T - n*.5}{\sqrt{n*.5(1-.5)}}$. Simplifying the denominator yields: $z = \frac{T - n*.5}{\sqrt{n*.5*.5}} = \frac{T - n*.5}{\sqrt{n}*.5}$. Multiplying the top and bottom of the fraction by two and simplifying the denominator yields

$$z = \frac{2}{2} \left[\frac{T - n*.5}{\sqrt{n}*.5} \right] = \frac{2T - n}{\sqrt{n}}.$$ (Eq. 7.A2)

In performing the normal approximation, one should apply a continuity correction by increasing or decreasing T by one-half unit. Because T is multiplied by two in the last term in Eq. 7.A2, this operation is equivalent to adding or subtracting one unit to the numerator, producing:

$$z = \frac{2T - n - 1}{\sqrt{n}},$$ (Eq. 7.A3a)

or

$$z = \frac{2T - n + 1}{\sqrt{n}}.$$ (Eq. 7.A3b)

The probability of z, using (7.A3a), gives the probability of obtaining that z or something less extreme. The correction decreases the probability by half a bar. One then needs to subtract this value from 1.00 to obtain the one-sided p value of obtaining this z or something more extreme (double the value to obtain a two-sided p value). The equation in 7.A3b should be used if one is performing a one-sided test and one thinks there could not be a positive effect (i.e., $H_a = \pi < 50\%$). This situation is unusual, so the equation in 7.A3a is reported in the main text as Eq. 7.4.

APPENDIX 7.3

EXAMPLE AND THEORY BEHIND THE WILCOXON SIGNED-RANKS TEST

This test is used with metric data with related samples when the distribution of difference scores is fairly symmetric. A normal approximation may be used with $n \geq 10$.

Theory Underlying the Normal Approximation

For this test, the test statistic is the sum of the signed ranks, $\sum R_i^s$. To calculate the z score, one needs the expected value and variance of the sampling distribution. Consider the very simple case of $n = 3$. (The small sample size is for illustrative purposes only and is of course not large enough to apply to an asymptotic test.)

Mean

The test statistic, $\sum R_i^s$, is a composite variable composed of three other random variables (for the case of $n = 3$), namely R_1^s, R_2^s, R_3^s (these are the signed ranks). Let R_i represent the unsigned ranks. Therefore, each R_i^s can take on one of two values: $1 * R_i$ or $-1 * R_i$, depending on the sign of the rank. So, for example, the third rank can be either 3 or -3. The sign of the rank is the random component. It is also random which rank gets assigned to which individual.

For any random variable, $E(X) = X_j * \text{Prob}(X_j)$, where j represents the different possible values of the random variable. In our example, X_j can be either the positive or negative rank. Under the null hypothesis, an increase is just as likely as a decrease, so the probability of X_j being positive (or negative) is 50%. Therefore, $E(R_i^s) = (R_i * 50\%) + (-R_{i*} * 50\%) = 0$. Therefore, the mean of each $R_i^s = 0$. Also, $E(R_i^s) = 0$ and $E(\sum R_i^s) = 0$.

Variance and Standard Error

For any RV, $Var(X) = \sum[(X_j - E(X_j))^2 * \text{Prob}(X_j)]$. So $Var(R_i^s) = \sum[(R_i^s - E(R_i^s))^2 * \text{Prob}(R_i^s)]$. Because $E(R_i^s) = 0, Var(R_i^s) = \sum[(R_i^s)^2 * \text{Prob}(R_i^s)]$, where R_i^s can be either positive or negative, and each of these has a 50% probability under the null hypothesis. Using the third rank as an example, it follows that the variance for the third rank is $(R_3^s) = 3^2 * 50\% + -3^2 * 50\% = 3^2 * 50\% + 3^2 * 50\% = 3^2$. More generally, $Var(R_i^s) = (R_i^s)^2$ and $Var(\sum R_i^s) = \sum Var(R_i^s) = (\sum(R_i^s)^2)$. The square root is the standard error.

Z value

By definition, $z = \frac{X - E(X)}{SD(X)}$. For the normal approximation, the test statistic is $T = \sum R_i^s$, where T is the random variable (X) in the equation. As shown above, the sampling distribution of T has a mean of 0 and a standard error of $\sqrt{\sum (R_i^s)^2}$. Standardizing, it follows: $z = \frac{\sum R_i^s - 0}{\sqrt{\sum (R_i^s)^2}} = \frac{\sum R_i^s}{\sqrt{\sum (R_i^s)^2}}$, proving Eq. 7.6.

Exact Tables

I will now use the exact tables to conduct the test of the data in Table 7.7. It is recommended that one use the normal approximation when there are ties in the data, but this example is given for illustrative purposes only. Consulting exact tables, such as Table 7.8, one would find that the lower critical value (LCV) as 14 (given $n = 12$ and $a = .025$) and an upper critical value (UCV) for $a = .975$ based on the formula UCV $= \frac{n(n + 1)}{2} - $ LCV. In the example, UCV $= \frac{12 * 13}{2} - 14 = 64$. Because the test statistic, $T^+ = 24.5$, is between the critical values (14 and 64), the results are not significant.

Theory

Conover (1999, p. 358) provided the following illustration. Suppose you have eight chips, numbered 1 through 8. The number is positive on one side of the chip and negative on the other side. For example, one chip has a 3 on one side, and a -3 on the other. In any trial, one flips all the chips and will obtain a sample of signed ranks ranging from 1 to 8 in absolute value.

For $T^+ = 1$, there is only one possible permutation. Remember that T^+ is the number of positive ranks, so the only permutation that has 1 positive rank is if the one chip comes up positive and all the other chips come up negative [$+1$, -2, -3, -4, -5, -6, -7, -8]. The total number of possible permutations is $2^8 = 256$, so Prob($T^+ = 1$) $= \frac{1}{256} = .004$. Likewise, there is only one permutation corresponding to $T^+ = 2$ [-1, 2, -3, -4, -5, -6, -7, -8], so Prob($T^+ = 2$) also equals .004. However, for $T^+ = 3$, there are two favorable permutations, one involving $+3$ [i.e., -1, -2, $+3$, -4, -5, -6, -7, -8], and one involving $+1$ and $+2$ [$+1$, $+2$, -3, -4, -5, -6, -7, -8]. Therefore, Prob($T^+ = 3$) $= \frac{2}{256} = .008$. All these scenarios are mutually exclusive, so add these probabilities together to obtain the cumulative probability of 3 or less: Prob($T^+ \le 3$) $= .004 + .004 + .008 = .016$. Likewise, Prob($T^+ \le 2$) $= .004 + .004 = .008$. Repeating this logic for different values of T^+ and for different sample sizes is how the exact table was constructed.

The logic of the exact tables does not afford any room for ties. (There were no chips for 2.5 or -2.5, for example.) The methods for modifying the tables to account for ties is very cumbersome. For this reason, it is recommended that the normal approximation be used instead of the exact tables when there are ties and $n \ge 10$.

References

Alock, J. E. (1974). Cooperation, competition, and the effects of time pressure in Canada and India. *Journal of Conflict Resolution, 18,* 171–197.

Arbuthnott, J. (1710). An argument for divine providence, taken from the constant regularity observed in the birth of both sexes. *Philosophical Transactions, 27,* 186–190.

Campbell, G., & Skillings, J. H. (1985). Nonparametric stepwise multiple comparison procedures. *Journal of the American Statistical Association, 80,* 998–1003.

Conover, W. J. (1999). *Practical nonparametric statistics* (3rd ed.). New York, NY: John Wiley & Sons.

Daniel, W. W. (1990). *Applied nonparametric statistics* (2nd ed.). Boston: PWS-Kent.

Dunn, O. J. (1964). Multiple comparisons using rank sums. *Technometrics, 6,* 241–252.

Haffajee, A. D., Thompson, M., Torresyap, G., Guerrero, D., & Socransky, S. S. (2000). Efficacy of manual and powered toothbrushes (I). *Journal of Clinical Periodontology, 28,* 937–946.

Harter, M. L., & Owens, D. B. (1970). *Selected tables in mathematical statistics* (Vol. 1). Chicago, IL: Markham.

Hawkins, D. M. (1990). *Identification of outliers.* New York, NY: Chapman and Hall.

Iman, R. L., Quade, D., & Alexander, D. A. (1975). Exact probability levels for the Kruskal-Wallis test. In H. L. Harter & D. B. Owen, *Selected tables in mathematical statistics* (Vol. III, pp. 329–384). Providence, RI: American Mathematical Society.

Karpinsky, A. (2007). *Graduate statistics and data analysis.* Retrieved from http://astro.temple.edu/~andykarp/psych522524.html

Lewis-Beck, M. S., & Tien, C. (1996). The future in forecasting: Prospective presidential models. *American Politics Quarterly, 24,* 468–491.

Marascuilo, L. A., & Dagenais, F. (1974). The meaning of the word "integration" to seniors in a multi-racial high school. *The Journal of Negro Education, 43,* 179–189.

Neave, H. R. (2011). *Statistical tables for mathematicians, engineers, economists, and the behavioural and management sciences.* New York, NY: Routledge.

Page, E. B. (1963). Ordered hypotheses for multiple treatments: A significance test for linear ranks. *American Statistical Association Journal, 58,* 216–230.

Pallant, J. (2010). *SPSS survival manual* (4th ed.). New York, NY: McGraw Hill.

Randles, R. H., & Wolfe, D. A. (1979). *Introduction to the theory of nonparametric statistics.* New York, NY: John Wiley & Sons.

Sawilowksy, S. S. (1990). Nonparametric tests on interaction in experimental design. *Review of Educational Research, 60,* 91–126.

Scheffé, H. (1959). *The analysis of variance.* New York, NY: John Wiley & Sons.

Sheshkin, D. J. (1997). *Handbook of parametric and nonparametric statistical procedures.* New York, NY: CRC Press.

Sprent, P. (1989). *Applied nonparametric statistical methods.* New York, NY: Chapman & Hall.

Syed, E. U., Siddiqi, M. N., Dogar, I., Hamrani, M. M., Yousafzai, A. W., & Zuberi, S. (2008). Attitudes of Pakistani medical students towards psychiatry as a prospective career: A survey. *Academic Psychiatry, 32,*160–164.

CHAPTER 8

ADVANCED RANK TESTS (FOR INTERACTIONS AND ROBUST ANOVA)

This chapter addresses nonparametric techniques for testing statistical interactions, specifically, *aligned rank transforms* (ARTs) and robust ANOVA-type statistics (ATSs). Most of the statistical tests that we have examined so far are for relatively simple research designs: Comparing two groups or examining pre-to-post changes. We have also considered somewhat more complex designs involving the comparison of three or more groups (or time points). However, many research designs require the examination of statistical interactions. A common design in experimental research is a 2 × 2 factorial design involving two independent variables (IVs) and examination of their interaction (i.e., for a positive interaction, is there an effect above and beyond the main effects of the two IVs added together?) There are also more complex two-way or multiway factorial designs where the same issues apply. Another common design is the repeated measures factorial design, which involves use of a control group and examination of whether there is an interaction of the treatment variable with time (i.e., is there a different pattern of growth in the experimental group than in the control group?) Figure 8.1 demonstrates such an interaction (indicated by the fact that the two growth lines are not parallel). This type of design is methodologically stronger than pretest–posttest designs (see Technical Note 8.1), so it is important to examine methods for examining time-by-group interactions.

In conducting nonparametric analysis of these designs, researchers often neglect to test formally for interactions. For example, in measuring growth over time, it is common practice for researchers to perform separate sign tests for the experimental and control groups, and if growth in the experimental group is significant and growth in the control group is not, to conclude that the treatment had an effect. This practice is tantamount to inferring an interaction effect but without actually testing for one. To see why this approach is problematic, consider the situation where the *p* value for a sign test in the experimental group is .049 and that for the control group is .051. Would one really conclude that the experimental group grew significantly more than the control group? The supporting logic is faulty because

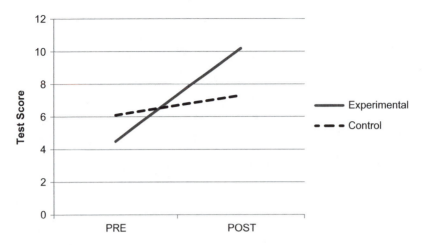

Figure 8.1 Graph illustrating an interaction effect in a pre–post design. The graph shows means by condition. The experimental group shows more growth than the control group. Participants in the experimental group grew, on average, by 5 test points, whereas those in the control group grew, on average, by only 1 point.

if a test of the control group's growth returns nonsignificant results, that does not mean one has to reject the alternative hypothesis, as there still might be some growth one did not have enough statistical power to detect.

There is a common misperception that there are no good nonparametric methods for testing interactions. There are in fact some nonparametric approaches for examining statistical interactions, but they are not widely known or understood. This situation is unfortunate, because testing interactions is an essential part of many research designs. This chapter examines two of these methods: (a) the ART procedure for metric data, and (b) ATS. The latter reflects work over the last few decades in the area of "robust statistics" (Wilcox, 2005). Robust statistics make few if any assumptions about the underlying population and sampling distributions, and are attractive for that reason as well as for the ability to test statistical interactions. The chapter's treatment of robust (nonparametric) statistics is not meant to be comprehensive but rather to introduce readers to this important area of research.

Aligned Rank Transform

This particular procedure is for metric data that are skewed or do not otherwise meet the normality assumptions of analysis of variance (ANOVA) and where a rank-based (nonparametric) alternative is desired. One approach is to take all the observations, rank them, and then perform an ANOVA on the ranks. This approach, known as the *rank transformation* (RT), has been found to be unreliable, especially in regard to estimating interactions. Significant interactions can be created that did not previously exist in the data, or conversely, an interaction that existed in the original data can be lost when the transformation is applied. This is because the RT makes the data less statistically independent; when ranks are computed, the values in one cell (i.e., ranks) are no longer independent from the values in another. A rank depends on the values of all the other observations.

EXPLORING THE CONCEPT

Table 8.1 presents hypothetical data from a 2 × 2 ANOVA. Examine the tables, especially the cells where Factors A and B are both 1. Do you see how the rank transformation exaggerates the magnitude of the interaction?

Table 8.1 Example of a Rank Transformation (RT) Distorting a 2 × 2 ANOVA Interaction Data

Factor A	Factor B	Metric data	Ranked data
0	0	11	10.5
0	0	10	5.5
0	0	9	1.0
0	1	10	5.5
0	1	10	5.5
0	1	10	5.5
1	0	10	5.5
1	0	10	5.5
1	0	10	5.5
1	1	10	5.5
1	1	11	10.5
1	1	12	12.0

Metric means

	A = 1	A = 0	Marginal mean
B = 1	11	10	10.5
B = 0	10	10	10
Marginal mean	10.5	10	

Rank means

	A = 1	A = 0	Marginal mean
B = 1	9.33	5.50	7.42
B = 0	5.50	5.67	5.58
Marginal mean	7.42	5.58	

Table 8.2 shows how the RT can also reduce the magnitude of an interaction. With the metric data, there is clearly an interaction; the cell mean when both factors are 1 is much higher than the other cell means (more than is explained by the main effects). With the ranked data, the cell mean is only slightly higher than the cell beneath it.

A much better method is the ART. This idea with the ART is to remove the main effects from the data before ranking. The values in each cell will then be less

Table 8.2 Example of Rank Transformation (RT) Reducing a 2 × 2 ANOVA Interaction Data

Factor A	Factor B	Metric data	Ranked data
0	0	11	6.0
0	0	10	3.5
0	0	9	1.0
0	1	10	3.5
0	1	10	3.5
0	1	10	3.5
1	0	20	9.0
1	0	20	9.0
1	0	20	9.0
1	1	15	7.0
1	1	25	11.0
1	1	35	12.0

Metric means

	A = 1	A = 0	Marginal mean
B = 1	25	10	17.5
B = 0	20	10	15
Marginal mean	22.5	10	

Rank means

	A = 1	A = 0	Marginal mean
B = 1	10	3.5	6.75
B = 0	9	3.5	6.25
Marginal mean	9.5	3.5	

dependent on the values in the other cells, and the information related to the inter-action will be preserved to a greater extent than with the RT.

The ART can be used as a remedy when the normality assumptions of ANOVA are not met, for example, with skewed or kurtotic data in one cell, or when one suspects that a nonparametric approach will be more powerful because of skew or kurtosis. (There are qualifications, however, and I will discuss research on ART later in the chapter.) Ranking will reduce the influence of skew and kurtosis. An ART test should not be used as a remedy for violations of the equal variance assumption among groups because the ART (being an ANOVA technique) makes this assumption too, and ranking will not necessarily reduce these problems. Furthermore, as with the Wilcoxon-Mann-Whitney (WMW) test, when there are distributional differences among cells, a statistically significant result might not be due to differences in the population means or medians (i.e., locations) but could be due to these other distributional differences, especially in small samples or when sample sizes are unequal (Beasley & Zumbo, 2009). In that event, more robust measures, such as relative treatment effects (presented in the next section) should also be used with an ART test to determine if one group has *stochastic dominance*, that is, tends to have larger or smaller scores than the other groups.

Conducting a 2 × 2 ART

To implement the ART, use the following steps. The data in Table 8.3 will be used as an example; the data reflect a 2 × 2 layout. To differentiate the two levels of each factor, I used dummy variable coding (e.g., A = 1, A = 0); one could also code these levels as A = 1 and A = 2. The ART should only be used with balanced designs (i.e., equal cell sizes).

1. Compute the overall grand mean ($\hat{\mu}_{\bullet\bullet}$) for all Y observations (1.27 in our example).
2. Compute the main effects for Factor A ($\hat{\alpha}_j$), one for each level of the factor. A main effect is the difference between the factor mean at a particular level and the grand mean ($\hat{\alpha}_j = \hat{\mu}_{j\bullet} - \hat{\mu}_{\bullet\bullet}$). If the number of cases at each level is equal, the main effects for a factor will sum to zero. So because there are two levels of Factor A and, using means in Table 8.3, the main effect for Factor A = 1 is 1.62 – 1.27 = 0.35, the main effect for Factor A = 0 will be −0.35.

   In SPSS, one can calculate the factor means and grand mean using the "Means," "Compare Means" option. If one uses EXCEL, a relatively easy way to calculate factor-level means is to use the "IF" function. For example, and as illustrated in Figure 8.2, for the mean when A = 0, I used a blank column to

Table 8.3 Aligned Rank Transform (ART) for 2 × 2 ANOVA (Example Data)

		Factor						Factor			
Case	A	B	Y	Y'	Rank Y'	Case	A	B	Y	Y'	Rank Y'
1	0	0	0.73	0.16	57	41	0	1	1.26	0.01	43
2	0	0	0.74	0.17	58	42	0	1	0.94	−0.31	17
3	0	0	0.65	0.08	49	43	0	1	0.90	−0.35	11
4	0	0	1.16	0.59	80	44	0	1	1.21	−0.04	38
5	0	0	0.22	−0.35	10	45	0	1	1.61	0.36	68
6	0	0	0.85	0.28	63	46	0	1	1.24	−0.01	39
7	0	0	1.12	0.55	78	47	0	1	1.64	0.39	71
8	0	0	0.88	0.31	64	48	0	1	0.91	−0.34	12
9	0	0	0.51	−0.06	33	49	0	1	1.00	−0.25	21.5
10	0	0	0.93	0.36	67	50	0	1	1.46	0.21	60
11	0	0	0.25	−0.32	16	51	0	1	1.31	0.06	46
12	0	0	0.71	0.14	54.5	52	0	1	1.36	0.11	51
13	0	0	0.58	0.01	42	53	0	1	0.83	−0.42	5
14	0	0	0.71	0.14	54.5	54	0	1	1.69	0.43	74
15	0	0	0.81	0.24	61	55	0	1	0.95	−0.30	18
16	0	0	0.21	−0.36	8.5	56	0	1	1.00	−0.25	21.5
17	0	0	0.41	−0.16	28	57	0	1	0.88	−0.37	7
18	0	0	0.30	−0.27	20	58	0	1	1.78	0.53	77
19	0	0	0.77	0.20	59	59	0	1	0.82	−0.43	3
20	0	0	0.21	−0.36	8.5	60	0	1	1.04	−0.21	24
21	1	0	1.17	−0.11	31	61	1	1	1.72	−0.24	23
22	1	0	1.70	0.42	72	62	1	1	2.39	0.43	73
23	1	0	1.30	0.02	44	63	1	1	1.84	−0.12	30
24	1	0	1.35	0.07	48	64	1	1	1.68	−0.28	19
25	1	0	1.16	−0.12	29	65	1	1	1.91	−0.05	36.5
26	1	0	1.27	−0.01	40	66	1	1	1.85	−0.11	32
27	1	0	1.22	−0.06	34.5	67	1	1	1.63	−0.33	14
28	1	0	1.65	0.37	69.5	68	1	1	1.95	−0.01	41
29	1	0	0.89	−0.39	6	69	1	1	2.20	0.24	62
30	1	0	1.10	−0.18	27	70	1	1	1.63	−0.33	14
31	1	0	1.43	0.15	56	71	1	1	2.00	0.04	45
32	1	0	1.22	−0.06	34.5	72	1	1	1.63	−0.33	14
33	1	0	1.07	−0.21	25	73	1	1	1.91	−0.05	36.5
34	1	0	1.65	0.37	69.5	74	1	1	2.30	0.34	66
35	1	0	1.08	−0.20	26	75	1	1	2.28	0.32	65
36	1	0	1.41	0.13	53	76	1	1	2.53	0.57	79
37	1	0	1.36	0.08	50	77	1	1	2.02	0.06	47
38	1	0	0.85	−0.43	4	78	1	1	2.48	0.52	76
39	1	0	0.76	−0.52	2	79	1	1	2.45	0.49	75
40	1	0	0.72	−0.56	1	80	1	1	2.08	0.12	52

Note: Figures based on unrounded values. $\hat{\mu}_{\bullet\bullet} = 1.267713$, $\hat{\mu}_{1\bullet} = 1.62100$, $\hat{\mu}_{\bullet 1} = 1.60767$.

	A	B	C	D	E	F	G	H	I	J	K
1	Case	A	B	Y		Y if A = 0					
2											
3	1	0	0	0.73		=IF(B3=0,D3,"")		Formulas shown for			
4	2	0	0	0.74		=IF(B4=0,D4,"")		first three rows.			
5	3	0	0	0.65		=IF(B5=0,D5,"")					
6	4	0	0	1.16		1.16		Values shown for			
7	5	0	0	0.22		0.22		remaining rows.			
8	6	0	0	0.85		0.85					
9	7	0	0	1.12		1.12					
10	8	0	0	0.88		0.88			Factor A Mean		
11	9	0	0	0.51		0.51			=AVERAGE(F3:F72)		
12	10	0	0	0.93		0.93			0.91		
13	11	0	0	0.25		0.25					
14	12	0	0	0.71		0.71					
15	13	0	0	0.58		0.58					
16	14	0	0	0.71		0.71					
17	15	0	0	0.81		0.81					
18	16	0	0	0.21		0.21					
19	17	0	0	0.41		0.41					
20	18	0	0	0.3		0.3					
21	19	0	0	0.77		0.77					
22	20	0	0	0.21		0.21					
23	21	1	0	1.17							
24	22	1	0	1.7							
25	23	1	0	1.3							
26	24	1	0	1.35							
27	[CASES 25 TO 35 NOT SHOWN]										
28	36	1	0	1.41							
29	37	1	0	1.36							
30	38	1	0	0.85							
31	39	1	0	0.76							
32	40	1	0	0.72							
33	41	0	1	1.26		1.26					
34	42	0	1	0.94		0.94					
35	43	0	1	0.9		0.9					
36	44	0	1	1.21		1.21					
37	45	0	1	1.61		1.61					
38	46	0	1	1.24		1.24					
39	47	0	1	1.64		1.64					

Figure 8.2 EXCEL data format for computing factor level mean (for A = 0). Figure shows formula for Row 3 (in cell F3), which was copied down the column (not all rows are shown). Values of Y are generated only when A = 0, allowing computation of the factor level mean (0.91 in cell I12).

create a new cell entry for the first row of the data set (which was row 3), using the syntax =IF(B3=0, D3, ""), which means if cell B3 (which contained the level of Factor A for the first individual) is 1, use the value in D3 (which contains the outcome value), otherwise use nothing ("" inserts blank text). The new value was placed in cell F3. I then copied this entry down the F column and used

"=AVERAGE(F3:F72)" to obtain the factor-level mean for level 0. I then calculated the main effect by subtracting off the grand mean. The main effects for other factor levels can be calculated in the same way (although in this example, one could multiply by -1).

3. Compute the main effects for Factor B ($\hat{\beta}_m$) in a similar fashion; $\hat{\beta}_m = 0.34$ in our example (when B = 1) and -0.34 (when B = 0).

EXPLORING THE CONCEPT

The note to Table 8.3 indicates that the grand mean rounds to $\hat{\mu}_{..} = 1.27$ and the marginal mean for when Factor B is 1 rounds to $\hat{\mu}_{.1} = 1.61$. Try calculating the main effects for Factor B yourself.

4. Take all the observations (Y_{ijm}) and from each one subtract the appropriate main effects and the grand mean. (The resulting aligned data—call each observation Y'_{ijm}—reflect just the interaction AB). These values are shown in Table 8.3. To illustrate, Case 61 in the example has $Y_{ijm} = Y_{61,1,1} = 1.72$. (The subscript j reflects the level of the first factor, and m the second.) Subtracting off the main effects and the grand mean yields $Y'_{61,1,1} = 1.72 - 0.35 - 0.34 - 1.27 = -0.24$.

   The above-described case was in the cell where Factor A = 1 and Factor B = 1. Case 1, on the other hand, has Factor A = 0 and Factor B = 0 ($Y_{ijm} = 0.73$). In this case, the main effects are negative, so $Y'_{1,1,1} = 0.73 - (-0.35) - (-0.34) - 1.27 = 0.15$ (0.16 without rounding error). Then rank all the cases to create the aligned ranks. This ranking can be performed in SPSS with "Transform" ➜ "Rank Cases" or in EXCEL with "RANK.AVG".

5. Next rank the Y_{ijm} and then using statistical software conduct an ANOVA on the aligned ranks (using the full factorial model with main effects and the interaction term). The main effects will be nonsignificant since you have already removed their effects. You want to see if the interaction is significant.

Except for the last step, I prefer to use EXCEL because there is not an option in IBM SPSS (as of Version 20) to align ranks. If one prefers, one could perform in SPSS some steps of the data transformation using basic commands such as "Compute Variable" and "Rank Cases." One can also find the aligned ranks using the web-based program, ARTweb (Wobbrock, Findlater, Gergle, & Higgins, 2011), found at http://faculty.washington.edu/wobbrock/art/artweb/.

   In our example, if one runs the aligned ranks through a full-factorial 2 × 2 ANOVA (using the SPSS "General Linear Model" procedure), one finds that $F(1,76) = 3.40$, $p = .07$, so the interaction effect is not quite significant. (Note: Results may differ somewhat depending on the rounding protocol used in the calculations if it affects the rankings; less rounding is preferable.) In contrast to the interaction effect, the p values for the main effects are .99 and .90, but these are not meaningful because we subtracted out the main effects. To properly test for main effects, say the main effect of Factor A, find the aligned ranks for Factor A. (Before the ranking, subtract off the main effect of Factor B, the interaction effect, and the grand mean.) These are not shown in Table 8.3 but were easily calculated using ARTweb. The main effects for Factors A and B were both highly significant—A: $F(1,76) = 131.36$, $p < .001$; B: $F(1,76) = 132.22$, $p < .001$.

Conducting a Repeated Measures ART Test

The following example (from Beasley & Zumbo, 2009) illustrates an ART test related to a memory list learning experiment: After committing a poem to memory, do slow learners increasingly forget more words from the poem after a 1-, 2-, or 3-week delay than fast learners (as in Gentile, Voelkl, Mt. Pleasant, & Monaco, 1995)? Table 8.4 presents and analyzes the data.

In the example, there are 18 participants (i is the subject number), two groups (denoted by $j = 1$ or $j = 2$), and three observations ($t = 1$, 2, or 3). The data are aligned by taking each value and subtracting the row mean and column mean and adding the grand mean of all the observations. (Note: for mathematical reasons, this procedure differs from the one for independent samples in that the grand mean is added, not subtracted.) So, for example, to obtain the aligned value of the first observation ($i = 1, j = 1, t = 1$), calculate

$$1.00 - 5.90 - 3.60 + 7.89 = -0.61.$$

Formally, this can be stated as

$$Y'_{ijt} = Y_{ijt} - \bar{Y}_{\bullet j \bullet} - \bar{Y}_{\bullet \bullet t} + \bar{Y}_{\bullet \bullet \bullet}. \tag{Eq. 8.1}$$

The aligned values are then ranked and a multivariate repeated measures ANOVA performed on the aligned ranks. Performing the ANOVA in SPSS produced results indicating a significant interaction. Beasley and Zumbo (2009) recommended using Hotelling's Trace [$H_{(A)} = 1.436$, $F(2,15) = 10.771$, $p = .001$]. I used a multivariate repeated measures approach in the example because it makes fewer assumptions; the population distributions among groups is assumed identical at each time period but not among time periods (as is assumed by a univariate repeated measures model). Sphericity is not assumed (*sphericity* refers to constant variances and covariances). Thus in SPSS, it is better to consult the multivariate statistics. Table 8.4 shows the variance, skew, and kurtosis values to be greater at Time 1, but because the values are similar for both groups, the distributional assumption appears to be met. However,

Table 8.4 Aligned Ranks Transformation (ART) for Repeated Measures (Example Data)

Data

Subject Time:	Original				Aligned data			Aligned ranks		
	1	2	3	(Row mean)	1	2	3	1	2	3
Slow learners:										
1	1.0	7.9	8.8	(5.90)	−0.61	0.60	0.00	16	41	29
2	2.4	9.2	10.1	(7.23)	−0.54	0.57	−0.03	17	40	28
3	2.2	10.1	11.8	(8.03)	−1.54	0.67	0.87	3	42	45
4	2.3	10.9	11.1	(8.10)	−1.51	1.40	0.10	4	50	33
5	3.1	10.1	13.2	(8.80)	−1.41	−0.10	1.50	5	25	51
6	3.3	9.9	12.1	(8.43)	−0.84	0.07	0.77	12	32	43
7	3.2	11.2	14.4	(9.60)	−2.11	0.20	1.90	2	35	52
8	4.4	12.3	13.1	(9.93)	−1.24	0.97	0.27	6	46	36
9	4.9	11.2	14.2	(10.10)	−0.91	−0.30	1.20	11	21	47
10	9.2	13.1	14.3	(12.20)	1.29	−0.50	−0.80	49	19	13
Fast learners:										
11	1.1	6.2	7.2	(4.83)	0.56	−0.03	−0.53	39	27	18
12	2.2	4.8	6.1	(4.37)	2.13	−0.96	−1.16	53	10	8
13	2.3	7.1	8.0	(5.80)	0.79	−0.10	−0.70	44	24	15
14	2.4	8.1	9.4	(6.63)	0.06	0.07	−0.13	30	31	23[a]
15	3.2	7.3	10.4	(6.97)	0.53	−1.06	0.54	37	9	38
16	3.4	9.3	10.5	(7.73)	−0.04	0.17	−0.13	26	34	22[a]
17	4.1	8.1	9.3	(7.17)	1.23	−0.46	−0.76	48	20	14
18	10.1	10.4	10.2	(10.23)	4.16	−1.23	−2.93	54	7	1
Column mean:	3.60	9.29	10.79	(7.89 = grand mean)						

Descriptive statistics (by group)

Slow learners:				
Mean	3.60	10.59	12.31	(8.83)
Median	3.15	10.50	12.60	(8.62)
Variance	4.60	2.02	3.23	(2.72)
Skew	1.85	−0.06	−0.63	(0.31)
Kurtosis	4.35	0.21	−0.50	(0.76)
Fast learners:				
Mean	3.60	7.66	8.89	(6.72)
Median	2.80	7.70	9.35	(6.80)
Variance	6.76	2.67	2.31	(2.95)
Skew	2.24	−0.06	−0.76	(0.75)
Kurtosis	5.64	0.09	−0.75	(1.08)

Note: Adapted from Beasley and Zumbo (2009). Figures based on unrounded data.
[a]Ranks for Subjects 14 and 16 are not tied when aligned data are not rounded.

we cannot be entirely sure of this point because sample moment estimates are unstable in small samples (Beasley & Zumbo, 2009). So although the means and medians for the slow learners do become more different than those for the fast learners over time, it is probably safest to interpret the results not in term of location parameters but in terms of stochastic dominance (the tendency for fast learners to have higher scores). Slow learners are not stochastically dominant at Time 1 but become increasingly more so over time. Because the stochastic dominance differs over time, Beasley and Zumbo call this establishing *stochastic heterogeneity*. Formulas for conducting specific contrasts are provided by Beasley and Zumbo (2009, p. 30).

Interpreting results in terms of stochastic heterogeneity is particularly necessary in small samples or when sample sizes among groups are unequal—as this makes ARTs especially sensitive to violations of equal variances among groups. Otherwise, unless the distributional assumptions are obviously violated, Beasley and Zumbo (2009) recommended that the results be interpreted as differences in location parameters (specifically medians), because aligned rank tests are especially sensitive to location differences and, therefore, significant results are most likely due to that. Because of the ranking, it is safer to interpret the differences in terms of medians, but the original data should be consulted in interpreting the magnitude of the effect.

EXPLORING THE CONCEPT

Figure 8.3 displays the medians for each group; there is about a three-unit difference at Times 2 and 3. The slow learners forget more rapidly than fast learners. Judgment must be used to interpret the magnitude of the effect. In your opinion, is this a large effect?

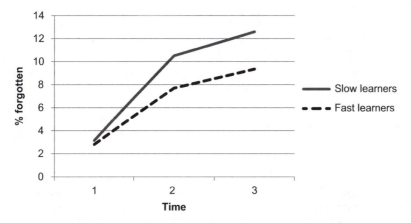

Figure 8.3 Medians by time (and group) for percentages of items forgotten (for data in Table 8.4). Illustrates magnitude of effect for aligned rank transform (ART).

The ART test described above for repeated measures was developed by Higgins and Tashtoush (1994). It relies on ranking by column. One can also use Friedman ranking and find ranks within each row. Beasley (2000) extended the Friedman test to the situation where there are two or more independent groups of subjects.

Research on the ART

There are some alternative aligned-rank procedures (see Hettmansperger, 1984; McSweeney, 1967; Puri & Sen, 1969), but these have not been found to have as desirable statistical properties as those described above, so the remarks in this section do not apply to these alternatives. The research studies to be described were conducted as Monte Carlo simulations, in which the researcher stipulates the population distributions, difference in means, variances, and sample sizes, and then a sample for each group is drawn from its respective populations and a statistical test performed. If the test is significant when there are no differences in the population means or medians, a Type I error is declared, and vice versa for Type II errors. The procedure is repeated thousands of time (e.g., 10,000) so that error rates can be determined empirically. A good test will have an empirical Type I error rate very close to .05 (the *nominal level*), although nominal levels of .01 or .001 can also be used. A good test will also have good statistical power (i.e., avoidance of Type II errors) compared to competing tests, and be robust, so that violation of any assumptions will not affect the error rates much. (Of course, the fewer assumptions made, the more robust the test.) To explore these properties, studies typically conduct sets of simulations on various distributions while varying different parameters such as sample size. The distributions discussed in this section are described in Table 8.5 and illustrated in Figure 8.4.

Table 8.5 Description of Selected Probability Distributions

Distribution	Description
Exponential	Skewed, nonlinear distribution with exponentially decreasing slope. Used to model such things as delays between events or learning curves.
Mixed normal	Normal distribution contaminated by another distribution, causing skew.
Double exponential	Symmetric distribution reflecting the difference between two exponentially distributed random variables.
Log normal	Typically skewed distribution, reflecting the logarithms of a normally distributed random variable. Used to model the joint occurrences (and multiplication) of a series of random variables, such as (in finance) the compounding of returns from a number of trades.
Cauchy	Symmetric distribution created by dividing one normally distributed random variable by another normally distributed one.

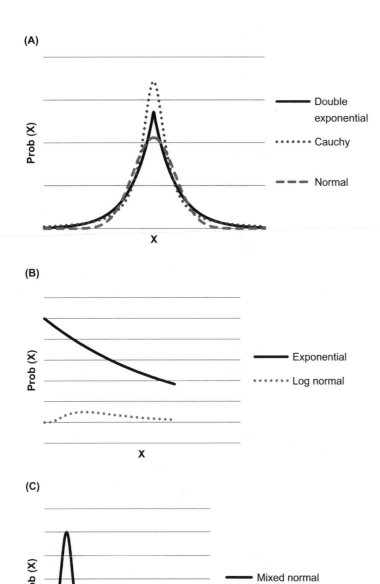

Figure 8.4 Selected probability distributions. (a) Symmetrical, kurtotic (double exponential and Cauchy) compared to normal; (b) skewed (exponential and lognormal); and (c) mixed normal (also skewed).

Toothaker and Newman (1994) conducted a simulation study with 5,000 replications per scenario; their results for the ART test described in this chapter are shown in Table 8.6. The table shows that if the underlying population distribution was normal or exponential, the ART produced good or slightly liberal empirical Type I error (α) rates. The authors defined slightly liberal as an empirical α that exceeds the nominal one by 1% (e.g., between .05 and .06), and slightly conservative as between .04 and .05. Slightly liberal tests are usable, but the p values must be interpreted with caution, especially if they are marginally significant (e.g., between .04 and .05). For the exponential distribution, the ART was consistently

Table 8.6 Results of Toothaker and Newman (1994) Simulation Study Comparing ART and ANOVA F

Population distribution	Factorial design	Cell sizes	Empirical α		Empirical power[a] Medium effect[b]		Large effect[c]	
			ART	F[d]	ART	F[d]	ART	F[d]
Normal	2×2	5	.0602	.0556	.8478	.8506	.9878	.9914
		10	.0554	.0540	.8354	.8480	.9856	.9896
	2×4	5	.0526	.0524	.8442	.8500	.9868	.9878
		10	.0462	.0474	.8294	.8432	.9884	.9916
	4×4	5	.0518	.0506	.8340	.8412	.9894	.9900
		10	.0482	.0490	.8494	.8608	.9908	.9928
Exponential	2×2	5	.0594	.0478	.9002	.8530	.9818	.9664
		10	.0570	.0518	.9394	.8428	.9946	.9740
	2×4	5	.0582	.0472	.9436	.8486	.9936	.9738
		10	.0562	.0506	.9694	.8464	.9992	.9846
	4×4	5	.0604	.0494	.9662	.8370	.9990	.9806
		10	.0504	.0444	.9878	.8528	.9998	.9858
Mixed normal	2×2	5	.0282	.0192	.9814	.8040	.9924	.9308
		10	.0368	.0156	.9974	.8144	.9998	.9516
	2×4	5	.1212	.0186	.9984	.8244	.9998	.9462
		10	.0882	.0238	1.0000	.8336	1.0000	.9642
	4×4	5	.1372	.0410	1.0000	.8212	1.0000	.9556
		10	.1030	.0372	1.0000	.8362	1.0000	.9740

Notes: Based on 5,000 trials per scenario. Figures are for the interaction effect only and nominal $\alpha = .05$. See Toothaker and Newman (1994) for main effects. They label the ART as RO.
[a]Empirical power is $1 - \beta$, where β is the probability of a Type II error.
[b]Medium population effect selected so power, given a normal distribution, was about .85.
[c]Large population effect selected so power, given a normal distribution, was about .99.
[d]F statistic based on ANOVA.

more powerful than the ANOVA F, with big differences when there was a medium population effect. On the other hand, consider the mixed normal distribution. A mixed normal distribution is a normal distribution with some extreme value that create skew. The empirical α's were acceptable but rather conservative for the 2×2 case; they were totally unacceptable for larger factorial designs.

Mansouri and Chang (1995) conducted a simulation study using a 4×3 design but did not investigate the mixed normal distribution. As shown in Table 8.7, they found the empirical α rates for the ART to be slightly liberal or good, except for the Cauchy distribution, which is a specific type of kurtotic distribution (where they recommend the unaligned RT). With regard to power, the ART was slightly more powerful with the exponential distribution with $n = 20$ and (although not shown in the table) substantially more powerful in large samples

Table 8.7 Results of Mansouri and Chang (1995) Simulation Study Comparing ART and ANOVA F, 4×3 Design

Population distribution	Cell sizes	Empirical α				Empirical power[b]			
		$c = 1.0$[a]		$c = 2.5$[a]		$c = 1.0$[c]		$c = 2.5$[a]	
		ART	F[e]	ART	F[e]	ART	F[e]	ART	F[e]
Normal	2	.059	.050	.058	.048	.145	.134	.634	.664
	5	.052	.057	.044	.045	.379	.394	.998	.999
	10	.052	.068	.055	.054	.728	.752	1.000	1.000
	20	.055	.087	.050	.051	.977	.981	1.000	1.000
	50	.058	.194	.053	.050	1.000	1.000	1.000	1.000
Exponential	20	.051	.045	.055	.051	1.000	.976	1.000	1.000
	50	.053	.052	.046	.043	1.000	1.000	1.000	1.000
Double exponential	20	.052	.051	.049	.046	.897	.775	1.000	1.000
	50	.052	.054	.047	.047	1.000	.996	1.000	1.000
Lognormal	20	.057	.040	.055	.042	.972	.442	1.000	.982
	50	.058	.047	.049	.041	1.000	.815	1.000	1.000
Cauchy	20	.366	.016	.367	.020	.536	.023	.929	.077
	50	.468	.016	.472	.018	.806	.024	.994	.082
Uniform	20	.052	.052	.051	.051	1.000	1.000	1.000	1.000
	50	.056	.054	.050	.048	1.000	1.000	1.000	1.000

Notes: Based on 5,000 trials per scenario. Figures are for the interaction effect only and nominal $\alpha = .05$.
[a]Assumes $c = \alpha_2 = \beta_1$ and $-c = \alpha_3 = \beta_2$, where α and β denote main effects.
[b]Empirical power is $1 - \beta$, where β is the probability of a Type II error.
[c]Connotes medium population effect. $c = \gamma_{11}$; and $-c = \gamma_{41} = \beta_1$, where γ denotes an interaction effect.
[d]Connotes large population effect, $c = \gamma_{11}$; and $-c = \gamma_{41} = \beta_1$.
[e]F statistic based on ANOVA.

when the distribution was log normal and the effects to be detected were small (i.e., $c = 0.50$; see footnotes to Table 8.7 for definition of c). In the latter circumstance, the ART was also somewhat more powerful with the exponential and double exponential distributions.

In summary, for 2×2 designs, the ART fares well (although it is sometimes slightly liberal) and is often more powerful when the underlying distribution is exponential. For larger designs, the ART is advantageous in kurtotic distributions, except Cauchy. For skewed distributions, ART is advantageous with the exponential and log-normal distribution but not a mixed-normal distribution.

In respect to using aligned rank methods for repeated measures, research by Beasley (2002) found stable α levels and robustness against assumption violations (see also Beasley, 2000), but only the normal, exponential, and double exponential distributions were investigated. Also, Rasmussen, Heumann, Heumann, and Botzum (cited in Beasely, 1996) found superior statistical power when the distributions were log normal or skewed (Beasely & Zumbo, 2003). These results are generally consistent with those for independent samples.

Of course, one does not usually know the shape of the population distribution. The sample distribution may provide some clues but may contain substantial sampling error, especially in small samples. Prior studies may provide some additional clues, and sometimes there are theoretical considerations pointing to a particular population distribution (e.g., the distribution may be exponential when modeling a learning curve). Nevertheless, given some sensitivity to the shape of the underlying population distribution, and the uncertainty surrounding this issue, we can conclude that ARTs are only moderately robust. (Rank-based methods have also been criticized because "ranked data generally will have unequal variances," Shah & Madden, 2004, p. 34.) We next examine a more robust and sophisticated method for examining nonparametric statistical interactions as well as main effects.

Robust ANOVA-Type Statistics

Just as the study of nonparametrics is a subfield of statistics, the study of robust statistics is another subfield (which intersects with the study of both parametric and nonparametric statistics). A *robust* statistical method is one that either makes fewer assumptions than other methods or is less affected by violation of the assumptions that are made. Therefore, violations of underlying assumptions—which can produce biased estimates and reduce statistical power—is, by definition, less of an issue with robust statistics. See Erceg-Hurn and Mirosevich (2008) for a concise treatment of robust statistics and Wilcox (2005) for a more extended treatment.

Most of the statistical tests that have been considered so far in this book relate to testing a null hypothesis concerning a population parameter: means, proportions, medians, or mean ranks. Many of these tests also attempt to estimate the standard deviation of these parameters in the population and the sampling distribution. Various assumptions are required for different tests regarding sample size and normality, statistical independence, equality of variances among groups (or the symmetry or identicalness of the distributions among groups), and the absence of outliers. In contrast to making hypotheses about population *parameters*, robust statistics often just make hypotheses about the population *distributions*. For example, the null hypothesis typically is that the population distribution in Group A is the same as in Group B. One does not need to make separate distributional assumptions (e.g., of homogenous variances) because these distributional features are implied by the null hypothesis; one just needs to assume the null. If the null hypothesis is rejected, one can conclude that our treatment has had some effect on the population distribution. One does not infer whether there was an effect on the population mean or median, so some specificity is sacrificed for greater robustness, but the shape of the empirical distributions for each group can be examined graphically to see how a treatment might be affecting the population distribution. There are also some effect-size measures that can be used.

The particular set of robust statistics to be examined in this section are *ANOVA-type statistics* (ATS), and specifically the ATS for the nonparametric analysis of ranks. Although nonparametric statistics are less affected by outliers than parametric statistics, nonparametric techniques can still be affected somewhat. Furthermore, ranked data often have unequal variances among groups (even if the data were originally derived from metric data with equal group variances), and traditional nonparametric tests like WMW or Kruskal-Wallis require that assumption unless a null hypothesis of no distributional differences is used. So does the ART method (see previous section), which also requires metric-level data. The nonparametric ATS is much more robust than other tests to outliers and violation of distributional assumptions. Moreover, we examine the ATS here in particular because it can be used to test for nonparametric statistical interactions.

If one is analyzing metric data, one should also consider using a parametric ATS if some of the assumptions of traditional parametric tests are violated. A parametric ATS can be an alternative to reducing the level of measurement down to the ordinal or nominal level and using a nonparametric test, as data reductions do sacrifice some statistical power. Other advantages of an ATS are that (a) it is robust to outliers, (b) it can be performed with missing values (assuming these are missing at random), and (c) although asymptotic, it still provides a good approximation in small samples (Shah & Madden, 2004). The statistics in small samples will be slightly biased, but the amount of bias is typically small and can be quantified and reported. It is also a useful alternative to using generalized linear models (described in Chapters 9–13) when there are many factors, factor levels, or ordered categories (Shah & Madden, 2004), or again when sample sizes are small. The downside is that an ATS may be more complex to conduct and to explain to others than traditional nonparametric techniques, although one can always refer those interested in an explanation to a relevant book chapter such as this one.

Comparing Two Groups

For purposes of illustration, I will assume that one is comparing two groups, as in a WMW or *t* test, and then will extend the discussion to the comparison of multiple groups, as in the Kruskal-Wallis test or ANOVA. The null hypothesis is that the *cumulative population distributions* of the two groups are identical.

When one is dealing with ranks, the overall distribution is uniform, as illustrated in Figure 8.5. For example, if there are 1,000 individuals, there will be 1,000 ranks, and each rank will appear once unless there are ties—but set that possibility aside for now. The probability of each rank will therefore be the same from rank to rank, as shown in Figure 8.5. (To make the figure easier to read, an *n* of 7, rather than of 1,000, is used.) Figure 8.5 also shows the cumulative probability of each rank, which is the probability of obtaining that rank or a lesser one if one randomly samples from the distribution.

If one randomly divides the individuals into two groups, then the probability distributions will not be uniform, as shown for Group A in Figure 8.6(A). The figure uses the data from Table 6.14 (see Chapter 6), which I used to illustrate the median and WMW tests in the previous chapter. Figure 8.6(B) shows the cumulative distributions. The cumulative distribution is easier to examine perceptually because

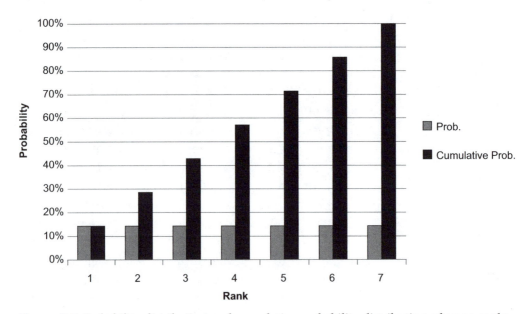

Figure 8.5 Probability distribution and cumulative probability distribution of seven ranks.

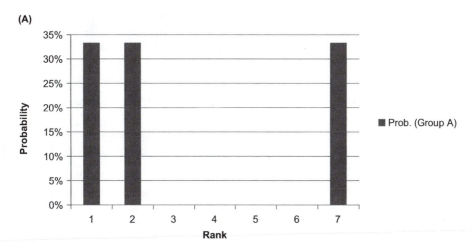

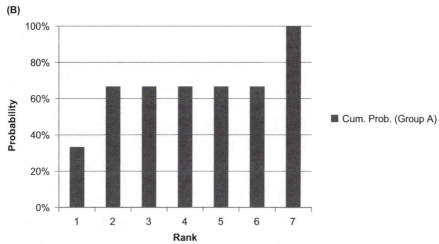

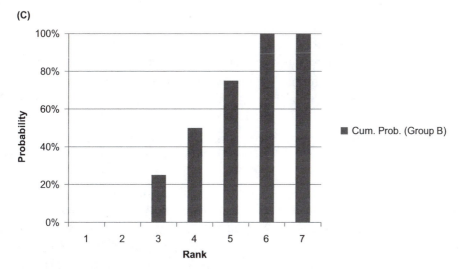

Figure 8.6 Probability distributions for data from Table 6.14 (median test). (a) Noncumulative distribution for Group A. (b) Cumulative distribution for Group A. (c) Cumulative distribution for Group B.

there are no values with zero frequencies that break up the distribution. This in turn facilitates comparisons. For example, Figure 8.6(C) shows the cumulative distribution for Group B, and that distribution clearly has a different shape than that for Group A (although we still need to test whether the difference is statistically significant; see Appendix 8.1).

This approach of comparing the shapes of the distributions was discussed briefly in the previous chapter in our discussion of the WMW test. The WMW test typically directly tests the differences in mean ranks between two groups, under the null hypothesis that the population distributions of the two different groups are identical. The ATS (Brunner, Domhof, & Langer, 2002; Brunner & Puri, 2001) takes a similar approach, but instead an ATS is based on a statistic known as the relative treatment effect:

$$\theta = \mathrm{Prob}(X_A < X_B) + \frac{1}{2} * \mathrm{Prob}(X_A = X_B),\qquad \text{(Eq. 8.2)}$$

which is the probability that a randomly chosen value in Group A will be less than a randomly chosen value in Group B. (I would write "less than *or equal to*," but θ is really a compromise between that and just "*less than*," because of the $\frac{1}{2}$ coefficient.) A good estimate of the relative treatment effect is given by the following equation:

$$\hat{p}_j = \frac{\bar{R}_{\cdot j} - .50}{N},\qquad \text{(Eq. 8.3)}$$

where $\hat{p}_j$ is an estimate of the population relative treatment effect for a specific group, $\bar{R}_{\cdot j}$ is the mean rank of the jth group, and N is the total sample size. Using the data in Table 6.14, the mean rank for Group A is 4.67, and so its estimated relative treatment effect is $\hat{p}_A = \frac{4.67 - .50}{7} = 0.60$. Likewise, the mean rank for group B is 3.5, and so $\hat{p}_B = \frac{3.50 - .50}{7} = 0.43$. A $\hat{p}_j$ greater than 0.50 suggests that there is a tendency for the distribution in group j to have values greater than the overall mean rank. In the case when there are only two treatments, one can also calculate $\hat{p}$ (without any subscript), which is the relative treatment effect of Group A relative to B (rather than the overall mean distribution). It is $\hat{p}$ that estimates θ in Eq. 8.2. It can be shown that

$$\hat{p} = \hat{p}_A - \hat{p}_B + .50.\qquad \text{(Eq. 8.4)}$$

In our example, $\hat{p} = .60 - .43 + .50 = .67$, which means that one estimates that there is a 67% probability that the (ranked) values in Group A will be greater than in Group B. If one inspects Table 6.14, one can see that 2 of the 3 values in Group A are in fact greater than all the values in Group B. The values in Group A are *tendentiously greater* than those in Group B. (This is the same concept as stochastic dominance discussed in the literature on the ART.) A probability of 50% indicates a lack of stochastic dominance.

EXPLORING THE CONCEPT

If the two groups are of equal size, then the two relative treatment effects would sum to 1.0. Suppose there was one additional data point (rank = 8) in Group A. Complete the following table.

	Group A	Group B	Row Sum
	8	5	
	7	4	
	6	3	
	1	2	
Sum	_____	_____	_____ (a)
Mean Rank ($\bar{R}_{\bullet j}$)	_____	_____	_____ (b)
$\bar{R}_{\bullet j} - 0.5$	_____	_____	_____ (c)
Relative treatment effects ($\hat{p}_j$)	_____	_____	_____ (d)

Verify that the value in (a) is the sum of ranks $\left(\frac{N(N+1)}{2}\right)$. Verify that the overall mean rank is $\frac{(N+1)}{2} = 4.5$. Verify that the value in (b) is $(N + 1)$. Why would the value in (b) be equal to $(N + 1)$? [Hint: (b) is the sum of two means.] Verify that the value in (c) is N and the value in (d) is $\frac{N}{N} = 1$. Does Group A tend to have values greater than the overall mean rank (4.5)? Is $\hat{p}_a > 0.5$? Calculate and interpret the relative treatment effect $\hat{p}$ from Equation 8.4 ($\hat{p} = \hat{p}_A - \hat{p}_B + .50$). It should be 75%. Are 75% of the values in Group A greater than those in Group B?

The null hypothesis is that the population distributions of the two groups are the same. ($H_0^F : F_A = F_B$), where F refers to the population distributions. According to Brunner et al. (2002), a somewhat weaker hypothesis is that the relative treatment effects of each group are equal ($H_0^p : p_A = p_B$), which can also be stated as $H_0^p : p = .50$. The first null hypothesis (H_0^F) implies the second (H_0^p), but the converse is not true: Relative treatment effects can be equal in the presence of distributional differences if the central tendencies of the two distributions are the same. For now, let us use the stronger null hypothesis of no distributional differences.

What we are doing here may not initially sound much different from the WMW test, which compares the mean ranks (and $\hat{p}_j$ is a function of the mean rank), but there are two important differences. First, as we shall see, the ATS formulation can be extended to multiple groups, including 2×2 designs or larger factorial designs, as well as repeated measures. Second, the null hypothesis H_0^p is formally stated in terms of relative treatment effects.

Associated with the ATS is an effect size measure, the probability of superiority (PS), which was introduced in Chapter 6 as an effect size measure when comparing two independent groups. The PS is the same as the relative treatment effect p.

Regarding minimum samples sizes, when comparing two groups there should be at least 20 observations in each group (Bathke & Brunner, 2003). Even with the required sample size, the statistic may be somewhat liberal (with simulated Type I error rates ranging from 5% to 7%).

Extensions to Multifactorial Designs

Previously I discussed the ATS for comparing two independent groups, but the procedure can be extended to more advanced designs. Suppose one has a two-way factorial design with two factors, A and B. Factor A has a levels ($j = 1 \ldots a$), and Factor B has b levels ($m = 1 \ldots b$). In a 2×2 design, $a = 2$ and $b = 2$, producing four cells. Each cell would have a cumulative rank distribution. Denote each of these distributions by a term F_{jm}. In our example, there are four cells with the following distributions, $F_{11}, F_{12}, F_{21}, F_{22}$. Similar to a contingency table, one can also examine the distribution just for Factor A by averaging across Factor B (these are the two marginal distributions $F_{1\bullet}$ and $F_{2\bullet}$) and likewise for Factor B ($F_{\bullet 1}$ and $F_{\bullet 2}$). Averaging two or more distributions produces an average distribution denoted by $\bar{F}_{j\bullet}, \bar{F}_{\bullet m}$, and $\bar{F}_{\bullet\bullet}$.

The test for a main effect of Factor A uses the null hypothesis

$$H_0^F(A){:}\ \bar{F}_{1\bullet} = \bar{F}_{2\bullet} = \cdots = \bar{F}_{a\bullet}. \tag{Eq. 8.5a}$$

The test for the main effect of Factor B uses the null hypothesis

$$H_0^F(B){:}\ \bar{F}_{\bullet 1} = \bar{F}_{\bullet 2} = \cdots = \bar{F}_{\bullet b}. \tag{Eq. 8.5b}$$

The test of the interaction effect between A and B uses the null hypothesis

$$H_0^F(AB){:}\ \bar{F}_{jm} = \bar{F}_{j\bullet} + \bar{F}_{\bullet m} - \bar{F}_{\bullet\bullet}, \tag{Eq. 8.5c}$$

or alternatively,

$$H_0^F(AB){:}\ \bar{F}_{jm} + \bar{F}_{\bullet\bullet} = \bar{F}_{j\bullet} + \bar{F}_{\bullet m}, \tag{Eq. 8.5d}$$

which is mathematically equivalent. These equations are further explained in Appendix 8.2, which also shows the derivation of Eq. 8.5c.

The notation is similar to that used in ANOVA (which is not surprising because the ATS is an ANOVA-type statistic). However, there is more to computing an ATS than just running the ranks through an ANOVA; this point is explained more fully in the Appendix 8.1, which presents the technical details of calculating and testing an ATS.

EXPLORING THE CONCEPT

Although the mean ranks are used in the calculation of an ATS, these are a function of the distributions in both groups (and these are in turn a function of the population distributions). So is the ATS ultimately a function of the population distributions? In interpreting an ATS, do we make inferences about the distributions or the locations (i.e., means or medians)? How does this differ from ANOVA?

In the two-way factorial design, one can estimate the relative treatment effects (p_{jm}) with Eq. 8.6, which is analogous to Eq. 8.3:

$$\hat{p}_{jm} = \frac{1}{N}\left(\bar{R}_{\bullet jm} - \frac{1}{2}\right). \tag{Eq. 8.6}$$

Eq. 8.6 gives the relative treatment effects for each cell. For main effects, the relative treatment effects for each level of Factor A (across all levels of Factor B) are given by

$$\hat{p}_{j\bullet} = \frac{1}{N}\left(\bar{R}_{\bullet j\bullet} - \frac{1}{2}\right), \tag{Eq. 8.7}$$

and likewise for Factor B (except the *j*s are summed over rather than the *m*s).

According to Bathke and Brunner (2003), with at least four factor levels, the ATS requires at least 10 replications per cell. (However, testing for covariate effects requires only seven per cell.) Fewer factor levels require larger samples.

Although software packages are constantly being updated, at the time of this writing the ATS procedure can be performed in SAS using "PROC MIXED" (ANOVAF option), but one must first download macros provided free by Brunner et al. (2002; http://www.ams.med.uni-goettingen.de/de/sof/ld/makros.html). These macros are needed to estimate the standard errors. The software *R* can also compute an ATS (see Noguchi, Gel, Brunner, & Konietschke, 2012).

An example of an ATS analysis is provided by Shah and Madden (2004), reanalyzing data from Krause, Madden, and Hoitink (2001). The health of certain plant seedlings was rated 42 to 60 days after planting on the following ordinal scale: 1 = *symptomless*; 2 = *mild root rot*; 3 = *mild root and crown rot*; 4 = *severe root and crown rot*; and 5 = *dead plant*. The seedlings were planted in one of nine different potting mixtures (Factor A), and some mixtures were fortified with biocontrol agents (Factor B). So Factor A had nine levels and Factor B had two ("fortified" and "not fortified"). The number of batches tested per condition ranged from 13 to 81.

The results of the ATS are shown in Table 8.8. The table shows a main effect of potting mix and an interaction between potting mix and fortification. This interaction means that fortification was beneficial for some mixtures but not others. For example, the relative treatment effects for dark peat moss that had been autoclaved

Table 8.8 ANOVA-Type Statistics (ATS) Results for Seedling Study, 2 × 9 Factorial Design

Effect	F	$df_{numerator}$	$df_{denominator}$	p value
Fortification	3.54	1.00	89.1	.063
Potting mix	49.26	6.83	89.1	< .001
Fortification × potting mix	2.49	6.83	89.1	.023

From Shah and Madden (2004).

(i.e., the soil was first heated under high pressure to sterilize it) were significantly different ($\hat{p}_{jm} = .67$ for natural, .37 for fortified) but not so for dark peat moss that had not been autoclaved ($\hat{p}_{jm} = .66$ for natural, .57 for fortified). Fortifying the soil with biocontrol agents only helped when the soil had first been sterilized.

Extension to Repeated Measures

Brunner et al. (2002) provided examples of the various longitudinal designs shown in Table 8.9. One of these examples relates to a stem-cell concentrate study, where stem-cell concentrates were extracted from individuals with skin cancer and were later injected back into the patients after the completion of chemotherapy to help restore blood production. In the meantime, the stem-cell concentrates were frozen. The research question was whether the freezing process reduced the beneficial quality of the stem-cell concentrates by reducing the amount of a chemical known as CFU-GM. For each sample the amount of this chemical was measured both before and after the freezing process. Between-group factors were gender and whether the patient was under low or high strain due to the amount of chemotherapy.

Brunner et al. (2002) call this design F2-LD-F1. The "F2" on the left of the "LD" indicates that there are two "between-subject" factors: Factor A (gender) and Factor B (strain). The LD stands for "longitudinal data." The "F1" to the right of the "LD" indicates that that there is one time factor.

To simplify matters, I will drop consideration of gender and just analyze the data for the male patients. This reduces the design to F1-LD-F1. When there are two groups and two time points, the design can be analyzed using an EXCEL worksheet. More complicated designs will require specialized software.

The original data for the example are in Brunner et al. (2002, p. 239). I used =RANK.AVG in EXCEL to produce the values shown in Table 8.10. The next step is to calculate the sample variances for each group and also for each time point. (For these equations, the first subscript i refers to the individual case, the second j to group, and the third t to the time point; make a note of these conventions so you can properly interpret the equations.) These particular equations are written for two

Table 8.9 Longitudinal Designs with an Applicable ANOVA-Type Statistic (ATS)

Number of groups (Factor A)	Groups stratified[a] (Factor B)	Time factor stratified[b] (Factor C)	Name of design
One	No	No	LD-F1
One	No	Yes	LD-F2
Several	No	No	F1-LD-F1
Several	Yes	No	F2-LD-F1
Several	No	Yes	F1-LD-F2
Several	Yes	Yes	F2-LD-F2

From Brunner et al. (2002).
[a]Stratified, for example, by gender, or a second treatment condition. This creates a second factor.
[b]Stratified, for example, by having a morning and an evening observation per time period.

time points, so t is instantiated as either 1 or 2. Now, the equation for calculating the group variances is

$$\hat{\sigma}_j^2 = \frac{\sum_{i=1}^{n_j}\left(R_{ij1} + R_{ij2} - \bar{R}_{\bullet j1} - \bar{R}_{\bullet j2}\right)^2}{n_j - 1}, j = 1, 2. \qquad \text{(Eq. 8.8)}$$

So for the low-strain group, $\sum_{i=1}^{n_j}(R_{ij1} + R_{ij2} - \bar{R}_{\bullet j1} - \bar{R}_{\bullet j2})^2 = 111{,}656.4$ (as shown in the "Sum" row at the bottom of the table), and $n_j = 30$ (for both groups), so $\hat{\sigma}_1^2 = \frac{111{,}656.4}{29} = 3850.22$. Likewise, $\hat{\sigma}_2^2 = 5668.68$. The time variances are

$$\hat{\tau}_j^2 = \frac{\sum_{i=1}^{n_j}\left(R_{ij1} - R_{ij2} - \bar{R}_{\bullet j1} + \bar{R}_{\bullet j2}\right)^2}{n_j - 1}, j = 1, 2. \qquad \text{(Eq. 8.9)}$$

So for the first group, $\sum_{i=1}^{30}(R_{ij1} - R_{ij2} - \bar{R}_{\bullet j1} + \bar{R}_{\bullet j2})^2 = 6{,}054.0$ (as shown in the table), so $\hat{\tau}_1^2 = \frac{6054}{29} = 208.76$. Likewise, $\hat{\tau}_2^2 = 447.26$.

The test statistics for the main effects are:

$$U_n^A = \frac{\bar{R}_{\bullet 11} + \bar{R}_{\bullet 12} - \bar{R}_{\bullet 21} - \bar{R}_{\bullet 22}}{\sqrt{\dfrac{\hat{\sigma}_1^2}{n_1} + \dfrac{\hat{\sigma}_2^2}{n_2}}} \text{ (group)}, \qquad \text{(Eq. 8.10)}$$

$$U_n^T = \frac{\bar{R}_{\bullet 11} - \bar{R}_{\bullet 12} + \bar{R}_{\bullet 21} - \bar{R}_{\bullet 22}}{\sqrt{\dfrac{\hat{\tau}_1^2}{n_1} + \dfrac{\hat{\tau}_2^2}{n_2}}} \text{ (time)}. \qquad \text{(Eq. 8.11)}$$

In our example, $U_n^A = 0.92$, and, $U_n^T = 8.30$. In large samples, the sampling distribution for U (used when there are only two time points) asymptotically follows the standard normal distribution, so one has $p = .36$ for the group effect and $p < .001$ for the time effect. So there appears to be a significant main effect of time. But one also needs to check for the interaction of group and time.

Table 8.10 ANOVA-Type Statistic (ATS) for Two Groups/Two Time Points: Stem-Cell Concentrate Study

| (A) Male ranks (N = 128) | | | | (B) For group effect $(R_{ij1} + R_{ij2} - \bar{R}_{\bullet j1} - \bar{R}_{\bullet j2})^2$ | | (C) For time effect $(R_{ij1} - R_{ij2} - \bar{R}_{\bullet j1} + \bar{R}_{\bullet j2})^2$ | |
| Low strain | | High strain | | | | | |
Pre	Post	Pre	Post	Low strain	High strain	Low strain	High strain
65	63	23	49	89.0	2,456.1	100.7	2,622.0
72	45	121	96	417.5	9,109.0	224.0	0.0
91	78	123	103	996.5	10,908.0	0.9	27.1
46.5	43	18	13	2,297.6	8,200.9	72.8	408.3
71	67	10	9	0.3	10,518.3	64.5	585.9
16	15	68	60	11,328.1	41.5	121.7	296.0
17	11	101	94	11,975.7	5,393.6	36.4	331.5
22	24	44	33	8,360.1	1,985.5	196.9	201.8
39	31	105	86	4,547.3	4,822.1	16.3	38.5
57	76.5	30	4	15.5	7,666.5	994.4	0.6
88	52	12	2	6.6	11,568.9	574.4	231.2
127	128	114	41	13,821.9	1,118.3	169.9	2,284.3
117	102	25	7	6,653.1	8,020.8	8.8	51.9
69	61	62	19	55.3	1,645.0	16.3	316.6
122	110	38	20	8,942.9	4,039.7	0.0	51.9
93	85	6	2	1,645.7	12,895.6	16.3	449.7
51	28	8	5	3,414.5	11,785.0	120.3	493.1
46.5	53	54	26	1,438.9	1,727.1	343.5	7.8
79	76.5	116	64	326.4	3,415.4	90.9	717.9
89	81	115	106	1,060.6	9,888.5	16.3	262.6
118	104	126	108	7,151.5	12,643.0	3.9	51.9
50	42	125	112	2,064.2	13,326.7	16.3	149.0
99	75	119	109	1,337.1	11,329.7	143.2	231.2
82	37	66	34	339.8	464.8	1,086.8	46.2
74	58	100	48	29.5	699.1	15.7	717.9
98	55	92	21	242.3	73.3	958.9	2,097.1
97	87	84	35	2,168.5	6.5	4.1	566.2
124	120	107	80	11,356.5	4,282.5	64.5	3.2
27	14	111	70	9,299.4	3,533.3	0.9	249.5
95	59	90	36	274.5	19.7	574.4	829.1
		40	2		6,329.6		163.7
		56	29		1,336.5		3.2
		113	83		5,541.5		23.0
		73	32		274.2		249.5
$\bar{R}_{\bullet jt}$ =74.74	62.70	73.38	48.18				
n_j = 30	30	34	34				
$\hat{p}_j$ = .58[a]	.49	.57	.37				
Sum				111,656.8	187,066.2	6,054.0	14,759.4

The test statistic for the interaction effect is

$$U_n^{AT} = \frac{\bar{R}_{\bullet 11} - \bar{R}_{\bullet 12} - \bar{R}_{\bullet 21} + \bar{R}_{\bullet 22}}{\sqrt{\dfrac{\hat{\tau}_1^2}{n_1} + \dfrac{\hat{\tau}_2^2}{n_2}}}. \tag{Eq. 8.12}$$

In our example, the numerator in Eq. 8.12 is: $74.73 - 62.70 - 73.38 + 48.18 = -13.17$. The denominator is $\sqrt{\frac{208.76}{30} + \frac{447.26}{34}} = 4.48$. So $U_n^{AT} = \frac{-13.17}{4.48} = -2.93$. The p value is therefore .003. (In EXCEL, $=$NORM.S.DIST(-2.93, 1) ➔ .0017; double this for the two-sided value of .0034.) The interaction effect is therefore significant.

Figure 8.7 graphs the relative treatment effects shown in Table 8.10, and it can be seen that freezing the blood samples has a worse effect on samples from patients with high levels of strain from chemotherapy; their samples are less "robust." Because the interaction effect is significant, one cannot interpret the main effect of time but must interpret the "decline" for each group separately. Is the decline for the high-strain group significant? To determine this, one takes the raw pre–post data just for this group ($n = 34$) and reranks the data from 1 to 34. The resulting data are shown in Table 8.11. The change scores are computed (post–pre), and these are used to compute the standard error for a LD-F1 design (with two time points) using Eq. 8.13:

$$SE = \sqrt{\frac{\sum[(\text{Change score}) + \bar{R}_{\text{pre}} - \bar{R}_{\text{post}}]^2}{n-1}}, \tag{Eq. 8.13}$$

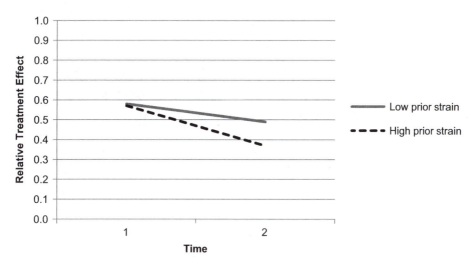

Figure 8.7 Relative treatment effects by time (and group) on amount of CFU-GM (for stem-cell concentrate data in Table 8.10). Illustrates a group-by-time interaction effect for an ANOVA-type statistic.

Table 8.11 ANOVA-Type Statistic (ATS) for One Group/Two Time Points: Stem-Cell Concentrate Study (High Strain Males)

Ranks ($N = 68$)		Change score	$\left[(Change\ score) + \bar{R}_{pre} - \bar{R}_{post}\right]^2$
Pre	Post		
17	32	15	770.62
65	49	−16	10.50
66	52	−14	1.54
13	12	−1	138.30
10	9	−1	138.30
39	35	−4	76.74
51	48	−3	95.26
30	23	−7	33.18
53	45	−8	22.66
21	4	−17	17.98
11	2	−9	14.14
61	29	−32	370.18
18	7	−11	3.10
36	14	−22	85.38
27	15	−12	0.58
6	2	−4	76.74
8	5	−3	95.26
33	19	−14	1.54
63	37	−26	175.30
62	54	−8	22.66
68	56	−12	0.58
67	59	−8	22.66
64	57	−7	33.18
38	24	−14	1.54
50	31	−19	38.94
47	16	−31	332.70
44	25	−19	38.94
55	42	−13	0.06
58	40	−18	27.46
46	26	−20	52.42
28	2	−26	175.30
34	20	−14	1.54
60	43	−17	17.98
41	22	−19	38.94
$\bar{R} = 40.88$	28.12		$\sum = 2{,}932.2$
$n_j = 34$	34	34	$SE = \sqrt{\frac{Sum}{n-1}} = \sqrt{\frac{2{,}932.2}{33}} = 9.43$
$\hat{p}_j = .59$[a]	.41		

[a] $\hat{p}_j$ is the relative treatment effect for each group. The overall estimated relative treatment effect ($\hat{p}$) is $.59 - .41 + .50 = .68$.

which the table shows to be 9.43. The test statistic, which is asymptotically normally distributed, is given by Eq. 8.14:

$$T_n^F = \frac{(\bar{R}_{\text{post}} - \bar{R}_{\text{pre}})}{SE} * \sqrt{n}. \tag{Eq. 8.14}$$

In our example, $T_n^F = \frac{(28.12 - 40.88)}{9.43} * \sqrt{34} = -7.89$. The one-sided p value is very small (in EXCEL =NORM.S.DIST(-7.89, 1) $= 1.51062 * 10^{-15}$). Doubling this result to obtain the two-sided p value, one still finds that $p < .001$. One can therefore reject the null hypothesis that the two population distributions are identical ($F_{\text{pre}} = F_{\text{post}}$). There is a significant tendency for the pre scores to be higher ($\hat{p} = .68$, from Table 8.11).

One could also plot a 95% confidence interval around the estimate using the SAS macro LC_CI (see Brunner et al., 2002, for details; also see Appendix 8.3). Brunner et al. (2002) report CIs for the stem-cell concentrate example of [.57, .62] for Time 1, and [.28 to .43] for Time 2.

If one uses the weaker null hypothesis of equal relative treatment effects ($H_0^p : p = .50$), use the following test statistic:

$$T_n^p = \frac{(\bar{R}_{\text{post}} - \bar{R}_{\text{pre}})}{2S_{n,p}} * \sqrt{n}, \tag{Eq. 8.15}$$

where $S_{n,p}$ is the standard error (see Technical Note 8.2).

In small samples the estimates will be biased, but the amount of bias can be estimated (see Brunner 2002, p. 90) and is typically small. (In the stem-cell concentrate example, the biases were 0.004 pre and -0.004 post.) In small samples, computer programs approximate the distribution of T_n^F with a t distribution. The minimum sample size is $n = 7$ for continuous data and $n = 14$ for discrete data (assuming there are not more than a couple ties). If one uses the weaker null hypothesis of equal relative treatment effects ($H_0^p : p = .50$), the sampling distribution for the test statistic T_n^p can be approximated in the same way, but a minimum sample size of $n = 15$ is needed for both continuous and discrete data. If there are more than five time points, however, the ATS becomes conservative, especially in small samples (see Brunner et al., 2002, p. 73). Also, in large samples (e.g., $n = 100$) there are other methods that are more statistically powerful (see Chapter 9). However, Brunner et al. (2002) argued that when using purely ordinal scales, use of the ATS is preferable. In those cases, use T_n^F with a normal approximation.

EXPLORING THE CONCEPT

Figure 8.8 presents the raw metric data in the form of cumulative probability distributions. (The H distribution is an average of the pre and post distributions and is used to compute the relative treatment effects for each group; see Appendix 8.1.)

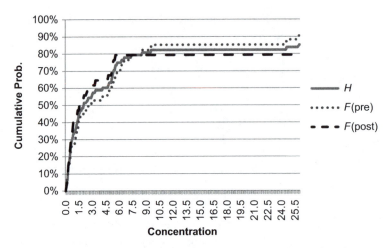

Figure 8.8 Cumulative probability distributions of raw data for stem-cell CFU-GM concentrate data (high prior strain males).

Is there a greater probability for blood samples in the estimated F_{post} distribution to have smaller concentrations of CFU-GM? Is there a greater tendency for the "pre" blood samples (that is, before being frozen) to have large concentrations? Is this pattern consistent with Figure 8.7?

Brunner et al. (2002) reported that for the stem-cell concentrate data, the raw metric data had a couple of large outliers. Because ATS tests involve ranking the data, these tests are robust to outliers and also make fewer assumptions than the other tests discussed in the prior two chapters. For instance, although the raw metric data for the high strain group could have also been analyzed with the Wilcoxon signed-ranks test, the ATS procedure does not assume that the difference scores are symmetrically distributed.

Research on the ATS

Some advocate using the ATS (when the sampling distribution is approximated by the F distribution; see Technical Note 8.3) if the normality or covariance homogeneity assumptions of ANOVA are violated and group sizes are unequal (Vallejo & Ato, 2012), although the ART is also a possibility.

Like the ART, a number of simulation studies of the ATS have been conducted that can give guidance regarding when it is best to use the technique. Brunner et al. (2002) simulated calculations of the ATS using a discrete uniform distribution for sample sizes from 8 to 100 and for four to eight time points. For four time points, the ATS was slightly liberal (5.9% Type I errors) for $n = 8$ and close to 5% for $n \geq 14$. For eight time points, the ATS was conservative in small samples ($n \leq 20$; see Brunner et al., 2002, p. 73).

As of 2003, Bathke and Brunner (2003) summarized the results of simulation studies using various distributions and concluded that the ATS was slightly liberal (with Type I error rates ranging from 5% to 7%) when there were at least four factor levels and $n \geq 10$; fewer factor levels requires larger sample size (e.g., $n \geq 10$ for two factor levels). Tests for covariate effects, on the other hand, were slightly conservative (if $n \geq 7$ per cell).

Using data generated from a multivariate normal distribution and with a common correlation coefficient between time points, Tian and Wilcox (2007) found the ATS method for repeated methods performed well in terms of Type I errors and power except when there were only two time points and $n = 10$. With unbalanced designs, Vallejo, Fernández, and Livacic-Rojas (2010) found that the ATS had satisfactory error rates with nonnormal data only when the distributions were not heavily skewed (heavy tails)—otherwise the ATS was too conservative. The authors did propose an adjustment to achieve robustness with heavily skewed data (see Vallejo, Ato, & Fernández, 2010). (The adjustment involves a transformation invented by Hall that makes the data symmetric than inverts the data back to asymmetric form.) Bathke, Harrar, and Madden (2007), working in a multivariate context with small sample sizes (e.g., $n = 6$), found the ATS to adequately control Type I error rates only when the sampling distribution was approximated by the F distribution and there were strong positive correlations between time points (with uncorrelated data, the test was too conservative).

Summary

The ability to use ATS tests in small samples, and to test for interactions when data are ranked, makes ATS tests an attractive alternative. (Also, for some recent developments since Brunner et al.'s (2002) book was published; see Technical Note 8.4.) The downside is that the ATS is not yet an option in IBM SPSS; however, Bathke, Schabenberer, Tobias, and Madden (2009) showed that under some situations, the ATS is equivalent to using the Greenhouse-Geisser F test, which is available in SPSS under the "General Linear Model" option. As previously noted, the ATS is an option in SAS with added macros. One other downside is that the ATS does not characterize precisely how population distributions differ (Serlin & Harwell, 2001, cited in Beasley & Zumbo, 2009), although we have seen that some inferences can be made about central tendencies. Furthermore, the ATS procedure can be supplemented with graphical methods.

Problems

1. A researcher investigates the effect, on a learning outcome measure, of providing feedback on a student's concept map (Factor A), and providing a second day of practice (Factor B). The data are in the EXCEL file *Concept Maps.8.Problem_1*.
    a) Using EXCEL or statistical software, find the grand mean, factor-level means for A = 1 and B = 1, and the four main effects (two for each factor).
    b) For each student, calculate Y' and the aligned ranks. (Optional: Check your calculations using ARTweb.)
    c) Using statistical software, run the aligned ranks through an ANOVA. (In SPSS, use "General Linear Model," "Univariate.") Report the F (with df) and p values for the interaction. Is the interaction significant?
    d) The raw data are negatively kurtotic: Overall $kurt = -1.19$ (with some cell kurtosis values of -1.79 and -1.96). Would you expect a nonparametric approach to be more statistically powerful here?
2. Use the same data as in problem 1 but change the outcome value for student 1 from 5.66 to 9.00, thereby creating an outlier. This change creates positive skew (2.0) and kurtosis (4.7) in the first cell. Answer the same questions as in problem 1(a–c).
3. Similar to Gustafson et al. (1999), a researcher evaluates the effectiveness of a home-based, computerized health-enhancement support system for HIV patients. The system provides patients with information and decision support, as well as connection to other patients and experts. Among other outcome variables, the researcher collected self-report data at three times on the number of times the patient telephoned health-care providers in the last 2 months, comparing 15 participants who received the computer-support system with 15 control group subjects. Because the data for accessing health care was skewed, the researcher tested for a time-by-group interaction using a nonparametric ART. The data are located in the EXCEL file *Health Support.8.Problem_3*. For "Condition," the experimental group is coded 1.
    a) Using EXCEL or statistical software, find the grand mean, row means (over the three time periods), and column means.
    b) For each student, calculate Y' and the aligned ranks.
    c) Using statistical software, run the aligned ranks through an ANOVA. (In SPSS, use "General Linear Model," "Repeated Measures," and define a three-level factor.) Report the multivariate Hotelling's Trace $(H_{(A)})$, F (with df), and p values for the interaction. Is the interaction significant?
4. Repeat the problems in question 3 but just using the data for "prescore" and "6 months" (i.e., two time points).
5. Bandelow et al. (1998) conducted a psychiatric clinical trial in which 16 patients suffering from panic disorder attacks were treated with the antidepressant

imipramine over an 8-week period. Every 2 weeks, observers rated the severity of each patient's disorder on an ordinal scale ranging from 2 to 8, with 8 reflecting a high frequency of attacks. The data for weeks 0, 4, and 8 are in the EXCEL file, *Panic Attack.8.Problem_5*. Using EXCEL, conduct an ATS-analysis to contrast weeks 0 and 4 only. (Because only simple analyses can be conducted in EXCEL, do not compare all three time points at once.) Be sure to rank the data before conducting the analysis. Find:

a)   the mean ranks (to 4 decimal places), n, and N;

b)   relative treatment effects ($\hat{p}_{pre}$, $\hat{p}_{post}$, and $\hat{p}$);

c)   the standard error (to 4 decimal places) from Eq. 8.13;

d) T_n^F from Eq. 8.14; and

e)   the two-sided p value, based on the t distribution with 15 df.

f)   Is it valid to compare the results to those of a traditional sign test? To the Wilcoxon signed ranks test? Why or why not?

6.   a)   Repeat the analysis in problem 5, contrasting weeks 0 and 8.

    b)   Optional: Perform a full analysis (involving weeks 0, 4, and 8) using SAS PROC MIXED, after downloading the LD_F1 and LD_CI macros from http://www.ams.med.uni-goettingen.de/sasmakr-de.shtml, following the instructions in Brunner et al. (2002, pp. 110–116). For sample SAS programs, see the electronic appendix to Shah and Madden (2004).

7.   Similar to Dickerson et al. (2010), a researcher randomly divides 48 young adults into two groups ($n = 24$); half consume the drug ketamine and half are given a placebo. (The drug ketamine blocks NMDA receptors in the brain and is therefore hypothesized to cause feelings of being intoxicated with alcohol.) After 45 minutes, participants rate their feelings of intoxication by estimating the number of alcoholic drinks that would cause a similar feeling. The data are in the EXCEL file *Intoxication.8.Problem_7*. (Data on thiopental are also included for Problem 8.) Because the data are skewed, the researcher uses a nonparametric ATS to analyze the interaction of drug with time. Using EXCEL, find

a)   the four mean ranks (to 4 decimal places), n_j (one for each treatment), and N;

b)   the four relative treatment effects $\hat{p}_{j*t}$ (*i.e.*, $\hat{p}_{group*time}$) from Eq. 8.3;

c)   the two time variances ($\hat{\tau}_j^2$) to 4 decimal places from Eq. 8.9, one for each treatment; and

d) U_n^{AT} (from Eq. 8.12), along with the significance level.

8.   Thiopental is another drug that can cause feelings of intoxication by affecting the levels of the neurotransmitter GABA in the brain (GABA is associated with inhibition). Does thiopental have a significantly greater effect than ketamine? For this comparison, answer the same questions as in the previous problem.

Technical Notes

8.1   Pretest–posttest one-group designs are subject to various *threats to internal validity* (Shadish, Cook, & Campbell, 2002), that is, threats to the validity of drawing conclusions about cause and effect. For example, rather than because of the treatment, students can become better at a task due to natural development (*maturation*), some other event besides the treatment that occurs between tests (*history*), or practice with taking tests (*testing*). However, if one includes a control group with participants that take the same tests, these threats can be attenuated because the control group will also experience maturation, testing effects, and history effects (unless only the experimental group is affected by the event). One might therefore observe an increase in the scores for the control group, but if the experimental group shows a significantly greater increase, that is evidence that the treatment has had a significant causal effect on the test scores, assuming random assignment. Even without randomization, if the pretest means of the two groups are close to one another, this provides some evidence that the two groups are similar to one another. One can also use the pretest information to examine individuals who dropped out of the study (*mortality threat*) and whether this caused bias.

8.2   The standard error of T_n^p is $S_{n,p}$, and can be derived from the variance, which is

$$S_{n,p}^2 = \frac{1}{n-1}\sum_{i=1}^{n}[(R_{i2} - R_{i2}^{(2)}) - (R_{i1} - R_{i1}^{(1)}) - (\bar{R}_{\bullet 2} - \bar{R}_{\bullet 1})]^2, \qquad \text{(Eq. 8.16)}$$

where $R_{ij}^{(i)}$ is the internal rank of each observation among the n_j observations in group j. (An internal rank is a ranking of just the observations in a given group.)

8.3   The ATS has an asymptotic chi-square sampling distribution. However, when the number of factors or factor levels is large, use of a chi-square distribution is not recommended because rank transformations do not preserve homogeneous variances (Brunner, Dette, & Munk, 1997). Furthermore, the moments of the distribution are unknown, as they are a function of unknown eigenvalues (Brunner et al., 2002). Based on a technique developed by Box (1954), one can approximate the sampling distribution, even in small samples ($n \geq 7$), with a chi-square distribution (with f degrees of freedom, where f is defined below), multiplied by a scale g. The first two moments of the approximate distribution $g * \chi_f^2$ tend to coincide with the first two moments of the actual sampling distribution. However, f and $f * g$ must be estimated; as a result, the sampling distribution is F shaped, with f degrees of freedom in the numerator. The test statistic involves linear algebra (see Chapter 9 for the basic principles) and is calculated with the following formula (from Tian & Wilcox, 2007, based on Brunner et al., 1997, 2002):

$$F_n(\mathbf{C}_2) = \frac{n}{N^2 trace(\mathbf{C}_2 \hat{\mathbf{V}}_n)} \sum_{j=1}^{k} \left(\overline{\mathbf{R}}_{\bullet j} - \frac{N+1}{2} \right)^2,$$

where n is the sample size of each group ($N = n * k$), k is the number of j groups, $\mathbf{C}_2 = \mathbf{I}_j - \frac{1}{k} \mathbf{K}_j, \mathbf{K}_j,$ is a k by k matrix of 1's, the covariance matrix $\hat{\mathbf{V}}_n = \frac{1}{N^2(n-1)} \sum_{i=1}^{n} (\mathbf{R}_i - \overline{\mathbf{R}})(\mathbf{R}_i - \overline{\mathbf{R}})'$, the matrix of ranks $\mathbf{R}_i = (R_{i1, \ldots}, R_{ik})$, and $\overline{\mathbf{R}} = \frac{1}{n} \sum_{i=1}^{n} \mathbf{R}_i$. (A trace is the sum of the main diagonal of a square matrix.) The null hypothesis is

$$H_o : \mathbf{C}_2 \mathbf{F} = 0 = \begin{pmatrix} F_1 - \overline{F} \\ \ldots \\ F_j - \overline{F} \end{pmatrix} = \begin{pmatrix} 0 \\ 0 \\ 0 \end{pmatrix}.$$

For the F distribution, the numerator df is $\hat{f} = \frac{[trace(\mathbf{C}_2 \hat{V}_n)]^2}{trace(\mathbf{C}_2 \hat{V}_n \mathbf{C}_2 \hat{V}_n)}$ and the denominator $df = \infty$. (In some cases, finite denominator df are estimated to improve the approximation.)

In Brunner et al. (2002), the F distribution is used when there are several groups (three or more) and several time points. When there are only two groups, the z or t distributions may be used, as in the chapter examples.

8.4 There have been some additional statistical developments since the Brunner et al. (2002) book was published. Konietschke, Bathke, Hothorn, and Brunner (2010) developed a multiple contrast test for longitudinal data and a procedure based on H_0^p for constructing confidence intervals. There has also been some work conducted on the properties of a modified ATS in "high dimensional" situations when the number of parameters exceeds the number of observations (Ahmad, Werner, & Brunner, 2008; Choopradit & Chongcharoen, 2011).

APPENDIX 8.1

TECHNICAL DETAILS ON THE ATS PROCEDURE

1. Each observation has a sampling distribution. We make no assumptions about the shape of these distributions or that they are identical. If there are n cases, there are $i = 1 \ldots n$ observations. If there are k groups (e.g., two conditions), there are $j = 1 \ldots k$ groups and $n \times k$ observations. If each case is observed at multiple time periods (t), there will be a total of $n \times k \times t$ observations, indexed by the subscripts i, j, and t.
2. When data are ranked, ties are assigned the average rank (i.e., "midrank").
3. If X is the RV, let lowercase x represent specific values of the RV. There are three types of cumulative distributions. In the right continuous version, the cumulative distribution is $F^+ = \text{Prob}(X \leq x)$. In the left continuous version, $F^- = \text{Prob}(X < x)$. (Note that the "less than" sign, not the "less than or equal to" sign, is used.) With continuous variables, these distributions will not differ, but they will for discrete distributions, so the averages of the left and right distributions are used, producing the *normalized distribution*, F. The use of the term *normalized* is a bit of a misnomer, because this procedure does not impose a normal distribution on the data; it is more akin to a continuity correction.
4. If there are k multiple groups, there will be F_j distributions. One does not need to estimate each separately, but only the weighted mean of these, which yields the average distribution. Eq. 8.17 gives the formula.

$$H(x) = \frac{1}{N} \sum_{j=1}^{k} n_j F(x)_j. \qquad \text{(Eq. 8.17)}$$

5. The relative treatment effect (p_j), describes the tendency of F_j relative to H. If the cases in F_j tend to be less than average (given by H), then p_j will be less than 50%; and vice versa if p_j is greater than 50%. The relative treatment effect is formally defined as $p_j = \int H dF_j$. This expression is an integral reflecting the area under a curve, specifically corresponding to the difference in area (probability). (The term dF_j is the first derivative of F_j, which is also the noncumulative probability density function of F; Shah & Madden, 2004.) It can be shown that Eq. 8.3 provides a very good estimate of the relative treatment effect.

   SAS Proc Mixed (with ANOVAF option) can provide estimates of the relative treatment effects, but one will also need to download macros (http://www.ams.med.uni-goettingen.de/sasmakr-de.shtml). The output will report the REs (relative effects), the amount of bias in the estimates, and the variance. The formula for the variances is

$$V_j = Var\left(\sqrt{N}\left(\hat{p}_j - p\right)\right), \qquad \text{(Eq. 8.18)}$$

where p is the true relative effect size. The formula for the standard error is

$$\mathrm{SE}(\hat{p}_j) = \sqrt{\frac{V_j}{N}}.$$ (Eq. 8.19)

The program then uses the standard error to construct a 95% CI around each estimated treatment effect (under the headings "lower" and "upper"). A CI for a significant effect will not bracket 50%.

PROC MIX will also provide mean ranks by condition under the heading "Least Square Means," along with estimates of the standard errors. Divide these by N for rough estimates of the standard errors for the relative treatment effects if the macros are not available (but these estimates will be less accurate than those discussed above).

For analysis of main effects and interactions, consult the output under ANOVA F.

6. For within-subject variables in a repeated-measures analysis, use an infinite number of degrees of freedom in the denominator ("infty") of the F statistic.
7. On any computer output, ignore the F values associated with Wald-type statistics. These are only reliable in very large samples. Use those for the ATS.

Source: Brunner, Domhof, and Langer (2002); Shah and Madden (2004). See also Brunner and Puri, 2001.

APPENDIX 8.2

DERIVATION OF EQ. 8.5C (DEFINITION OF A TWO-WAY INTERACTION EFFECT)

In ANOVA, each observation is denoted by Y_{ijm}, where j reflects the level of one factor and m reflects the level of the second. Each observation is a linear function of the main and interaction effects:

$$Y_{ijm} = \mu + \alpha_j + \beta_m + (\alpha\beta)_{jm} + \varepsilon,$$

where α_j and β_m are the main effects, $(\alpha\beta)_{jm}$ is the interaction effect, and ε is an error term reflecting deviations from the predicted value, which is the cell mean (μ_{jm}). The error term can be dropped by replacing Y_{ijm} with μ_{jm}:

$$\mu_{jm} = \mu + \alpha_j + \beta_m + (\alpha\beta)_{jm}.$$

Because the main effects are defined as $\alpha_j = \mu_j - \mu$ and $\beta_m = \mu_m - \mu$, it follows that

$$\mu_{jm} = \mu + \mu_j - \mu + \mu_m - \mu + (\alpha\beta)_{jm},$$

which simplifies to

$$\mu_{jm} = \mu_j + \mu_m - \mu + (\alpha\beta)_{jm}.$$

Under the null hypothesis of no interaction, $(\alpha\beta)_{jm} = 0$, therefore,

$$H_0(AB): \mu_{jm} = \mu_j + \mu_m - \mu.$$

With ANOVA-type statistics (ATS), the equations are written in terms of cumulative distribution functions, so Eq. 8.5c is

$$H_0^F(AB): \bar{F}_{jm} = \bar{F}_{j\bullet} + \bar{F}_{\bullet m} - \bar{F}_{\bullet\bullet} \quad \text{QED.}$$

APPENDIX 8.3

CONFIDENCE INTERVALS FOR RELATIVE TREATMENT EFFECTS

Let the relative treatment effect for the jth group and tth time point be denoted by p_{jt}. It can be shown that the relative treatment effect is contained in the interval $\frac{n_j}{2N}, 1 - \frac{n_j}{2N}$. In the chapter example (Table 8.11), $N = 2n = 68$, so the maximum bounds for the first group are .25 and .75. One then applies a procedure known as the delta method (or δ-method; Rao, 1973, p. 385ff), with the following steps.

1. Transform to a 0, 1 scale using the following formula, where $\hat{p}_{jt}^*$ are the transformed values, also written as $g_2(\hat{P}_{jt})$.

$$g_2\left(\hat{P}_{jt}\right) = \hat{p}_{jt}^* = (N\hat{p}_{jt} - \frac{n_j}{2}) / (N - n_j).$$ (Eq. 8.20)

For the data in Table 8.11, this yields .68 (Time 1) and .32 (Time 2). Because there is only one group in the example, one can drop the subscript for group (j). Eq. 8.20 then simplifies to

$$\hat{p}_t^* = \frac{N\hat{p}_t - \dfrac{n}{2}}{N - n} = 2\hat{p}_t - .5, t = 1, 2.$$ (Eq. 8.21)

This equation produces the same result as did Eq. 8.20.

2. Apply the logit transformation (see Chapter 10), $ln\left(\frac{\hat{p}_{jt}^*}{1-\hat{p}_{jt}^*}\right)$, producing .75 for Time 1 and $-.75$ for Time 2. This makes the scale unbounded.

3. Estimate the variance, $\hat{\sigma}_{jt}^2$. With just one group and two time points, find $\hat{\sigma}_t^2$, which is equal to $\frac{S_{n,p}^2}{4n^2}$. Using the formula in Technical Note 8.2, one finds that $S_{n,p}^2 = 29.28$, so $\hat{\sigma}_t^2 = \frac{29.28}{4*34^2} = .006$.

4. Estimate the variance $\left(\hat{\sigma}_t^*\right)^2 = \frac{N^2\hat{\sigma}_t^2}{N-n} = 4\hat{\sigma}_t^2$. In our example, $\left(\hat{\sigma}_t^*\right)^2 = .024$ and the standard error $\hat{\sigma}_t^* = \sqrt{.024} = .155$.

5. For a two-sided, $(1-\alpha)$ CI (e.g., 95%), find the $u_{1-\alpha/2}$ quantile of the standard normal distribution (e.g., 1.96).

6. Find the bounds of the CI for the transformed effect. For the lower bound, $$p_{jt,L}^g = \text{logit}\left(\hat{p}_{jt}^*\right) - \frac{\hat{\sigma}_{jt}^*}{\hat{p}_{jt}^*\left(1-\hat{p}_i^*\right)\sqrt{n}} * u_{1-\frac{\alpha}{2}}$$ (for the upper bound, add rather than subtract). In our example, for the first time period, $p_{j1,L}^g = .75 - \frac{.155}{.68(1-.68)\sqrt{34}} * 1.96 = .51$. $p_{j2,U}^g = .75 + \frac{.155}{.68(1-.68)\sqrt{34}} * 1.96 = .99$. Calculations for the second time period are left to the reader as an exercise.

7. Conduct an inverse transformation: $p_{jt,L} = \frac{n_j}{2N} + \frac{N-n_j}{N} * \frac{\exp(p_{jt,L}^g)}{1+\exp(p_{jt,L}^g)}$ and $p_{jt,U} = \frac{n_j}{2N} + \frac{N-n_j}{N} * \frac{\exp(p_{jt,U}^g)}{1+\exp(p_{jt,U}^g)}$; this will provide the end points of the CI. With one group and two time points, these expressions simplify to $p_{t,L} = \frac{1}{4} + \frac{1}{2} * \frac{\exp(p_{t,L}^g)}{1+\exp(p_{t,L}^g)}$ and $p_{t,U} = \frac{1}{4} + \frac{1}{2} * \frac{\exp(p_{t,U}^g)}{1+\exp(p_{t,U}^g)}$. In our example, $p_{1,L} = .56$ and $p_{1,U} = .62$.

According to Brunner et al. (2002, p. 61), "Simulation studies showed that this method provides intervals with an approximate confidence probability $1 - \alpha$, even for small sample sizes ($n_j \geq 10$)."

References

Ahmad, M. R., Werner, C., & Brunner, E. (2008). Analysis of high dimensional repeated measures designs: The one sample case. *Computational Statistics & Data Analysis, 53,* 416–427.

Bandelow, B., Brunner, E., Broocks, A., Beinroth, D., Hajak, G., Pralle, L., & Rüther, E. (1998). The use of the Panic and Agoraphobia Scale in a clinical trial. *Psychiatry Research, 77,* 43–49.

Bathke, A., & Brunner, E. (2003). A nonparametric alternative to analysis of covariance. In M. G. Akritas & D. N. Politis (Eds.), *Recent advances and trends in nonparametric statistics.* San Diego, CA: Elsevier Science.

Bathke, A. C., Harrar, S. W., & Madden, L. V. (2007). How to compare small multivariate samples using nonparametric tests. *Computational Statistics and Data Analysis, 52,* 4951–4965.

Bathke, A., Schabenberger, O., Tobias, R. D., & Madden, L. V. (2009), Greenhouse-Geisser adjustment and the ANOVA-type statistic: Cousins or twins? *The American Statistician, 63,* 239–246.

Beasley, T. M. (1996, October). *Parametric and nonparametric test of interaction in the split-plot design under conditions on nonnormality and covariance heterogeneity.* Paper presented at the annual meeting of the Mid-Western Educational Research Association, Chicago, IL.

Beasely, T. M. (2000). Nonparametric tests for analyzing interactions among intrablock ranks in multiple groups repeated measures designs. *Journal of Educational & Behavioral Statistics, 25,* 20–59.

Beasley, T. M. (2002). Multivariate aligned rank test for interactions in multiple group repeated measures designs. *Multivariate Behavioral Research, 3,* 197–226.

Beasley, T. M., & Zumbo, B. D. (2003). Comparison of aligned Friedman rank and parametric methods for testing interactions in split-plot designs. *Computational Statistics & Data Analysis, 42,* 569–593.

Beasley, T. M., & Zumbo, B. D. (2009). Aligned rank tests for interactions in split-plot designs: Distributional assumptions and stochastic heterogeneity. *Journal of Modern Applied Statistical Methods, 8,* 16–50.

Box, G.E.P. (1954). Some theorems on quadratic forms applied in the study of analysis of variance problems, I: Effect of inequality of variance in the one-way classification. *Annals of Mathematical Statistics, 25,* 290–302.

Brunner, E., Dette, H., & Munk, A. (1997). Box-type approximations in factorial designs. *Journal of the American Statistical Association, 92,* 1494–1503.

Brunner, E., Domhof, S., & Langer, F. (2002). *Nonparametric analysis of longitudinal data in factorial experiments.* New York, NY: John Wiley & Sons.

Brunner, E., & Puri, M. L. (2001). Nonparametric methods in factorial designs. *Statistical Papers, 42*, 1–52.

Choopradit, B., & Chongcharoen, S. (2011). A test for two-sample repeated measures designs: Effects of high dimensional data. *Journal of Mathematics and Statistics, 7*, 332–342.

Dickerson, D., Pittman, B., Ralevski, E., Perrino, A., Limoncelli, D., Edgecombe, J., . . . Petrakis, I. (2010). Ethanol-like effects of thiopental and ketamine in healthy humans. *Journal of Psychopharmacology, 24*, 203–211.

Erceg-Hurn, D., & Mirosevich, V. M. (2008). Modern robust statistical methods. *American Psychologist, 63*, 591–601.

Gentile, J. R., Voelkl, K. E., Mt. Pleasant, J., & Monaco, N. M. (1995). Recall after relearning by fast and slow learners. *Journal of Experimental Education, 63*, 185–197.

Gustafson, D. H., Hawkins, R., Boberg, E., Pingree, S., Serlin, R. E., Grazianoa, F., & Chan, C. L. (1999). Impact of a patient-centered, computer-based health information/support system. *American Journal of Preventive Medicine, 16*(1), 1–9.

Hettmansperger, T. P. (1984). *Statistical inference based on ranks.* New York, NY: John Wiley & Sons.

Higgins, J. J., & Tashtoush, S. (1994). An aligned rank transform test for interaction. *Nonlinear World, 1*, 201–211.

Konietschke, F., Bathke, A. C., Hothorn, L. A., & Brunner, E. (2010). Testing and estimation of purely nonparametric effects in repeated measures designs. *Computational Statistics and Data Analysis, 54*, 1895–1905.

Krause, M. S., Madden, L. V., & Hoitink, H. A. J. (2001). Effect of potting mix microbial carrying capacity of biological control of Rhizoctonia damping-off of radish and Rhizoctonia crown and root rot of Poinsettia. *Phytopathology, 91*, 1116–1123.

Mansouri, H., & Chang, G. H. (1995). A comparative study of some rank tests for interaction. *Computational Statistics & Data Analysis, 19*, 85–96.

McSweeney, M. (1967). An empirical study of two proposed nonparametric test for main effects and interaction. *Dissertation Abstracts International, 28*(10), 4005.

Noguchi, K., Gel, Y. R., Brunner, E., & Konietschke, F. (2012). nparLD: An R software package for the nonparametric analysis of longitudinal data in factorial experiments. *Journal of Statistical Software, 50*, 1–23.

Puri, M. L., & Sen, P. K. (1969). A class of rank order tests for a general linear model. *Annals of Mathematical Statistics*, 1325–1343.

Rao, C. R. (1973). *Linear statistical inference and its applications* (2nd ed.). New York, NY: John Wiley & Sons.

Shadish, W., Cook, T., & Campbell, D. (2002). *Experimental and quasi-experimental designs for generalized causal inference.* Boston, MA: Houghton Mifflin.

Shah, D. A., & Madden, L. V. (2004). Nonparametric analysis of ordinal data in designed factorial experiments. *Phytopathology, 94*, 33–43.

Tian, T., & Wilcox, R. (2007). A comparison of two rank tests for repeated measures designs. *Journal of Modern Applied Statistical Methods, 6*(1). Retrieved from http://www.uh.edu/~ttian/rank.pdf

Toothaker, L. E., & Newman, D. (1994). Nonparametric competitors to the two-way ANOVA. *Journal of Educational and Behavioral Statistics, 19*, 237–273.

Vallejo, G., & Ato, M. (2012). Robust tests for multivariate factorial designs under heteroscedasticity. *Behavior Research,44*, 471–489.

Vallejo, G., Ato, M., & Fernádez, P. (2010). A robust approach to analyzing unbalanced factorial designs with fixed levels. *Behavior Research Methods, 42*, 607–617.

Vallejo, G., Fernádez, P., & Livacic-Rojas, P. (2010). Analysis of unbalanced factorial designs with heteroscedastic data. *Journal of Statistical Computation and Simulation, 80*, 75–88.

Wilcox, R. R. (2005). *Introduction to robust estimation and hypothesis testing* (2nd ed.). San Diego, CA: Academic Press.

Wobbrock, J. O., Findlater, L., Gergle, D., & Higgins, J. J. (2011). "The aligned rank transform for nonparametric factorial analyses using only ANOVA procedures." *Proceedings of the ACM Conference on Human Factors in Computing Systems (CHI '11).* New York, NY: ACM Press.

CHAPTER 9

LINEAR REGRESSION AND GENERALIZED LINEAR MODELS

The remaining portion of this book presents statistical models where one or more of the independent variables are metric. This chapter briefly reviews linear regression, highlighting topics that will be important in later chapters (dummy variables and functional form) as well as the assumptions of the general linear model. It then introduces the concept of a *generalized* linear model (GLM), which forms the basis of logistic, ordinal, and Poisson regression. (As described in Chapters 10–13, these methods are applicable when the dependent variable [DV] is nominal or ordinal or a count.) The final section of this chapter presents GLMs using linear algebra.

Presented at the end of the book (in Chapter 15) is Table 15.1. This table is a test matrix summarizing the various statistical techniques covered in this book. It may be useful for readers to review the table now to gain a sense of the ground still to be covered.

Review of Linear Regression

Basics of Ordinary Least Squares

As shown in Figure 9.1, the ordinary least squares (OLS) technique calculates the intercept and slope of the best-fitting line, specifically, coefficients that minimize the sum of the squared residuals. The basic concepts are listed below. Proofs of most of these equations can be found in Appendix 9.1.

1. The equation of a line is $\hat{Y}_i = a + bX_i$, where $\hat{Y}_i$ is the predicted value for the *i*th observation. (A hat sign always indicates an estimated value.)
2. A residual is the difference between an observed and predicted value $(Y_i - \hat{Y}_i)$; the smaller the residual, the closer a data point will be to the regression line. Residuals are also sometimes referred to as "errors," although, technically, the

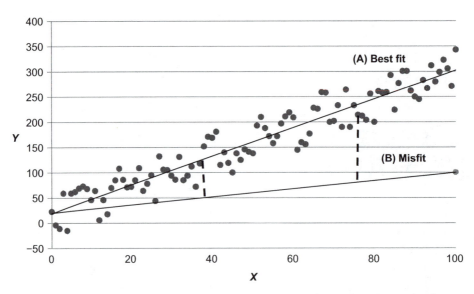

Figure 9.1 Best-fitting line (Line A) compared with misfitting line (Line B). The dashed lines are examples of Line B residuals and become larger as X increases.

term *error* relates to the population regression line, not the regression line in the sample (see Item 5).

For the best-fitting line, it can be shown (see Appendix 9.1) that

$$b = \frac{Cov_{X,Y}}{Var_X}, \text{ and} \qquad\qquad\qquad\qquad\qquad \text{(Eq. 9.1)}$$

$$a = \bar{Y} - b\bar{X}. \qquad\qquad\qquad\qquad\qquad\qquad\qquad \text{(Eq. 9.2)}$$

The slope b is an estimate of how much Y will change from a one-unit increase in X.

3. The coefficient of determination (R^2) measures the amount of variance in Y that is explained by the regression equation. Specifically, $R^2 = \frac{\sum(\hat{Y}-\bar{Y})^2}{\sum(Y-\bar{Y})^2}$. R^2 is also the square of the Pearson correlation coefficient, r, between X and Y.

4. The coefficients a, b, and R^2 are derived from a sample. If one performed a regression using all people or objects in the population, then one would discover the "true relation" between X and Y devoid of sampling error. The equation for the population regression line is

$$Y_i = \alpha + \beta X_i + \varepsilon_i, \qquad\qquad\qquad\qquad\qquad \text{(Eq. 9.3)}$$

where the subscript i indexes the value of X along the real number line (e.g., $X_4 = 4$, $X_{4.01} = 4.01$). The error term ε_i reflects stochastic error (not sampling error); stochastic error occurs from unmeasured or otherwise omitted variables that affect Y.

5. The equation for the regression line estimated from a sample of data is sometimes written as

$$Y_i = \hat{\alpha} + \hat{\beta} X_i + e_i,$$ (Eq. 9.4)

where the subscript i ($i = 1 \ldots n$) indicates the observation number, $\hat{\alpha} = a$, and $\hat{\beta} = b$.

6. $\hat{\beta}$ has a sampling distribution (as does $\hat{\alpha}$). It can be shown that the mean of the $\hat{\beta}$ sampling distribution is β [i.e., $E(\hat{\beta}) = \beta$] and that the standard error is

$$SE(\hat{\beta}) = \sqrt{\frac{\sigma^2}{\sum(X_i - \bar{X})}}, \text{where } \sigma^2 \text{ is estimated by } s^2 = \frac{\sum(Y_i - \hat{Y}_i)^2}{n - 2}.$$ (Eq. 9.5)

7. In a large sample, or when Y is normally distributed, the sampling distribution of the test statistic (t) will follow the t distribution. (It does not follow the normal distribution because one is estimating two parameters, β and the standard error.) Under the null hypothesis, $\beta = 0$, and because $E(\hat{\beta}) = \beta$, it follows that $E(\hat{\beta}) = 0$ under the null. The t statistic therefore is

$$t(n-2) = \frac{\hat{\beta} - 0}{SE(\hat{\beta})} = \frac{\hat{\beta}}{SE(\hat{\beta})}.$$ (Eq. 9.6)

From the t statistic one can derive a p value.

For the data on which Figure 9.1 were based, I performed a regression analysis in SPSS using "Analyze" ➜ "Regression" ➜ "Linear." Suppose X reflects scores (percent correct) on a cognitive abilities exam taken at age 20, and Y reflects annual income at age 50 (in thousands). The coefficients of the estimated regression line were found to be $\hat{\alpha} = 19.082$, $\hat{\beta} = 2.836$, $R^2 = 90.3\%$. For $\hat{\beta}$, $t(98) = 30.38$ and $p < .001$, so we can reject the null hypothesis that $\beta = 0$. The results imply that each additional point on the cognitive exam predicts an increase in income of $2.8 thousand.

The equations above use one predictor, but there can be two or more predictors if using multiple regression. For two predictors, we can rewrite Eq. 9.3 as

$$Y_i = \beta_0 + \beta_1 X_i + \beta_2 Z_i + \varepsilon_i.$$ (Eq. 9.7)

A multiple regression with two predictors consists of finding the best-fitting plane in a three-dimensional space. The parameter β_1 reflects the association between X and Y while holding Z constant. The t tests for the significance of the slope estimates are based on $n - k - 1$ df, where k is the number of predictors.

EXPLORING THE CONCEPT

In what ways is multiple regression similar to the Mantel-Haenszel test described in Chapter 5?

Dummy Variables

In multiple regression, predictors can be nominal as well as metric. A dichotomous nominal variable can be represented by a *dummy variable*, which is a variable coded 0 or 1. For example, the variable gender can be coded 0 for females and 1 for males. Suppose one hypothesizes that there is a different relationship between experience at a job and monthly salary (in thousands) for females than for males. One could fit two different regression lines (one for each gender), but one would not know if the estimates differed significantly for males and females. One can make this determination by fitting one regression line using the model

$$Y_i = \hat{\beta}_0 + \hat{\beta}_1 X_i + \hat{\beta}_2 D_i + \hat{\beta}_3 X_i D_i + e, \quad \text{(Eq. 9.8)}$$

where D_i is the dummy variable, and $\hat{\beta}_3 X_i D_i$ is an interaction term between the metric and nominal predictors. (An interaction term reflects how the relationship between two variables is moderated—i.e., altered—by changes in a third variable.) Suppose one obtained the following:

$$Y = 1.6 + 0.7\,Gender + 0.5\,Experience + 2.0(Gender * Experience) + e. \quad \text{(Eq. 9.9)}$$

For females, *Gender* = 0, and Eq. 9.9 reduces to

$$Y = 1.6 + 0.5\,Experience + e. \quad \text{(Eq. 9.10)}$$

For males, *Gender* = 1, and Eq. 9.9 simplifies to:

$$\begin{aligned} Y &= 1.6 + 0.7 + 0.5\,Experience + 2\,Experience + e, \\ &= 2.3 + 2.5\,Experience + e. \end{aligned} \quad \text{(Eq. 9.11)}$$

One can therefore derive two individual regression equations, one for males and one for females, and infer from the regression results for Eq. 9.9 whether the coefficients of the two lines are significantly different by testing the significance of the dummy variable terms, involving gender.

In comparing Eq. 9.10 and Eq. 9.11, one sees that the coefficient for *Gender* (0.7) is the amount that the intercepts differ and the coefficient for the interaction term *Gender * Experience* (2.0) is the amount that the slopes differ. Some students mistakenly think that these coefficients represent the intercept and slope of one of the regression lines, but this interpretation is incorrect.

Dummy variable coding can also be used with multinomial variables, such as political party affiliation. Suppose that individuals are classified into three groups: Democrats, Republicans, and Other. The regression equation for some DV could be specified as

$$Y_i = \hat{\beta}_0 + \hat{\beta}_1 D_{i1} + \hat{\beta}_2 D_{i2} + e, \qquad \text{(Eq. 9.12)}$$

where $D_1 = 1$ if Republican, 0 otherwise, and $D_2 = 1$ if party is "Other," 0 otherwise. The coefficients for the dummy variables will indicate how the means for each category differ from the mean of the reference group, Democrats. If one were to include a dummy variable for Democrats as well, it would create a technical problem known as perfect multicollinearity (see Technical Notes 9.1 and 9.2), so such a variable is omitted. If one were to also include a metric independent variable (IV), one would typically also include interaction terms with each of the dummy variables.

Dummy variable coding can also be used with experimental data (e.g., 1 = *experimental*, 0 = *control*). The equivalent of analysis of variance (ANOVA) can be performed with linear regression by using a different dummy variable for each treatment level except the reference category. Also including a metric variable in the model is equivalent to performing analysis of covariance (ANCOVA). Because ANOVA and ANCOVA can be performed with regression equations, the linear regression model became known as the *general* linear model. Regressions, ANOVAs, and ANCOVAs can all be conducted in SPSS with the general linear model procedure. Unlike the regression procedure, the general linear model procedure will automatically create dummy variables for each category of a multinomial variable; just place the variable name in a factor box. One can obtain the parameter estimates by checking that box under "Options"; however, the sign of the estimates can be confusing to interpret because SPSS uses $D = 1$ as the reference category. One should therefore still double check one's interpretation of directionality, especially for interaction terms, by also using the regression procedure on the initial or final model specified, or the GLMs procedure, which gives the option of choosing the reference category to use.

Functional Form

Sometimes the relationship between an IV and DV is not linear. For example, the relationship between test anxiety and test performance is curvilinear (e.g., having an inverted U-shape), with small amounts of anxiety having a positive effect on performance and large amounts a negative effect. A curvilinear relationship can be modeled by including a quadratic term (X^2) in the model along with X.

Another type of nonlinear relationship is exponential. I focus here on exponential growth in some detail because these concepts will be used when discussing logistic regression in Chapter 10. Consider the following scenario: You put $10,000 in a savings account with an interest rate of 6%, compounded annually. The amount in the bank account will grow exponentially over time, with a growth rate of 6% (or more specifically, 1.06). How much will the bank account be worth after 10 years? I simulated this problem in EXCEL, producing the values in Table 9.1. Under "Amount," each value was derived by multiplying the previous one, contained in the cell above it, by 1.06. It is assumed that there are no further deposits or withdrawals. The amount grows exponentially year after year. After 10 years, the amount increased to $17,908 simply from compound interest. Figure 9.2 shows the exponential growth curve over a century.

Table 9.1 Exponential Growth Example: Bank Account Amount

Year	Amount interest at 6% annually	Ln(amount)
0	$10,000	9.21
1	10,600	9.27
2	11,236	9.33
3	11,910	9.39
4	12,625	9.44
5	13,382	9.50
6	14,185	9.56
7	15,036	9.62
8	15,938	9.68
9	16,895	9.73
10	17,908	9.79

Note: $Amount = e^{10,000 + 1.06 Year}$. $Ln(amount) = 9.21 + 0.06 * Year$.

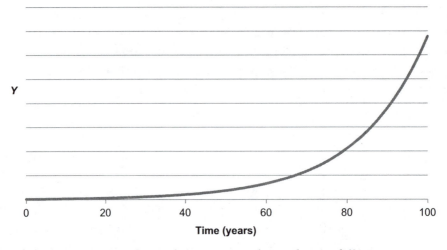

Figure 9.2 Exponential growth curve: Annual growth rate of 6%.

Typically with exponential growth problems one will not know the growth rate but will need to estimate it from the data. The equation for exponential growth is

$$\hat{Y}_i = e^{\alpha + \beta X_i}, \qquad (\text{Eq. 9.13})$$

where α is the logarithm of the starting amount and β is the logarithm of the growth rate. Taking logs of both sides and using one of the rules of logarithms (see Table 9.2 for a review), Eq. 9.13 can be rewritten as

$$ln\left(\hat{Y}_i\right) = \alpha + \beta X_i \ ln\left(e\right) = \alpha + \beta X_i, \qquad (\text{Eq. 9.14})$$

since $ln(e) = 1$. Because the log transformation makes the equation linear, one can perform an OLS regression using $ln(Y_i)$ as the DV.

Performing the regression in SPSS produced the following results: $a = 9.21$, $b = .058$. These are the estimates of the parameters of the population regression line using $ln(Y)$ as the DV, in this case, ln(amount). The reader can confirm from visual inspection of Table 9.1 that ln(amount) increases linearly by .058 per year (.06 with rounding). But one is ultimately interested in estimating the starting value and growth rate. The estimate a is just the estimated logarithm of the starting value, and b is just the estimated logarithm of the slope, so calculating the antilogarithms of these values will produce the starting value and growth rate. One can find antilogarithms through exponentiation. More specifically, $e^{9.21} = exp(9.21) = 10,000$ and

Table 9.2 Rules of Logarithms

---

1. **Definitions.** Common logarithms are the exponents of 10 that produce a certain number (e.g., $log(100) = 2$). Natural logarithms are the exponents of e ($e = 2.71828$) that produce a certain number (e.g., $ln(7.39) = 2$, because $e^2 = 7.39$). All the following rules apply to both common and natural logarithms.

2. **Antilogarithms.** An antilogarithm is the reverse of the logarithm. For example, the common antilog of 2 is 100 ($10^2 = 100$). The natural antilog of 2 is 7.39 ($e^2 = 7.39$). The latter is also referred to as *exponentiation* (in EXCEL, $=$EXP(2) $= 7.39$).

3. **Product rule: $log(xy) = log(x) + log(y)$.**
   For example, $log(2 * 3) = log(2) + log(3)$.
   Note that $log(2) = 0.30$, $log(3) = 0.48$, $log(6) = .78$.
   Because logarithms are exponents, the rule follows from the rule of exponents: $a^x + a^y = a^{x+y}$. (In the example, $10^{0.30} * 10^{0.48} = 10^{0.78}$.)

4. **Quotient rule: $log\left(\frac{x}{y}\right) = log\left(x\right) - log\left(y\right)$.**
   For example, $log\left(\frac{6}{3}\right) = log\left(6\right) - log\left(3\right) = log(2)$.
   This rule follows from the previous rule.

5. **Power rule: $log(x^k) = k * log(x)$.**
   For example, $log(10^3) = 3 * log(10) = 3 * 1 = 3$. The power rule follows from the product rule. To see why, note that $log(10^3) = log(10^1 * 10^1 * 10^1) = log(10^1) + log(10^1) + log(10^1) = 3 * log(10^1)$.

$e^{.058} = exp(.058) = 1.06$. (These calculations can be easily performed in EXCEL using the =EXP command.)

If one were to perform instead an OLS regression on the data in Table 9.1 without taking the logarithm of Y, one would obtain a value of $b = \$787.6$. This value is a biased estimate of the relationship between X and Y because the wrong functional form has been used: linear rather than exponential.

EXPLORING THE CONCEPT

$787.60 is the average annual amount that the account value grew over the 10-year period. If we used the b to predict the amount in year 11, why would the prediction be biased downward?

This example does not include an error term, but there would be stochastic error if there were other random factors that influenced how much one had in the bank account over time, such as occasional deposits or withdrawals.

There are a number of other functional forms that are sometimes used in addition to those discussed here (exponential and curvilinear). In SPSS, go to "Regression"➔ "Curve Fitting" for an overview.

The General Linear Model

In conducting statistical hypothesis tests of beta estimates, one makes a number of assumptions that constitute the normal *general linear model*, summarized in Table 9.3.

Table 9.3 Assumptions of the General Linear Model

Specification assumptions	Distributional assumptions
1. **Linear functional form:** $Y_i = \beta_0 + \beta_1 X_{i1} + \ldots + \beta_k X_{ik} + \varepsilon_i$.	4. **Homoscedasticity:** Error variance the same for each value of X_i [$\text{Var}(\varepsilon_i) = \sigma^2$].
2. **Fixed (nonstochastic) X_i:** X_i used to predict Y_i (hence Y_i is conditional on X_i); no influence of Y_i or of ε_i on X_i.	5. **Errors statistically independent/ nonautocorrelation:** $\text{Cov}(\varepsilon_i, \varepsilon_j) = 0, i \neq j$.
3. **Errors average zero:** $E(\varepsilon_i) = 0$.	6. **Normally distributed errors:** Errors at each value of X_i follow a normal distribution ($\varepsilon_i \sim N$).

Notes: The general linear model reflects the assumptions underlying ordinary least squares (OLS). In the population, the subscript i represents the different levels of X along the real number line. In a sample, i subscripts the observation number.

Specification Assumptions

The first three assumptions, which are known as the *specification* assumptions, are used in the proof that $E(\hat{\beta}) = \beta$ (see Appendix 9.1). This equation asserts that $\hat{\beta}$ is an unbiased estimate of β, and that β is the mean of the sampling distribution. Violations of the specification assumptions will therefore bias the beta estimates. Two common types of violations are (a) use of the wrong functional form, which violates the first assumption, and (b) omitting relevant variables that are correlated with both X and Y, which violates the third assumption, for the following reason. Suppose there is an omitted third variable Z that is correlated with both X and Y. Omitting Z from the statistical model would cause a confound. For example, research has shown that there is some positive correlation between class size and student achievement (Lazear, 1999). (We will assume the relationship is linear although it most likely is not.) Class sizes are often smaller in high socioeconomic (SES) areas, and SES is known to be a strong predictor of achievement. Therefore, any correlation between class size and achievement could be due, at least in part, to the effect of SES. Including SES in the model would control for its effects, but if omitted from the model, the effect of SES would show up in the error term (which reflects the effect of all omitted variables). There would therefore be a correlation between class size (X) and the population error term. The expected value of the error term would not be zero, because error would increase or decrease with X.

Because there often are many potentially relevant variables, how does one decide which variables to include in the model? The best approach is to use theoretical considerations. A *theoretical model* is a set of propositions that explain and predict a causal pattern of associations among variables. For example, the reason that class size reduction is often hypothesized to increase achievement is that students can receive more individualized attention from their teachers. A competing theoretical model is that class size reduction has no effect on achievement and that any relationship is spurious (e.g., caused by SES). The point is that theoretical considerations can be used to help identify relevant variables, such as SES. One could also use statistical considerations, for example, by examining a correlation matrix of various variables in a data set, but some of the correlations (especially if small) may not reflect real causal relationships. Also, statistical considerations are limited to only those variables one has data on, whereas theoretical considerations help guide decisions as to what data to collect.

Distributional Assumptions

Returning to Table 9.3, the second set of assumptions (the "distributional assumptions") are used either to derive the standard errors or, in the case of Assumption 6, to establish the sampling distribution as t shaped. The normality assumption is illustrated in Figure 9.3. To make the graph easier to read, a discrete IV is used. For

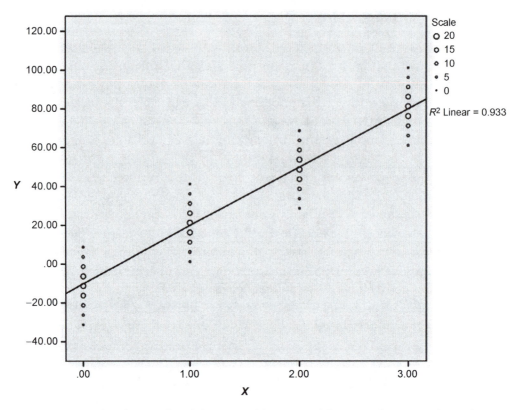

Figure 9.3 Graphical examples of the general linear model using a discrete independent variable. Points have been "binned" so that the size of the circle indicates number of cases. Values of Y follow a normal distribution for each level of X with a constant variance. Values of ε (distanced from the regression line) also follow a normal distribution.

each level of X, the errors follow a normal distribution (this situation is known as *conditional normality*). The majority of errors are small, but there are some larger ones, as shown in the figure.

Suppose for simplicity that we only take a sample of two points, as shown in Figure 9.4. We are more likely to obtain a regression line similar to the solid line, with the slope only slightly different than the population regression line, than one like the dashed line, where the slope is more deviant. After all, the dashed line involves sampling a more extreme point, and there are fewer of those because of conditional normality. The sampling distribution of $\hat{\beta}$ consists of all the possible slopes that we can sample, and if the errors are normally distributed, so will be the sampling distribution of $\hat{\beta}$, because the latter is a function of the errors. (The sampling distribution of the t statistic is t shaped because the numerator, $\hat{\beta}$, has a normal sampling distribution but the denominator, which is the standard error, has a chi-square sampling distribution.)

If conditional normality is not met, the sampling distribution of $\hat{\beta}$ is not guaranteed to be normal. The central limit theorem shows that in a large enough sample,

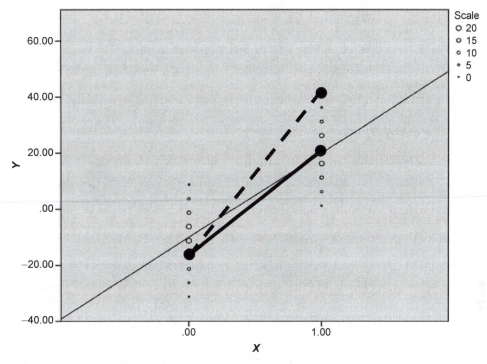

Figure 9.4 Two estimated regression lines from a sample of $n = 2$. The probability of drawing the solid line is much higher than the dashed line.

the sampling distribution will be approximately normal, but how large the sample needs to be is unclear. Most minimum sample size guidelines, such as the ones in Tabachnick and Fidell (2001); that is, $n \geq 104 + k$, are based on considerations of statistical power, not validity, and also assume a lack of significant skew in the DV. It is therefore safest to transform a skewed DV into a more normally shaped one using some sort of logarithmic or Box-Cox transformation (see Technical Note 9.3), or to use a GLM (described in the next section). With a skewed DV, conditional normality is unlikely to be met. It is also important to note that skew also can substantially decrease statistical power, as was explained in Chapter 6. So normality is important on both power and validity grounds.

Outliers

An important part of model building is checking for outliers and deciding whether to downgrade, delete, or retain outliers in the dataset. The presence of outliers can violate both specification and distributional assumptions and can reduce statistical power.

As was discussed in Chapter 6, cases with standardized $|z|$ scores greater than 3.29, can be considered univariate outliers (Tabachnick & Fidell, 2001), because the probability of such scores is less than .001. Cases with standardized or *studentized* residuals (see Technical Note 9.4) greater than $\approx|3|$ can be considered multivariate outliers (Hawkins, 1990), although it is more customary to use Mahalanobis distances (see Technical Note 9.5) or other distance measures. However, outliers can alter the mean and standard deviations of variables, reducing the power of these methods to identify all outliers.

EXPLORING THE CONCEPT

(a) If there is a positive univariate outlier, will that make the mean larger or smaller? Will it make the standard deviation larger or smaller? Will these considerations make its z score larger or smaller? How would this affect the power of the $|3.29|$ criterion to detect outliers?

In regard to multivariate outliers, some statistical programs calculate "deleted residuals," where the residual for each observation is calculated based on the predicted value from a regression with that observation deleted from the analysis. (SPSS calculates studentized deleted residuals in the regression procedure.) Use of deleted residuals is, however, not that robust when there are multiple outliers, because it only deletes one observation at a time. One might therefore use the median absolute deviation (MAD) method described in Chapter 6. Other methods for detecting outlier are discussed by Barnett and Todd (1994), Freund and Wilson (1998), Hawkins (1990), and Huber (1981).

As was outlined in Chapter 6 (see Figure 6.1), an outlier can substantially alter the slope of the regression line, especially if it is in the tail of one of the distributions. Why can this create bias? There may be something qualitatively different about the outlying case from the rest of the cases. For example, in psychological or educational research, there might be one student who is unmotivated to complete a questionnaire and answers randomly. Another example is a student who is an exceptional genius or somehow more advanced than the other study participants. In these examples, the outliers are likely caused by a different data generating process than the other observations (i.e., drawn from a different population distribution) and are said to *contaminate* the distribution of the DV. These outliers should have their own population regression line with its own intercept and slope. In principle, one could include a dummy variable for geniuses as well as an interaction term between genius status and the treatment; however, it would be impossible to properly estimate the parameters of these variables given only one or a few cases. Researchers often therefore delete these cases from the regression analysis; not doing so could cause the same specification bias as omitting a relevant (dummy) variable. Contaminants can also reduce statistical power by pulling all the other observations away from the true population regression line, thereby increasing the standard errors.

It is also possible that, rather than contaminants, the outliers are part of the same probability distribution as the other cases but that the probability distribution has heavy tails associated with skew or kurtosis in the population (Hawkins, 1990). (Note that a distribution may be heavy in just one of the tails due to skew; see Technical Note 9.6.) In this situation, Hawkins (1990) argued that one should not delete the outliers but can downgrade their influence through various means (data transformation, robust methods using trimmed means, etc.).

Generalized Linear Models

Many of the issues discussed in this chapter (nonlinear functional forms, lack of normality) can be addressed by using a *generalized* linear model (GLM). The GLM approach was introduced by McCullagh and Nelder (1998) and provides more elegant and powerful solutions to these problems than previously developed remedies. As a case in point, not all distributions can be made normal through a data transformation (especially where the mode is zero or near zero; Bradley, 1982). Such distributions may be more easily modeled with a GLM. Also, the beta coefficients of such models are often easier to interpret than those based on a Box-Cox transformation (see Sakia, 1992, for controversies surrounding the latter).

GLMs are somewhat less appropriate for small samples, because the estimates may be biased. (However, there are some bias correction formulas; see Cordeiro & McCullagh, 1991; Firth, 1993; Kosmidis & Firth, 2009.)

Components of a GLM

A GLM consists of three elements:

1. Specification of a set of IVs (which is the *systematic* component of the model);
2. the random component, which involves identifying a DV and the shape and parameters of its probability density function; and
3. specification of a *link function* that connects the IV(s) and DV.

For example, the general linear model underlying OLS has the following form:

$$Y_i = \beta_0 + \beta_1 X_{i1} + \beta_2 X_{i2} + \cdots + \beta_k X_{ik} + \varepsilon_i, \varepsilon_i \sim \text{i.i.d. } N(0, \sigma^2), \tag{Eq. 9.15}$$

where there are k independent variables. The abbreviation, "i.i.d." stands for "independently and identically distributed."

As was the case with exponential growth, GLMs often involve nonlinear functions that are transformed into linear ones. The DV of the linear equation is denoted by η_i. In the case of the general linear model, $\eta_i = Y_i$, which is known as the identity link. In the case of exponential growth, $\eta_i = ln(Y_i)$, which is known as the log link.

In regard to other types of GLMs, all require correct specification (e.g., inclusion of all relevant IVs), so the real differences among GLMs are in the types of distributions used and the link functions. For example, logistic models assume a binomial or Bernoulli distribution and require that the link between the IVs and DV is a logit function (see Chapter 10). Poisson models used with count data (see Chapter 11) use a Poisson distribution and a log-link function. It is possible to use a different link function, for example, a linear-link function (i.e., identity link) with a Poisson-distributed random variable. The most common and customary link function for a particular distribution is typically the *canonical link function,* which uses a special term in the distribution's probability density function, known as the *natural parameter,* as the basis for the link function (see Technical Note 9.7). The canonical link function has special properties.

In IBM SPSS, "Generalized Linear Models" is an option under the "Analyze" menu. Clicking on "Generalized Linear Models" again brings up a series of nine tabs, the first of which is "Type of Model." If one clicks on "Custom," one has the choice of nine distributions and several link functions for each distribution; these are summarized in Table 9.4. However, instead of a custom model, one can just choose certain distributions and use the recommended link function. For example, for a Poisson model, under "Counts," click on "Poisson log linear," which will automatically invoke the canonical log-link function. For the normal general linear model, just click on "Linear."

The other tabs under GLMs in SPSS are

- *Response* (for identifying the dependent variable);
- *Predictors* (for identifying IVs that may be used in the analysis; list nominal variables under "factors" and metric variables under "covariates");
- *Model* (for specifying which predictors and interactions to use in the model);
- *Estimation* (for selecting a parameter estimation method; see Chapter 14);
- *Statistics* (for determining what type of statistics to print in the output);
- *EM Means* (for determining what expected marginal means of nominal variables to print and what multiple comparisons to perform);
- *Save* (for determining what type of residuals and related statistics to save in the database); and
- *Export* (for exporting model parameter and correlations to other applications).

There are a lot of choices to be made, but users can typically use the default settings for the last few tabs and focus on choosing the type of model and the variables

Table 9.4 Generalized Linear Models: Distributions and Link Functions (Used in IBM SPSS)

Distribution	Description
Binomial	Probability distribution for a series of binary events (nominal RVs).
Gamma	Continuous, positively skewed, bounded at zero. Shape and scale (or rate) parameters used to model such things as waiting times or insurance claims (with some very high values).
Inverse Gaussian	Continuous, positively skewed, bounded at zero. Shape becomes more normal as shape parameter approaches infinity. Inverse of RV describes a random walk. Used to model such things as Brownian motion (i.e., motion of particle suspended in a fluid), stock prices, repair times.
Multinomial	Probability distribution for nominal RVs with multiple categories.
Negative binomial	For overdispersed count data; also probability of k successes before r failures.
Normal	Normal (aka Gaussian, bell-shaped curve).
Poisson	Distribution for count data; generally positively skewed with lots of zeros.
Tweedie	Mixture of Poisson distribution (for frequency) and gamma distribution (for severity). Used to measure such things as rainfall or insurance claims.
Link function	
Cauchit	aka inverse Cauchy. Based on tangent of probabilities; used when there are many extreme values.
Identity	Linear relationship.
Log	For modeling an exponential relationship.
Logit	"S"-shaped function used in logistic regression for modeling binary outcomes.
Log-log complementary	Asymmetric "S"-shaped function which increases slowly at first then rapidly at large values of X. Used to model binary or ordinal outcomes with very small or large probabilities, and with "interval censored" survival analysis when X cannot be estimated precisely (but only in intervals).
Negative log-log	Asymmetric "S"-shaped function which increases rapidly at first then more slowly at larger values of X. Used in ordinal regression when the probability of lower response categories is high.
Power	Variable base raised to a fix exponent (i.e., cx^a), in contrast to an exponential function (fixed based raised to variable exponent). Power functions are used to model growth and decay and learning curves (alternative to the log-link function for exponential growth).
Probit	"S"-shaped function that is the inverse of the normal cumulative distribution, for modeling binary outcomes.

to be included in the GLM. As will be described in Chapters 10 through 12, some of the models are available under other SPSS options.

Matrix Formulation

This section addresses how GLMs may be expressed in matrix terms and parameters estimated using linear algebra. Readers might choose to skip this section on their first reading of this chapter, but the material will be important in the last two chapters of the book. The basic principles of linear algebra are shown in Table 9.5.

Table 9.5 Basic Principles of Linear Algebra

1. **Matrix addition.** In general, if $\mathbf{A} = \begin{bmatrix} a & c \\ b & d \end{bmatrix}$ and $\mathbf{B} = \begin{bmatrix} e & g \\ f & h \end{bmatrix}$, then $\mathbf{A} + \mathbf{B} = \begin{bmatrix} a+e & c+g \\ b+f & d+h \end{bmatrix}$.

2. **Scalar multiplication.** A scalar is a matrix with dimensions 1×1. If λ is a scalar, then $\lambda \begin{bmatrix} a & c \\ b & d \end{bmatrix} = \begin{bmatrix} \lambda a & \lambda c \\ \lambda b & \lambda d \end{bmatrix}$.

3. **Matrix multiplication.** If $\mathbf{A} = \begin{bmatrix} a & c \\ b & d \end{bmatrix}$ and $\mathbf{B} = \begin{bmatrix} e & g \\ f & h \end{bmatrix}$, then $\mathbf{AB} = \begin{bmatrix} ae+cf & ag+ch \\ be+df & bg+dh \end{bmatrix}$.

4. **Transposition.** To transpose a matrix, list the column values as rows and vice versa. For example, if $\mathbf{A} = \begin{bmatrix} a & c \\ b & d \end{bmatrix}$, then $\mathbf{A^T} = \begin{bmatrix} a & b \\ c & d \end{bmatrix}$.

5. **Square and diagonal matrices.** A matrix is *square* if it has the same number of rows as columns (**A** and **B** are both square). A *diagonal* matrix is a matrix where all elements off the main diagonal are zero. For example, $\begin{bmatrix} 2 & 0 \\ 0 & 3 \end{bmatrix}$ is diagonal. Most diagonal matrices are square.

6. **Identity matrices.** An identity matrix is a diagonal matrix with only ones on the diagonal, e.g., $\mathbf{I} = \begin{bmatrix} 1 & 0 \\ 0 & 1 \end{bmatrix}$. Analogous to the fact that in ordinary multiplication, $k * 1 = k$, in matrix multiplication, $\mathbf{AI} = \mathbf{A}$. For example, $\begin{bmatrix} 2 & 0 \\ 0 & 3 \end{bmatrix} * \begin{bmatrix} 1 & 0 \\ 0 & 1 \end{bmatrix} = \begin{bmatrix} 2 & 0 \\ 0 & 3 \end{bmatrix}$.

7. **Inverting matrices.** Analogous to the fact that in ordinary multiplication, $k * \dfrac{1}{k} = 1$, in matrix algebra, $\mathbf{A} * \mathbf{A^{-1}} = \mathbf{I}$. The inverse of a matrix is the matrix that satisfies this equation. For example, the inverse of $\begin{bmatrix} 2 & 0 \\ 0 & 3 \end{bmatrix}$ is $\begin{bmatrix} 0.5 & 0 \\ 0 & 0.33 \end{bmatrix}$ because the product of the two matrices is $\begin{bmatrix} 1 & 0 \\ 0 & 1 \end{bmatrix}$.

OLS

The regression equation in OLS estimated from the data is

$$\mathbf{Y}_{n\times 1} = \left(\mathbf{X}_{n\times p}\right)*\left(\hat{\boldsymbol{\beta}}_{p\times 1}\right) + \boldsymbol{e}_{n\times 1}.$$ (Eq. 9.16)

Expressions in bold refer to matrices. The **X** matrix (also known as the *design* matrix) refers to the data, with one row for each case and one column for each IV. The first column consists of all ones; these are the values of X_0 in the expression $\beta_0 X_0$, which relates to the intercept.

EXPLORING THE CONCEPT

Why do we use the term β_0 in the general linear model instead of $\beta_0 X_0$? Are these terms equal to one another, given that all $X_0 = 1$?

Let p represent the number of parameters. So if there are two predictors, there are three parameters because β_0 is considered a parameter. The dimensions of **X** are ($n \times p$; the number of rows is always listed first, then the number of columns, a fact that must be committed to memory.) The **β** matrix is a column vector with dimensions ($p \times 1$).

To multiply two matrices together, multiply the first row of the first matrix by the first column of the second (each corresponding term is multiplied and the terms added). So for example,
$$\begin{bmatrix} 1 & 4 & 4 \\ 1 & 3 & 5 \\ 1 & 3 & 3 \\ 1 & 1 & 3 \end{bmatrix} * \begin{bmatrix} 3 \\ 2 \\ 1 \end{bmatrix} = \begin{bmatrix} 15 \\ 14 \\ 12 \\ 8 \end{bmatrix}.$$
If one multiplies the first row of the first matrix by [3, 2, 1], which is the first (and only) column of the second matrix transposed, one obtains:

$$(1 * 3) + (4 * 2) + (4 * 1) = 15,$$

which is the first entry of the final matrix. Likewise, the second entry is:

$$(1 * 3) + (3 * 2) + (5 * 1) = 14.$$

To multiply two matrices together, the column dimension of the first matrix must match the row dimension of the second. For example, consider two hypothetical matrices, $\mathbf{A}_{m \times n}$ and $\mathbf{B}_{o \times q}$. To multiply the two matrices, one takes each row of **B** and multiplies it by a value in each column of **A**, summing the results. To be able to perform the operation, n must equal o or there will be terms in one matrix without a corresponding partner in the other. If $n = o$, the matrices are said to conform to one another for purposes of multiplication. For example, the matrices $\mathbf{A}_{3 \times 7}$ and $\mathbf{B}_{7 \times 2}$ conform, multiplication produces a matrix with the two outer dimensions, $\mathbf{A}_{3 \times 2}$. On the other hand, the matrices $\mathbf{A}_{2 \times 3}$ and $\mathbf{B}_{4 \times 2}$ do not conform and so can't be multiplied together.

EXPLORING THE CONCEPT

In the general linear model, we have $\mathbf{X}_{n \times p}$ and $\mathbf{B}_{p \times 1}$. Do the dimensions conform? What would be the dimensions of the resulting $\hat{\mathbf{Y}}$ matrix?

As the above exercise suggests, $\mathbf{Y}$ is a column vector with dimensions $n \times 1$. An important part of reading matrix equations is understanding how the dimensions conform.

Matrix multiplication can be performed in EXCEL. First, calculate the dimensions of the output array. Second, highlight blank cells for this array. Third, type the EXCEL command =MMULT(array$_1$, array$_2$), where *array* is the cell addresses of each matrix being multiplied. Fourth, press *control-shift-enter*. If one needs to first transpose or invert a matrix, use =TRANSPOSE(array) or =MINVERSE(array). It is recommended that readers practice multiplying the matrices shown above.

Suppose the matrices shown above are from a small data set of four cases and two predictors, where $(\mathbf{X}_{4 \times 3}) * (\hat{\boldsymbol{\beta}}_{3 \times 1}) = \hat{\mathbf{Y}}_{4 \times 1}$ (i.e., the first matrix is the design matrix and the second the beta estimates). Suppose the observed DV is $\mathbf{Y} = \begin{bmatrix} 16 \\ 14 \\ 11 \\ 6 \end{bmatrix}$. Note that in the sample, $\mathbf{Y}_{4 \times 1} - \hat{\mathbf{Y}}_{4 \times 1} = e_{4 \times 1}$, so $e = \begin{bmatrix} 1 \\ 0 \\ -1 \\ -2 \end{bmatrix}$. (In adding or subtracting matrices, the matrices must have the same dimensions; one simply adds or subtracts the individual values from one another.) The term e is a vector of residuals, one for each case. The goal of OLS is to minimize the sum of the squared residuals, that is,

$$\left(e_{1 \times n}^T * e_{n \times 1} \right) = \left(\boldsymbol{\mathit{Residual\ Sum\ of\ Squares}} \right)_{1 \times 1}, \tag{Eq. 9.17}$$

where the superscript $\mathbf{T}$ means the matrix has been transposed (rows turned into columns and columns turned into rows). Matrix multiplication automatically sums the squared residuals. It can be shown that the following formula for $\hat{\beta}$ minimizes the residual sum of squares:

$$\hat{\boldsymbol{\beta}}_{p \times 1} = \left(\mathbf{X}_{p \times n}^T * \mathbf{X}_{n \times p} \right)^{-1} \mathbf{X}_{p \times n}^T * \mathbf{Y}_{n \times 1}. \tag{Eq. 9.18}$$

GLM Formulations

The general linear model can be written in matrix terms as

$$\mathbf{Y} = \mathbf{X}\boldsymbol{\beta} + \boldsymbol{\varepsilon}, \; \boldsymbol{\varepsilon} \sim N_n\left(\mathbf{0}, \sigma_\varepsilon^2 \mathbf{I}_n\right),$$

(Eq. 9.19)

where $\mathbf{I}_n$ is the variance-covariance matrix: $\sigma_\varepsilon^2 \mathbf{I}_n = \begin{bmatrix} \sigma_\varepsilon^2 & 0 & \cdots & 0 \\ 0 & \sigma_\varepsilon^2 & \cdots & 0 \\ \vdots & \vdots & \ddots & \vdots \\ 0 & 0 & \cdots & \sigma_\varepsilon^2 \end{bmatrix}.$

Other generalized linear models can be written as

$$\boldsymbol{\eta} = \mathbf{X}\boldsymbol{\beta}, \; E(\mathbf{Y}) = \boldsymbol{\mu} = \mathbf{g}^{-1}(\boldsymbol{\eta}),$$

(Eq. 9.20)

where $\mathbf{g}^{-1}(\boldsymbol{\eta})$ is the inverse of the link function, and where $\mathbf{Y}$ is assumed to be generated from a particular probability distribution [for example, $\mathbf{Y} \sim \text{binomial}(\boldsymbol{\pi})$]. The distribution should be from the exponential family (see Chapter 10). The distribution reflects the random component of the model and is typically included in the model in lieu of an error term. These GLMs will be covered in the next five chapters.

Summary

This chapter has reviewed the basic meaning of intercept and slope parameters, as well as how to interpret dummy variable parameter estimates (which reflect differences in intercept and slope parameters). Nonlinear functional forms were also discussed, particularly exponential growth, which is related to logistic growth (to be discussed in the next chapter). The assumptions underlying OLS regression (i.e., the specification and distributional assumptions) were also presented, as were methods for identifying outliers.

Because multiple regression can be used to perform t tests, ANOVA, and ANCOVA, Eq. 9.19 is referred to as the general linear model. However, the normal general linear model is actually not general enough because it assumes, in the population, conditional normal error distributions and linear relationships. Therefore, McCullagh and Nelder (1998) proposed generalized linear models (GLMs) with alternative distributional forms and link functions, of which the normal general linear model is a subset. The remaining chapters of this book present various GLMs specifically appropriate for categorical data analysis.

Problems

1. As in Bennett, Elliott, and Peters (2005), researchers examine the behavior problems of 14,000 kindergartners. A 20-item externalizing behavior scale was completed for each student by their teacher (metric DV). The IVs were (a) being from a single-parent home (nominal) and (b) parent involvement in school (metric), as measured by the number of school events the parent(s) had attended over the past year. Using multiple regression, the researchers estimated this equation: $\hat{Y}_i = 2.80 - 0.10 * Parental\ Involvement + 0.50 * Single\ Parent\ Dummy - 0.3 * (Parent\ Involvement * Single\ Parent\ Dummy)$. All estimates were significant.

   a) Rewrite the regression equation as two equations, one each for single-parent and two-parent households. Use EXCEL to add and subtract coefficients.

   b) Using EXCEL, for each equation in part (a), calculate the $\hat{Y}_i$'s, one for each possible value of parental involvement (0 to 7 discrete).

   c) Graph the effect of parental involvement on behavior problems, showing the two regression lines. (If using EXCEL, highlight data and use "Insert" ➔ "Line Chart." If using IBM SPSS, create the parental involvement dummy variable and use "Graphs" ➔ "Chart Builder" ➔ "Line" ➔ "Multiple Lines.")

   d) Interpret the results.

2. In a study on the effects of teaching students a particular learning strategy (concept mapping), the DV was scores on an achievement test (ranging from 0 to 100%) and the IVs were IQ scores (ranging from 0 to 145) and whether a participant received training on a learning strategy (concept mapping). The estimated regression equation was: $\hat{Y}_i = 30 + 20 * Strategy\ Training + 0.3 * IQ - 0.2 * (IQ * Strategy\ Training)$. Answer the same questions as in the previous problem, using strategy training as the dummy variable.

3. A researcher investigates the relationship of fear appeals and topic engagement, using 90 eighth-grade students as her sample and a randomized 2×2 factorial design. Students are given 1 week of class time (in a social studies class) to research and write a paper on one of two topics. One topic is of high personal relevance (banning soda sales from schools) and another of less immediate relevance (campaign finance reform). The researcher tries to induce fear in some students by giving them a pamphlet on harmful effects related to their issue; the students in the control group do not receive a pamphlet. Topic engagement is measured by the number of pages that each student writes. The researcher uses regression to analyze the results, using a dummy variable for fear appeal (1 = if a fear appeal) and a dummy variable for topic (1 = high relevance). She obtains the following equation:

$$\hat{Y}_i = 4.00 - 2.0 * \left(Fear_{Appeal} \right) + 2.8 * High_{Rel} + 4.0 * \left(Fear\_Appeal * High\_Rel \right).$$

a) Rewrite the regression equation as two equations, one for the low-relevance condition and one for the high-relevance condition.

b) Using EXCEL, for each equation in part (a), calculate the predicted means for the experimental (fear) and control (no fear) groups.

c) Create a graph displaying the two regression lines.

d) Interpret the results, assuming all effects are significant.

4. Nussbaum and Kardash (2005) conducted a study of argumentative essay writing, using a 2 × 2 factorial (ANOVA) design, $n = 77$. One factor was asking students to persuade the reader versus not providing a persuasion goal; the other factor was giving participants a dual-positional text summarizing potential arguments and counterarguments on the issue (vs. providing no text). The DV was the number of counterclaims in each student's essay (i.e., claims counter to their stated position). The estimated regression line was $\hat{Y}_i = 0.3 + 0.1 * Persuade + 0.7 * Text - 0.88 * (Text * Persuade)$.

a) Rewrite the regression equation as two equations, one for the no-persuade condition and one for the condition with a persuasion goal.

b) Using EXCEL, for each equation in part (a), calculate for the four cells the predicted cell means. (Hint: The cell means are equal either to the intercept of one of the equations or to the intercept plus slope value.)

c) Create a graph displaying the two regression lines.

d) The researchers only found the *Text * Persuade* term to be significant. Under what condition did the provision of a dual-positional text encourage students to consider counterarguments in their essays?

5. Enter the data from the file *Curve Estimation.9.Problem_5* into your statistical software package.

a) Regress Y_i onto X_i. Report the estimated equation (e.g., $\hat{Y}_i = 1 + 2 * X_i$), the R^2, and the p value for the estimate of beta. Is the estimate significantly different from zero?

b) Using your statistical software or EXCEL, generate a scatterplot. What do you notice about the shape of the scatterplot?

c) Calculate the square of X. (You can do this in SPSS by using "TRANSFORM" ➜ "COMPUTE," and computing X * X.) Then perform a multiple regression, regressing Y on both X and X^2. Report the estimated equation (e.g., $\hat{Y}_i = 1 + 2 * X_i + 3 * X_i^2$), the R^2, and the p values for $\hat{\beta}_1$ and $\hat{\beta}_2$. Do the estimates significantly differ from zero?

d) Compare your results from the two regressions. Why are the results so different?

e) Repeat both regressions and create scatterplots of the standardized residuals as the DV and the predicted values as the IV. (In SPSS, use "Analyze" ➜ "Regression" ➜ "Linear" ➜ "Plot," using ZRESID and ZPRD.) What do you notice about the differences?

    f)   If you have access to SPSS, use "Analyze" ➔ "Regression" ➔ "Curve Estimation" to fit linear, quadratic, and cubic functional forms to the data (check "plot models"). Which functional form provides the greatest fit to the data?

6.   For this problem, use the data from Problem 5. If you have software for Generalized Linear Models, regress Y on X and X^2. In SPSS, use "Type of Model" ➔ "Linear," "Response": Y, "Predictors" ➔ "Covariates": X, X^2, and complete the model tab. What is the Pearson chi-square statistic? (This statistic is a measure of fit; the squared residuals are used to compute a chi-square statistic.)

7.   Jennifer puts $10,000 in a bank account (at Time 0). It grows at a constant rate for 30 years, as shown by the data in the file *Money.9.Problem_7*.

    a)   Copy the data into EXCEL and using any two adjacent Y values, calculate the percentage rate of growth per year.

    b)   What is the growth coefficient? For example, the coefficient for a 5% rate of growth would be 1.05.

    c)   The function describing these data is $\hat{Y}_i = e^{(9.21 + \beta X_i)}$. Based on your answer in part (a), what is β? Remember β is the natural logarithm of the growth coefficient.

    d)   Verify your answer by computing $ln(Y)$ and regressing it on time. Report the regression equation.

    e)   Use EXCEL ("Insert" ➔ "Scatterplot") or statistical software to graph the data.

    f)   Using "Account Balance" as the DV, in SPSS run "Regression" ➔ "Curve Estimation." Choose linear, exponential, and quadratic. What are the respective R^2s and which is highest?

8.   A researcher conducts a study on whether background knowledge predicts how much students in a history class participate in whole-class discussions throughout a semester. The data are in the file *Discussions.9.Problem_8*. The IV is a pretest assessing knowledge of the time period (France from 1600 to 1900). The DV is the number of times students speak during the semester (during discussions videotaped for observation).

    a)   Fit a linear and exponential curve to the data. (In SPSS, run "Curve Estimation," and check "plot models.") What are the respective R^2s, and which is highest?

    b)   Inspect a graph fitting both models. Does the graph that is generated show that the exponential curve fits better? Attach the output.

    c)   Perform a linear regression of $ln(Y)$ on X. Do you get the same parameter estimate for the slope as you did for the exponential model in Part (a)?

    d)   A one-unit increase in X will predict what percent increase in Y (speech turns)? Remember that you will need to exponentiate the parameter estimate.

e) Generate a plot of *ln(Y)* on *X* and save the residuals (in SPSS use "Curve Estimation"). Do the errors look homoscedastic (i.e., having equal variances at different values of *X*)?

f) Visually inspect your plot in Part (e). Which value appears to be an outlier?

g) Repeat the linear regression from Part (c) (in SPSS, use the linear regression procedure), and save the studentized deleted residuals (or if that option is not available, just the studentized or standardized residuals). (See Technical Note 9.4 for an explanation of studentized residuals.) Does the suspected outlier qualify as one using statistical criteria?

9. For this problem, use the hypothetical data below.

a) Find the skew and kurtosis for Y_i. Optional: Use your statistical software to create a histogram of Y_i (e.g., in SPSS, "Analyze" → "Descriptive Statistics" → "Frequencies" → "Charts," check histogram).

b) A gamma distribution can be used to model skewed metric distributions. In EXCEL, find the gamma probabilities associated with the numbers $x = 1$ through 10, using =GAMMA.DIST(*x, alpha, beta,* 0). Set the shape parameter *alpha* = 2.0 and the scale parameter *beta* = 1.0. (Enter these values in a cell and, when referring to them in the above formula, press F4 so the cell addresses will not change.) For graphic purposes, add a row into the dataset for $x = 0.5$, Prob(*x*) = 8%, to account for the probabilities of $x < 1.0$ and to make the total add up to 100%. Use "Insert" → "Scatter" (with smooth lines) to plot the probability distribution. Experiment with changing the parameter values to see how the distribution's shape and scale changes.

c) If available, use the "Generalized Linear Model" procedure in IBM SPSS (or similar software) to regress Y_i on X_i. Report the Pearson χ^2 for the following three models: linear, gamma with log link, gamma with identity link (in SPSS, click on "Custom" under the "Type of Model" tab and then specify this model type). Which model predictions best fit the data?

d) Rerun the linear model and save the residuals and the "Predicted value of linear predictor." Create a scatterplot of the former against the latter. (In SPSS, use "Graphs" → "Chart Builder" → "Scatter/Dots" under the gallery tab, and choose "Simple Scatter;" then drag the variables onto the appropriate axes.) Can you detect the skew in the error distributions from this plot? What is the skew value of the residuals?

e) What is the skew value of the residuals from the best-fitting model in Part (c)?

Case	X	Y	Case	X	Y	Case	X	Y	Case	X	Y	Case	X	Y
1	1	15	5	5	9	9	9	22	13	13	25	17	17	18
2	2	1	6	6	9	10	10	33	14	14	26	18	18	19
3	3	17	7	7	40	11	11	13	15	15	57	19	19	40
4	4	28	8	8	6	12	12	14	16	16	17	20	20	21

10. Analyze the functional form of the relationship between X and Y for the following hypothetical data.
    a) Create a scatterplot. Does the relationship look linear or nonlinear?
    b) Assuming normal errors, use the following custom link functions to fit a GLM and report the Pearson chi-square values: Identity, Log, and Power (-2). Which model fits the data best?

Case	X	Y	Case	X	Y	Case	X	Y	Case	X	Y	Case	X	Y
1	1	.25	5	5	.21	9	9	.08	13	13	.07	17	17	.10
2	2	.23	6	6	.16	10	10	.14	14	14	.15	18	18	.07
3	3	.22	7	7	.17	11	11	.14	15	15	.15	19	19	.07
4	4	.16	8	8	.13	12	12	.13	16	16	.08	20	20	.15

11. For practice conducting matrix multiplication in EXCEL (and related operations), complete Problems 5 and 6 from Chapter 14.

Technical Notes

9.1 Perfect multicollinearity occurs when two or more of the predictors are perfectly correlated. The condition makes it impossible to find the inverse $(\mathbf{X^TX})^{-1}$ in the $\hat{\boldsymbol{\beta}}$ formula and in the formula for the standard error. Coding a multinomial IV with a dummy variable for each category creates perfect multicollinearity between the IV (which will always have a value of 1 for one of the categories) and X_o, which consists of a series of ones. For this reason, one of the dummy variables must be omitted and that category designated as the reference category.

9.2 The problem of perfect multicollinearity is different from that of partial multicollinearity, where two IVs are partially correlated. Creating partial multicollinearity, if not too severe, may be necessary to correctly specify a model if both IVs are relevant (i.e., correlated with each other and the DV). Partial multicollinearity does not violate any assumptions of the normal linear model but does reduce statistical power.

9.3 The Box-Cox power transformation (Box & Cox, 1964) is based on the equation, $Y_i^{(\lambda)} = \dfrac{\left(Y_i^{\lambda} - 1\right)}{\lambda}$ if $\lambda \neq 0$, otherwise $Y_i^{(\lambda)} = ln(Y_i)$. One can use different values of λ to determine which value best removes skew from the data (or see Osborne, 2013). To aid interpretation, note that a λ of 2 is equivalent to Y^2, 0.5 to $\sqrt{Y}$, 0 to log(Y), −1.0 to $\frac{1}{Y}$, and 0.5 to $\frac{1}{\sqrt{Y}}$.

9.4 Both standardized and studentized residuals divide the raw residual by an estimate of the standard error of the residuals (based on the square root of the mean-squared error), but studentization adjusts the latter value by the leverage of each observation [specifically, $MSE * (1 - h_{ii})$]. Cases farther away from

$\bar{X}$ have greater leverage on the slope of the regression line, and therefore the residuals for those values tend to be smaller and less variable. For this reason, sample residuals—unlike the population errors under homoscedasticity—may not all have the same standard errors. Because they follow a t distribution, studentized residuals greater than 3.39 in absolute value (as opposed to 3.29 for standardized residuals) should be inspected, although some authors, such as Freund and Wilson (1998) use more liberal standards (e.g., 2.5 in absolute value).

9.5 Mahalanobis distances (D^2) can also be used to identify outliers when all variables are metric. The observations in a multiple regression data set form an ellipsoid (or hyperellipsoid) in at least a three-dimensional space. D^2 measures the distance of an observation from the center of the ellipsoid, standardized by the standard deviation of the points in the direction of the observation in question. The measure also takes into account the covariances between the IVs.

In SPSS, the appropriate function can be found under the "Save" tab under the regression and general linear model procedures. Assuming a chi-square distribution, calculate the probability of each distance using "Transform" → "Compute Variable," and then by typing: SIG.CHISQ(MAH_1, df) where df is equal to k (k is the number of IVs). Cases with probabilities less than .001 should be considered outliers.

9.6 A distribution is considered "heavy tailed" if one or both tails decay more slowly than an exponential function.

9.7 The canonical link function expresses the dependent variable of a GLM in its linear form, η, in terms of what is known as the natural parameter. Most commonly used distributions, such as the normal, Poisson, or binomial have probability density functions (PDFs) that can be written in exponential form, $\text{Prob}(Y = y) = exp\left\{\frac{y\theta - b(\theta)}{\phi + c(y,\,\phi)}\right\}$, where ϕ is a scale parameter (to be discussed in Ch 10) and θ is the natural parameter. So, for example, it can be shown that the PDF of the binomial distribution $\left[f(k) = \binom{n}{k}\pi^k(1-\pi)^{n-k}\right]$ can be rewritten as $f(k) = \binom{n}{k}exp\left[k * log\left(\frac{\pi}{1-\pi}\right) + nlog(1-\pi)\right]$, making $log\left(\frac{\pi}{1-\pi}\right)$ the natural parameter, θ. (The second term in the parentheses, $nlog(1-\pi)$, is a function, $b(\theta)$, of the natural parameter $log\left(\frac{\pi}{1-\pi}\right)$; the ϕ here can be ignored.) In the binomial case, the canonical link function is $\eta = log\left(\frac{\pi}{1-\pi}\right) = \mathbf{X\beta}$, showing how the canonical link is expressed in terms of the natural parameter. For the normal distribution, the natural parameter $\theta = \mu$, so the canonical link function is just $\eta = \mu = \mathbf{X\beta}$ (the identity link function). Canonical link functions have special mathematical properties; for example, different estimation procedures produce identical results (see Chapter 14).

APPENDIX 9.1

PROOFS OF SELECTED OLS THEOREMS

Proof of Eq. 9.1 $b = \dfrac{Cov(X,Y)}{Var(X)}$

1. $e_i = Y_i - \hat{Y}_i$. $\hat{Y}_i = a + bX_i$, so $e_i = Y_i - a - bX_i$.

2. To simplify the calculations, let $x_i = X_i - \bar{X}$. Substituting x_i for X_i (this will alter a but not the slope b) yields: $e_i = Y_i - a - bx_i$. Making a similar substitution for $y_i = Y_i - \bar{Y}$ eliminates the intercept term, so $e_i = y_i - bx_i$.

3. Find the value of b (and a) that minimizes $\sum e_i^2 = \sum (y_i - bx_i)^2$.

4. The slope of the function is given by the partial derivative: $\dfrac{\partial}{\partial b} \sum (y_i - a - bx_i)^2$
   $= \sum 2(-x_i)(y_i - a - bx_i) = -2\sum x_i (y_i - a - bx_i)$.

5. Setting the partial derivative to zero to find the minimum point and dividing both sides of the equation by -2 yields $\sum x_i (y_i - a - bx_i) = 0$.

6. Therefore, $\sum x_i y_i - a\sum x_i - b\sum x_i^2 = 0$.

7. Because it can be shown that $\sum x_i = 0$ (sum of the positive and negative deviations cancel out one another), the middle term is eliminated, yielding $\sum x_i y_i - b\sum x_i^2 = 0$.

8. Therefore, $b\sum x_i^2 = \sum x_i y_i$, and $b = \dfrac{\sum x_i y_i}{\sum x_i^2} = \dfrac{Cov(X,Y)}{Var(X)}$. QED.

Proof of Eq. 9.5 $SE(\hat{\beta}) = \sqrt{\dfrac{\sigma^2}{\sum (X_i - \bar{X})}}$, where σ^2 is estimated by $s^2 = \dfrac{\sum (Y_i - \hat{Y}_i)^2}{n-2}$

1. $b = \sum (X_i - \bar{X})(Y_i - \bar{Y}) / \sum (X_i - \bar{X})^2$ per Eq. 9.1. If one lets x_i represent the mean deviated value $(X_i - \bar{X})$, and likewise $y_i = (Y_i - \bar{Y})$, then $b = \dfrac{\sum x_i y_i}{\sum x_i^2}$.

2. Let $k = \sum x_i^2$. Then $b = \dfrac{\sum x_i y_i}{k} = \sum \left(\dfrac{x_i}{k}\right) y_i$, per the distributive property of addition.

3. Let the weight $w_i = \dfrac{x_i}{k}$. Therefore b is a weighted sum $b = \sum w_i y_i$.

4. Recall that in general, Var($A + B$) = Var(A) + Var(B) + 2Cov(A,B), and Var(kA) = k^2Var(A) (see Chapter 3). Therefore, assuming that all observations are statistically independent so that the covariances are zero, it follows that $Var(b) = Var(\sum w_i y_i) = w_1^2 Var(y_1) + w_2^2 Var(y_2) + \cdots + w_n^2 Var(y_n) = \sum w_i^2 Var(y_i) = \sum w_i^2 \sigma^2$.

5. Because $w_i = \dfrac{x_i}{k}$, $Var(b) = \sum \dfrac{x_i^2}{k^2}\sigma^2 = \dfrac{\sigma^2}{k^2}\sum x_i^2$, per the distributive law.

6. Because $k = \sum x_i^2$ (see step 2), $Var(b) = \dfrac{\sigma^2}{\left(\sum x_i^2\right)^2}\left(\sum x_i^2\right) = \dfrac{\sigma^2}{\sum x_i^{2}}$.

7. Because $b = \hat{\beta}$, $SE(\hat{\beta}) = \sqrt{\dfrac{\sigma^2}{\sum x_i^{2}}}$. QED.

Proof of $Cov(X_i, \varepsilon_i) = 0$

1. $Cov(X_i, \varepsilon_i) = E\left[(X_i - \bar{X})(\varepsilon_i - E(\varepsilon))\right]$. Because $E(\varepsilon) = 0$ per Table 9.3, $Cov(X_i, \varepsilon_i) = E\left[(X_i - \bar{X})(\varepsilon_i)\right]$. Because $x_i = (X_i - \bar{X})$, $Cov(X_i, \varepsilon_i) = E(x_i * \varepsilon_i)$.

2. Because X_i and x_i are assumed to be fixed variables per Table 9.3, $E(x_i * \varepsilon_i) = x_i E(\varepsilon_i)$ for the ith value of X. Because we also assume $E(\varepsilon_i) = 0$ per Table 9.3, $E(x_i * \varepsilon_i) = 0$ for the ith value of X.

3. From step 1, $Cov(X_i, \varepsilon_i) = E(x_i * \varepsilon_i)$. Therefore, $Cov(X_i, \varepsilon_i) = 0$. QED.

Proof of $E(b) = \beta$

1. Let $x_i = (X_i - \bar{X})$ and $y_i = (Y_i - \bar{Y})$. Then $b = \dfrac{Cov(X_i, Y_i)}{Var(X_i)} = \dfrac{\sum x_i * y_i}{\sum x_i^2}$.

2. Because $y_i = \beta x_i + \varepsilon_i$, then $b = \dfrac{\sum x_i * (\beta x_i + \varepsilon_i)}{\sum x_i^2} = \dfrac{\sum \beta x_i^2 + \sum x_i * e_i}{\sum x_i^2}$.

3. Because of the distributive law, $b = \dfrac{\beta \sum x_i^2 + \sum x_i * e_i}{\sum x_i^2} = \beta + \dfrac{\sum x_i * e_i}{\sum x_i^2}$.

4. $E(b) = E\left(\beta + \dfrac{\sum x_i * e_i}{\sum x_i^2}\right) = \beta + \dfrac{\sum E(x_i * e_i)}{\sum E(x_i^2)}$, per Eq. 3.6.

5. Because $E(x_i * \varepsilon_i) = 0$ per step 2 of previous proof, $E(b) = \beta$. QED.

References

Barnett, V., & Todd, T. (1994). *Outliers in statistical data* (3rd ed.). New York, NY: John Wiley & Sons.

Bennett, P., Elliot, M., & Peters, D. (2005). Classroom and family effects on children's social and behavioral problems. *The Elementary School Journal, 105*, 461–480.

Box, G.E.P., & Cox, D. R. (1964). An analysis of transformations. *Journal of the Royal Statistical Society, Series B, 26*, 211–234.

Bradley, J. V. (1982). The insidious L-shaped distribution. *Bulletin of the Psychonomic Society, 20*, 85–88.

Cordeiro, G. M., & McCullagh, P. (1991). Bias correction in generalized linear models. *Journal of the Royal Statistical Society, Series B, 53*, 629–643.

Firth, D. (1993). Bias reduction of maximum likelihood estimates. *Biometrika, 80*, 27–38.

Freund, R. J., & Wilson, W. J. (1998). *Regression analysis: Statistical modeling of a response variable*. San Diego, CA: Academic Press.

Hawkins, D. M. (1990). *Identification of outliers*. New York, NY: Chapman & Hall.

Huber, P. J. (1981). *Robust statistics*. New York, NY: John Wiley & Sons.

Kosmidis, I., & Firth, D. (2009). Bias reduction in exponential family nonlinear models. *Biometrika, 96*, 793–804.

Lazear, E. P. (1999). *Educational production* [Working paper 7349]. Cambridge, MA: National Bureau of Economic Research.

McCullagh, P., & Nelder, J. A. (1998). *Generalized linear models* (2nd ed). New York, NY: Chapman & Hall/CRC.

Nussbaum, E. M., & Kardash, C. M. (2005). The effect of goal instructions and text on the generation of counterarguments during writing. *Journal of Educational Psychology, 97,* 157–169.

Osborne, J. W. (2013). *Best practices in data cleaning: A complete guide to everything you need to do before and after collecting your data.* Thousand Oaks, CA: Sage.

Sakia, R. M. (1992). The Box-Cox transformation technique: A review. *The Statistician, 41,* 169–178.

Tabachnick, B. G., & Fidell, L. S. (2001). *Using multivariate statistics* (4th ed.). Boston, MA: Allyn & Bacon.

CHAPTER 10

BINARY LOGISTIC REGRESSION

This chapter considers the situations where there is a binary dependent variable (DV). Examples of such DVs include whether an individual will or will not drop out of college, contract a certain disease, or vote in the next election. Logistic regression is appropriate when the DV is nominal and the independent variables (IVs) are metric. If the nominal DV is dichotomous, binary logistic regression can be used. As will be discussed in the next chapter, when the DV contains more than two categories, multinomial logistic or ordinal regression is applicable.

The generalized linear models (GLM) approach provides a unified framework for these different techniques. So, for example, in IBM SPSS, the GLM procedure can be used to perform various logistic or ordinal regressions. However, there are also separate options in SPSS for performing these techniques under the regression menu. These options are available in older versions of SPSS (as well as newer versions) and also contain some features that are not available when using GLM, so users should become familiar with both.

Basics of Logistic Regression

When one has a binary DV or IV, it should typically be coded as a dummy variable. In the case of a binary IV, one can use ordinary least squares (OLS), as discussed in the previous chapter. However, using OLS with a binary DV will violate several underlying assumptions, of OLS, specifically linearity, homoscedasticity, and normality. Figure 10.1(A) displays the best-fitting (OLS) line for a binary DV (Y). It can be seen that for values of X greater than 7, the values of $\hat{Y}$ exceed 1.0. Likewise, for values of $X < -7$, the values of $\hat{Y}$ are less than 0. These are nonsensical values: The values of $\hat{Y}$ should be between 0 and 1. A value of $\hat{Y}$ between 0 and 1 is meaningful; for example, $\hat{Y} = .70$ means that there is a 70% chance that a case with $X = 2$ will have $Y = 1$. But a $\hat{Y}$ in excess of one or less than zero is not meaningful, implying that we are using the wrong functional form (i.e., link

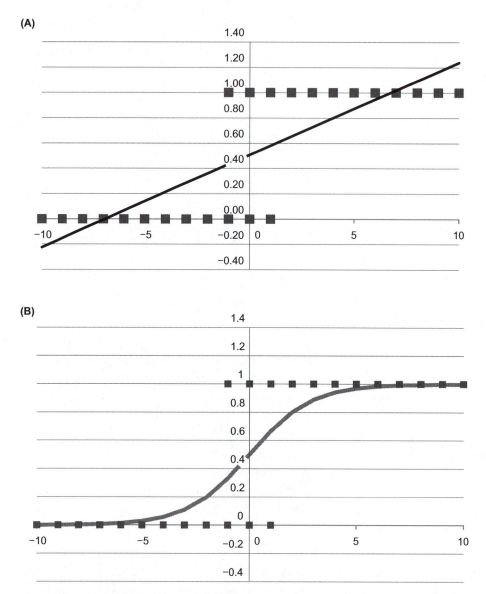

Figure 10.1 (a) Linear regression line for a binary outcome variable. (b) Logistic regression line for binary outcome variable.

function). A more sensible functional form is shown in Figure 10.1(B); the figure displays the logistic curve, to be discussed shortly. Use of the (incorrect) linear functional form can result in implausible estimates. For example, an increase of $10,000 in annual income would not have the same effect on the likelihood of buying a home for someone currently making $20,000 per year as it would for someone making $50,000, but OLS assumes that the effect would be uniform across incomes. The relationship between income and house buying is better modeled with a logistic curve.

Returning to Figure 10.1(A), there are also two other flagrant violations of the general linear model displayed in the figure. First, the residuals are not normally distributed for any value of X. For example, where $X = 0$, the residuals are only 0.5 and -0.5, with nothing in between. The error distribution is actually binomial. Second, the residual error variances are not constant. There are two residuals in the middle, where $X = 0$ in this example, but smaller residuals near where $X = 7$ or -7. This heteroscedasticity occurs because the variance of a binomial distribution $[n\pi(1 - \pi)]$ is a function of the mean $(n\pi)$, and hence the mean (conditioned on different values of the IV) changes with the IV. The variance is greatest when $\pi_i = 50\%$, which occurs in the figure where $X = 0$.

These violations of the specification and distributional assumptions of OLS will cause parameter estimates (of β's and standard errors) to be biased, potentially resulting in Type I and II errors.

The Logistic Curve

I will explain, in due course, how the logistic regression procedure addresses these various issues, beginning with the issue of nonlinearity.

The Basic Logistic Curve

Figure 10.1(B) displays a logistic curve. The curve originated in the field of population biology but was found useful in categorical data analysis. In population biology, the curve is related to exponential growth. Inspection of the figure shows that the first part of the curve is exponential. However, in population biology, there are limits to how many animals a particular habitat can support. For example, if the population of a certain species exceeds the carrying capacity of the environment, there will not be enough food and some animals of the species will die of starvation or become more vulnerable to predation. On the other hand, if the population is below the carrying capacity, the drive to reproduce will increase the population until it reaches equilibrium. In the context of logistic growth, we could consider the Y values in Figure 10.1(B) to represent the population size in proportion to the total population that the habitat can support.

Statisticians appropriated use of the logistic growth curve to model something entirely different: the growth in the probability that a nominal variable (Y) will equal one as X increases by one unit. The probability can be written as $\pi_i = \text{Prob}(Y_i = 1 \mid X_i)$. (For ease of exposition, I will at times leave off the i subscript.)

The formula for the logistic curve is

$$\pi = \frac{e^{(\alpha + \beta X)}}{1 + e^{(\alpha + \beta X)}},$$

(Eq. 10.1)

where e is the number 2.7183. It should be noted that according to Eq. 10.1, the Y intercept is not α but rather $\frac{e^{\alpha}}{1+e^{\alpha}}$.

EXPLORING THE CONCEPT

(a) Will the denominator in Eq. 10.1 always be larger than the numerator? How does this prevent π from ever exceeding 1.0? (b) Using EXCEL, and assuming that $\beta = 0$, what is the value of π if $\alpha = 0$? $\alpha = -3$? $\alpha = -6$? $\alpha = -9$? What is the lower asymptotic value of π? If the numerator of Eq. 10.1 is close to 0, what will be the value of the denominator? Of π? (c) Suppose the value of the numerator is very small, say .02 or even .0002. What will be the value of π? You will see that it is close to the value of the numerator. Why does this make the lower part of the logistic growth curve reflect exponential growth?

The reader is encouraged to use Eq. 10.1 (and EXCEL) to generate values of Y for discrete X values ranging from -10 to 10 and an associated scatterplot (experimenting with different values of α and β).

The Transformed Curve

To construct a GLM, Eq. 10.1 needs to be transformed into a linear equation. It can be shown (see Appendix 10.1) that the odds $\frac{\pi}{1-\pi} = e^{(\alpha+\beta X)}$. Taking the natural logs of each side yields $ln\left(\frac{\pi}{1-\pi}\right) = ln\left(e^{(\alpha+\beta X)}\right)$, which simplifies to:

$$ln\left(\frac{\pi}{1-\pi}\right) = \alpha + \beta X. \tag{Eq. 10.2}$$

The term $ln\left(\frac{\pi}{1-\pi}\right)$ represents the *log odds,* also known as the *logit.* The right-hand side of the equation is now a linear equation. Beta reflects the amount that the log odds will change from a one unit change in X. One can move off the log scale onto the original scale through exponentiation. (This same procedure was used in the previous chapter in finding rates of exponential growth.) If X is itself a dummy variable (coded 0, 1), $exp(\beta)$ will just be an odds ratio, that is, the ratio of the odds that $Y = 1$ when $X = 1$ to the odds that $Y = 1$ when $X = 0$ (see Chapter 4). Subtracting 1.0 from the odds ratio yields the percentage increase in the odds of a hit from a one unit increase in X (see Chapter 4). If X is metric, the interpretation is for *any* one unit increase of X.

In logistic regression, $exp(\hat{\beta}) - 1$ is the predicted percentage increase in the odds of a hit ($Y = 1$) from a one unit increase in X.

For example, if $\hat{\beta} = 0.5$, then $exp(0.5) = e^{0.5} = 1.65$, which means that the odds that $Y = 1$ increases by 65% from a one unit increase in X. (Alternatively, one could say that the odds that $Y = 1$ are 1.65 times the odds when X is one unit less.) Note that the increase here is in terms of the odds, $\frac{\pi_i}{1-\pi_i}$, not in terms of the probabilities, π_i.

EXPLORING THE CONCEPT

Suppose $\pi_1 = 60\%$. The odds of $Y = 1$ when $X = 1$ are $\frac{60\%}{40\%} = 1.5$. A 65% increase in the odds from a one unit increase in X means that if $X = 2$, we would predict the odds to be $1.5 * 1.65 = 2.48$, which corresponds to $\pi_1 = 71.3\%$ (see Eq. 4.7). In percentage terms, have the odds increased more than the probabilities? Why is it important to use the word *odds* rather than *chance* in writing up the results of a logistic regression?

The Logistic Regression Model

In the binary logistic regression model, although Y is dichotomous, suppose that there is a latent (i.e., unobserved) continuous variable, Y^*, that corresponds to Y, such that if $Y* > 0$, $Y = 1$, and if $Y* \leq 0$, $Y = 0$. This assumption of a latent continuous variable is not critical but is useful for understanding the logistic model. Because $Y*$ is metric, it can be modeled using an ordinary multiple regression model:

$$Y_i^* = \gamma_0 + \gamma_1 X_{ij} + \cdots + \gamma_k X_{ik} + \sigma\varepsilon_i, \qquad \text{(Eq. 10.3)}$$

where σ is an adjustment factor that allows the variance of the residuals to be adjusted downwards or upwards (Williams, 2009). The logistic model for Y_i (the observed nominal variable) is

$$ln\left(\frac{\text{Prob}(Y_i = 1)}{1 - \text{Prob}(Y_i = 1)}\right) = \beta_0 + \beta_1 X_{ij} + \cdots + \beta_k X_{ik}. \qquad \text{(Eq. 10.4)}$$

Although logistic regression procedures estimate the βs and not the γs, the two are related because it has been shown that $\beta_j = \frac{\gamma_j}{\sigma}$ (Amemiya, 1985).

Underlying Assumptions

Let us now return to the consideration of other underlying assumptions. The model in Eq. 10.4 meets the linearity assumption of regression. Note that there are no positive bounds on the size of odds; for example, the odds of $\pi = .9999$ are 9999 to 1 and could potentially increase to infinity. On the other hand, odds cannot be equal to or less than zero, but the log odds are not so bounded; if the odds are very small (say, .0001), then the log odds will be very negative (e.g., -9.21).

EXPLORING THE CONCEPT

Because the DV (i.e., logit) in Eq. 10.4 is a function of π, there will be intermediate values of the DV between 0 and 1. Inspect Figure 10.1(B). What is the value of π for $X = 0$? For $X = 2.5$?

The fact that the DV is on an unbounded continuous scale does not, however, mean that the errors are now normally distributed with constant variance. These issues will be resolved by estimating the parameters, not with OLS but with a very important and widely used approach developed by Sir Ronald Fisher, known as *maximum likelihood estimation* (MLE). The MLE approach, to be presented later in this chapter, will assume a Bernoulli or binomial distribution and does not assume homoscedasticity.

Data Generation Example

To better understand the logistic model, I will now generate some data using the model, assuming certain parameter values. Generating data is not typically part of empirical research, where the goal is to estimate parameter values, but it is a common practice in statistical research involving computer simulations and is also useful instructionally.

Suppose a drug manufacturer has invented a new drug, call it Trimone, for treating depression. As part of the clinical trials, Trimone is administered to 4,000 outpatients at varying doses (10−40 mg) plus to 1,000 outpatients who receive a placebo (sugar pill) with 0 mg Trimone. The DV is whether or not a patient's psychiatrist notes any improvement in symptoms after a 2 week period. Let the parameter values be $\beta_0 = -5.0$ and $\beta_1 = 0.2$.

Table 10.1 illustrates the data generation process. For each value of X_i, the linear equation $\eta_i = -5.0 + 0.2X_i$ is used to generate a logit. (η_i is the value of the linear predictor(s) in a GLM and in this case denotes the logit.) So, for example, $X_3 = 20$ mg, so $\eta_2 = -5.0 + 0.2 * 20 = -1.0$. Because the logit is also known as the log odds, we can find the odds just by exponentiating $exp(-1.0) = 0.37$. From the odds, one can then derive the probability using the following equation:

$$\text{Prob}(x) = \frac{odds(x)}{[1 + odds(x)]}. \qquad \text{(Eq. 10.5)}$$

Therefore, $\text{Prob}(Y = 1 \mid X_3 = 20) = \frac{0.37}{1 + 0.37} = 27\%$. Now it is assumed in the population that the Y_i values at a given value of X_i follow a Bernoulli distribution because Y_i equals either 1 or 0. However, because in this example there is more than one case at each value of X_i, the total number of hits at a given value of X_i will follow a binomial distribution with a mean given by Eq 10.5. So for $X_3 = 20$, $\mu_3 = 27\%$. To

Table 10.1 Generation of Hypothetical Data for Trimone Clinical Trials

Group	Dosage (X_i)	Logit($\eta_i = -5.0 + 0.2X_i$)	Odds $= \exp(\eta_i)$	Means/probabilities $= \frac{odds}{[1 + odds]}$	Simulated no. hits	Simulated no. misses
1	0	−5	0.01	0.01	8	992
2	10	−3	0.05	0.05	43	957
3	20	−1	0.37	0.27	298	702
4	30	1	2.72	0.73	732	268
5	40	3	20.09	0.95	952	48

Note: For each group, $\eta_i = 1,000$, $\beta_0 = -5.0$, $\beta_1 = 0.2$. The number of hits was simulated using the binomial distribution.

generate an actual frequency count, I used SPSS "Transform" → "Compute Variable" and used the following command: RV.BINOM(1000, "Means"), where "Means" was the variable name for the results calculated from Eq. 10.5. The calculation and final results are shown in Table 10.1.

Later, we shall see that if we perform a logistic regression using the DV values in the last two columns of Table 10.1, the procedure estimates $\hat{\beta}_0 = -4.874$ and $\hat{\beta}_1 = 0.197$, which are close to the population values from which I generated the data.

The Problem of Unobserved Heterogeneity

Note that the logistic model (Eq. 10.4) does not contain an error term. As shown in Table 10.1, the model predicts a logit, from which we can derive the means of the binomial distributions, one for each X_i. The binomial (or Bernoulli) distribution is the random component of the logistic GLM, from which the observed values of Y_i are generated. (The Bernoulli distribution is used if there is at most just one observation for each value on the X axis.) There are still errors, because the observed values of Y_i will differ from the predicted (mean) values. In the example, for X_3 the predicted number of hits is $0.27 * 1,000 = 270$ and the actual value is 298, producing a residual of 29. For each observation in Group 3, the predicted probabilities are all 27% and the actual probabilities are either 100% or 0% (depending on whether $Y = 1$ or 0), producing residuals of 0.73 or −0.27, respectively. The major point is that the presence of stochastic errors is implicit in the model. The errors cause unexplained variance.

The error term in Eq. 10.3 (the model for the latent continuous variable) is assumed to follow the logistic distribution, which is a distribution with a probability density function similar to the normal curve but with slightly thicker tails and a wider variance. When standardized, the error variance is fixed at 3.29. As noted by Mood (2010), any increase in the explained variance causes the total variance of the dependent variable $[Y_i^*]$ to increase, as the total variance is the sum of the explained and unexplained variance and the latter is fixed at 3.29. As a result, the scale of Y_i^* increases. "When the scale of the dependent variable increases, b_1 must also increase since it now expresses the change in the dependent variable in another metric" (Mood, 2010, p. 69). The implication is that when there is unexplained variance, the beta estimates are biased downward and will tend to increase as more relevant predictors are added to the model. Unlike OLS, this process occurs even if the new predictors are not correlated with the existing predictors.

Mood (2010) gave the example, using artificial data, of whether high school graduates transition to college or not ($Y_i = 1$ or 0). Model 1 contains only the predictor IQ ($\hat{\beta}_1 = 0.80$). Model 2 adds the predictor gender (female = 1), which is uncorrelated with IQ. The results of a logistic regression using model 2 were $\hat{\beta}_1 = 0.99$ and $\hat{\beta}_2 = 2.00$. The beta estimate for IQ ($\hat{\beta}_1$) has increased because more significant variables have been added to the model. The amount of variance (or what Mood calls *unobserved heterogeneity*) has increased.

Of course, if any omitted predictors are correlated with the existing predictors and the DV, adding the omitted predictors will cause the beta estimates of the existing predictors to increase or decrease (depending on the pattern of correlations), so the net effect on the beta estimates of these two mechanisms (explaining more variance and controlling for confounds) is uncertain.

In logistic regression, one has to be careful when one compares two different beta estimates, as those estimates are affected by both the population betas and the amount of unobserved heterogeneity. To validly compare two different beta estimates, one must assume that the amount of unobserved heterogeneity in both estimates is the same. (See Mood, 2010, for description of several statistics that are robust to this assumption.)

Maximum Likelihood Estimation

Overview

As noted above, logistic regression uses MLE, an extremely important approach invented by Sir Ronald Fisher in 1922 (Aldrich, 1997). The essence of maximum likelihood estimation is as follows:

Definition of MLE. A maximum likelihood (ML) estimator of a parameter θ is the value of $\hat{\theta}$ that maximizes the likelihood of obtaining the observed results.

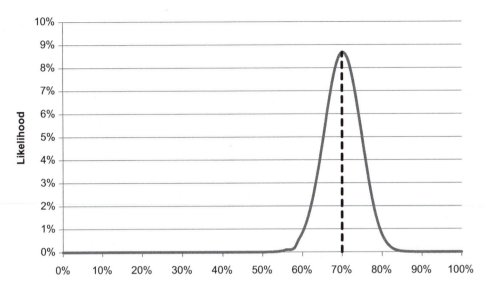

Figure 10.2 Likelihood function for a proportion. The data ($n = 100$) indicated $X = 70$. The likelihood of obtaining the data, given different values of the parameter π, is given by the binomial formula and results are shown in the figure. The maximum of the likelihood function corresponds to the maximum likelihood estimate of 70%, which is also the sample proportion.

The likelihood of obtaining the observed results, given a certain value of $\hat{\theta}$, is given by a *likelihood function*, so the ML estimator is the $\hat{\theta}$ that maximizes this function. It can be shown that the ML estimator of π is the sample proportion P. Suppose that $\pi = 70\%$ and that x (the number of hits) is 65 ($n = 100$). The probability of obtaining 70 hits given $\pi = 70\%$ is given by the binomial formula (Eq. 1.10), and equals 8.7%. P in this case is $\frac{70}{100} = 70\%$, so $l(P = 70\% \mid \pi = 70\%) = .05$. However, one typically does not know the value of π but will need to estimate it. Figure 10.2 shows the likelihood of obtaining the observed data ($P = 70\%$) for different values of π. It can be seen graphically that the likelihood is maximized when $\pi = 70\%$. So the ML estimate of π is just 70%.

More generally, the likelihood function for estimating a population proportion is the binomial formula:

$$l(P \mid \pi) = \binom{n}{x} \pi^x (1 - \pi)^{n-x}. \qquad \text{(Eq. 10.6)}$$

In Appendix 10.1, it is shown using calculus that $l(P \mid \pi)$ is maximized when $\pi = P$. So P is the ML estimator of π.

ML estimates are not in general unbiased but they are *consistent*, meaning that the amount of bias approaches zero as the sample size approaches infinity. Assuming correct model specification to control for unobserved heterogeneity, ML estimates are essentially unbiased in large samples, but there may be some bias in small samples. The bias is of the order n^{-1}, so for $n = 100$, it would be of the order of 1%. See Firth (1993) for one method for correcting the bias in small samples. Allison (1999) recommended that in small samples, higher levels of statistical significance (i.e., smaller p values) should be required to compensate for bias, or that *profile likelihood confidence intervals* should be used (see Technical Note 10.1).

MLE in Logistic Regression

Reflecting the Bernoulli distribution, the probability of obtaining a certain Y value for any given case in logistic regression is

$$\text{Prob}(Y_i = y) = \pi^y(1 - \pi)^{1-y}, \tag{Eq. 10.7}$$

where y is either 1 or 0.

EXPLORING THE CONCEPT

If $y = 1$, then $1 - y = 0$. Therefore, $(1 - \pi)^0 = 1$. According to Eq. 10.7, what would Prob $(Y_i = y)$ equal? What if $y = 0$?

Eq. 10.7 expresses the stochastic component of the logistic GLM for individual points. The link function is given by Eq. 10.1. Substituting Eq. 10.1 into Eq. 10.7 yields

$$\text{Prob}(Y_i = y) = \left(\frac{e^{(\alpha+\beta X)}}{1 + e^{(\alpha+\beta X)}}\right)^y \left(1 - \frac{e^{(\alpha+\beta X)}}{1 + e^{(\alpha+\beta X)}}\right)^{1-y}. \tag{Eq. 10.8}$$

It is shown in Appendix 10.1 that Eq 10.8 simplifies to

$$\text{Prob}(Y_i = y) = \frac{\left(e^{(\alpha+\beta X)}\right)^y}{1 + e^{(\alpha+\beta X)}}. \tag{Eq. 10.9}$$

From Eq. 10.9 one can derive the likelihood equation for logistic regression. The likelihood is the probability of obtaining the observed data given specific values of α and β. Eq. 10.9 gives the likelihood for one data point; the likelihood for the entire dataset is the product of the likelihoods for each point. The symbol $\Pi_{i=1}^n$ is a multiplication operator (analogous to the summation operator, $\Sigma_{i=1}^n$). The likelihood equation is hence:

$$l = \Pi_{i=1}^n \frac{\left(e^{(\alpha+\beta X)}\right)^y}{1 + e^{(\alpha+\beta X)}}. \tag{Eq. 10.10}$$

EXPLORING THE CONCEPT

Why is the likelihood based on a multiplicative product? Because the likelihood reflects a joint probability, what assumption is being made concerning statistical independence?

In ML estimation, the goal is to find the values of α and β that maximize the likelihood. Calculus was employed in the previous example where P was used to estimate π (see Appendix 10.1), but Eq. 10.10 is more complicated, and it would become even more complicated if multiple IVs were used. In general, many of the likelihood functions used in MLE are too complex to differentiate, so instead, *iterative* methods are used to find the maximum of the likelihood function. Iterative methods begin with an initial guess of the parameter values (often derived from OLS), use those to derive predicted values and calculate residuals, and then update the parameter estimates by an incremental amount that reduces the size of the residuals and produces a better fit to the data. The process is repeated until there is no further improvement in the fit. ML estimation algorithms are based either on the *Newton-Raphson* or the *Fisher scoring methods,* or a hybrid of the two. These methods are described in more detail in Chapter 14.

The Fisher scoring method is mathematically equivalent to iterative reweighted least squares, which is a method used to correct for heteroscedasticity when using the general linear model. This fact is one reason heteroscedasticity is no longer a concern here. The MLE approach does require some assumption regarding the shape of the error distribution, but unlike OLS, normality is not necessarily assumed. Therefore, the problems associated with using OLS estimation are avoided with MLE.

Statistical Hypothesis Tests

One beneficial consequence of using MLE is that once the regression parameters are estimated, we can also calculate l (the likelihood of obtaining the data given those parameter estimates). Computer programs simply plug the *ML* parameter estimates into the likelihood equations. The *likelihood ratio* (LR) is defined as

$$\text{LR} = \frac{l(\textit{Reduced Model})}{l(\textit{Fuller Model})}, \tag{Eq. 10.11}$$

where the reduced model is nested in the fuller model. *Nesting* means that the fuller model will have all the same terms as the reduced model plus some additional terms.

EXPLORING THE CONCEPT

If the likelihoods are the same, the fuller model will have no more explanatory power than the reduced model, which implies that the additional term or terms in the larger model are not statistically significant. What will the value of the LR be?

The LR provides one indication of how well one model fits the data relative to another model, but a better fit statistic is the $-2LL$, which stands for the *negative two log likelihood*. The $-2LL$ is a better statistic for two reasons. First, the use of logarithms simplifies mathematical calculations; one can add terms (i.e., log likelihoods) rather than multiplying likelihoods together; the latter can lead to unmanageably small numbers. Second, the sampling distribution of the $-2LL$ is known: Wilks (1938) showed that changes in the $-2LL$ followed a chi-square distribution.

By way of illustration, suppose we fit a model with one IV and $l_1 = .001$. (Maximum likelihoods tend to be small because there are so many possible data configurations.) We then add a second IV and the l increases to $.01$. Intuitively, the second model seems better than the first, but is the difference statistically significant? We need to calculate the $-2LLs$. First, we calculate the natural logs, $L_1 = -6.91$, and $L_2 = -4.61$. (The capital L stands for *log* likelihood.) The logs are negative because likelihoods, being probabilities, are always less than 1.0. To make the values positive, we multiply by -2. (We could also multiply by -1, but then the results of Wilks's theorem would not apply.) In the example, the $-2LLs$ are 13.82 (reduced model) and 9.21 (fuller model), and the difference is 4.61. (Note that a lower $-2LL$ indicates a better fit; Model 2, which contains an additional term and fits the data better, has a lower $-2LL$ than Model 1.) The change in the $-2LL$ is 4.61, which is known as the *likelihood ratio test statistic* and has a chi-square sampling distribution with 1 degree of freedom, because there is only a one-parameter difference between the two models. Therefore, $\chi^2_{LR}(1) = 4.61$, $p = .03$. We can therefore reject the null hypothesis and infer that Model 2 fits the data significantly better than Model 1. In other words, adding the second IV significantly improves the model fit. This approach for testing the significance of variables is known as the *likelihood ratio method*. The likelihood ratio test statistic is

$$G = \chi^2_{LR} = -2[L(M_1) - L(M_2)], \text{ where } M_1 \text{ is nested in } M_2. \quad \text{(Eq. 10.12)}$$

The letter G (or G^2) is typically used to denote the likelihood ratio chi-square. The equation shows that we can also calculate G simply by taking the difference in the log likelihoods and multiplying by -2. In the example, $-6.91 - (-4.61) = -2.30$, and $-2 * -2.30 = 4.60$. This value differs from the previous one only due to rounding error. The formula for G is also sometimes written as

$$G = -2ln\left(\frac{l(Reduced\ Model)}{l(Fuller\ Model)}\right).$$

(Eq. 10.13)

EXPLORING THE CONCEPT

Notice that the model with the higher likelihood has a lower $-2LL$. Why is this the case? What is the effect of multiplying by a negative number?

A likelihood ratio $\chi^2(1)$, or G statistic, can be calculated for each predictor in a model by comparing the $-2LL$ of the model with the one omitting that term.

A second method for testing statistical significance is the *Wald method*. This method assumes that all relevant variables have been specified and so L cannot be increased further (i.e., one is at the maximum of the log likelihood function). Wald showed that under this situation, the statistic $z_k = \beta_k / SE_{\beta_k}$ was normally distributed under the null hypothesis that $\beta_k = 0$. Wald's z is analogous to the t statistic in OLS, except that the Wald test uses the z distribution instead. (The square of the z value is Wald's χ^2.) Technically, the Wald method should only be used to answer research questions pertaining to the predictive validity of IVs and not for determining which variables should be included or dropped from the model (i.e., model building); however, the two different methods often give similar if not identical p values, so this prohibition is often ignored. However, the Wald method is less statistically powerful than the LR methods and can yield biased estimates when the data are sparse for some values of an IV (Cohen, Cohen, West, & Aiken, 2003). A third method of testing significance (*Fisher scoring*) will be described in Chapter 14.

The Trimone Example

In IBM SPSS, there is an option under the regression menu for performing logistic regression, but it can also be performed using the GLM procedure. I focus on the latter here because this SPSS procedure is more recent.

To conduct the logistic regression using the clinical trial data in Table 10.1, I inputted the five different daily drug doses as a scale (i.e., metric) variable. The DV "Improvement" (Y_i) was either 1 or 0, weighted by the frequency counts shown in Table 10.1. Under the GLM procedure, I selected "Binary Logistic" (under the "Type of Model" tab). On the "Response" tab, I designated "Improvement" as the DV and clicked on "Reference Category" to designate "First (lowest category)" as the reference category (in my opinion, this choice makes interpretation easier). On the "Predictors" tab, I designated "Dose" as the covariate as well as on the "Model" tab. Under the

"Statistics" tab, I chose "Likelihood ratio" for the chi-square statistic and also checked "Include exponential parameter estimates." Finally, I clicked "OK" to run the model.

The SPSS output, slightly condensed, is shown in Figure 10.3. The model information box shows the probability distribution was binomial and the link function was the Logit. The Goodness of Fit box displays various statistics (described in the next section) measuring how well the specified model fits the data. Note that the

Generalized Linear Models

Model Information

Dependent Variable	Improve[a]
Probability Distribution	Binomial
Link Function	Logit

a. The procedure models 1.00 as the response, treating .00 as the reference category.

Goodness of Fit[a]

	Value	df	Value/df
Deviance	3.186	3	1.062
Scaled Deviance	3.186	3	
Pearson Chi-Square	3.104	3	1.035
Scaled Pearson Chi-Square	3.104	3	
Log Likelihood[b]	−16.319		
Akaike's Information Criterion (AIC)	36.638		
Finite Sample Corrected AIC (AICC)	36.640		
Bayesian Information Criterion (BIC)	49.672		
Consistent AIC (CAIC)	51.672		

Dependent Variable: Improve
Model: (Intercept), Dosage[a]
a. Information criteria are in small-is-better form.
b. The full log likelihood function is displayed and used in computing information criteria.

Omnibus Test[a]

Likelihood Ratio Chi-Square	df	Sig.
3538.863	1	.000

Dependent Variable: Improve
Model: (Intercept), Dosage[a]
a. Compares the fitted model against the intercept-only model.

Tests of Model Effects

Source	Type III		
	Likelihood Ratio Chi-Square	df	Sig.
(Intercept)	3345.488	1	.000
Dosage	3538.863[a]	1	.000

Dependent Variable: Improve
Model: (Intercept), Dosage
a. The validity of the likelihood ratio chi-square is uncertain because log-likelihood convergence was not achieved for the constrained model. Results shown are based on the last iteration.

Figure 10.3 SPSS logistic regression output for Trimone antidepressant data. The DV is improvement in patient symptoms.

Parameter Estimates

Parameter	B	Std. Error	95% Profile Likelihood Confidence Interval		Hypothesis Test	
			Lower	Upper	Wald Chi-Square	df
(Intercept)	−4.874	.1369	−5.148	−4.612	1268.546	1
Dosage	.197	.0053	.187	.208	1382.092	1
(Scale)	1[a]					

Parameter Estimates

Parameter	Hypothesis Test	Exp(B)	95% Profile Likelihood Confidence Interval for Exp(B)	
	Sig.		Lower	Upper
(Intercept)	.000	.008	.006	.010
Dosage	.000	1.218	1.205	1.231
(Scale)				

Dependent Variable: Improve
Model: (Intercept), Dosage
a. Fixed at the displayed value.

Figure 10.3 Continued.

log likelihood is −16.319 so the −2LL = 32.638. The omnibus test is significant, indicating that model fits the data significantly better than the intercept-only model. Under the tests of model effect, the dosage variable is highly significant (under the LR test), but there is a warning message that the estimates failed to reach convergence, so in this case it is better to test significance of individual predictors with the Wald test. In the parameter estimates box, note that B for dosage is .197 and the standard error is .0053. Wald's z would therefore be $\frac{0.197}{0.0053} = 37.17$; squaring produces a Wald $\chi^2 = 1381.6$, which is very close to the value shown on the output (the difference is due to rounding error). The associated p value is < .001. Note that *exp*(B) for dosage is 1.218, implying the odds for improvement increase by 21.8% for every 1-milligram increase in the Trimone dosage.

Grouped Versus Ungrouped Data

The Trimone example used grouped data, as there were multiple observations for each value of X_i. It is often the case when using metric predictors, however, that there will be only one observation (if any) for different values of X. Such data are *ungrouped*. The distinction between grouped and ungrouped data is important because some statistics and procedures suitable for grouped data are not appropriate for ungrouped ones.

It is important at this point to introduce the notion of a *covariance pattern*, which refers to a distinct set of predictor values (Hosmer & Lemeshow, 2013). Let J denote the number of covariate patterns. If there is only one predictor with 50 different values of X_i, then $J = 50$. ($J = n$ in this case; $J < n$ if two or more cases have the same X_i.) With two predictors, $J < n$ if there are two or more cases with the same IV values on both predictors—for example, $X_1 = 1$ and $X_2 = 2$ for two or more observations. $X_1 = 1$ and $X_2 = 2$ is one covariate pattern.

Data where there are a small number of covariate patterns and a large number of cases can be considered grouped. Categorical predictors involve grouped data, but so can metric predictors if there are a small number of IV levels (as in the Trimone example).

Fit Statistics

The Deviance and Pearson's X_2

As noted previously, the likelihood ratio method for testing significance is based on the $-2LL$. A related statistic, known as the *deviance* (*D*), involves a comparison of a model's $-2LL$ with that of a "saturated" model. (In a saturated model, there is a term for each observation or, when categorical predictions are used, for each cell.) The deviance is defined as

$$D = -2\ln\left[\frac{l(Current\ Model)}{l(Saturated\ Model)}\right].$$
(Eq. 10.14)

If the current model fits the data as well as the saturated model, the LR will be 1.0. The deviance will therefore be 0 [because $ln(1) = 0$].

Unlike when one is testing individual terms for significance, it is desirable for the deviance to be nonsignificant, because this indicates that the current model does not deviate much from one with a maximum fit. In contrast, the *omnibus test* contrasts the current model with the *null model* which contains only the intercept term. Typically, one hopes the omnibus test is significant, implying that the predictors, taken together, have some predictive power. Note that a significant result could be caused by just one of the predictors or by a combination of several. The omnibus test is useful when there is multicollinearity that makes it difficult to establish predictors as individually significant.

Returning to the deviance, when the data are grouped, the log likelihood for the saturated model is given by the following formula (Collett, 2003):

$$L_{saturated} = \Sigma_{i=1}^n \left\{ \ln\binom{n_i}{Y_i} + Y_i\ln(\tilde{p}_i) + (n_i - Y_i)\ln(1 - \tilde{p}_i) \right\},$$
(Eq. 10.15)

where Y_i is the number of hits for each of the i levels of X, n_i the number of cases, and $\tilde{p}_i$ is the observed proportion of hits ($\tilde{p}_i = Y_i/n_i$). Eq. 10.15 was derived directly from Eq. 10.6 by taking logarithms and summing. It reflects a binomial distribution of points for each level of X (or when there are multiple predictors, for each covariate pattern).

For ungrouped data, the deviance test is not valid. An asymptotic approximation for each covariate pattern—based on the central limit theorem—is not reasonable, as each covariate pattern may have only one or two observations (Hosmer & Lemeshow, 2013). Also, "the number of parameters in the saturated model increases with sample size" (Allison, 1999, p. 52), which violates a condition of asymptotic theory.

EXPLORING THE CONCEPT

For the Trimone data, one can obtain $L_{saturated}$ in SPSS by making "Dose" a factor rather than a covariate; doing so will create a dummy variable for each group except the reference category. Through this procedure, we find that $L_{saturated}$ is -14.726. What is the $-2LL$ for the saturated model, and by how much does it differ from the $-2LL$ for the current model (32.638)? This value is the deviance.

The deviance is also the sum of the squared deviance residuals, which are basically the raw residuals placed on a -2 log scale. See Technical Note 10.2 for the specific formula.

A statistic that is similar to the deviance is Pearson's X^2. The statistic is based on the difference between the observed and expected values of each subgroup (see Chapter 4). It is defined for cell ij as

$$X^2 = \Sigma_{i=1}^n \frac{\left(Y_i - n_i\hat{\pi}_i\right)^2}{n_i\hat{\pi}_i\left(1-\hat{\pi}_i\right)},$$ (Eq. 10.16)

where i is the index for a subgroup and Y_i the number of hits in the subgroup. The value of X^2 is also the sum of the squared Pearson residuals. The formula for a Pearson residual is $r_i = \frac{Y_i - n_i\hat{\pi}_i}{\sqrt{n_i\hat{\pi}_i(1-\hat{\pi}_i)}}$. The Pearson residual is the raw residual divided by the conditional standard deviation. The division corrects for the natural heteroscedasticity of these variances. According to Cohen et al. (2003), it is still generally preferable to use deviance residuals because (a) the latter will tend to be more normally distributed, and (b) Pearson residuals are unstable at values of $\hat{\pi}_i$ close to 0 or 1. Nevertheless, Pearson's X^2 is sometimes a preferable fit measure over the deviance, as will be noted below.

When using grouped data, one can test the significance of the deviance or Pearson's X^2 using a chi-square test with $[J - (k + 1)]$ df, where k is the number of predictors, excluding the intercept. J refers to the number of covariate patterns. If the test is nonsignificant, one can conclude that the predicted values do not differ significantly from the actual ones. So for example, if there were two binary predictors and therefore four covariate patterns, then $df = [4 - (2 + 1)] = 1$. Suppose $D = 1.5$, then $\chi^2(1) = 1.5$, $p = .22$, so the model would adequately fit the data. For the

Trimone data, $D = 3.186$. There are 3 degrees of freedom because $J = 5$, $k = 1$, and $J - (k + 1) = 3$. The test result is therefore $\chi^2(3) = 3.186$, $p = .36$, indicating an adequate fit. Likewise, Pearson's $X^2 = 3.104$, $p = .38$.

These tests assume, however, that none of the cells are sparse (i.e., having very few if any observations) so that an asymptotic approximation will be reasonable. One rule of thumb that is sometimes used is that—for a test of X^2—no more than 20% of the cells should have expected frequencies less than 5 (Powers & Xie, 2000). Tests of the deviance (D) may be biased in small samples (where n divided by the number of cells is less than 5; Agresti, 2002; Koehler, 1986). According to Koehler and Larntz (1980), the chi-square approximation for the deviance "gives conservative critical levels when most expected frequencies are smaller than 0.5 and liberal levels when most expected frequencies are between 1 and 5" (p. 341). For Pearson's X^2 goodness of fit, a chi-square approximation is liberal if most expected frequencies are less than 1. These guidelines can be used in deciding whether to report the test of X^2 versus the deviance, or nothing at all, as it is not mandatory to conduct an overall goodness-of-fit test.

Hosmer-Lemeshow Test

What about metric predictors? If the data are grouped, then the deviance may be used as an overall measure of fit, but not if the data are ungrouped. Hosmer and Lemeshow developed a fit test for ungrouped data where the observations are divided into about 10 equal groups along the X axis or based on the predicted values ($\pi_i's$) if there are multiple predictors. The idea is to create "cells," each with enough observations that a chi-square test can be performed. The specific formula is

$$\hat{C} = \Sigma_{k=1}^{g} \frac{(O_k - n_k' \bar{\pi}_k)^2}{n_k' \bar{\pi}_k (1 - \bar{\pi}_k)}, \tag{Eq. 10.17}$$

where O_g is the number of observed hits for the kth decile group, $n_k' \bar{\pi}_k$ is the expected number of hits, $\bar{\pi}_k$ is the average estimated probability, and n_k' is the total number of subjects in the kth group. $\hat{C}$ asymptotically has a chi-square distribution with $n - 2$ degrees of freedom. The number of groups can be adjusted to avoid having any with too few observations.

The GLM procedure in SPSS does not report $\hat{C}$, but it can be calculated with the binary logistic regression procedure.

O'Connell (2006) reported that the Hosmer-Lemeshow (H-L) test is fairly robust and therefore recommends its use in small samples or when there are sparse cells (see Technical Note 10.3). (One can also resolve the problem of sparseness by combining categories, but this will reduce statistical power and, if one is not careful, theoretical interpretability.) In general, the H-L test is appropriate when the number of covariate patterns is close to n. The test is conservative, so except in very large samples ($n \geq 400$), it may not detect a lack of fit to some of the data that other diagnostics might (Hosmer & Lemeshow, 2013).

On the other hand, if the test is significant, but only because of small departures between the observed and expected values, Hosmer and Lemeshow (2013) suggested partitioning the data at random into two subsets, using one subset to develop the model and the other to cross-validate it. Because the subsets are half the size of the total data set, the problem of the H-L test having "too much power" is reduced.

Even with this refinement, use of the H-L test is controversial. Allison (2013) noted that the results of the test can be sensitive to the number of groups used, and that the p value of the test can increase when significant predictors are added to the model (and likewise decrease when nonsignificant predictors are added). Hosmer and Lemeshow (2013) argued that a nonsignificant p value of the test should not be used to decide that one model is better than another (i.e., it would be inappropriate to adopt Model A over Model B because the first had a p value of .5 and the second .7) but, in my opinion, the usefulness of the test is also questionable if both p values are somewhat close to .05 (e.g., .04 vs. .09) and give conflicting decisions regarding significance.

Pseudo-R²

In OLS regression, R^2 is a measure of the percentage of variance explained. The formula for R^2 is not valid in the context of logistic regression (see Technical Note 10.4), but some quasiequivalent measures have been devised known as pseudo-R^2.

The Cox and Snell pseudo-$R^2 = 1 - \left[\dfrac{L(M_{null})}{L(M_{full})}\right]^{2/n}$. The statistic compares the log likelihoods of the null model (containing only the intercept) and the full model. It

reflects the proportional reduction in the deviance from adding the predictors. A limitation is that if the DV is nominal, Cox and Snell's pseudo-R^2 has a theoretical maximum less than 100%. Nagelkerke's R^2 therefore adjusts the Cox and Snell R^2 by dividing it by the theoretical maximum, $1 - [L(M_{null})]^{2/n}$. Nagelkerke's R^2 can equal 100% where there is a perfect fit, so its use is preferred over Cox and Snell's.

Information Criteria

The SPSS output for the Trimone data also reports Akaike's information criterion (AIC = 36.368) and the Bayesian information criterion (BIC = 49.672). These criteria are not meant as overall measures of fit but are used to compare different statistical models (to determine which model fits the data best). A smaller information criterion indicates a better fit (although some software packages use a higher-is-better form). The formula for the AIC is

$$\text{AIC} = 2 * p + (-2\text{LL}),$$
(Eq. 10.18)

where p is the number of parameters (Akaike, 1973). Because one can improve fit just by adding more parameters to a model, the AIC contains a penalty for adding parameters. The formula for the BIC (Raftery, 1995) is

$$\text{BIC} = p * ln(n) + (-2\text{LL}). \tag{Eq. 10.19}$$

The BIC has a stronger penalty for adding parameters; the penalty is also a function of sample size. (In larger samples, one might detect more small and erroneous effects.) The BIC can actually be negative, with negative values implying better fit than positive ones.

Although one can also use the -2LLs for comparing models (and perform a likelihood ratio test to determine if differences are significant), LR tests are only useful in comparing nested models. The AIC and BIC statistics, on the other hand, can be used for comparing nonnested as well as nested models. Use of these statistics can (a) quantify the strength of evidence for a null hypothesis (Raftery, 1995) and (b) result in more parsimonious models. Parsimony counteracts the tendency to *overfit* models, that is, to add parameter estimates that improve fit but which are really irrelevant predictors associated with Type I errors. Models that overfit the data often do not predict well when applied to new data sets.

The other information criterion reported by SPSS (AICC and CAIC) contains even stronger penalties and therefore result in even more parsimony (see Technical Note 10.5). While I would recommend the use of these statistics over the AIC and BIC, this book focuses on the latter because the AIC and BIC are so widely used.

Unlike the LR test, there are no statistical tests to determine if a change in the AIC or BIC is significant. For the BIC, however, Raftery (1995) suggested the following guidelines: (a) a BIC difference between models of 0 and 2 constitutes weak evidence for the better-fitting model (the probability of the model, given the data, is between 50% and 75%); (b) a difference of 2 to 6 constitutes positive evidence (reflecting a probability of from 75%–95%); (c) a difference of 6 to 10 constitutes strong evidence (probability of 95%–99%); and (d) a difference >10 constitutes very strong evidence (probability > 99%).

Percentage of Cases Classified Correctly

The percentage of cases classified correctly is conceptually most similar to the percentage of variance explained (rather than, for example, percentage of deviance reduced). To calculate it, one forms a contingency table with the DV binary category as one variable and the cases correctly versus incorrectly classified as the other variable. One can then use some measure of association, such as ϕ or $\hat{\tau}$ (see Chapter 6) to determine if there is a significant difference. (There should be if the model fits adequately.) Table 10.2 provides an example. The number of cases classified correctly (shown on one of the diagonals of the table) is 63; the percentage correctly

Table 10.2 Percent of Cases Correctly Classified ($n = 100$)

Observed	Predicted		
	1	0	Total
1	39	21	60
0	16	24	40
Total	55	45	100

Note: Percent correctly classified: $\frac{(39+24)}{100} = 63\%$. $\tau_b = .25, p = .012$.

classified is 63%. The percentage is statistically different from 50% according to a test of Kendall's $\hat{\tau}_b$. An advantage of this fit statistic is that it does not require other assumptions to be met (like having a large sample, no sparse cells, etc.). However, a weakness is that some information is thrown away by dichotomizing the predicted probabilities into just two groups (< 50% and ≥ 50%). Furthermore, a model may fit the data quite well but predict poorly if many predicted values are close to 50% (e.g., predicted probabilities of 49% and 51% may fit the data well but result in opposite classification predictions). Many authors (Hosmer & Lemeshow, 2013; King, 2008) therefore recommend only focusing on correct classification statistics when the primary goal is to make classification predictions. (When the primary goal is theory development, the procedure might provide some useful adjunct information but should not be of primary concern.)

Even when the goal is classification, correct classification statistics are not that useful when dealing with rare events. For example, if only 20% of the cases are hits, but the model predicts all cases as misses, 80% will be correctly classified. That high rate of correct classification is potentially misleading as to the actual predictive power of the model to identify rare events.

Model Building and Refinement

Model Building Steps

Building a logistic regression model involves a number of steps. The following list of steps is based primarily on the recommendations of Hosmer and Lemeshow (2013), with some modifications. Bear in mind that model building is an iterative process and so some or all steps may need to be repeated until the researcher is satisfied that all underlying assumptions are met. The researcher also needs to be flexible in regard to the order that these steps are implemented.

1. Verify adequacy of sample size. According to Hosmer and Lemeshow (2013), there should be a minimum of 10 cases per IV; however, 20 to 1 is preferred (50 to 1 for stepwise regression). Other authors (e.g., Long, 1997) recommend

a minimum of 100 cases for simple analyses to avoid MLE small-sample bias. Furthermore, sufficient power may require substantially larger samples, so a power analysis should be performed before data are collected (e.g., using G∗Power software, or formulas in Hosmer & Lemeshow, 2013, pp. 401−408). Exact methods for conducting logistic regressions in small samples (such as *conditional logistic regression*) are available in SAS and some other statistical software. If there are fewer hits than misses on the DV, simulation studies by Vittinghof and McCulloch (2006) suggested that there should be at least 5 to 9 hits per model parameter (intercept plus number of predictors) for accurate estimation of the standard errors. If there are more hits than misses, the rule applies to the misses.

2. Specify a set of predictors. It is best to use theory to choose relevant predictors. Statistical/stepwise regression can also be used, but can lead to the omission of relevant variables and the inclusion of irrelevant variables reflecting Type I errors. The technique is most appropriate when prediction and not theory development is the primary goal. Stepwise regression is not an option in the SPSS GLM procedure but is available using the logistic regression procedure under the regression menu (see Technical Note 10.6 for further details). Bursac, Gauus, Williams, and Hosmer (2008) presented an alternative to stepwise regression, *purposeful selection*, which performed better in simulation studies than stepwise regression (see Technical Note 10.7). Finally, Mood (2010) noted that it is important to include in the final model all major covariates (even if uncorrelated with one another) to minimize bias associated with unobserved heterogeneity.

3. Determine the extent of multicollinearity among predictors. Multicollinearity, which is correlation among predictors, does not violate any model assumptions, but it does reduce statistical power. If theory development is the goal, relevant predictors should not be dropped as this can cause statistical confounds; the remedy is to collect more data. Short of that, the presence of severe multicollinearity should be reported so the effect on statistical power can be taken into account when interpreting the results. Procedures to detect multicollinearity are not available in SPSS with the logistic regression options, but one can use ordinary linear regression to detect excessive correlation among the IVs. In SPSS, go to "Regression" ➔ "Linear," click on "statistics" and then "collinearity statistics." Small eigenvalues (close to zero), and two (or more) variance components on the same dimension greater than .5 are indicative of potential problems (Kleinbaum & Klein, 2010). A condition index of > 15 suggests a potential problem, and one > 30 a serious problem.

4. Examine the univariate distributions of the predictors and check for any extreme cases (e.g., |z| scores greater than 3.29 are extreme if the variable is normally distributed). The DV need not be checked because it can only assume values of 0 or 1. A skewed predictor is not inherently problematic but might contribute to other problems, such as sparse cells.

5. For categorical predictors, construct contingency tables with the DV and identify any cells that are sparse (e.g., using the chi-square criteria outlined in the previous section). Sparse or empty cells can be eliminated by combining some categories, but do so only if theoretically meaningful.

6. Conduct an initial logistic regression. Examine measures of overall fit such as the deviance or Pearson's X^2 (if the cell sizes are adequate). Also examine the beta estimates and standard errors. If they are very large, this is indicative of one or more problems with the data (multicollinearity, zero cell counts, or quasicomplete separation, which is discussed later in the chapter).

7. Check the residuals for deviant and influential cases. Outliers should not be deleted at this point, as other revisions to the model may improve the fit of deviant observations, but checking will provide information as to where the model may not be adequately fitting. Even if the summary statistics indicate an overall good fit, parts of the model may not fit well. How to check for outliers is discussed later in the chapter.

8. Consider whether any nonsignificant predictors should be dropped from the model. Dropping these simplifies the model and analysis. However, the researcher may wish to retain predictors that bear directly on a specific research question. Also, a nonsignificant variable could potentially become significant later in the analysis if additional changes are made to the model. Therefore, if the variable is eliminated at this point, the researcher may wish to try adding it back into the model later on.

9. Check for linearity in the logits. As implied by Eqs. 10.2 and 10.4, logistic regression assumes a linear relationship between the logits (i.e., log odds) and any metric predictors. One can perform a Box-Tidwell test by including the term $X * ln(X)$ in the model where X is a metric predictor; a significant term indicates nonlinearity. This test may be overly powerful in large samples and may lack power in small samples. King (2008) therefore recommended grouping the predictor into categories (four or more) and graphing the log odds of the DV mean for each category against the IV. (See Technical Note 10.8 for a similar approach suggested by Hosmer & Lemeshow, 2013.) Squaring or using the logs of X may help satisfy the linearity assumption (see Chapter 9). However, the modifications should make some theoretical sense—otherwise, one may be capitalizing on chance and overfitting the data.

10. Consider the inclusion of interaction terms. Interaction terms involve the product of two or more other terms included in the model, often a nominal and metric variable. Because there are often a large number of potential interaction terms, stepwise regression can be useful in finding those that are statistically significant. However, only include interaction terms that make some theoretical sense. Remember that differences in beta estimates for different levels of the nominal predictor can also reflect differences in unobserved heterogeneity rather than real differences.

11. Consider alternative link functions. Some alternatives to the logit link are shown in Table 10.3. The complementary log-log or negative log-log link should be fitted if the overall proportion of cases where $Y = 1$ is very different from the proportion where $Y = 0$, for example, 80% versus 20%. So what exactly is the complementary log-log (cloglog) function? Mathematically, it is $\eta = ln[-ln(1 - \pi)]$, where η is the predicted value of the function. So for example, if $\pi = .30$, then $\eta = ln[-ln(1 - \pi)] = -1$. The predicted ηs can then be used to predict the πs. As shown in Figure 10.4, the logit curve is symmetric around .50, whereas the cloglog function is not. Furthermore, the cloglog curve approaches a 100% probability more rapidly than the logistic. The cloglog link is often useful when there are many more hits than misses on the DV (see Technical Note 10.9 for more details on the cloglog link). The negative log-log link is the flip side of cloglog one and should be fitted if there are many more misses than hits on the DV, especially for low values of X. See Technical Note 10.10 for explanation of the other link functions shown in Table 10.3.

Table 10.3 Alternative to the Logit Link Function

Link function name	Equation	Situation used for
Complementary log-log	$ln[-ln(1 - \gamma)]$	Higher categories more probable.
Negative log-log	$-ln[-ln(\gamma)]$	Lower categories more probable.
Probit	$\Phi^{-1}(\gamma)$	Normally distributed latent variable.
Cauchit	$tan[\pi(\gamma - 0.5)]$	Latent variable with many extreme values.

Note: The symbol γ reflects probabilities or cumulative probabilities; reflects the cumulative normal probability density function.

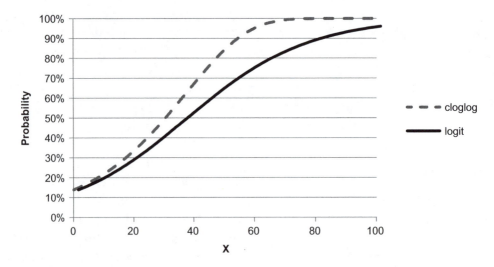

Figure 10.4 Complementary log-log (cloglog) versus logit link.

In choosing a link function, Hosmer and Lemeshow (2013) suggested using a Stukel test (1988).

a) Perform a standard logistic regression. Save the estimated logits (call them $\hat{g}_i$), which in the IBM SPSS GLM procedure are the "predicted values of the linear predictor" (XB). Also save the predicted probabilities ($\hat{\pi}_i$), which in IBM SPSS are called "Predicted value of mean of response."

b) Copy and paste the logits into EXCEL, and compute the following two new variables: $z_{1i} = 0.5 * g_i^2 * I(\hat{\pi}_i \geq 0.5)$ and $z_{2i} = -0.5 * g_i^2 * I(\hat{\pi}_i < 0.5)$, where $I(arg) = 1$ if arg is true and 0 if false. As an example, the EXCEL code for z_{1i} might look like "=0.5 * C3 * IF(D3 >= 0.5, 1, 0)" where C3 and D3 are the cell addresses for the estimated logits and probabilities, respectively.

c) Add z_{1i} and/or z_{2i} to the model and test their significance with likelihood ratio tests. (A score test—described in Chapter 14—should be used instead if available in your statistical software.) Do not include a variable if the values are all zero.

d) If the new covariates are significant, use their estimated beta coefficients (and CIs) as a guide to choosing an alternative link function. If both coefficients are about 0.165, the lighter-tailed probit link might be considered (see Technical Note 10.10). A cloglog link is supported by coefficients about 0.620 and -0.037 for z_{1i} and z_{2i}, respectively, and a negative log-log link by coefficients about -0.037 and 0.620. If the CIs contain these values, Hosmer and Lemeshow (2013) argue that the Stukel test provides one piece of evidence for using an alternative-link function, but theoretical plausibility and ease of interpretation should also enter into the final decision of what link function to use. Pregibon (1980) also proposed a goodness-of-link (GOL) test that can provide additional evidence that a particular link function is or is not suitable (see Appendix 10.2).

12. Rescreen for influential cases and outliers. If outliers are deleted, the previous model-building steps should be repeated to determine if prior decisions are still valid.

13. Check for over- or underdispersion. Overdispersion occurs in grouped data when there is more variability in the data than there should be. This issue will be discussed later in the chapter.

An Extended Example: Model Specification, Fit, and Diagnostics

The file *Proficiency Exam.10* contains hypothetical data on whether 168 students passed a mathematics proficiency exam (DV) and on two predictors: student socioeconomic status (as measured on a 100-point scale) and gender (1 = *female*, 0 = *male*).

Univariate Analysis

The first step listed in the previous section is to verify that the sample size is adequate. The sample size is sufficient, as $n = 168$. Second, one needs to select predictors if there are a large number to choose from in a database; here we will just use gender and socioeconomic status (SES). Third, assess multicollinearity. Because one of the predictors is nominal, I calculated Kendall's $\hat{\tau}_b$ between the two predictors; the value was extremely small (.038, $p = .56$) so multicollinearity was not indicated. For completeness, however, I generated collinearity statistics from an OLS regression; the highest condition index (4.95) was well below 15. Fourth, using SPSS "Descriptives," the univariate statistics for SES showed its distribution was somewhat flat (*kurt* $= -1.05$) but not skewed (*skew* $= -0.122$). The standardized values ranged from -1.98 to 1.79, so none were univariate outliers (greater than 3.29 in absolute value), assuming a normal distribution. (The distribution was not quite normal, but a histogram, created using SPSS "Frequencies," did not visually show any univariate outliers, only a mode in the 80s that skewed the distribution.) Fifth, a contingency table with the nominal predictor (gender) as one variable and "passing" as the other did not indicate any sparse cell. The smallest cell frequency was 23 and expected frequency (calculated with Eq. 4.3) was 30.5.

Initial Logistic Regression

Having conducted a preliminary univariate analysis, we are now ready to run an initial logistic regression using the GLM procedure in IBM SPSS. In addition to the steps described previously for the Trimone data, on the "Predictors" tab, designate "Female" as the factor and "SES" as the covariate. Under the "Factor" box, click on the "Options" button and choose "Descending" for "Category Order for Factor." Under the "Model" tab, designate "Female" and "SES" as main effects. Under the "Statistics" tab, choose "Likelihood ratio" for the chi-square statistic and also check "Include exponential parameter estimates." Under the "EM Means" tab, check "Female," select the arrow to move the variable to the "Display Means for" box, and choose "Pairwise" from the "Contrast" menu. Then select "OK" to run the model.

EXPLORING THE CONCEPT

Download the proficiency exam data from the book's website (in file *Proficiency Exam.10*) and run the GLM procedure as outlined above. Are your results the same as reported below?

Fit of Initial Model

From the output (not shown), it can be seen in the goodness of fit box that the model deviance is 154.41, but the data are ungrouped, so the deviance should not be interpreted as a measure of overall fit. Instead, we should use the H-L statistic calculated with the binary linear regression procedure: $\hat{C} = \chi^2_{HL}(8) = 12.93, p = .11$. The results indicate that the predicted values do not differ significantly from the actual ones, according to this test. The model "passes" the H-L test, but just barely (and it is a conservative test)—so there is reason to check other diagnostic indicators carefully.

Returning to the GLM output, the AIC is 187.348 and the BIC is 196.720. These are not summary measures of overall fit but will be used to compare models. Note also that the log likelihood is −90.674, therefore the −2LL = 181.35. Recall from Eq. 10.18 that AIC = 2 ∗ p + (−2LL), and so in this case, because there are three parameters (for intercept, females, and SES), (2 ∗ 3) + 181.35 = 187.35.

EXPLORING THE CONCEPT

(a) If there were only two parameters instead of three, would the AIC be lower? (In SPSS, a smaller AIC indicates a better fit.) Do you see how the formula contains a penalty for adding parameters to the model? (b) Use the formula in Eq. 10.19 to calculate the BIC.

Displayed next is the result of the omnibus test, which compares the fitted model with an intercept-only model ($LR\ \chi^2_{omnibus} = 6.776$). Because the fitted model has three parameters and the intercept-only model has one, the difference has 2 degrees of freedom, and the p value is therefore .034, a significant result.

Test of Model Effects

The test of model effects is examined next. There is a likelihood ratio chi-square for each variable, which reflects an LR test for the fitted model against a reduced model with that variable eliminated. The output shows that SES is not significant. The output also displays the parameter estimates and the results of the Wald test: For "females," Wald $\chi^2(1) = 5.54$, $p = .019$. Note that the LR test result is $p = .017$. The result is very close to the Wald p value, but the LR test is more valid because we have not yet settled on a final fitted model. Note also that, as a general rule, the LR test is slightly more powerful. If one omits SES from the model, one finds that the results of Wald and LR tests converge at $p = .017$.

Interpreting the Beta Coefficient

The next step is to interpret the beta estimate for females (0.776). $Exp(0.776) = 2.17$, which means that the predicted odds of females passing the proficiency test are 217% those of males (or that the estimated odds increase by 117% if the value of the predictor changes from males to females).

In interpreting the sign of coefficients in logistic regression and other categorical analysis procedures, it is easy to make mistakes regarding directionality, so it is a good practice to generate a measure of association between a nominal IV and DV just to verify one's conclusion. In the present example, $\phi = .186$, which confirms the positive relationship between being female and being more likely to pass the proficiency test.

Although the coefficient for SES is not significant, I will nonetheless interpret it to show how coefficients for metric predictors should be explained. The coefficient is 0.006, $exp(0.006) = 1.01$. In the sample, the odds of passing increase by 1% from each unit increase in SES. Anderson and Rutkowski (2008) noted that it is sometimes more informative to report an odds ratio for a one-standard-deviation increase in the IV. The standard deviation for SES was 26.58. We need to calculate $(exp(\hat{\beta}\sigma) = exp(0.006 * 26.58) = 1.17$. Therefore, a one standard deviation increase in SES will result in a 17% increase in the odds.

Another alternative to using the standard deviation is to rescale the IV into larger units that are more meaningful (e.g., thousands of dollars rather than single dollars when measuring income). An increase in thousands is likely more substantively interesting than an increase of a dollar.

Contrasts

The last part of the output shows that the estimated marginal mean for females is .73 and for males is .55. These are probabilities and can be converted into odds and an odds ratio as follows: $\frac{.73/.27}{.55/.45} = \frac{2.70}{1.22} = 2.21$, which is very close to the previous estimate for gender. There are about even odds of males passing the exam (i.e., 1.22), but for females the odds are almost 3:1 (i.e., 2.70).

The difference between the mean predicted probabilities of .73 and .55 is .18; the contrast is significant (Wald $\chi^2(1) = 5.83$, $p = .016$). Recall that in SPSS we had set the contrast menu to "pairwise." There is only one pair in this example, but if a nominal IV with three categories had been used, there would be three pairwise comparisons. The other types of contrast available for nominal variables are as follows: (a) Simple: Compares the mean of each level with the mean of a reference group (e.g., control group); (b) Deviation: Compares each mean with the grand mean (.64 in the example); (c) Difference: Compares the mean at each level with the mean for all previous levels; (d) Helmert: Compares the mean at each level with the mean of all subsequent levels; (e) Repeated: Compares the mean at each level with the mean of the subsequent level; and (f) Polynomial: Tests for linear and quadratic trends in the means (which is particularly useful for DVs that are ordered categories). The type of contrast to use will depend on the research questions and the type of data involved.

If making more than one contrast, an adjustment for multiple comparisons can be made. The options in IBM SPSS (see the lower left part of the "EM Means" tab) are Least Significant Difference (unadjusted), Bonferroni, Sequential Bonferroni, Sidak, and Sequential Sidak. See Technical Note 10.11 for an explanation of these adjustments.

Screening Residuals for Outliers and Influential Cases

In SPSS, residuals can be generated through the "save" command (or tab). We have already discussed the deviance residuals; using these is preferable to the raw residuals because their distribution will be closer to normal. The Pearson residuals can also be examined as these tend to be larger than deviance residuals, sometimes making it easier to detect problematic cases. *Likelihood residuals* are deleted residuals (see Chapter 9), which are calculated by omitting the corresponding case in the calculation of the beta estimates and then calculating the deviance; doing so makes it easier to identify influential cases as outliers. The square of a likelihood residual gives the effect on the deviance from omitting that particular observation. (The difference between the squares of the likelihood residuals from two different models is useful in identifying cases that most affect the difference in deviance; see Collett, 2003.) Likelihood residuals turn out to be weighted averages of the Pearson and deviance residuals, with greater weight given to the deviance residuals.

According to Coxe, West, and Aiken (2009), with categorical data analysis (CDA) there is less consensus than with OLS regression about agreed upon cutpoints for eliminating cases as outliers. Based on assuming normality, some authors recommend using 2.5 or 3 in absolute value as a cutpoint (see Technical Note 10.12 for a formal test), whereas others (e.g., Hosmer & Lemeshow, 2013) argue that one cannot assume that the sampling distributions of the residuals are normal unless the data are grouped. Coxe et al. (2009) recommend focusing on one or at most several cases with the highest values on a diagnostic measure, such as the standardized or studentized deviance residuals or the DfBetas for the most theoretically important variables. (A DfBeta is a measure of how much a beta estimate would change if the observation were deleted; see Technical Note 10.13.) If eliminating a case (or collectively a set of cases) makes a large difference to the fit statistics, then it should be investigated further. Only if there is a reason to believe that the case is qualitatively different from the other cases should it then be eliminated from the analysis. This standard is somewhat higher than the one outlined in Chapter 9, because in OLS regression there is more consensus on the statistical criteria for outliers. Also, in CDA, extreme values can be caused by other diagnostic problems and remedied in ways other than removing them. These cases might also just be natural features of the data, especially if nonnormal distributions are used.

A case can be an outlier but be hard to detect if it is influential and therefore has a large effect on the coefficients (creating a better fit); this outcome will reduce the size of the residual. Measures of influence should therefore also be examined, including leverage, Cook's D, and/or DfBetas. In OLS regression, a point that is far from the center of the data has high leverage, meaning that it has high potential to change the results. However, in logistic regression, extreme values on the IV only have high leverage if the corresponding $\hat{\pi}_i$ is between 10% and 90%. Because of the S-shaped nature of the logistic curve, points at the far extremes have less leverage. A case with high leverage should be assessed for influence, which means that the estimate(s) of one or more beta coefficients would change substantially if the case were removed. DfBetas provide this information for each case and parameter estimate (but are only calculated in SPSS by the logistic regression procedures). Cook's D—a weighted average of the DfBetas—can be generated by GLM in IBM SPSS (Kleinbaum & Klein, 2010).

For the proficiency data, I saved various residuals and diagnostic statistics and then sorted the cases by the likelihood residuals. I repeated this procedure for leverage and Cook's D. No cases stood out from the others.

Diagnostic Graphs

Graphical methods should always be part of the identification and interpretation of outliers as well as other diagnostic problems. One can graph various diagnostic indicators (standardized or studentized residuals, leverage values, Cook's D, or DfBetas) against the predicted logits to determine if there is a poorer fit for some parts of the data than others. This checking can be performed in IBM SPSS by going to the "Save" tab in the GLM procedure and checking "Predicted Value of the Linear Predictor" (XBPredicted) and "Deviance Residuals." In all versions of SPSS, one can also use the logistic regression procedure; click on "Save" and then check the "Deviance" box under residuals and "Probabilities" under "Predicted values." When using the regression option, one would then need to use the "Transform" ➜ "Compute Values" procedure to transform the probabilities into logits.

Next, construct a scatterplot by going to "Graph" ➜ "Chart Builder." Select "Scatter/Dot," select a simple scatterplot (generally the first option), and drag "Standardized Deviance Residuals" to the y-axis and "Predicted Value of the Linear Predictor" to the x-axis, and then click "OK." On the output, double click on the graph; this will bring up the chart editor. Then click on the icon for "Add Fit Line at Total." Under fit method, choose to add a "Loess" fit line (Cleveland, 1979).

The loess line (aka "lowess line") does not assume any particular functional form (e.g., linear, quadratic) and instead is created by taking a small group of neighboring points (e.g., 6 points) and fitting a regression line to each of these "windows" (see Cohen et al., 2003, p. 114, for computational details and additional examples).

For grouped data, if the model adequately fits the data, the fit line should be flat. A lack of fit can be caused by outliers, omission of a major relevant predictor, or by some of the problems discussed below (nonlinearities or overdispersion). Note that homoscedasticity is not an assumption of the logistic regression model.

The proficiency data are not grouped, but loess lines—if interpreted properly—can still help identify areas of poor fit. Figure 10.5(a) plots the likelihood residuals against the predicted logits (which SPSS calls "the predicted values of the linear predictor"). The loess line is U-shaped, indicating a large number of positive residuals at both the low and high ends. To aid interpretation, I also copied the DV and the predicted category into an EXCEL worksheet to determine which predictions were successful [using a formula such as =IF(A1 = B1, 1, 0), which is read, "if A1 = B1, then 1, otherwise 0"]. The resulting values were then copied back into SPSS with the "success" variable coded as "scale" to enable creation of the loess line. Figure 10.5(b) shows a high concentration of successful predictions at the high end of the scale; these would still generate positive residuals because the predicted $\pi_i's$ are less than 1.0. There is room to improve the fit, but of more concern is the concentration of unsuccessful predictions at the low end of the scale; these pull the loess line down in Figure 10.5(b) and upward in Figure 10.5(a). These are essentially low SES individuals who pass the proficiency test even though it is predicted that they would not, thereby generating positive residuals.

A useful graph (specifically for identifying outliers) is the absolute value of the likelihood residuals against the leverages. Plotting residuals against the values of an IV is also useful, as areas of poor fit may suggest nonlinearities.

Checking for Nonlinearity

Before dropping the nonsignificant SES term from the model, we should check for nonlinearities, interactions, and the need for alternative link functions. Constructing and adding an SES*ln(SES) term to the model ("Box-Tidwell" test) produced $\chi^2_{LR}(1) = 2.47$, $p = .116$. The test of the Box-Tidwell term is not significant but is close, so there still may be an aspect of the model we could improve.

Inclusion of Interaction Term

Next, consider including an interaction term, here SES*Female; it is significant when included in the model ($\chi^2_{LR}(1) = 6.52$, $p = .01$).

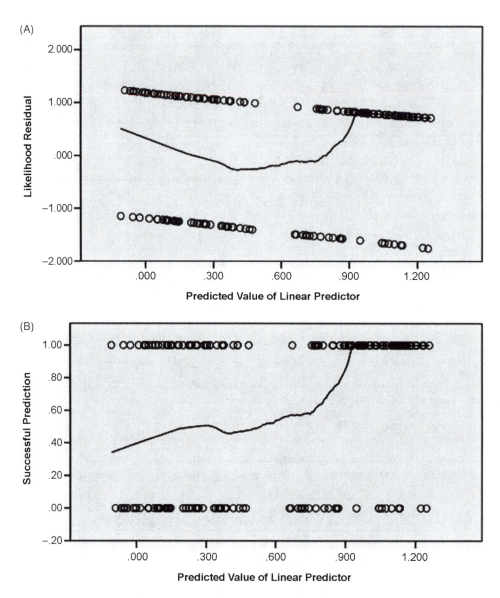

Figure 10.5 Loess lines for proficiency data of (a) likelihood residuals plotted against predicted logits and (b) successful predictions plotted against these logits.

Table 10.4 displays the parameter estimates. When [female = 0] (i.e., for males), the interaction term "disappears" and the relationship of SES to passing rates is negative but nonsignificant. However, for females, SES levels have a positive effect. The "slope" of SES is .032 + (−.009) = .023. Recall that the standard deviation of SES is 26.6, so a one standard deviation change is associated with a change in the DV (η) of .023 ∗ 26.6 = .61, $exp(.061)$ = 1.84. So a one standard deviation increase in SES increases the odds of passing by 84%.

Table 10.4 Parameter Estimates for Proficiency Test Model with Interaction Term (SPSS Output)

Parameter	B	Std. Error	95% Wald Confidence Interval		Hypothesis Test			Exp(B)	95% Wald Confidence Interval for Exp(B)	
			Lower	Upper	Wald Chi-Square	df	Sig.		Lower	Upper
(Intercept)	.643	.4967	−.316	1.647	1.675	1	.196	.643	.4967	−.316
[Female = 1.00]	−.843	.7167	−2.267	.557	1.384	1	.239	−.843	.7167	−2.267
[Female = .00]	0[a]							0[a]		
SES	−.009	.0086	−.026	.008	1.042	1	.307	−.009	.0086	−.026
[Female = 1.00] * SES	.032	.0129	.007	.058	6.252	1	.012	.032	.0129	.007
[Female = .00] * SES	0[a]							0[a]		
(Scale)	1[b]							1[b]		

Note: Dependent Variable: Pass Model: (Intercept), Female, SES, Female * SES
[a]Set to zero because this parameter is redundant.
[b]Fixed at the displayed value.

One next needs to ask whether the interaction makes any theoretical sense. If not, one can drop the interaction term on the grounds that one may be overfitting the data and finding significant terms that are really Type I errors (given that one is conducting multiple statistical tests). On the other hand, one might be able to make a plausible case for the interaction. In the present example, we need to explain why many low-SES males pass the exam whereas low-SES females do not. Perhaps upon investigation we find the proficiency exam contained several problems related to baseball and one related to constructing a fence around a garden. These problem contexts may be more familiar to low-SES males than females, giving them an advantage. We can therefore retain the interaction term in the model because it is theoretically meaningful. Otherwise, it should be dropped.

Alternative Link Functions

Using the current model with the interaction term, I saved the predicted logits and probabilities and performed the Stukel test. Adding z_1 and z_2 to the model reduced the log likelihood from -87.42 to -86.85, a reduction of 0.47. Doubling this value to perform an LR test with two degrees of freedom (because two parameters were added to the model) yielded $\chi^2_{LR}(2) = 0.92$, $p = .63$. The nonsignificant result means that there does not appear to be any need to consider alternative link functions.

Outliers and Influential Observations Revisited

Now that the model is almost final, we should check again for outliers or influential values. There are two observations with somewhat deviant values of Cook's D: .066 and .057 (the next lower value is .030). The cases are more influential than the others. Bollen and Jackman (1990) have suggested that cases with Cook's D values $> \frac{4}{n}$ should be examined, and both are $> \frac{4}{168} = .024$. (A higher standard is $D > 1.0$; Cook & Weisberg, 1982.) Consulting the raw data, we see that these cases correspond to high-SES females who failed the proficiency exam. Deletion of these cases did not substantively change the results, however, and there were a lot of other high-SES females who failed the test, so it is hard to make an argument that these cases are somehow unusual and should be removed.

In summary, interpretation of the final model shows that, at least for females, the probability of passing the proficiency test is positively predicted by SES level.

Overdispersion

Overdispersion occurs in grouped data when there is more variability in the data than there should be. It is often caused by outliers, model misspecification, or violations of statistical independence among observations. The latter can be caused by a clustering of observations in the population, for example, in families or neighborhoods. A classic example is that puppies from the same litter are more similar to one another than puppies from different litters, which can affect the variability in the data. The situation where statistical independence is violated reflects *genuine overdispersion* (also known as *extrabinomial variation*), as opposed to *apparent overdispersion* caused by outliers and model misspecification. Genuine overdispersion can also involve variance heterogeneity, that is, if there is significantly more variation within some clusters than others.

Let the scale statistic (ϕ) be defined as X^2 (or D) divided by df:

$$\phi = \frac{X^2}{df} \text{ or } \frac{D}{df}. \tag{Eq. 10.20}$$

In SPSS, the scale statistic is reported as "Value/df" corresponding to X^2 (or D).

The scale statistic should be close to 1.0. The rationale for this standard is because, with grouped data and an adequate model fit, the residual deviance should have a chi-square distribution with $(n - p)$ degrees of freedom. Recall from Chapter 3 that the mean (expected value) of a chi-square distribution is given by the number of degrees of freedom, so in a well-fitting model one would expect the residual deviance to equal the number of degrees of freedom. Therefore the ratio of the two should be 1.0 on average (Collett, 2003). For example, for the Trimone data, $\phi = 1.062$ using D and 1.035 using X^2 (see Figure 10.3). Scale values substantially greater than 1.0 suggest overdispersion, and lower scale values suggest underdispersion (which can also be problematic). Over- and underdispersion are only problems for *grouped* data; high or low scale values for ungrouped data can be ignored.

Also keep in mind that while a ϕ value of one is the expected value, there is also variation (recall from Chapter 3 that the variance of a chi-square distribution is $2 * df$), so a ϕ value of two, for example, is not necessarily problematic; consider whether the model fit is adequate.

EXPLORING THE CONCEPT

(a) Violations of statistical independence can cause observations to become more extreme (see Chapter 1). For example, if one is studying students who are nested in classrooms, there will be less variability within classrooms and more variability among classrooms than there should be. How could this contribute to overdispersion? (b) Suppose one omits a relevant third variable that is correlated with both X and Y. Could this cause the proportion of hits in each subgroup to become more extreme as X increases? This situation would also contribute to overdispersion.

A formal test for overdispersion is the Pearson X^2 test; simply calculate a p value for a chi-square value based on the following expression:

$$\chi^2(df) = X^2.$$ (Eq. 10.21)

So, for example, if Pearson's $X^2 = 22.00$ on 10 df, $\chi^2(10) = 22$, and so $p = .015$. One would reject the null hypothesis of no overdispersion. Note that in this example, $\phi = 2.2$. Simulation studies of the Pearson X^2 test by O'Hara Hines (1997) showed that it has good statistical properties (e.g., keeping the Type I error rate to about 5%) unless there are a large number of small predicted probabilities involved, in which case she recommends use of a test developed by Dean (1992; see Appendix 10.3). She found Dean's test was less reliable than Pearson's X^2 in other situations.

Overdispersion biases the standard errors, making them too low (thereby increasing Type I error rates). If dispersion problems are indicated, one should first try to remedy any problems with outliers and model misspecification. Only if a dispersion problem remains should one conclude that the data are genuinely overdispersed. It is a good practice to test for overdispersion with a full model that includes the full set of variables (to reduce the probability that relevant variables have been omitted), as well as on the "final model." However, this practice does not eliminate the possibility of model misspecification due to nonlinearities, outliers, or use of an incorrect link function. For this reason, I listed checking for overdispersion in the final part of "steps for model building." (However, the steps are iterative; they should not be implemented in a rigid linear fashion but may require revisiting previous steps—for example, to revisit what variables should be included in the model after adjusting for overdispersion.) Overdispersion can also be checked throughout the model-building process.

Underdispersion also biases the standard errors but has the opposite effect, increasing Type II error rates. Genuine underdispersion is rare but can be caused, for example, by violations of statistical independence when the correlation among cases is negative and produces a better fit to the data. As discussed in the next section, too good a fit will reduce power.

One remedial measure for overdispersion or underdispersion is to multiply the variance–covariance matrix by ϕ, which in effect multiplies the standard errors by ϕ. This technique is known as *using a scale parameter*. In the SPSS GLM procedure, go to the "estimation" tab and select the "scale parameter" method. One can choose either "Pearson's chi-square" or "deviance." Allison (1999) notes that both methods usually yield similar results but that the theory of quasilikelihoods (discussed in Chapters 13 and 14) suggests that it is better to base the adjustment on Pearson's X^2 (see also McCullagh & Nelder, 1989). The X^2 statistic is consistent (bias decreases as sample size increases), whereas the deviance estimate is not.

If a scale parameter is introduced into the analysis, other model fitting decisions (e.g., variables included/excluded, shape of link function) should be reexamined and various models again compared to one another. The deviance and X^2 should also be scaled by ϕ, and the scaled values used to compare models.

Complete and Quasicomplete Separation

One problem that sometimes occurs with logistic regression is quasicomplete or complete separation of the data. This situation occurs when one or more IVs perfectly classify the cases into groups. For example, if one uses gender to predict whether a student attends college, and all the males do go to college, there would be complete separation (Anderson & Rutkowski, 2008). As is the case with multicollinearity, the amount of available variance on which to perform the calculations is very low, leading to large standard errors or a complete failure of MLE algorithms to reach convergence. SPSS will issue a warning if a problem is detected. One can address this problem by simplifying the model or obtaining more data. One can also use advanced techniques such as exact logistic regression, which is available in some statistical packages.

Related Statistical Procedures

There are several other statistical procedures that are sometimes used, instead of logistic regression, when the DV is nominal. As mentioned previously, log-linear analysis can be used if most of the predictors are nominal. Probit regression can be used if one assumes that the value of the DV is determined by a latent, normally distributed continuous variable that reflects the probability of a hit. (Probit regression is often used in dose-response studies.) Probit regression typically gives similar results to logistic regression, but predictions at or near the 50% probability level may be more accurate. Discriminant analysis is used to find a linear function that best divides the cases into two groups on the DV. The technique has more restrictive assumptions than logistic regression (i.e., normally distributed errors and homogenous variance–covariance matrices for each group) and cannot handle categorical predictors (Klecka, 1980). For this reason, King (2008) and others recommend using logistic regression instead.

If using logistic regression, there are specialized methods for handling missing data, using *propensity scores* (a method used when assignment to treatment groups is not random), performing mediation analyses, or conducting Bayesian analyses. These special topics are covered in Hosmer and Lemeshow (2013).

Summary

In this chapter I have addressed logistic regression for when the DV is binary. The logistic curve is transformed into a linear one by using the log odds as the DV. Logistic regression uses maximum likelihood estimation (MLE) to estimate parameters; the estimated likelihoods of models with and without a certain variable can then be used to perform a likelihood ratio (LR) tests of statistical significance. Other methods of testing significance include the Wald and score tests. I described various fit statistics, including the deviance, Pearson's X^2, and the AIC. I then outlined some recommended steps for building and refining logistic models, which included steps for checking for outliers, nonlinearity in the logits, alternative-link functions, and overdispersion with grouped data.

Problems

1. A market researcher collects data on the daily amount of e-mail received by 100 randomly chosen young adults and uses the variable to predict whether or not each individual has a Facebook account. The data are in the file *FB.10.Problem_1*. The dataset was designed so that the IV ranges uniformly from 1 to 100.

   a) Using statistical software (e.g., "Descriptive Statistics" ➔ "Descriptives" procedure in SPSS), find the skew and kurtosis of the IV, and generate standardized z values. Report the minimum and maximum z values. Do any of the descriptive statistics appear problematic?

   b) Perform a logistic regression, regressing "Facebook account" on "E-mail." (In SPSS, use the GLM procedure if available; otherwise, use binary logistic regression.) Set the reference category to be the first ("lowest category"). On the "Statistics" tab, under "Chi-square Statistics," check "Likelihood ratio." Report the log likelihood, AIC, and BIC. Calculate the $-2LL$.

   c) Exponentiate the log likelihood to obtain the likelihood. What does the likelihood represent?

   d) From the log likelihood, use Eqs. 10.18 and 10.19 to calculate the AIC and BIC. Show your work.

   e) For the omnibus test, what is the value of the LR chi-square and p value? Why is there only 1 degree of freedom? Use the χ^2 value to derive the $-2LL$ of the intercept-only model.

   f) Is the IV (e-mail) statistically significant? Report both the LR and Wald chi-squares and associated p values. Which tends to be more statistically powerful? Which makes fewer assumptions?

   g) What is the estimated beta parameter (B) for the e-mail variable? The standard error? Without rounding, divide B by its standard error and report Wald's z.

   h) Interpret $exp(B)$.

   i) Verify the directionality of B by generating Kendall's $\hat{\tau}_b$.

   j) Check for linearity in the logits by conducting a Box-Tidwell test. (Note: In SPSS, the predicted logits are "the Predicted Values of the Linear Predictor," or "XB predicted.")

k) Test the adequacy of the link function with a Stukel test and report the LR results.

l) Using the logistic link, rerun the regression. Save the following diagnostic values: Predicted value of mean response, predicted category, predicted value of linear predictor (XB), Cook's distance, leverage value, standardized Pearson residuals, standardized deviance residuals, and likelihood residuals. What are the values of all of these for the first observation?

m) What is the predicted probability of the 24th observation (i.e., predicted value of mean response)? Why is this case predicted not to have a Facebook account?

n) Using descriptive statistic procedures, find the maximum and minimum values of the diagnostic statistics in part (m), starting with Cook's D.

o) Sort the database by the likelihood residuals (in SPSS, use "Data" → "Sort Cases"). Are any cases potential outliers (because the values of the residuals or other diagnostic statistics are quite different from the rest)? Should any cases be deleted?

p) Plot the DV (Facebook account) against the IV. Also generate the absolute values of the likelihood residuals and graph against the leverages. Which graph better displays the outliers?

2. As in Del Prette, Del Prette, de Oliveira, Gresham, and Vance (2012), a researcher classifies 100 elementary school students into those who display learning problems (LP = 1) and those who do not (LP = 0). On the theory that those who have weaker social skills are more likely to display learning problems, data are also collected from each student's classmates, on a 20-point scale, on the degree that the student is socially collaborative. The data are in the file *Learning Problems.10.Problem_2*.

a) For collaboration, report skew, kurtosis, and maximum and minimum z values.

b) Using logistic regression, regress LP on collaboration. Report the log likelihood, AIC, BIC, and calculate the $-2LL$.

c) From the log likelihood, use Eqs. 10.18 and 10.19 to calculate the AIC and BIC. Show your work.

d) For the omnibus test, what is the χ^2_{LR} and p value? Use the chi-square value to derive the $-2LL$ of the intercept-only model.

e) Is the IV statistically significant? Report both the LR and Wald chi-squares and associated p values.

f) What is the B for collaboration? The standard error? Without rounding, divide B by its standard error and report Wald's z.

g) Interpret $exp(B)$.

h) Verify the directionality of B by generating Kendall's $\hat{\tau}_b$.

i) Check for linearity in the logits by conducting a Box-Tidwell test.

    j)   What are the maximum and minimum values of Cook's D, leverage, standardized Pearson and deviance residuals, and the likelihood residuals?

    k)   Are any cases potential outliers (because the values of the residuals or other diagnostic statistics are quite different from the rest)? Should any cases be deleted? Generate the absolute values of the likelihood residuals and graph against the leverages.

3.   Males using a computerized dating service are administered an assertiveness questionnaire, with the scale ranging from 1 to 100. Afterward, the woman is asked whether the data were a success or not. Does assertiveness (X) predict success (Y)? The data are in the file *Dating.10.Problem_3*.

    a)   Conduct a binary logistic regression. Report B, χ^2_{LR}, and p value.

    b)   Conduct a Box-Tidwell test. Are nonlinearities indicated?

    c)   Copy the data for X and Y into EXCEL. Sort by X. Divide X into five categories of 10 cases each and compute the mean Y *(sum of Ys divided by 10)*. What are the five mean Ys? What are the five logits (log odds)? Graph these if possible. What sort of relationship is indicated?

    d)   Include a quadratic term (X^2) in the model. Report B, χ^2_{LR}, and p value. Is the quadratic term significant?

4.   A researcher administers an online training program to certify information technology specialists and, after the training program, instructs participants to study for the certification exam. Participants are tested on their knowledge after a delay of from 1 to 10 days. The DV is whether students pass the test (the data are below and in the file *Certification.10.Problem_4*). Conduct a logistic regression of "Pass" on "Days." The latter should be included in the model as a covariate, not a factor.

    a)   Report B, the χ^2_{LR}, and the AIC and BIC.

    b)   Conduct a Box-Tidwell test. Are nonlinearities indicated?

    c)   Copy the data for "Days" and "Pass" into EXCEL. Compute the proportion of passes for each value of "Days." What are the 10 values? What are the 10 logits (log odds)? Graph these if possible. What sort of relationship is indicated?

    d)   Compute the logarithm of "Days," and use that variable in the model rather than "Days" for this and the remaining subproblems. Report its estimated coefficient, standard error, and p value. How have the AIC and BIC changed?

    e)   What is the deviance divided by the degree of freedom? Do the data appear overdispersed? What are the results of a Pearson X^2 test of overdispersion using Eq. 10.21?

    f)   Rerun the model using a X^2 scale parameter. (In IBM SPSS, GLM procedure, go to the "estimation" tab and set the scale parameter method to "Pearson chi-square.") What is the standard error for ln(days), and how has it changed from the previous model?

Days	No. pass	No. fail	Days	No. pass	No. fail	Days	No. pass	No. fail
1	132	868	5	354	146	9	418	82
2	104	396	6	289	211	10	382	118
3	236	264	7	360	140			
4	255	245	8	358	142			

Technical Notes

10.1 Profile likelihood confidence intervals involve iterative evaluations of the likelihood function (Allison, 1999) and provide better approximations in small samples. If the CI does not bracket the null hypothesized value (e.g., $\beta = 0$), then the value is significant.

10.2 The formula for the deviance residuals (from Lloyd, 1999) is

$$d_i = sign(Y_i - \hat{\pi}_i) \sqrt{\left[\left(2y_i ln\left(\frac{Y_i}{\hat{\pi}_i}\right) + 2(1 - Y_i) ln\left(\frac{1 - Y_i}{1 - \hat{\pi}_i}\right)\right)\right]}, \qquad \text{(Eq. 10.22)}$$

where $sign(Y_i - \hat{\pi}_i) = 1$ if the raw residual is positive, -1 if negative, and 0 if $(Y_i - \hat{\pi}_i) = 0$. For example, suppose $\hat{\pi}_i$ for one observation is .70 and $Y_i = 1$. The raw residual is .30 but the deviance residual is 0.85. Specifically, $sign(Y_i - \hat{\pi}_i) = sign(.30) = 1$ Therefore, $d_i = +1\sqrt{\left[\left(2*1*ln\left(\frac{1}{.70}\right) + 2*(1-1)ln\left(\frac{1-1}{.30}\right)\right)\right]} = +\sqrt{2*[(.36+0)]} = 0.85$. Note that when the residual is positive, the second term is eliminated and the deviance is a function of the log of the reciprocal of $\hat{\pi}_i$. Also note that, by the quotient rules of logarithms, $ln\left(\frac{1}{\hat{\pi}_i}\right) = ln(1 - \hat{\pi}_i)$. The last term is the raw residual on the log scale. If $Y_i = 0$, then the raw residual would be $-.70$ and the deviance residual would be $-\left[2*ln\left(\frac{1}{.30}\right)\right] = -2.41$.

10.3 Cohen et al. (2003) argued that the Hosmer-Lemeshow test is only valid if there are large expected frequencies in all cells and that the test does not have adequate power for sample sizes under 400 (Hosmer & Lemeshow, 2013); however, others (e.g., Anderson & Rutowski, 2008) pointed out that the test is more robust than others to violations of sample size requirements.

10.4 Pseudo-R^2s are used as fit statistics in generalized linear models rather than actual R^2s. The latter in OLS regression reflects percentage of variance explained, which is based on partitioning the total sum of squares into explained and residual sum of squares with the following equation: $\sum(Y_i - \bar{Y})^2 = \sum(Y_i - \hat{Y}_i)^2 + \sum(\hat{Y}_i - \bar{Y})^2$. However, when nonlinear link functions are used, the cross-product term $2\sum(Y_i - \hat{Y}_i)(\hat{Y}_i - \bar{Y})$ will not equal zero, so R^2 cannot be used (Cameron & Trivedi, 1998).

10.5 AICC is the AIC "corrected" for finite sample size: $AICC = -2LL + \frac{2pn}{n-p-1}$. Compared to the AIC, the AICC contains an additional penalty for including more parameters and helps avoid a model that "overfits" the data because of the inclusion of too many parameters but is less likely to generalize to new data. Use AICC when n is small or k is large (Burnham & Anderson, 2002). The CAIC is the Consistent AIC $[-2LL + p(ln(n) + 1)]$ and is a slight modification to the BIC reflected in the inclusion of the 1, which increases the penalty for adding parameters.

10.6 For stepwise regression, SPSS uses either the Wald or LR test to determine which variable to drop at each step (a score test is used to determine which variable to add; see Chapter 14). Many authors recommend setting a liberal entry criterion (e.g., $\alpha = .15$ or $.25$) to protect against suppressor effects and loss of power due to multicollinearity.

10.7 Bursac et al. (2008) recommended a purposeful selection procedure for model building. The procedure involves (a) performing univariate regressions (with each predictor individually) and retaining those with p values less than .25, or those which are theoretically important; (b) performing a multiple logistic regression with all retained variables simultaneously, and retaining those with p values less than .05, or those which are theoretically important; (c) performing another multiple regression with the smaller model and examining any variable whose beta coefficient changed by more than 20% (one or more of the variables excluded from the smaller model may correlate with this included variable and may need to be entered back into the model to control for the confound); and (d) reentering one at a time those variables excluded in the first step to check their significance at the .05 level. One then checks for nonlinearities and interactions.

10.8 One approach proposed by Hosmer and Lemeshow (2013) for checking for nonlinearities involves dividing a metric predictor into four categories based on quartiles and using dummy variables for each category (except the lowest). Plot the estimated coefficients for each dummy variable against the midpoint of each quartile (plot zero against the reference category). The plot may suggest the shape of any nonlinearities. If there are different parameter values for different categories but no obvious ordering, consider including both $ln(X)$ and $ln(1 - X)$, which reflects a beta distribution (Hosmer & Lemeshow, 2013).

10.9 The cloglog function can result from truncating an observed or latent metric variable into a discrete response (e.g., reducing "number of accidents" into "no accidents" and "some accidents; see Piegorsch, 1992). Cloglog models are also used in survival analysis for modeling the length of time until some event occurs when time is measured in discrete units, such as years. It can be shown that the model—if one reverses the sign of the location parameter and the order of the Y categories—is equivalent to a proportional hazard model using Cox regression, and exp(B) is equivalent to a hazard ratio, where a hazard can be thought of as an instantaneous propensity for an event to occur (Allison, 1999). For example, a hazard ratio of 2 might indicate that individuals in one group die at twice the rate as the other. Hazards multiplied by a period of time equal the probability of the event occurring during that time period.

10.10 The probit function is the inverse of the standard cumulative normal distribution. The coefficients of a probit model are usually close to those of a logit model but are less easy to interpret. The Cauchit function (also known as the inverse Cauchy) is $g(u) = \tan[\pi(u - 0.5)]$ and is sometimes useful when there are a few large outliers.

10.11 In SPSS, in performing R comparisons, the Bonferroni correction adjusts the p value by R (specifically, $p_i^* = Rp_i$), where p_i^* is the adjusted value. The adjustment is equivalent to using the unadjusted value p_i and dividing the alpha level by R. The goal is to maintain the familywise error rate at .05 or some other designated level. The Bonferroni adjusted p value is an approximation, and the exact p values are given by the Sidak correction, $p_i^* = 1 - (1 - p_i)^R$. For example, if $R = 2$ and $p_i = .05$, $p_i^* = 1 - .95^2 = .0975$ (whereas $p_i^* = .10$ with the Bonferroni correction). Note that the logic behind the Sidak correction is that the probability of obtaining a significant result (p_i^*) is 100% minus the probability of obtaining R nonsignificant results. The Bonferroni correction is easier to compute and was widely used before the advent of personal computers. The Sidak adjustment is preferable, especially if the number of comparisons is large, but it assumes the comparisons are statistically independent. Sequential Bonferroni and Sequential Sidak adjustments can reduce the number of comparisons, thereby increasing statistical power (see the discussion in Chapter 8 on the stepwise, stepdown procedures). The least significant difference procedure makes no adjustments.

10.12 When using grouped data, a formal test of whether a case is an outlier is whether the largest likelihood residual (in absolute value) is greater than $\Phi^{-1}\left\{1 - \left(\frac{\alpha}{2n}\right)\right\}$, where Φ^{-1} is the inverse of the standard cumulative normal distribution (=NORM.S.INV in EXCEL), and n is the total number of observations (Collett, 2003).

10.13 A DfBeta value is calculated for each observation and predictor; it is a measure of how much the beta estimate for the predictor would change if that observation were deleted from the data set, scaled by the standard error with the observation deleted. DfBeta values greater than 1.0 flag influential observation (a cutoff of $\frac{2}{\sqrt{n}}$ is also sometimes used). Cook's D is a weighted average of the DfBetas for a particular case and is often used instead.

APPENDIX 10.1

SELECTED PROOFS

Derivation of the Logistic Generalized Linear Model (Eq. 10.2)

1. Nonlinear logistic model is $\pi = \dfrac{e^{(\alpha+\beta X)}}{1+e^{(\alpha+\beta X)}}$.

2. Let $z = \alpha + \beta X$. Therefore $\pi = \dfrac{e^z}{1+e^z}$ and odds are $\dfrac{\pi}{(1-\pi)} = \dfrac{\frac{e^z}{1+e^z}}{1-\left(\frac{e^z}{1+e^z}\right)}$.

3. Replacing the 1 in the denominator on the right with $\dfrac{1+e^z}{1+e^z}$ yields $\dfrac{\frac{e^z}{1+e^z}}{\frac{1}{1+e^z}} = e^z$.

4. Log odds are therefore $ln\left(\dfrac{\pi}{1-\pi}\right) = ln\left(e^z\right) = ln\left(e^{\alpha+\beta X}\right) = \alpha + \beta X$. QED.

Derivation of Logistic Likelihood Function (Eq. 10.10)

1. Let $z = \alpha + \beta X$. Therefore, from Eq. 10.8, $Prob(Y_i = y) = \left(\dfrac{e^z}{1+e^z}\right)^y \left(1-\dfrac{e^z}{1+e^z}\right)^{1-y}$.

2. Last term $\left(1-\dfrac{e^z}{1+e^z}\right)^{1-y} = \left(\dfrac{1}{1+e^z}\right)^{1-y}$, so $Prob(Y_i = y) = \left(\dfrac{e^z}{1+e^z}\right)^y \left(\dfrac{1}{1+e^z}\right)^{1-y} = \left(e^z\right)^y$
$\left(\dfrac{1}{1+e^z}\right)^y \left(\dfrac{1}{1+e^z}\right)^{1-y} = \left(e^z\right)^y \left(\dfrac{1}{1+e^z}\right)$ (from adding exponents).

3. Rewrite as $Prob(Y_i = y) = \left(e^{\alpha+\beta X}\right)^y \left(\dfrac{1}{1+e^{\alpha+\beta X}}\right)$ (Eq. 10.9).

4. Eq. 10.10: $l = \Pi_{i=1}^n \dfrac{\left(e^{(\alpha+\beta X)}\right)^y}{1+e^{(\alpha+\beta X)}}$ follows as joint probability of n independent events.

Proof that P is the Maximum Likelihood Estimator of π

1. $l(P\,|\,\pi) = \binom{n}{x}\pi^x (1-\pi)^{n-x}$ (from Eq. 10.6).

2. Because $x = nP, \dfrac{\partial l(P\,|\,\pi)}{\partial \pi} = \dfrac{\partial}{\partial \pi}\left[\binom{n}{nP}\pi^{nP}(1-\pi)^{nQ}\right]$, where $Q = 1 - P$.

3. $\dfrac{\partial}{\partial \pi}\left[\binom{n}{nP}\pi^{nP}(1-\pi)^{nQ}\right] = nP(1-\pi) - \pi nQ$. [Note: $\binom{n}{nP}$ is a constant.]

4. $nP(1-\pi) - \pi n(1-P) = 0$ at maximum point, so $nP(1-\pi) = \pi n(1-P)$.

5. Simplifying expression, at point of maximum likelihood, $\dfrac{P}{(1-P)} = \dfrac{\pi}{(1-\pi)}$, therefore odds$(P)$ = odds(π), therefore $P = \pi$ (at maximum point). QED.

APPENDIX 10.2

GOODNESS OF LINK TEST

Pregibon (1980) proposed that two link functions could be compared if they both could be embedded in a family of link functions—for example $g\left(p_i;\,\alpha\right)=ln\left\{\dfrac{\left(1-p_i\right)^{-\alpha}-1}{\alpha}\right\}$, which reduces to the logit link if $\alpha=1$ and the cloglog link as α approaches 0 (Collett, 2003). If one constructs the following variable (z_i) with $\alpha=1$ and adds it to a logistic model, if z_i is significant based on an LR test, it indicates that the logistic link does not fit well (Pregibon, 1980). The formula for z_i is

$$z_i = \frac{ln\left(1-\hat{p}_i\right)}{\left(1-\hat{p}_i\right)^{\alpha-1}}-\alpha^{-1}. \tag{Eq. 10.23}$$

This procedure is called a *goodness-of-link* (GOL) test. Likewise, one can test the goodness of a cloglog link by setting α to a very small value and adding z_i to a cloglog model.

Note that this test is sensitive to sample size: There could be lack of significance in a small sample or large significance in a large sample, so other considerations (e.g., reductions in deviances, visual inspections, theoretical interpretability, or ease of interpretation) should also be taken into consideration when choosing a link function.

Underlying Theory

The GOL test is an LR test, which can only be used to compare nested models. As noted above, Pregibon (1980) proposed that two link functions could be compared if they both could be embedded in a family of link functions, such as the Box-Cox family of power transform (see Chapter 9), where $\lambda=1$ results in the identity link and $\lambda=0$ in the log link (Breslow, 1995). Likewise, the family $g\left(p_i;\,\alpha\right)=ln\left\{\dfrac{\left(1-p_i\right)^{-\alpha}-1}{\alpha}\right\}$ yields the logit link when $\alpha=1$ and the cloglog link when α approaches 0. More specifically, when $\alpha=1$, then $\dfrac{\left(1-p_i\right)^{-\alpha}-1}{\alpha}=\dfrac{1}{\left(1-p_i\right)}-1=\dfrac{1-\left(1-p_i\right)}{\left(1-p_i\right)}=\dfrac{p_i}{\left(1-p_i\right)}$, which is the odds, so $ln\left\{\dfrac{\left(1-p_i\right)^{-\alpha}-1}{\alpha}\right\}$ is the log odds (logit). Now $\lim\limits_{\alpha\to0}\dfrac{\left(1-p_i\right)^{-\alpha}-1}{\alpha}=-ln(1-p_i)$ (Collett, 2003), so $g(p_i;\,\alpha)=ln(-ln(1-p_i))$, which is the cloglog function.

In actuality, one does not know the true value of α, but it has a hypothesized value, α_0. The correct model $[g(p_i; \alpha)]$ can be approximated by the first two terms of a Taylor expansion (see Appendix 14.1), specifically:

$$g(p_i; \alpha) \approx g(p_i; \alpha_0) + (\alpha_0 - \alpha) \frac{\partial g(p_i; \alpha)}{\partial \alpha} \bigg|_{\alpha_0} .$$

The correct model can therefore be approximated by:

$$g(p_i; \alpha_0) = \eta_i + \gamma z_i,$$

where η_i is the value of the linear predictor given the hypothesized link function, $\gamma = (\alpha_0 - \alpha)$, and $z_i = \dfrac{\partial g(p_i; \alpha)}{\partial \alpha} \bigg|_{\alpha_0}$. The variable z_i is a constructed variable that can be shown to be equal to the expression in Eq. 10.23; γ is its regression parameter, which will be significant if there is a large discrepancy between α_0 and α (implying the hypothesized link function is incorrect).

APPENDIX 10.3

DEAN'S TEST FOR OVERDISPERSION

A formal test for detecting overdispersion is reviewed by Dean (1992, pp. 455–456). To perform this test, one needs to save the predicted means ($\hat{\pi}_i$) and perform the following calculation manually (e.g., in EXCEL):

$$N_B = \frac{\sum\left\{[\hat{\pi}_i(1-\hat{\pi}_i)]^{-1}\left[(Y_i - m_i\hat{\pi}_i)^2 + \hat{\pi}_i(Y_i - m_i\hat{\pi}_i) - Y_i(1-\hat{\pi}_i)\right]\right\}}{\sqrt{\{2\sum m_i(m_i - 1)\}}}, \qquad \text{(Eq. 10.24)}$$

where Y_i is the response (0 or 1) and m_i is the cell size. This test assumes that while Y_i has a binomial distribution conditional on a particular probability, π_i^*, the π_i^*s follow a beta distribution on the interval (0, 1) with $Var(\pi_i^*) = \tau \, \pi_i^* (1 - \pi_i^*)$. The null hypothesis is that $\tau = 0$, in which case $Var(\pi_i^*) = 0$. For any given value of x_i, there will just be one π_i^* as in the standard logistic model (i.e., no overdispersion). Under the null hypothesis, the test statistic N_B will have a normal distribution, but one can conduct a one-sided test (e.g., if $N_B = 1.88$, then $z = 1.88$, one-tailed $p = .03$; one would reject the null hypothesis and conclude there is overdispersion).

Another model of overdispersion is the *logistic linear alternative model*, which is of the form $log\left(\frac{\pi_i}{1-\pi_i}\right) = \beta_0 + \beta_1 X_i + z_i$. This model simply adds an error term z_i to the basic logistic model, with $Var(z_i) = \tau$. The null hypothesis is again that $\tau = 0$. The test statistic N_A is more complicated to compute. For the logistic linear alternative model of overdispersion (from Dean, 1992),

$$N_A = \frac{\sum\left\{(Y_i - m_i\hat{\pi}_i)^2 - m_i\hat{\pi}_i(1-\hat{\pi}_i)\right\}}{\hat{V}}. \qquad \text{(Eq. 10.25)}$$

To compute $\hat{V}$ first compute all of the following: $W_{1i} = m_i\pi_i(1 - \pi_i)$, $W_{2i} = \frac{1}{2}m_i\pi_i(1 - \pi_i)(1 - 2\pi_i)$.

Both N_A and N_B will give very similar results when the p_is are between 20% and 80%. When many of the p_i's are more extreme, then both tests should be performed (Dean, 1992). In small samples, a correction factor should be applied that speeds convergence to normality (see Dean, 1992, p. 456, for specific formulas).

References

Agresti, A. (2002). *Categorical data analysis* (2nd ed.). Hoboken, NJ: Wiley.

Akaike, H. (1973). Information theory and an extension of the maximum likelihood principle. In B. Petrov & F. Csaki (Eds.), *Second International Symposium on Information Theory* (pp. 267–281). Budapest: Akademini Kiado.

Aldrich, J. (1997). R. A. Fisher and the making of maximum likelihood 1912–1922. *Statistical Science, 12*, 162–176.

Allison, P. D. (1999). *Logistic regression using the SAS system: Theory and application.* Cary, NC: SAS Institute.

Allison, P. D. (2013). Why I don't trust the Hosmer-Lemeshow test for logistic regression. *Statistical Horizons.* Retrieved from http://www.statisticalhorizons.com/hosmer-lemeshow

Amemiya, T. (1985). *Advanced econometrics.* Cambridge, MA: Harvard University Press.

Anderson, C. J., & Rutkowski, L. (2008). Multinomial logistic regression. In J. W. Osborne (Ed.), *Best practices in quantitative methods* (pp. 390–409). Thousand Oaks, CA: Sage.

Bollen, K. A., & Jackman, R. W. (1990). Regression diagnostics: An expository treatment of outliers and influential cases. In J. Fox & J. S. Long (Eds.), *Modern methods of data analysis* (pp. 257–291). Newbury Park, CA: Sage.

Breslow, N. E. (1995, June). *Generalized linear models: Checking assumptions and strengthening conclusions.* Prepared for the Congresso Nazionale Societa' Italiana di Biometria, Centro Convegni S. Agostino, Cortona, Italy.

Burnham, K. P., & Anderson, D. R. (2002). *Model selection and multimodel inference: A practical information-theoretic approach* (2nd ed.). New York, NY: Springer-Verlag.

Bursac, Z., Gauss, H. C., Williams, D. K., & Hosmer, D .W. (2008). Purposeful selection of variables in logistic regression. *Source Code for Biology and Medicine, 16*, 3–17.

Cameron, A. C., & Trivedi, P. K. (1998). *Regression analysis of count data.* New York, NY: Cambridge University Press.

Cleveland, W. S. (1979). Robust locally weighted regression and smoothing scatterplots. *Journal of the American Statistical Association, 74*, 829–836.

Cohen, J., Cohen, P., West, S. G., & Aiken, L. S. (2003). *Applied multiple regression/correlation analysis for the behavioral sciences* (3rd ed.). Mahwah, NJ: Erlbaum.

Collett, D. (2003). *Modeling binary data* (2nd ed.). Boca Raton, LA: Chapman & Hall/CRC.

Cook, R. D., & Weisberg, S. (1982). *Residuals and influence in regression.* New York, NY: Chapman & Hall.

Coxe, S., West, S. G., & Aiken, L. S. (2009). The analysis of count data: A gentle introduction to Poisson regression and its alternatives. *Journal of Personality Assessment, 9*, 121–136.

Dean, D. B. (1992). Testing for overdispersion in Poisson and binomial regression models. *Journal of the American Statistical Association, 87*, 451–457.

Del Prette, Z.A.P., Del Prette, A., de Oliveira, L. A., Gresham, F. M., & Vance, M. J. (2012). Role of social performance in predicting learning problems: Prediction of risk using logistic regression analysis. *School Psychology International, 33*, 615–630.

Firth, D. (1993). Bias reduction of maximum likelihood estimates. *Biometrika, 80*, 27–38.

Hosmer, D. W., & Lemeshow, S. (2013). *Applied logistic regression* (3rd ed.). New York, NY: John Wiley & Sons.

King, J. E. (2008). Binary logistic regression. In J. W. Osborne (Ed.), *Best practices in quantitative methods* (pp. 358–384). Thousand Oaks, CA: Sage.

Klecka, W. R. (1980). *Discriminant analysis.* Beverly Hills, CA: Sage.

Kleinbaum, D. G., & Klein, M. (2010). *Logistic regression: A self-learning text* (3rd ed.). New York, NY: Springer.

Koehler, D. (1986). Goodness-of-fit test for log-linear models in sparse contingency tables. *Journal of the American Statistical Association, 81*, 483–493.

Koehler, D., & Larntz, K. (1980). An empirical investigation of goodness-of-fit statistics for sparse multinomials. *Journal of the American Statistical Association, 75*, 336–344.

Lloyd, C. J. (1999). *Statistical analysis of categorical data.* New York, NY: John Wiley & Sons.

Long J. S. (1997). *Regression models for categorical and limited dependent variables.* Thousand Oaks, CA: Sage.

McCullagh, P., & Nelder, J. A. (1989). *Generalized linear models* (2nd ed). New York, NY: Chapman & Hall/CRC.

Mood, C. (2010). Logistic regression: Why we cannot do what we think we can do, and what we can do about it. *European Sociological Review, 26*, 67–82. doi://10.1093/esr/jcp006

O'Connell, A. A. (2006). *Logistic regression models for ordinal response variables.* Thousand Oaks, CA: Sage.

O'Hara Hines, R. J. (1997). A comparison of tests for overdispersion in generalized linear models. *Journal of Statistical Computation and Simulation, 58*, 323–342.

Piegorsch, W. W. (1992). Complementary log regression for generalized linear models. *The American Statistician, 46*, 94–99.

Powers, D. A., & Xie, Y. (2000). *Statistical methods for categorical data analysis.* San Diego, CA: Academic Press.

Pregibon, D. (1980). Goodness of link tests for generalized linear models. *Applied Statistics, 29*, 15–24.

Raftery, A. E. (1995). Bayesian model selection in social research. *Sociological Methodology, 25*, 111–163.

Stukel, T. A. (1988). Generalized logistic models. *Journal of the American Statistical Association, 83*, 426–431.

Vittinghof, E., & McCulloch, C. E. (2006). Relaxing the rule of ten events per variable in logistic and Cox regression. *American Journal of Epidemiology, 165*, 710–718.

Wilks, S. S. (1938). The large-sample distribution of the likelihood ratio for testing composite hypotheses. *The Annals of Mathematical Statistics, 9*, 60–62.

Williams, R. (2009). Using heterogeneous choice models to compare logit and probit coefficients across groups. *Sociological Methods & Research, 37*, 531–559.

CHAPTER 11

MULTINOMIAL LOGISTIC, ORDINAL, AND POISSON REGRESSION

This chapter considers regression techniques where the dependent variable (DV) consists of multiple categories or counts. Multinomial or ordinal logistic regression is used when the DV consists of categories (unordered and ordered, respectively); Poisson regression is used for count data. For all three techniques, the independent variable or variables (IVs) can be metric or nominal. Ordinal IVs can be used in certain situations.

Multinomial Logistic Regression

The Multinomial Logistic Model

While with binary logistic regressions, the conditional population distributions of the Y_is are assumed to be binomial, the stochastic component of a multinomial model reflects the multinomial probability distribution. The probabilities can be computed using the multinomial formula.

The Multinomial Formula

This formula is similar to the binomial formula (and reduces to the binomial when the number of categories, J, is equal to 2). Let lowercase j denote the number of each category, so that j runs from 1 to J. The multinomial formula is

$$\text{Prob}\left(x_1 \ \& \ x_2 \ \& \ldots \& \ x_J\right) = \left(\frac{n!}{x_1! \ * \ x_2! \ * \ldots * x_J!}\right) * \left(\pi_1^{x_1} \ * \ \pi_2^{x_2} \ * \ldots * \pi_J^{x_J}\right), \qquad \text{(Eq. 11.1)}$$

where x_j is the number of hits in category j, and π_j is the probability of a hit in category j. In addition, $\sum_{j=1}^{J} \pi_j = 100\%$.

Each case is classified into one of the J categories and so will reflect a hit in one of the categories. For example, if $n = 10$, and there are three categories with $\pi_1 = .25$, $\pi_2 = .40$, and $\pi_3 = .35$, the probability that four of the cases will be classified in Category 1, two in Category 2, and four in Category 3 is $\frac{10!}{4!2!4!} * .25^4 * .40^2 * .35^4 = .03$.

EXPLORING THE CONCEPT

Suppose there are only two categories (e.g., passing or failing an exam). The probability of passing is 30% and that of failing 70%. Using the multinomial formula, what is the probability that out of 10 students, 6 will pass the exam and 4 will fail? What is the probability according to the binomial formula? Does the multinomial formula reduce to the binomial when $J = 2$?

Multinomial Logistic Regression

To conduct a regression analysis, one category must be designated as the reference category. For ease of interpretation, the reference category should be the category that is most different from the others. (In SPSS, there is an option for specifying whichever category you choose.) Let category J denote the reference category. In a multinomial logistic regression, one predicts the log odds that a case will be in the jth category as opposed to the Jth category. (There are $J - 1$ sets of odds.) For simplicity, suppose there is just one IV, then, similar to Eq. 10.2 for the binomial case, the multinomial logistic model can be written as

$$ln\left(\frac{\pi_j}{\pi_J}\right) = \alpha_j + \beta_j X, \; j = 1, ..., J - 1. \tag{Eq. 11.2}$$

Note that in this model there are separate regression equations for each of the $J - 1$ categories. Therefore, if there are three categories, there would be two separate regression equations. The following likelihood equation is used to estimate the parameters:

$$L = \sum_{j=1}^{J} \sum_{i=1}^{n_j} ln(\pi_{ij}). \tag{Eq. 11.3}$$

The second term in Eq. 11.3 is a summation over all cases that fall into category j. The overall likelihood of obtaining the data is the sum of the likelihoods for the individual categories. (The likelihood is expressed as a sum rather than a product in Eq. 11.3 because of use of the log scale.) One could also run separate binary logistic regressions for each of the $J - 1$ categories, but the multinomial approach is more statistically powerful.

In IBM SPSS, multinomial logistic regression can be conducted using either that option under the regression menu or using the Generalized Linear Models

(GLM) procedure. With the latter, on the "Type of Model" tab, choose "Custom," "Multinomial" for the distribution, and "Cumulative logit" for the link function.

Voting Example

Kotler-Berkowitz (2001) used multinomial linear regression to predict the effect of economic factors and religion on voting behavior during the 1992 British General Election ($N = 6,750$). The DV reflected what party (Labour, Liberal Democrat, or Tory) each citizen in the sample voted for in the 1992 general election. Table 11.1 shows a subset of the results.

Note from the table that *Tory* is the reference category for the DV. Two regression equations were estimated, one for Labour versus Tory and one for Liberal Democrat versus Tory. The last column, for Labour versus Liberal Democrat, is—with some rounding error—simply the values in the first column minus those in the second.

The table shows a significant negative effect of the "economic assessment" variable. Respondents completed a survey with one item asking them to assess their current economic situation on a 5-point scale, from *difficult* (low) to *comfortable* (high). Respondents who were economically comfortable were less likely to vote for the Labour or Liberal Democratic parties and therefore more likely to vote Tory. The estimated odds ratio (OR) for the former comparison, for example, was

Table 11.1 British General Election, 1992: Multinomial Logistic Regression Results

Independent variables	Voting behavior (DV)		
	Labour vs. Tory	Liberal Democrat vs. Tory	Labour vs. Liberal Democrat
Constant	0.21	−0.84**	−0.63*
Economic variables			
Economic assessments	−0.14**	−0.10**	−0.03
Unemployment benefits	0.04	0.10	−0.06
Metric religious variables			
Religious behavior	−0.08**	0.12**	−0.20
Religious belief	−0.06	−0.29**	0.23
Religious belonging			
Catholic	0.57***	−0.04	0.62**
Dissenting Protestant	0.21	0.35**	−0.14
Church of Scotland	−0.34	−0.34	0.00
Other religion	1.09**	0.26	0.83*
Secular	0.40***	0.19	0.22*

Notes: Reference category for voting behavior was "Tory" and for religious belonging was "Anglican." *$p < .05$. **$p < .01$. ***$p < .001$.

$exp(-0.14) = 0.87$, and $exp(-0.10) = 0.90$ for the latter. The table also shows that voters who attended church more often (with a higher score for "religious behavior") were less likely to be members of the Labour party (compared to either other party) but were more likely to be Liberal Democrats than Tory. Religious belief refers to agreement, on a 3-point scale, with the statement that the Bible is literally the word of God. Such voters were less likely to vote Liberal Democratic. Catholics were more likely to vote for the Labour Party, other variables being held constant. In general, the results contradict the hypothesis that religion is not a factor affecting how people vote in Britain.

EXPLORING THE CONCEPT

What is the sign of the first coefficient for religious behavior? Would a devout Catholic who attends church regularly be more or less likely to vote for Labour than a less devout Catholic?

Fit Statistics in Multinomial Logistic Regression

When assessing whether one's final model adequately fits the data, the Pearson or likelihood ratio chi-square statistics are often used. Because some of the multinomial fit indices are chi-square statistics, the usual requirements regarding cell sizes apply (see Chapter 10). When metric predictors are used, there will be many empty cells. One could combine cells, but that would discard information. C. J. Anderson and Rutkowski (2008) therefore recommend conducting a supplementary analysis where one conducts separate binary logistic regressions for each DV category against the reference category and computes the Hosmer-Lemeshow (H-L) statistic for each regression. (There is not an H-L statistic for multinomial regression.) Although the H-L procedure uses a chi-square test, it is more robust to violations of the sample size assumptions than the other fit statistics (C. J. Anderson & Rutkowski, 2008).

Ordinal Regression

Ordered Categories

If the categories of the dependent variable can be ordered, then it is typically preferable to use ordinal rather than multinomial logistic regression. The ordering provides additional information that can improve the "efficiency" (i.e., statistical power) of the estimates. Ordinal regression is only appropriate when the DV consists of ordered categories; if the DV consists of ranks, then the nonparametric tests presented in

Chapters 6 through 8 are more appropriate. The procedure is known by a number of other names, including cumulative logistic regression, cumulative logits, the polytomous universal model (PLUM), or the proportional odds model. Ordinal regression is available in IBM SPSS under the regression menu or through the GLM procedure.

Ordinal regression is ideally used when the DV consists of about three to six categories and none of the categories are sparse. If there is a category with only a handful of observations, consider combining the category with one of the adjacent categories. A subsequent section on minimum cell sizes gives more specific criteria.

Three-Category Example

For simplicity, assume that there are three categories, high, medium, and low, and that these are coded as 3, 2, and 1, respectively. Furthermore, let

- π_1 be the probability of a case falling into the high category;
- π_2 be the probability of a case falling into the medium or high categories; and
- π_3 be the probability of a case falling into the low, medium, or high categories.

The probabilities here are cumulative, with $\pi_1 < \pi_2 < \pi_3$. Because all cases are classified into one of the three categories, $\pi_3 = 100\%$, so one does not need to estimate π_3. One could estimate π_1 with a binary logistic regression, coding a case with a 1 if it falls into the high category or a 0 if it falls into the medium or low categories. One could again use binary logistic regression to estimate π_2, coding a case as 1 if it falls into the medium or high categories and 0 if it falls into the low category. However, it is more statistically efficient and parsimonious if one only estimates one regression equation. This can be done if we assume the two regression lines discussed previously have the same slope (i.e., are parallel). (Of course, the logistic functions actually define curves, but these are mathematically transformed into lines with a fixed slope.) We now only have to estimate one slope parameter for each independent variable. The parameter represents the log odds of being in a higher category if the IV is increased by one unit.

Indexing the category numbers with j, ranging from 1 to J (the total number of categories), the ordinal regression model can be expressed as

$$ln\left(\frac{\hat{\pi}_j}{1 - \hat{\pi}_j}\right) = \alpha_j - \beta X. \tag{Eq. 11.4}$$

A negative sign is used in the model so that larger coefficients will reflect larger scores. Recall that $\pi_1 < \pi_2 < \pi_3$. Therefore, π_1, which is associated with the high category, will have the lowest probability, so the negative sign is used in Eq. 11.4 to reverse the relationship. Doing so in effect makes the low category the baseline category. Note that the IBM SPSS GLM procedure allows one to make the high category the baseline category by choosing the "Descending" option for "Category order" on the "Response" tab. Descending is the approach also used in SAS.

Argumentation Example

Suppose a researcher provides a group of college students instruction on how to integrate arguments and counterarguments into an overall final solution when writing an argumentative essay, as in Nussbaum (2008). Students in the control group did not receive this instruction but did write the same number of essays as those in the treatment group. All essays were categorized into a high, medium, or low category, depending on their amount of argument–counterargument integration.

Inspection of the frequencies or proportions in Table 11.2 shows that there was a higher concentration of treatment group participants in the high category. The

Table 11.2 Ordinal Regression: Argument-Counterargument Integration Data

Amount of integration	Frequency		
	Treatment group	Control group	Total
High	44	11	55
Medium	45	46	91
Low	29	44	73
Total	118	101	219
	Proportions		
	Treatment group	Control group	Total
High	37.3	10.9	25.1
Medium	38.1	45.5	41.6
Low	24.6	43.6	33.3
Total	100.0	100.0	100.0
	Cumulative proportions		
	Treatment group	Control group	Total
High	37.3	10.9	25.1
Medium or high	75.4	56.4	66.7
	Odds		
	Treatment group	Control group	OR
High	0.59	0.12	4.86
Medium or high	3.07	1.29	2.38

cumulative proportions and cumulative odds derived from these are also shown. Inspection of the ORs shows that the treatment increased the odds of a case falling into the high or medium categories.

The ORs further indicate that the treatment increased the odds of being in the high category (OR = 4.86) and the odds of being in the medium or high category (OR = 2.38). An ordinal regression should be performed to determine if these differences are significant. In IBM SPSS, I prefer using the ordinal regression procedure over GLM because the former allows for a test of the parallel lines assumption. Go to "Analyze" ➜ "Regression" ➜ "Ordinal," and click on "Output . . ." Check "Test of parallel lines." I also recommend checking "Cell information" as this will provide information useful in interpreting the beta estimates. Conducting the ordinal regression in SPSS indicated that the location parameter (i.e., slope) was −1.13 for [condition = control] and therefore was 1.13 for [condition = experimental]. $Exp(1.13) = 3.10$. Note that this estimated OR is about in between the two cited previously. The estimate is significant ($p < .001$). The threshold estimates are intercepts and typically not of interest.

According to Hosmer and Lemeshow (2013), next perform any stepwise selection of main effect, check for nonlinearities and then for interactions. None of these are applicable to this example. One should then conduct the parallel lines test (i.e., test of proportional odds).

Parallel Lines Test

For the parallel lines test, $\chi^2(1) = 3.71$, $p = .054$. We fail to reject the null hypothesis of parallel lines, but just barely. In fact, the goodness-of-fit tests indicate that the model only barely fits the data (Pearson $p = .056$, Deviance $p = .054$). (See Williams, 2006, for technical details on the parallel lines test.)

Remedial Measures

However, if one's data were to fail the test, there are five types of remedial options. First, one could conduct two different logistic regressions, one for each cumulative nonbaseline category, as was described previously. The results would be to estimate the two ORs shown in Table 11.2. (The estimates would differ slightly than the actual values.) O'Connell (2006) recommended that separate regressions be routinely performed even if one reports the results of the overall ordinal regression so as to obtain a better sense of any patterns in the data. Furthermore, if the beta estimates for the different regressions are close together, we could accept the parallel lines assumption even if our model fails the parallel lines test. This practice is recommended because the test is overly sensitive to small departures from the null when (a) the sample is large, (b) there are a large number of predictors, and/

or (c) metric predictors are used (Allison, 1999; Clogg & Shihadeh, 1994; Peterson & Harrell, 1990).

A second option is to fit a multinomial model; however, this option discards information about how the categories are ordered, resulting in less statistical power. A third option is to combine some categories to determine if parallelism can be reached that way; however, combining also throws away information. A fourth option is to fit a *partial proportional odds model* where the parallel lines assumption is made only for those variables where it is justified (Williams, 2006); this option, however, while available in some software packages (e.g., Stata) is not currently available in SPSS. The fifth option (and the one I would try first) is to fit a model with an alternative link function. Table 10.3 (in Chapter 10) summarizes these functions.

Alternative Link Functions

In SPSS, alternative link functions can be found in the ordinal regression procedure under "Options . . ." (there is a drop-down menu for "Link:" at the bottom of the "Options" dialog box) or in the GLM procedure under "Custom" (for type of model) with a multinomial distribution. Two useful link functions are the cumulative complementary log-log (cloglog) function and the cumulative negative log-log function. As an example, consider the hypothetical data in Table 11.3. As in Zayeri, Kazemnejad, Khansfshar, and Nayeri (2005), 900 newborn babies ("neonates") were classified into three categories depending on body temperature (normal, mild hypothermia, severe hypothermia). Some had a risk factor that I will call Risk Factor X (e.g., mother had multiple pregnancies, is overweight). One can see from Table 11.3 that the cumulative ORs are positive, indicating that Risk Factor X does increase hypothermia, but the two ORs (3.83, 6.79) are substantially different. It is not surprising, therefore, that the logit model fails the parallel lines test ($\chi^2(1) = 8.03$, $p = .005$); however, a negative log-log model passes ($\chi^2(1) = 0.094$, $p = .76$).

For the negative log-log regression, the location parameter estimate for [Risk Factor X = 1] was 1.27 ($p < .001$). The estimate was positive, meaning that having Risk Factor X increases the odds of having hypothermia. However, it is also important to understand the magnitude of the effect. It is therefore best to create a table of proportions (as in Table 11.3) so that readers can get a sense of the direction and magnitude of the relationship. One needs to be careful in using alternative link functions because the beta parameters are no longer *ln*(ORs).

Minimum Cell Sizes

Because ordinal regression relies on chi-square tests, none of the cells should have a zero count and, as a rule of thumb, 80% should have a count of 5 or more; 10 is preferable (Allison, 1999). Therefore, before running an ordinal regression, use a

Table 11.3 Ordinal Regression: Neonatal Hypothermia Data

	Data		
	X present	X absent	Total
Severe (3)	106	46	152
Mild (2)	197	132	329
Normal (1)	84	335	419
Total	387	513	900
	Proportions		
	X present	X absent	Total
Severe	27.4	9.0	16.9
Mild	50.9	25.7	36.6
Normal	21.7	65.3	46.6
Total	100.0	100.0	100.0
	Cumulative proportions		
	X present	X absent	Total
Severe	27.4	9.0	16.9
Mild or severe	78.3	34.7	53.4
	Odds		
	X present	X absent	OR
Severe	0.38	0.10	3.83
Mild or severe	3.61	0.53	6.79

Note: X refers to Risk Factor X.

crosstabs procedure to ensure this assumption is satisfied. This check only needs to be performed for nominal predictors, not metric ones. One should also check the different levels of the DV to ensure that no level is sparse; otherwise, combine categories or collect more data.

Metric Predictors

The previous examples of ordinal regression all involved nominal predictors, but, more typically, one or all of the predictors will be metric. In that event, the location parameter (if the logit link is used) will be a log of an OR only in the sense that it represents the percentage increase in odds from a one-unit increase in the predictor variable. In SPSS, a metric variable should be included in the "Covariates" box, not the "Factors" box.

Table 11.4 Ordinal Regression with Metric Predictor: Argument-Counterargument Integration as Predicted by IQ

Model Fitting Information

Model	−2 log likelihood	Chi-square	df	Sig.
Intercept Only	254.878			
Final	19.381	235.496	1	.000

Link function: Logit.

Pseudo-R^2

Cox and Snell	.832
Nagelkerke	.962
McFadden	.892

Link function: Logit.

Parameter Estimates

		Estimate	Std. Error	Wald	df	Sig.	95% Confidence Interval	
							Lower Bound	Upper Bound
Threshold	[Arg_Integration = 1.00]	13.892	4.024	11.918	1	0.001	6.005	21.779
	[Arg_Integration = 2.00]	28.88	8.187	12.444	1	0.000	12.834	44.926
Location	IQ	0.708	0.2	12.494	1	0.000	0.315	1.1

Link function: Logit.

Test of Parallel Lines[a]

Model	−2 log likelihood	Chi-square	df	Sig.
Null Hypothesis	19.381			
General	19.351	.030	1	.862

Note: The null hypothesis states that the location parameters (slope coefficients) are the same across response categories.
[a]Link function: Logit.

For example, suppose one is predicting argument–counterargument integration (high, medium, or low) in essays by 219 individuals by IQ score. Table 11.4 shows a portion of the output from an ordinal regression using hypothetical data. Not shown is an error message: "There are 138 (63.9%) cells (i.e., dependent variable levels by combinations of predictor variable values) with zero frequencies." The empty cells result from there being about 70 different levels of the IV crossed with 3 levels of the DV. A contingency table of the data would thus have about 210 cells, whereas here $n = 132$, so there will be a lot of empty cells. This situation is not problematic because, with metric predictors, we're not basing the analysis on a contingency table. But do avoid using the deviance or Pearson's X^2 as overall fit measures.

Next, diagnostic tests should be conducted. The Box-Tidwell test is nonsignificant (B $= -0.14$, $p = 0.8$). The result of the parallel lines test is also nonsignificant ($\chi^2(1) = .03$, $p = .86$), confirming the parallel lines assumption. What about checking for outliers? In SPSS, one can examine Pearson residuals by checking "Cell Information" under "Output . . ." in the ordinal regression procedure; and then eyeball the output. In the example, the largest Pearson residual was 2.11. With ungrouped data, one cannot assume the residuals are normally distributed, but the next highest residual was 1.76 (not too distant) and, more importantly, graphical methods did not indicate anything unusual about this observation. Other diagnostic measures, such as deviance residuals or likelihood residuals or Cook's distances, are not directly available, but one could generate these by conducting separate logistic regressions for each DV category (high vs. medium/low and high/medium vs. low). Doing so, however, is labor intensive.

Finally, even if our interest is not primarily in prediction, it is useful to examine the predictive performance of the model to identify observations or clusters of observations that do not fit the model well. With the SPSS ordinal regression procedure, if one clicks the "Output . . ." button and saves the predicted categories, then one can use "Analyze" ➔ "Descriptive Statistics" ➔ "Crosstabs" to make a contingency table of the predicted values against the actual ones (which are given by the DV). Table 11.5 shows the results. Summing the diagonal of the table shows that 125 cases out of 132 (94.7%) were correctly classified. Furthermore, none of the off diagonal cells were particularly large, so there did not appear to be any systematic bias.

Finding clusters of observations that do not fit the model well could be due to various factors (e.g., nonlinearities, omitted relevant variables, choice of link function). Also, consider whether the DV categories are ordered correctly; sometimes

Table 11.5 Argument Integration and IQ: Predicted Versus Observed Categories

Actual response category	Predicted response category			Total
	1	2	3	
1	24	2	0	26
2	1	31	2	34
3	0	2	70	72
Total	25	35	72	132

the presumption underlying an assumed ordering is incorrect. One might also add a scale component to the model for certain variables (e.g., for gender if there is more variation on the DV for males than females).

Once we settle on a final model, the location parameter(s) can be interpreted. For the current model, the estimate is significantly different from zero [Wald $\chi^2(1) =$ 12.49, $p < .001$, $exp(0.708) = 2.03$]. (In SPSS, the exponentiation is not provided in the output and so it must be performed manually.) The result can be interpreted as follows: Each additional IQ point approximately doubles the odds that an essay will have a greater degree of argument–counterargument integration (falls into a higher category).

Ordinal Independent Variables

We have considered cases where the IVs are nominal or metric, but ordinal regression can also be performed when the IV consists of ordered categories. Although if both X and Y are ordinal, testing for statistical independence can be performed with a *tau* or *partial tau* test (see Chapter 6). Using ordinal regression also allows one to more easily control for third variables. If the third variable is nominal, one can employ the following logit model for an $I \times J \times K$ ordered contingency table (Agresti, 2002):

$$logit\left[\text{Prob}\left(Y \leq j \mid X = i, Z = k\right)\right] = a_j + \beta X_i + \beta_k^Z,$$

where $\{X_i\}$ are ordered scores. The model reflects k partial tables, one for each category of Z. The term β_k^Z reflects differences among the tables (the z superscript is not an exponent but identifies the coefficient as being assigned to Z, the third variable).

This approach is an extension of the Mantel-Haenszel test for $2 \times 2 \times k$ tables (see Chapter 5) but extended to larger, ordered tables. One can test for conditional independence between X and Y (controlling for Z) in this model by examining whether there is a common linear trend between X and Y among the partial tables. One simply conducts an ordinal regression and tests whether the beta estimate in the βX_i term is significantly different from zero. We do not need to assume that X is metric, only ordinal. If there is an association between X and Y, then β will not be zero.

There is still an issue of whether β should be given a precise quantitative interpretation. According to Agresti (2002), treating ordinal variables in a quantitative manner is only sensible "if their categorical nature reflects crude measurement of an inherently continuous variable" (p. 292). In this case, β could be interpreted as reflecting the probability of a one *category* increase in Y given a one *category* increase in X. If X is purely ordinal, then one should refrain from interpreting the magnitude of $\hat{\beta}$ but can still conduct a test of conditional independence.

On the other hand, sometimes measurements are accurate enough that scores for the ordered categories could be considered metric. For example, the hypothermia categories normal, mild, and severe reflect an underlying continuous variable, body temperature. If the assigned scores also represent equally spaced points on this scale, they could actually be considered metric and one could use certain *log-linear association models* that will be described in the next chapter. For example, categories for income (under $50,000, $50,001 to $100,000, $100,001 to $150,000) represent the same range of income ($50,000), so scores for the categories such as 1, 2, and 3 could be considered metric rather than ordinal. As will be explained in Chapter 12, some of these models have also been applied to cases involving ordinal variables.

Adjacent-Categories Model and Survival Analysis

In the adjacent-categories logistic model, the noncumulative probability of a case being in a certain category (j) is compared with the probability of being in the next lower category ($j - 1$), rather than with the probability of being in the reference category (as in the multinomial model). In contrast with the ordinal regression model described previously, the adjacent categories model avoids the use of cumulative probabilities and may be more meaningful in certain situations. However, there is not a specific option for this model in IBM SPSS. Parameter estimates can still be easily derived from a multinomial model (see Hosmer & Lemeshow, 2000), or from a log-linear model (see Technical Note 11.1). Whether or not to use an adjacent-categories model over a proportional-odds model should depend more on how one wishes to present the results than on considerations of data fit; typically, the results are about the same. Another alternative model is the continuation ratio model, which is applicable when individuals move through stages of development (O'Connell, 2006).

Survival analysis addresses the probability of an individual surviving into the next time period (or higher category) and in the social sciences can be applied to such things as college retention or human development (i.e., entering the next stage). Good treatments of survival analysis are provided by Klein and Moeschberger (2003), Kleinbaum and Klein (2012), and Le (1998). These models build on those considered in this chapter, particularly the complementary log-log model (see Technical Note 10.9).

Poisson Regression

Researchers often count things (e.g., the number of teachers leaving the profession in a given year, the number of people who vote, or the number of people in a given year who contract cancer). Counts are considered discrete, metric data. Metric data can be either interval or ratio; count data are specifically ratio in nature because a score of zero truly reflects the absence of something (e.g., no one contracting a rare disease in a given year). Although count data are often analyzed using ordinary least squares (OLS), the values have a lower bound of zero; therefore, using a linear model creates misspecification and heteroscedasticity (since the error variances become truncated and nonnormal near the zero point). Furthermore, count data are often skewed, especially if there are many zero counts. This condition reduces statistical power.

In IBM SPSS, Poisson regression can be conducted using the GLM procedure. Under the "Type of Model" tab, choose "Poisson log linear." The predictors can consist of factors (i.e., nominal variables) or covariates (i.e., metric variables). The response variable will be the number of positive cases observed per unit of time or space (e.g., the number of people per year who contract cancer in the United States). The total population of the United States per year must also be taken into account; this variable is known as the *offset* and is specified under the "Predictors" tab in SPSS (enter the logarithm of the offset variable). The total population is a predictor, because an increase in the number of people with cancer could just be due to there being more people, so one should control for the effect of this variable in the model. The number of negative cases (e.g., people without cancer) is just the count minus the offset.

In some Poisson models, an offset variable is used to reflect the time or duration that individuals have been exposed to an IV. For example, the duration that individuals have been exposed to a carcinogen might vary among time periods (or geographic areas). One can control for variations in exposure duration by including it in a Poisson model as an offset.

It is important to note that when using the GLM procedure, one should not weight the data by the counts. The count is not a weighting variable but is the DV. However, when using the log-linear procedures described in the next chapter, one should weight the data by the counts, as is standard practice with contingency tables.

The Poisson Distribution

The Poisson distribution was discovered in the 19th century by the French physicist and mathematician Siméon-Denis Poisson. Prior to its discovery, counts were modeled using the binomial distribution.

Binomial Distribution

As explained in Chapter 1, when working with frequencies, the mean of the binomial distribution is $n\pi$, where π is the probability of a hit. If one tosses 100 coins in the air, the most likely outcome is that 50 will come up heads. The actual probability of different outcomes is given by the binomial formula (see Eq. 1.10 in Chapter 1), which involves the factorial term $n!$. So, for example, for $n = 5$, $n! = 5 * 4 * 3 * 2 * 1 = 120$. For $n = 100$, $n! = 9.3326 * 10^{157}$. The second example shows that the term $n!$ becomes unmanageably large as n increases (EXCEL will not calculate the factorial for any number in excess of 170). How then could one possibly calculate the probability of a certain number of hits with a relatively large n?

Poisson Approximation of the Binomial

Poisson showed that as n approaches infinity and π approaches zero, then the binomial distribution approaches the following distribution (which became known as the Poisson distribution):

$$\text{Prob}(Y = y) = \frac{\lambda^y e^{-\lambda}}{y!}, \qquad\qquad \text{(Eq. 11.5)}$$

where y is the number of hits for a given time period and λ is the average number of hits over all time periods. (See Appendix 11.1 for the derivation.) The parameter lambda (λ) can be estimated by collecting data for a large number of time periods (e.g., the number of annual cancer cases over a 30-year period). Lambda (λ) is a naturally occurring number (e.g., the annual number of new cases of cancer in the United States from 1975 to 2005 was about 1.25 million, annually). Given that the annual U.S. population was about 249.4 million during this time period, the average annual rate was about 0.5%. One could therefore estimate $\pi = .5\%$, $n = 249.4$ million, and $\lambda = 1.25$ million. The binomial distribution only becomes equivalent to the Poisson distribution when $\pi = 0$ and $n = \infty$, but the Poisson distribution provides a reasonable approximation to the binomial when n is very large and Y is very small.

The Poisson distribution therefore is often used to find the probability of rare events in large populations, where λ is very low. Figure 11.1 illustrates the Poisson distribution for different values of λ. All the distributions shown are positively skewed, but the skew is particularly pronounced for lower values of λ. (Note: Because λ is an average, it need not be discrete.) There may be many time periods in which no positive cases (i.e., hits) are observed, that is, many values of 0. As λ increases, the distribution becomes more normal and, as a rule of thumb, a normal approximation is reasonable if $\lambda > 10$ (Coxe, West, & Aiken, 2009).

Suppose $\lambda = 0.25$. According to Eq. 11.5, the probability of having one incident of the event in a year is $\text{Prob}(Y = 1) = \frac{0.25^1 e^{-0.25}}{1!} = 19.5\%$. This can also be calculated with the EXCEL formula =POISSON.DIST(y, λ, [0 = noncumulative, 1 = cumulative]). In the previous example, =POISSON.DIST(1,0.25,0) ➔ 19.5%. Likewise, the probability of observing a 0 is 77.9%.

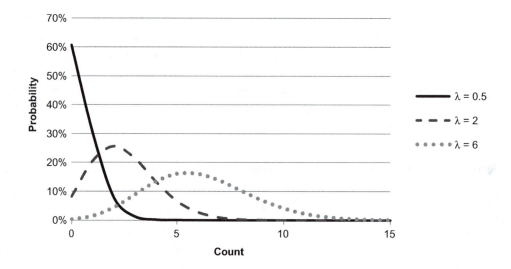

Figure 11.1 The Poisson distribution for different mean values (λ). To facilitate visual comparison, the probability distributions are drawn as continuous but are actually discrete. The amount of skew declines with greater values of λ.

EXPLORING THE CONCEPT

Why do the two probabilities sum to 97.4% rather than 100%? Using EXCEL, find the probability of $Y = 2$ and $Y = 3$.

Poisson Processes

A *Poisson process* is a causal process that generates cases, up to a few per time period. The resulting data will have a Poisson distribution. Nussbaum, Elsadat, and Khago (2008, p. 309) give the following example, adapted from Rosenfeld (2002):

> Suppose we have 20 orange trees, all of which look alike. . . . During the harvest season, we put a basket under each tree hoping to catch a few oranges each day. As long as each tree is similar, so that the same processes apply to each tree, and as long as each orange falls into the basket as a separate and independent event, this situation will represent a Poisson process and can be described by the Poisson distribution. The distribution will tell us how many baskets we can expect to contain 0, 1, 2, 3, 4, or more oranges each day.

In this example, λ is an average over both time (i.e., days) and space (i.e., basket). Agresti (2002) gave the example, taken from Brockmann (1996), of counting the number of horseshoe crabs per square meter on the beach at a given point of time; here the data reflect counts per unit of space rather than time. Other examples of rare phenomena for which the Poisson distribution has been applied include neuron firing, alpha emissions from radioactive sources, the distribution of stars in space, the frequency of accidents along a given stretch of highway per year (Antelman, 1997), the number of customers who arrive at a bank in a given time period (Padilla, 2003), or the number of rebuttals generated in argumentative essays (Nussbaum & Schraw, 2007).

Note that in some of these examples the event is quite rare, and in others, only semirare. For example, the frequency of accidents along some stretches of road per year is very low but is not that low along other stretches. The Poisson distribution is not just applicable to extremely rare events. Within 50 years of Poisson's discovery of the distribution, mathematicians began using the Poisson distribution in its own right, not just as an approximation to the binomial (Larsen & Marx, 1990). However, the events should be fairly low in frequency; otherwise, the distribution becomes approximately normal.

N *as a Source of Random Error*

When λ (and π) are not extremely low and n is not inordinately large, when should one use the Poisson distribution rather than the binomial distribution? The key consideration is whether n is a random variable or not. In calculating the probability of 100 coin tosses coming up heads, we can use the binomial distribution because we fixed n at 100. But in other situations, the n will naturally vary among units of time (or space) and is therefore represented by the offset variable. For example, the number of students who drop out of school in a particular school district per year is a function of both the probability of dropping out and the number of students enrolled per year ("the offset"). In the coin-tossing example, if the number of coins tossed was randomly determined, then the probabilities should be calculated using the Poisson distribution. For example, one might reach into a jar of coins and randomly pick a number to pull out and toss. One would need to do this multiple times to obtain an estimate of λ, that is, the average number of heads obtained per toss. The number of hits per time period (Y) would be a function of both π (50% for a fair coin) and n (number of coins scooped out per time period). The parameter λ would also be a function of these elements, because it is a function of Y (the number of hits). A Poisson distribution has the additional random element contributed by n.

Variance Equal to Mean

Poisson showed that the variance of the Poisson distribution is given by λ (see Appendix 11.1). It is a special feature of the Poisson distribution that the variance is equal to the mean. With simply a binomial distribution, the mean count when predicting frequencies is $n\pi$, which can be estimated by nP. Because λ is defined as the mean count, $\hat{\lambda} = nP$. The variance of a binomial distribution can be estimated by nPQ, where P is the probability of a hit and Q a probability of a miss ($Q = 1 - P$). The variance is typically less than the mean because of the Q term. For example, if $n = 100$ and $P = 5\%$, then $Q = 95\%$. The mean count would be estimated as 5 and the variance as $100 * 5\% * 95\% = 4.75$. With a Poisson distribution, if the mean were 5, the variance would also be 5. There is additional random variation because n is random.

EXPLORING THE CONCEPT

Poisson discovered his distribution by allowing the probability to approach zero. If $\pi = 0\%$, to what value would $(1 - \pi)$ be equal? (This value is estimated by Q.) Would it now be the case that the variance would equal the mean (i.e., $n\pi(1 - \pi) = n\pi$? (Note: Although $\pi = 0$, the mean and variance do not become equal to zero because $n = \infty$.)

Assessing Distributional Shape

One can assess whether one's data follow a Poisson distribution by performing a Kolmogorov-Smirnov (KS) test (see Chapter 3). In SPSS, go to "Analyze" ➔ "Nonparametric Tests" ➔ "One Sample." (In IBM SPSS, go to the "Settings" tab, "Customize tests," and "Compare observed distribution to hypothesized.")

In large samples, the KS test can be overly powerful, detecting small departures from the hypothesized distribution. Another option is to create a Q-Q plot, which compares the quantiles of the actual distribution with the hypothesized distribution; a straight line on the plot where $X = Y$ indicates a perfect fit. (SPSS has an option for Q-Q plots, but unfortunately not specifically for the Poisson distribution.) Finally, one should construct a histogram of the data (see Technical Note 11.2 for a graphical method of evaluating the distribution). Although these tests assess whether the marginal distribution of Y is Poisson, the Poisson model only assumes that the conditional distribution of each Y_i (i.e., conditional on specific values of X_i) are Poisson. So tests of the marginal distribution are helpful but not essential for justifying use of Poisson regression.

The Poisson Regression Model

Lambda (λ) is the mean parameter of the Poisson distribution. The underlying Poisson process that generates the data will determine λ. Variables that alter the Poisson process will change the value of λ. The goal of Poisson regression is to model how these variables alter λ and, in turn, the value of the DV (i.e., the counts). As explained by Nussbaum et al. (2008),

> using our previous example of orange trees, we assumed that each tree generated oranges at the same average rate, with a mean count of λ_1. Suppose, however, that the last row of trees in the orchard receives more sunlight (Rosenfeld, 2002). In that case, the row of trees will generate oranges at a higher rate, with a mean count of λ_2. The mean count (λ) will therefore be affected by an independent variable, the amount of sunlight.
>
> (p. 311)

Amount of sunlight could then be an IV in the regression. IVs can be metric or nominal.

The Poisson Generalized Linear Model

The Poisson regression model is a generalized linear model (GLM), as described in Chapter 9. For the stochastic component, it is assumed that each conditional distribution of Y_i is Poisson. The canonical link function is the log link, which is used for exponential relationships (e.g., exponential growth; see Chapter 9). The justification for use of the log link is that a count cannot be less than zero; a linear relationship might predict a negative value on the DV, whereas an exponential relationship would not. Figure 11.2 illustrates an exponential relationship between the mean counts for different values of X (three are shown).

EXPLORING THE CONCEPT

Will exponentiating a large negative number always produce a nonnegative number? What is $e^{-10,000}$?

The GLM procedure in IBM SPSS does allow one to use other link functions, including the identity link, if those are preferred for theoretical reasons. In some cases, a linear function might result in better predictions or be more theoretically interpretable when the predictions are positive values. However, because the log link is the canonical link function for the Poisson (see Chapter 9, Technical Note 9.7), it has certain desirable mathematical properties.

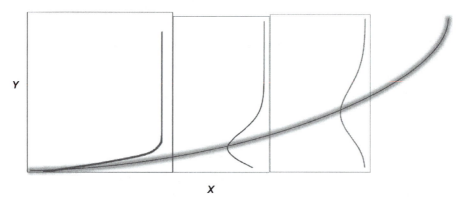

Figure 11.2 Poisson GLM: Exponential relationship between means for different values of X.

The Poisson regression model can be written as:

$$\lambda_i = s_i * \left(e^{\alpha + \beta X_i}\right),$$ (Eq. 11.6)

where λ_i is the predicted value of Y_i for the ith level of X, and s_i is the offset. (If no offset is used, s_i is set equal to 1.0.) Taking the logarithms of both sides yields

$$ln(\lambda_i) = ln(s_i) + \alpha + \beta X_i.$$ (Eq. 11.7)

Eq. 11.7 shows why it is necessary to create a new variable that is the logarithm of s_i when specifying the offset in the IBM SPSS GLM "Predictors" tab. Note that if the $ln(s_i)$ term is moved to the left-hand side of the equation, the regression equation predicts $ln(\lambda_i) - ln(s_i)$, which is equal to $ln\left(\frac{\lambda_i}{s_i}\right)$, the log of a rate (which can also be interpreted as a probability). For example, if in a school district there are 500 students enrolled at the beginning of the year and 131 leave, the dropout rate for that year is $\frac{131}{500} = 26.2\%$. This value can also be thought of as the probability that a student will leave school. The Poisson regression model can therefore be rewritten as

$$ln(\pi_i) = \alpha + \beta X_i.$$ (Eq. 11.8)

Exp(α) is the probability that a student will leave school when $X_i = 0$ and can be written as π_0. If X_i is dichotomous, *exp(β)* reflects how this probability will increase when $X_i = 1$. Because $\beta = ln(\pi_i) - ln(\pi_0)$, $exp(\beta) = \frac{\pi_1}{\pi_0}$. This last term is not a ratio of odds but a ratio of probabilities, known as the *relative risk* (RR), also known as the *incident risk ratio*. For example, if the drop-out rate for boys is 120% of the rate for girls, then RR = 1.2.

EXPLORING THE CONCEPT

In a student cohort, there are (a) 70 boys, and 15 drop out of school, and (b) 60 girls, and 11 drop out of school. Calculate the drop-out rate for boys and girls, and show that the RR = 1.169.

Note that if we subtract 1.0 from the ratio, $\frac{\pi_1}{\pi_0} - 1.0$ can be interpreted as the percentage increase in the probability (e.g., of dropping out) from a one-unit increase in X (e.g., 20%).

In this example, the IV is nominal, but if the IV were metric, $exp(\beta) - 1.0$ can be interpreted as a percentage increase in the probability from a one-unit increase in the IV. For example, if $exp(\beta) = 1.15$, then we could interpret the beta parameter as indicating a 15% increase in the chance of something (e.g., dropping out) from a unit increase in the IV (e.g., number of courses failed). The logic is similar to logistic regression except that the location parameter reflected percentage increases in the odds, not probabilities. (See Chapter 4 for an explanation of the differences between odds and probabilities.) This fact makes the interpretation of the beta coefficients in a Poisson regression more straightforward.

It is important to note that this interpretation is only valid if the log link function is used. If the identity link is used, then β should be interpreted in the same manner as in OLS regression, specifically as the expected increase in the DV from a one-unit increase in the IV. Poisson regression can still afford more statistical power than OLS—especially if the data are skewed—because the error term is more accurately modeled.

Figure 11.2 illustrates visually a Poisson regression model with the canonical log link. The means $(\lambda_i s)$ fall on the population regression line. Note that as the IV increases, so does λ_i, and because λ_i is also the variance for a given value of the IV, the variances increase as well. For an OLS model, this fact will not only cause heteroscedasticity but will also create high standard errors for the beta coefficient(s) (the greater variation at the higher values of the IV are given more weight). With Poisson regression (and assuming the data are generated through a Poisson process), the standard errors will tend to be lower because the error distributions are narrower for higher values of the DV, thereby constraining the number of possible regression lines that can be estimated from various samples (Nussbaum et al., 2008).

It is also the case that when OLS is used on count data with low means, there may be heteroscedasticity. At low values of the IV, the estimates of λ_i will be low assuming a positive relationship, and therefore so will the variance. At high values of the IV, the estimates of λ_i will be high and so will the variance. Coxe et al. (2009), for example, point out that the average number of children per woman in industrialized countries is much lower than in countries with economies based on subsistence agriculture, and so therefore are the variances. For example, the mean as of 2007 in Germany was 1.40, in Japan 1.23, and in the UK 1.66; in contrast, the mean in Laos was 4.59, Nigeria 5.45, and Uganda 6.84, indicating more variability. With OLS, heteroscedasticity can bias standard errors (see Chapter 9), often increasing the probability of Type I errors.

Diagnostics

The ratio of the deviance to degrees of freedom is the dispersion value ϕ. In a Poisson regression, when the predicted values are mostly greater than 5 and the number of units of time (or space) is fixed, the deviance and Pearson chi-square statistics will have an approximate chi-square sampling distribution with $n - p$ degrees of freedom (Agresti, 1996) and can be used in an overall goodness-of-fit test. (One tests whether the fit statistic is significantly different from zero.) Even if the prerequisite conditions are not met, and a formal test is not possible, a deviance substantially greater than the number of degrees of freedom indicates a poor fit (Allison, 1999).

To identify portions of the data that are not well fit by the model, plot the standardized deviance or likelihood residuals against the linear predictor ($\eta = XB$). Also plot the absolute values of these against η. For both graphs, add a best-fitting OLS line and loess line. If the model is adequate, these lines should be fairly flat. If the residuals increase with increases in the predicted values, this will indicate that the model is inadequate in some way. First, there could be an important, relevant predictor omitted from the model. A second possibility is nonlinearity. Just as with logistic regression, where there needs to be linearity in the logits, with Poisson regression there needs to be a linear relationship between the log of the counts and any metric predictors. The Box-Tidwell test can be conducted (see Chapter 10), but because the procedure is overly sensitive in large samples, minor nonlinearities can be safely ignored if the relationship looks linear graphically. If nonlinearity is a problem, quadratic and log terms can be added to the model. Sometimes the problem is the omission of an interaction term. These terms are more easily interpreted in terms of their effect on the log (counts) rather than the counts themselves.

In the SPSS GLM procedure, the predicted values can be saved by going to the "Save" tab and checking "Predicted values of linear predictor." These are the predicted log(counts). (Checking the box "Predicted value of mean of response" and exponentiating the results produces the predicted counts.) One can also go to the "Save" tab and save the "Deviance residuals" and other diagnostic statistics. It is important to examine the same diagnostic statistics that were discussed for logistic regression in Chapter 10, especially those relating to overdispersion.

Overdispersion

Overdispersion occurs in Poisson regression if there is more variation in the data than that predicted by the standard Poisson model. The standard Poisson model assumes that the conditional variance (i.e., variance at each level of an IV) is equal to the conditional mean (λ_i). This assumption is known as *equidispersion*.

Overdispersion can occur for all the reasons that were discussed in Chapter 10. In addition, the Poisson model does not explicitly contain an error term; rather, for the random component of the GLM, it is just assumed that the distribution of Y_i, conditioned on a particular level of X_i, follows a Poisson distribution (with mean λ_i). This point is illustrated in Figure 11.3(a). The figure shows simulated data that were generated by calculating the linear prediction $ln(\lambda_i)$ for each level of X_i using the specified parameters for β_0 and β_1, and then exponentiating the results to produce the variable "lambda." In SPSS, "Transform" ➔ "Compute Variable" ➔ "Count = RV.POISSON(lambda)" was used to model the random component and generate the observed values shown in Figure 11.3(a). An error term was not explicitly included, although residuals still could be derived from the data. Note from the figure that the variances increase as X_i increases because the predicted means ($\hat{\lambda}_i$) increase, and the variances equal the means. In the Poisson model, the counts can have variation around their means due to the effect of omitted variables, but the variance is limited to the mean.

In reality, however, the influence of omitted variables might be larger than what is recognized by the Poisson model (especially if there are violations of statistical independence among observations). If there are omitted variables that have an influence on the population counts, then different cases at the same level of X_i might have different λ_is, causing deviation from the Poisson distribution and a conditional variance greater than the mean.

EXPLORING THE CONCEPT

In Poisson regressions, omitted variables can cause overdispersion, which violates a model assumption. This holds even if the omitted variable is not correlated with the IVs. Why does this not cause a problem in OLS? Does the error term in OLS reflect the influence of omitted variables?

In practice, count data are frequently overdispersed. Consider the data in Figure 11.3(b). The variances of the higher values of X are greater than those shown in Figure 11.3(a). For example, for $X = 10$, I calculated the empirical variance to be 11.9 for the Poisson and 779 for the negative binomial.

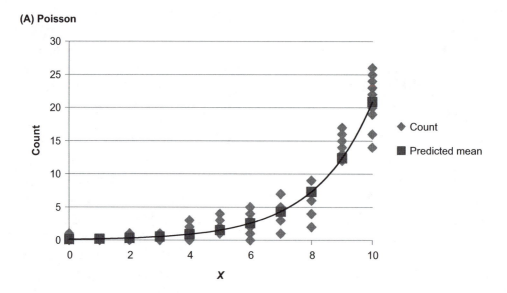

(A) Poisson

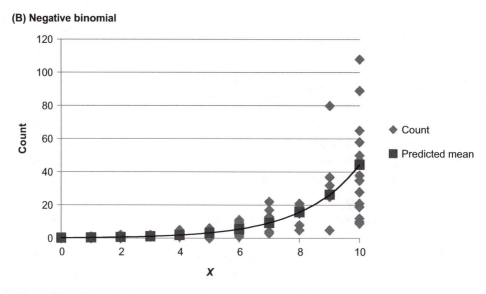

(B) Negative binomial

Figure 11.3 Simulated count data using log link with random component (a) Poisson, and (b) negative binomial. The latter illustrates extra-Poisson variation (overdispersion). (Parameter values used in simulation were $\beta_0 = -2.0$, $\beta_0 = 0.5$, $k = 2$.)

Negative Binomial Models

Although one can use a scaling factor to correct for overdispersion, it is more statistically powerful to diagnose and remediate overdispersion by fitting a *negative binomial* (NB) *model* (Allison, 1999). The nature of this model is described in more detail in Appendix 11.2 but briefly—like the Poisson distribution—it is a discrete probability distribution (positively skewed) that has zero at its lower bound. However, unlike the Poisson distribution, the variance can exceed the mean. The NB distribution can therefore be used when there is overdispersion. The NB model allows λ_i (the conditional mean count) to vary for individuals at the same level of X_i. This is modeled by adding an explicit error term where $exp(\varepsilon)$ has a standard gamma distribution. You may recall from Chapter 9 that the gamma distribution is a continuous distribution that is positively skewed and bounded at zero. The gamma distribution has a dispersion parameter (here set to 1.0) and a shape parameter, k.

The reciprocal of the shape parameter, k^{-1}, is called the *ancillary parameter* and measures the degree of overdispersion. As k approaches infinity, the gamma distribution approaches a Poisson distribution and k^{-1} approaches 0, indicating no overdispersion. The data in Figure 11.3 were generated using $k = 2$.

EXPLORING THE CONCEPT

The specific steps used to generate the data in Figure 11.3 were (a) an error term was created in SPSS using "Transform" ➜ "Compute Variable" ➜ "Error = RV.GAMMA(2, 1)," (b) predicted means were calculated using the linear equation, $ln(\lambda_i) = -2.0 + 0.5 * X + ln(e)$ and then exponentiating, and (c) observed values were generated using "Compute Variable" ➜ "Count = RV.POISSON(λ_i)." Try generating your own data using these steps. The data will not be identical to those in Figure 11.3(b) because of random error. Do you see how the NB distribution is a mixture of the gamma and Poisson distributions? The gamma distribution is continuous; why is λ_i considered so?

The NB distribution has a variance that exceeds the mean. The conditional variances are given by the following formula:

$$\mu + \frac{\mu^2}{k}, \tag{Eq. 11.9}$$

where μ is the mean of the NB distribution. In Figure 11.3(b), the predicted mean for $X = 10$ was 42.51, so the conditional population variance should be $42.51 + \frac{(42.51^2)}{2} = 946$ (the variance in the sample was 779). Here, the ancillary parameter was $k^{-1} = 0.5$.

Fitting NB Models

In IBM SPSS, one may fit an NB model using GLM by selecting, on the "Type of Model" tab, "Negative binomial with log link," or by selecting "Negative binomial" on the custom menu. The latter gives one a choice of alternative link functions and of specifying a value for the ancillary parameter or allowing the program to estimate it. Using the first option ("Negative binomial with log link") fixes $k^{-1} = 1$; how-ever, it is more sensible to allow the program to estimate k^{-1}. The value of k^{-1} will appear on the output under "Parameter Estimates" in the line marked "(Negative Binomial)" and the value should be significantly > 0. Because the Poisson model assumes $k^{-1} = 0$, it can be considered nested in the NB model and therefore an LR test (with 1 df) can be used to test the significance of k^{-1}. Simply compute the differ-ences between the deviances or negative two log likelihoods ($-2LLs$); the sampling distribution is $\chi^2(1)$. Simonoff (2003) noted that one should conduct a one-sided test because k^{-1} cannot be less than zero, and therefore the p value should be cut in half. (Although a chi-square distribution is one-tailed, its mass reflects the two tails of a z distribution combined.)

For the simulated data in Figure 11.3, the Poisson model, fit to the data in Panel A, had a dispersion parameter of $\phi = 0.86$ for the deviance, so there was no evidence of overdispersion. On the other hand, $\phi = 5.09$ for the data in Panel B, with a devi-ance of 498.64 on 98 df, strongly suggesting overdispersion. For the NB model, however, the deviance was 86.40 on 97 df (so $\phi = 0.89$), a reduction of 412.24. Thus, $\chi^2_{LR}(1) = 412.24$, $p < .001$, indicating that the estimated ancillary parameter ($k^{-1} = 0.32$) was significant, and overdispersion was present.

One should note that it is possible to estimate and add a scale factor to a NB model to determine if that would also reduce overdispersion and improve the fit (using a likelihood ratio test). That is not the case here, as $\phi < 1.0$.

Unlike logistic regression, overdispersion can be a problem with count data even if the data are ungrouped on the IV(s). This situation is illustrated in Figure 11.4. The data were generated using two beta parameters, one for X_1 ($\beta_1 = 1.0$) and one for X_2 ($\beta_2 = 0.5$), plus an intercept ($\beta_0 = 0$) and a Poisson random variable using the expected λ_is. In making estimates, if X_2 is omitted from the model, the deviations of the observed values from the predicted ones are much greater; with X_2 included, the predicted values are very close to the actual ones (the deviations are not even visible in the figure). As a result, the deviance is much greater (663.76 vs. 1.26), and when the deviances are divided by the degrees of freedom, the value of the disper-sion parameters (ϕ) are 165.94 versus 0.42. The former model, with the omitted variable, is clearly overdispersed.

Zero-Inflated Models

Sometimes overdispersion can be caused by the data having more zeros than would be predicted by the standard Poisson or negative binomial models. These excess zeros could be *structural zeroes*, cases that could not theoretically have a count or perform the behavior in question. For example, the number of drinks consumed per day by social drinkers could potentially be zero (i.e., be *sampling* zeroes), but the counts of those who do not drink at all should be treated differently, as structural

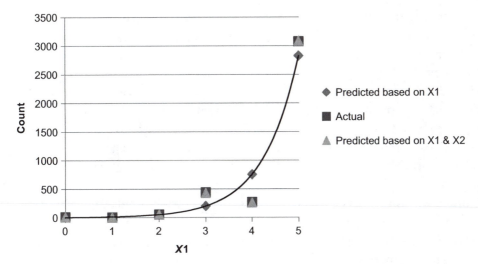

Figure 11.4 Poisson overdispersion from an omitted variable. When the variable (X2) is omitted, there are greater deviances of the actual values from the predicted. In this example, there is just one count observation per level of X1, so the data are ungrouped.

zeroes. Although there are zero-inflated models that contain an additional parameter to explain structural zeroes, these are not currently available in IBM SPSS. (One would need to use software such as Stata or SAS.) If one knows which cases are structural zeroes, however, one can eliminate them from the analysis, or fit a Poisson log-linear model with a cell structure, as will be described in the next chapter.

Example of Poisson Regression

The online supplementary materials to Coxe et al. (2009) contains simulated data that are similar to that collected by DeHart, Tennen, Armeli, Todd, and Affleck (2008) in a study of college student drinking. The study predicted the number of drinks consumed by 400 college students on a particular Saturday night from gender and sensation-seeking scores on a personality inventory. (No offset was used in this study, so its value was set to 1.0.) Fitting a full factorial model (Model 1) using IBM SPSS suggested possible overdispersion, because "Value/df" was 2.408 for the deviance (953.07) and 2.263 for Pearson's X^2. (We will formally test the assumption shortly.) For Model 2, setting the scale parameter method to "Deviance," we find that the scaled deviance is 396.00, a reduction of 557.07. However, Allison (1999) recommended using the Pearson X^2 instead (see Chapter 10). Setting the scale parameter method to Pearson X^2 yields a scaled Pearson X^2 of 396.00. In regard to the beta estimates, the one for the interaction term is now nonsignificant $[\chi^2_{LR}(1) = 2.65, p = .11]$.

Dropping the interaction term from the model produces Model 3, with both gender and sensation seeking as significant terms. For gender (with males coded as 1), the coefficient for [gender = 0] is −0.839, so the coefficient for [gender = 1] is 0.839. $Exp(0.839) = 2.31$, which is the relative risk (RR). Males were likely to have 2.31 as many drinks as females $[\chi^2_{LR}(1) = 83.54, p < .001]$. For sensation seeking, which is a metric variable, the beta estimate is 0.261, and $exp(0.261) = 1.30$ $[\chi^2_{LR}(1) = 20.21, p < .001]$, so increasing sensation seeking by 1 point (on a 5-point scale) is associated with a 30% increase in the number of drinks consumed.

One can also use a negative binomial model to correct for overdispersion (Model 4). Using the custom option on the "Type of Model" tab in SPSS ("Negative binomial with log link") and letting the program estimate the ancillary parameter α, we find—with the interaction term included and a scale adjustment excluded— that $\alpha = 0.51$, which is greater than 0 (thus confirming that there is overdispersion in the Poisson model—Model 1). The deviance is 460.14 on 395 df, so $\phi = 1.17$, indicating no need to include a scale parameter in the current model. However, the interaction term is again nonsignificant $[\chi^2_{LR}(1) = 2.32, p < .13]$, so dropping it (producing Model 5) yields a slightly lower deviance of 459.71.

Table 11.6 shows some of the output for Model 5. Akaike's information criterion (AIC) for the NB model (1,707.023) is lower than that for the Poisson with a scale adjustment (1,888.795), and so is the Bayesian information criterion (BIC), so it would be best to report the results for the NB model. (Recall that AIC and BIC are used to compare nonnested models, and smaller is better.) Proceeding to interpret Model 5, we find that estimated coefficient for [gender = 1.0] is 0.822 and for

Table 11.6. Selected Negative Binomial SPSS Regression Output (Model 5)

Generalized Linear Models

Goodness of Fit[a]

	Value	df	Value/df
Deviance	459.774	396	1.161
Scaled Deviance	459.774	396	
Pearson Chi-Square	381.395	396	.963
Scaled Pearson Chi-Square	381.395	396	
Log Likelihood[b]	−849.512		
Akaike's Information Criterion (AIC)	1707.023		

Dependent variable: y
Model: (Intercept), gender, sensation[a]

[a] Information criteria are in small-is-better form.
[b] The full log likelihood function is displayed and used in computing information criteria.

(Continued)

Table 11.6. Continued.

Tests of Model Effects

| Source | Type III | | |
	Likelihood Ratio Chi-Square	df	Sig.
(Intercept)	.506	1	.477
Gender	67.068	1	.000
Sensation	14.191	1	.000

Dependent variable: y

Model: (Intercept), gender, sensation

Parameter Estimates

| Parameter | B | Std. Error | 95% Profile Likelihood Confidence Interval | | Hypothesis Test | |
			Lower	Upper	Wald Chi-Square	df
(Intercept)	−.644	.3329	−1.299	.011	3.737	1
[gender = 1]	.822	.0956	.635	1.011	73.991	1
[gender = 0]	0[a]	.	.	.	.	.
Sensation	.235	.0616	.114	.356	14.559	1
(Scale)	1[b]					
(Negative binomial)	.510	.0699	.386	.662		

Parameter Estimates

| Parameter | Hypothesis Test Sig. | Exp(B) | 95% Profile Likelihood Confidence Interval for Exp(B) | |
			Lower	Upper
(Intercept)	.053	.525	.273	1.011
[gender = 1]	.000	2.276	1.887	2.747
[gender = 0]	.[a]	.	.	.
Sensation	.000	1.265	1.121	1.428
(Scale)				
(Negative binomial)				

Dependent variable: y
Model: (Intercept), gender, sensation
[a]Set to zero because this parameter is redundant.
[b]Fixed at the displayed value.

sensation seeking is 0.235. In the output, the estimate for "(Negative binomial)" is that of the ancillary parameter: $\hat{k}^{-1} = 0.510$.

One should next perform diagnostic tests. Performing a Box-Tidwell test to test for a nonlinear relationship between $ln(Y)$ and sensation seeking [by adding $X * ln(X)$ to the model] produces a deviance of 459.795, which is not an improvement over Model 5 [and the $X * ln(X)$ term is nonsignificant], so nonlinearity is not indicated. Figure 11.5 graphs the likelihood residuals against the predicted values (of the linear predictor, in SPSS "XBPredicted") and does not provide any evidence of outliers. The best-fitting OLS line is flat [$B = .075$, $t(400) = 0.743$, $p = .46$]. The loess fit line is also fairly flat, so the model appears adequate.

EXPLORING THE CONCEPT

(a) Inspect the lines in Figure 11.5. Are the points making up the line generally equal to zero? (b) Is it sensible that there would be some large residuals above the line, given the shape of the Poisson and NB distributions?

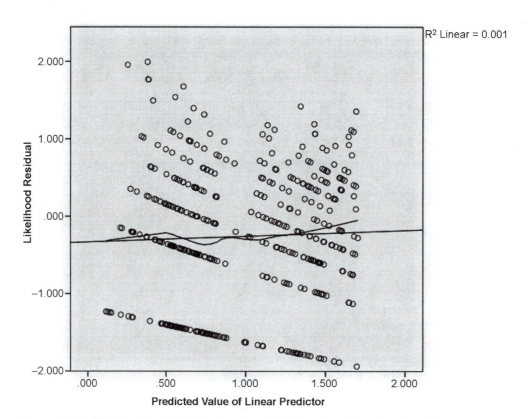

Figure 11.5 Likelihood residuals for student drinking data plotted against the predicted values of the linear predictor.

As noted previously, no outliers were identified by graphical methods. To be thorough, however, one should also use statistical methods to check for outliers. Coxe et al. (2009) noted that, unlike OLS, there are no recommended cut-off scores for diagnostic indicators when using nonlinear models. Instead, they recommend examining one or at most several cases with the highest scores on diagnostic variables (e.g., likelihood or standardized deviance residuals and/or Cook's D). In the present example, the two cases with the highest likelihood residuals (and Cook's D) were cases 336 and 78, but the likelihood residuals were not inordinately high (1.99 and 1.96, respectively), and removing these cases made no substantive difference to the analysis (both predictors were still significant). If one or both of these cases did make a difference, one would still need to investigate the potential outliers further to make a case for removing them.

Summary

This and the previous chapters have presented regression methods for categorical and count DVs when the IVs are metric, ordinal, or nominal. If the DV and IVs are entirely nominal, then an alternative to a GLM model is to use log-linear analysis, discussed in the next chapter.

Problems

1. A sociologist followed 100 non-self-employed women over a 10-year period to determine who received jobs or promotions into upper tier occupations, using measures similar to Wilson (2012). The DV was Mobility: One DV category was Professional/Technical (PT), one was Managerial/Administrative (MA), and the reference category was None (no upper tiered job was obtained). All the women were either African American (AA) or Caucasian, and ethnicity and years of education beyond high school (ED) were used as predictors. The data are in the file *Mobility.11.Problem_1*.

    a) Conduct a logistic, multinomial regression. In SPSS, go to "Analyze" ➔ "Regression" ➔ "Multinomial logistic." (Do not use GLM as it makes restrictive assumptions.) Select "Mobility" as the DV, click on "Reference Category," and choose "Custom" (Type "None" to make that category reference, do not include quotation marks). Select AA as a "Factor" and ED as a "Covariate." Under "Model," "Main Effects" should be left selected. Under "Statistics," select all options under "Model" except cell probabilities and monotonicity measures. Then press "Continue" and "OK." For model fit, report AIC, BIC, $-2LL$, and the results of the omnibus LR test against the null model. Also report Nagelkerke pseudo-R^2 and percent correctly classified (in SPSS, is at end of printout).

b) Report the results of the LR tests for the ED and AA parameters. Why are there two degrees of freedom per test?

c) When significant at $\alpha = .05$, interpret the parameter estimates $exp(Bs)$ for the ED and AA variables.

d) Include an interaction term and report the results of the LR test for the interaction term. (In SPSS, under "Model" → "Custom/Stepwise," specify the two main effect terms and the interaction term.) Do not include the interaction term in subsequent regressions.

e) For diagnostic purposes, conduct separate binary logistic regressions (omit interaction terms). In SPSS, for the "Professional" category, select out the Managerial cases by going to "Data" → "Select Cases" → "If condition is satisfied": "DV" $\sim=$ "MA"; Then perform a binary logistic regression for the Professional category (you could do so with the SPSS multinomial logistic regression procedure, but the binary logistic regression procedure provides more diagnostic information, so use the latter.) Designate "AA" as a categorical covariate and "ED" as a covariate. Save the predicted probabilities, all influence statistics, and the deviance residuals. Under "Options," select classification plot, Hosmer-Lemeshow, and casewise listing of residuals. Then click OK. Report the Hosmer-Lemeshow statistic. Is there an adequate fit to the data?

f) For PT, sort the data by the deviance residuals and consult the various diagnostic statistics. Report the largest positive and negative deviance residuals. Sort by Cook's D and report the largest value.

g) Conduct a binary logistic regression for the managerial/administrative data. What two cases have the largest deviance residuals (in absolute value) and Cook's D?

h) Repeat the multinomial logistic regression but without Cases 11 and 12. (In SPSS, select cases with the criteria: "case $\sim=$ 11 & case $\sim=$ 12.") Remember to omit an interaction term. How does deleting these cases affect the AA beta estimate and p value (for MA)?

2. Similar to Tsitsipis, Stamovlasis, and Papageorgiou (2012), a researcher administered a written test to 200 ninth-grade students assessing their understanding of the nature of matter as well as instruments assessing students' logical thinking (LTh), and another assessing a cognitive style variable known as field dependence (FD). The dependent variable is each student's answer to a multiple-choice question about the structure of water: "If you view a picture of a water droplet blown up a billion times, what would you see?" In addition to the right answer (a picture of liquid structure; R), the incorrect answers related to pictures of "nothing exists" (N), a uniform-continuous substance (U), and a gas structure (G). The data are in the file *Matter.11.Problem_2*.

a) Use the cognitive variables and multinomial logistic regression to predict students' misconceptions about the nature of matter. (FD was measured as a nominal variable.) Use R as the reference category (make dummy variable for "R"). Do not include interaction terms as the researcher did not find them significant. For model fit, report AIC, BIC, $-2LL$, Nagelkerke pseudo-R^2, percentage correctly classified, and the results of the omnibus LR test.

b) Report the results of the LR tests for LTh and FD. Why are there three degrees of freedom per test?

c) Interpret the parameter estimates $exp(Bs)$.

d) For diagnostic purposes, create a separate dummy variable for the gas (G) response. (In SPSS, use "Transform" ➔ "Recode into Different Variables" and be sure to check "Convert numeric strings to numbers" under "Old and New Values.) Then select only those cases where the DV was R (the reference category) or G. In SPSS, select "Data" ➔ "Select Cases" ➔ "If condition is satisfied" and "DV = R | G" as the condition. Then perform a binary logistic regression requesting results of the Hosmer-Lemeshow test. (Note: the researcher found nonsignificant results of the H-L test for the other categories.)

e) Why does the model not adequately predict G responses? Does the model predict any cases to be G? Are the other responses always more likely? (Note: This problem could be potentially remedied by adding an IV that better predicted G responses, just as FD predicts U responses.)

3. Similar to Sherkat (2008), a researcher surveyed 8,060 adult Americans on whether they believe in God. He hypothesized that the certainty of an individual's belief in God grows stronger over time due to continued participation in and affiliation with social groups that reinforce one's beliefs. The DV consisted of answers to an item: "Which statement comes closest to expressing what you believe about God?" The answers were considered ordered categories ranging from atheism (score of 1) to theistic certainty (score of 6), specifically (1) *don't believe*, (2) *agnostic*, (3) *believes in god sometimes*, (4) belief in a higher power (but not a personal god), (5) *belief with doubt*, and (6) *certain belief*. Following are data on the distribution of beliefs by age category (also see file *Belief in God.11.Problem_3*). The researcher used the midpoint of each age category to form a metric predictor.

a) Conduct an ordinal regression. In SPSS, weight the data by frequency (as the data are in table form) and go to "Analyze" ➔ "Regression" ➔ "Ordinal Regression." Under "Output," click on "Test of Parallel Lines." Report the deviance, Nagelkerke's pseudo-R^2, B for "Age" (i.e., the location parameter), and $exp(B)$. Interpret the latter (and verify directionality through computing Kendall's $\hat{\tau}_b$).

b) Report the results of the parallel lines test. Is this assumption met?

c) Examine the frequency of responses by DV category (see column totals in the following table). Are there any grounds for using an alternative link function?

d) Repeat the ordinal regression, but with the negative log-log link and then again with the cloglog link. (In SPSS, go to "Options" to change the link function.) Report the deviances and Nagelkerke pseudo-R^2s. Which of the three link functions produces the best fit and why?

e) For the cloglog model, report the B and $exp(B)$; interpret the latter. What are the results of the parallel lines test?

f) The test of parallel lines may be overly sensitive in large samples. Individual cloglog regressions for each cumulative category using dummy variables yielded the following B estimates for Categories 2 through 6: 0.008, 0.01, 0.008, 0.007, 0.01. Is it reasonable to make the parallel lines assumption and why?

g) Generate the first value in the previous subproblem (0.008) by creating a dummy variable with the value 1 reflecting responses in categories 2 through 6 and the value 0 reflecting responses in Category 1. (In SPSS, use "Transform" ➔ "Recode into Different Variables.") Conduct a binary logistic regression using the cloglog link (and multinomial distribution) in GLM. Report B (for age) and the results of the associated LR test. Save the residuals and diagnostic statistics. What is the largest likelihood residual in absolute value? Is there any evidence of outliers?

Age category	Age category midpoint	1	2	3	4	5	6	Total
18–24	21	34	57	103	59	217	580	1,050
25–29	27	26	43	67	39	187	527	889
30–34	32	23	32	72	30	163	543	863
35–39	37	17	49	73	24	158	494	815
40–44	42	23	29	79	40	130	491	792
45–49	47	18	34	69	35	119	493	768
50–54	52	23	20	57	25	91	448	664
55–59	57	11	18	47	27	64	402	569
60–64	62	9	22	39	17	79	323	489
65–69	67	11	9	33	16	56	284	409
70–74	72	4	7	15	12	38	260	336
75–79	77	4	6	14	11	27	162	224
80–84	82	4	5	10	5	14	99	137
85 and older	87[a]	1	1	4	1	6	42	55
Total	—	208	332	682	341	1,349	5,148	8,060

[a] Technically there is not a midpoint because there is no upper bound but—given the low frequencies—the loss of information through reducing extremely high potential ages was not considered important.

4. Denham (2011) utilized data from a yearly survey of American youth on alcohol and marijuana use. Following are data from 1,045 female 12th graders on how frequently they use marijuana, by ethnicity. Because the DV consisted of categories that do not have the same range, Denham used ordinal regression to analyze the table. The DV consisted of the categories numbered from 1 to 7.

    a–d) Perform an ordinal regression on the data, using the procedures (and reporting the statistics) specified in the previous problem, Subparts (a)–(d). Report estimates for both Hispanic and Black. In lieu of computing Kendall's $\hat{\tau}$, compare the means for Black, White, and Hispanic.

    e) Denham hypothesized that, controlling for ethnicity, competitive athletes would use marijuana and alcohol as a coping mechanism and release from competitive pressure and to obtain the same "high" as from playing sports. For example (and confirming Denham's hypothesis), for female marijuana use, $B = 0.90$ ($SE = 0.17$, $p < .001$) for a dummy variable regarding playing competitive volleyball ($X = 1$); $B = 0.65$ ($SE = 0.20$, $p < .001$) for playing soccer. Interpret these beta estimates. Why do you think the B for the anaerobic sport (volleyball) was greater than for the aerobic one (soccer)?

Race/Ethnicity	DV							Total
	1	2	3	4	5	6	7	
Black	125	10	12	2	6	4	2	161
White	450	83	45	30	25	18	46	697
Hispanic	141	26	7	2	4	3	4	187
Column totals	716	119	64	34	35	25	52	1,045

For DV, 1 = 0 times (per week), 2 = 1–2 times, 3 = 3–5 times, 4 = 6–9 times, 5 = 10–19 times, 6 = 20–39 times, 7 = 40 or more times.

5. As in Nussbaum (2008), a researcher conducts a study on the effect of a graphic organizer for arguments for and against a particular issue on how well college students ($n = 150$) can write an argumentative opinion essay. The data are in the file *Rebuttals.11.Problem_5*. Each participant wrote one essay, and the DV was the number of rebuttals (refutation of counterarguments) included in each essay. There were two conditions: use of the graphic organizer and without. There were also IVs on essay length (measured in number of pages) and each participant's score on the Need for Cognition (NC) scale (Cacioppo, Petty, & Kao, 1984), which measures the disposition to engage in effortful thinking.

    a) What is the mean, variance, skew, and kurtosis of the DV (rebuttals)? (In SPSS, go to "Analyze" → "Descriptive Statistics" → "Frequencies" and request these statistics along with a histogram.) What features of the

Poisson distribution (i.e., counts, zeroes, skew, relationship of variance to mean) do the data display?

b) Conduct a Kolmogorov-Smirnov (KS) test, testing the fit of the rebuttal distribution to the Poisson as well as the normal distribution. (In SPSS, go to "Analyze" ➜ "Nonparametric Statistics" ➜ "One Sample.") Report the z and p values. Does the empirical distribution differ significantly from the Poisson? From the normal?

c) Conduct a Poisson regression. Go to "Generalized Linear Models" and select Poisson log-linear for the type of model. For the DV, use number of rebuttals. For predictors, use "graphic organizers" as a factor and NC as a covariate (do not include an interaction term, as it will not be found significant). Use the logarithm of essay length as an offset. In SPSS, use the GLM procedure and select Poisson log linear for type of model. Be sure to click on options under the "Predictors" tab and choose descending; also include the offset on the "Predictors" tab. Run a full factorial model and save the predicted values of the linear predictor and the likelihood residuals. Plot the latter against the former and insert a loess line. (For the loess line, in SPSS click on the scatterplot and then the icon to insert a fit line at total; select "Loess.") Describe what you see.

d) Report the log likelihood and deviance. Compute the $-2LL$.

e) Explain the logic of using essay length as an offset.

f) Report ϕ for the deviance and Pearson's X^2. Does there appear to be over-dispersion in the data? Why or why not?

g) Formally test for overdispersion by fitting a negative binomial model to the data. (In IBM SPSS, on the "Type of Model" tab, check custom and choose negative binomial with a log link function. Choose "Estimate Value.") Report the $-2LL$ and deviance. Compared to the Poisson model, is there any reduction in the deviance? What is the estimated ancillary parameter for the negative binomial model?

h) Report and interpret the beta estimates and LR p values for the Poisson model.

i) Also conduct a pairwise contrast for the nominal predictor. (In IBM SPSS, go to the EM means tab and click on the blank white space; choose "Pairwise." Also click the box at the bottom of the screen "to display over-all means.") What are the means for the two conditions? Compute the relative risk by dividing one mean by the other. How does this actual RR differ from the estimated one?

j) Rerun the Poisson regression, saving the residuals and diagnostic values. Is there any evidence of outliers?

k) Rerun the regression using the identity link (use the type of model custom option in IBM SPSS) and report the deviance, AIC, and BIC for this model and the previous Poisson regression. Which model fits the data better?

6. As in Segal (2012), a researcher collected data for a certain year on U.S. homicide rates by state per 100,000 individuals. From these data and values on the population for each state, she derived the number of murders by state. States were also rated on whether or not they had broad criteria for involuntarily committing individuals to an inpatient mental health facility (ICC = 1 if yes). The research question was whether involuntary civil commitment is associated with lower homicide rates. The data are in the file *Homicides.11. Problem_6*. Use *ln*(pop) as an offset.

  a–g) Answer the same questions as in the previous problem for Parts (a) through (g).

  h) Report the ICC beta estimate and LR *p* value for the negative binomial model.

  i) Rerun the current model with a Pearson X^2 scale parameter; is the change in the $-2LL$ significant (using 1 *df*)?

  j–k) Answer the same question as in parts (j) through (k) above but using a negative binomial model with a Pearson X^2 scale parameter. For (k), do not delete any outliers.

Technical Notes

11.1 Adjacent-category models with categorical predictors can be estimated in SPSS with the log-linear procedure (see Chapter 12) if the DV is represented by two variables: a nominal variable and numeric one (e.g., 1, 2, 3). The nominal variable would be coded in the same way as the numeric one but included as a factor, not a covariate, and used as a main effect term, allowing for different intercepts for the different categories (Allison, 1999). The covariate would be used in any interaction terms with other variables.

11.2 Ord (cited in Johnson, Kotz, & Kemp, 1992, p. 167) proposed a simple graphical technique for testing whether an empirical distribution follows a Poisson distribution. Let X^* be the value of the variable (0, 1, 2, 3, etc.), and f_x the frequency of each value. Let $\mu_x = \frac{X * f_x}{f_{x-1}}$ and then plot μ_x against X. For a Poisson distribution, the plotted line should be horizontal with $\mu_x = \lambda$ (Nussbaum et al., 2008, p. 311).

APPENDIX 11.1

POISSON REGRESSION PROOFS

Derivation of the Poisson Distribution From the Binomial

1. Binomial Formula: $\text{Prob}(k) = \binom{n}{k}\pi^k(1-\pi)^{n-k}$.

2. Because $\lambda = n\pi, \pi = \frac{\lambda}{n}$. Substituting for π in binomial formula

$$\text{Prob}(k) = \binom{n}{k}\left(\frac{\lambda}{n}\right)^k\left(1-\frac{\lambda}{n}\right)^{n-k}.$$

3. Because $\frac{n!}{(n-k)!} = n(n-1)(n-2)...(n-k+1)$, $\binom{n}{k} = \frac{n!}{(n-k)!k!} = $

$$\frac{n(n-1)(n-2)...(n-k+1)}{k!}. \text{ Also, } \left(\frac{\lambda}{n}\right)^k = \frac{\lambda^k}{n^k}.$$

4. It follows from Steps 2 and 3 that

$$\text{Prob}(k) = \frac{n(n-1)(n-2)...(n-k+1)}{k!} * \frac{1}{n^k}\lambda^k\left[1-\frac{\lambda}{n}\right]^{n-k}.$$

5. Switching the placement in the denominator of $k!$ and n^k, and breaking the last

term into two: $\text{Prob}(k) = \frac{n(n-1)(n-2)...(n-k+1)}{n^k} * \frac{1}{k!}\lambda^k\left[1-\frac{\lambda}{n}\right]^n * \left[1-\frac{\lambda}{n}\right]^{-k}.$

6. Because $\binom{x}{x} = 1$, and therefore $\binom{\infty}{\infty} = 1$, the first term to the right of the

equals sign approaches 1 as $n \to \infty$. The last term also approaches 1 because

$\frac{\lambda}{n}$ approaches 0. Therefore, in the limit: $\text{Prob}(k) = \frac{1}{k!}\lambda^k\left[1-\frac{\lambda}{n}\right]^n.$

7. From calculus, $\lim_{n\to\infty}\left(1-\frac{1}{\lambda}\right)^n = e^{-\lambda}$. Therefore, $\text{Prob}(k) = \frac{1}{k!}\lambda^k e^{-\lambda}$ [Eq. 11.5].

Proof that Variance Equals Mean

1. Note that $x(x-1) + x = x^2$.

2. Therefore, $E(k^2) = E[k(k-1) + k] = E[k(k-1)] + E(k) = E[k(k-1)] + \lambda$.

3. $E[k^2] - \lambda = E[k(k-1)] = \sum_{k=0}^{\infty}[k(k-1) * \text{Prob}(k)]$.

4. Because if $k = 0$ or 1, $k(k-1) * \text{Prob}(k) = 0$, $E[k^2] - \lambda = \sum_{k=2}^{\infty}[k(k-1) * \text{Prob}(k)]$.

5. Because $\text{Prob}(k) = \frac{1}{k!}\lambda^k e^{-\lambda}$ from Eq. 11.5, $E[k^2] - \lambda = \sum_{k=2}^{\infty}k(k-1)\frac{1}{k!}\lambda^k e^{-\lambda}$.

6. Because tautologically, $\frac{k(k-1)}{k!} = \frac{k(k-1)}{k(k-1)*[(k-2)!]} = \frac{1}{(k-2)!}$,

$E[k^2] - \lambda = \sum_{k=2}^{\infty}\frac{\lambda^k e^{-\lambda}}{(k-2)!}.$

7. Let $y = k - 2$. Then $E[k^2] - \lambda = \sum_{y=0}^{\infty} \frac{\lambda^{y+2} e^{-\lambda}}{y!} = \sum_{y=0}^{\infty} \frac{\lambda^y * \lambda^2 * e^{-\lambda}}{y!} = \lambda^2 \sum_{y=0}^{\infty} \frac{\lambda^y e^{-y}}{y!}.$

8. $\sum_{y=0}^{\infty} \frac{\lambda^y e^{-y}}{y!}$ is the probability mass function for the Poisson distribution (see Eq. 11.5), which sums to 100%; therefore, $E[k^2] - \lambda = \lambda^2 * 1 = \lambda^2$.

9. Adding λ to both sides of the equation yields: $E[k^2] = \lambda^2 + \lambda$

10. It can be shown that $Var(X) = E(X^2) - [E(X)]^2$, so given the result in Step 9, $Var(k) = \lambda^2 + \lambda - [E(k)]^2$. Because $E(k) = \lambda$, $Var(k) = \lambda^2 + \lambda - \lambda^2 = \lambda$. QED.

APPENDIX 11.2

THE NEGATIVE BINOMIAL DISTRIBUTION

The *negative binomial distribution* (NBD) is a discrete probability distribution of the number of failures in a sequence of Bernoulli trials before a specified number (r) of successes occurs. The NBD is also used to fit overdispersed Poisson distributions (with variances greater than their means). If the NBD fits the data better than the Poisson, that is also evidence of overdispersion.

In a series of Bernoulli (yes/no) trials, the NBD is the probability that a certain number of failures (j) will occur before a certain number of successes (r), hence the name *negative* binomial. An example would be the probability of approaching j customers before three have bought a timeshare ($r = 3$). Note that the NBD can also be used to model the number of successes before a certain number of failures; one would just "mathematically" define a failed approach as a "success" in the following formulas. Examples include the number of days an alcoholic is sober before she experiences a relapse ($r = 1$) or, in customer service, the number of clients successfully served without complaint before three complaints are received ($r = 3$).

The formula for the regular binomial distribution is $\text{Prob}(j$ failures out of n trials$) = \binom{n}{j}\pi^j(1-\pi)^{n-j}$. In the NBD, n is equal to $j + r - 1$, which is the number of failures and successes minus 1. It is necessary to subtract one because we are looking at the number of failures *before r* successes. The NBD probability formula is therefore

$$\text{Prob}(j \text{ failures before } r \text{ successes}) = \binom{j+r-1}{r-1}\pi^r(1-\pi)^j,$$

where π is the probability of a success. Note that j is the random variable, while r is fixed.

Suppose there are three unsuccessful trials out of 10 ($j = 3, n = 10$, therefore r is 7). If $\pi = 0.5$, then the noncumulative probability of the results is $\binom{9}{6}0.5^3(0.5)^7 = 8\%$ (with the NBD) and 12% with the binomial (counting number of failures). The former probability is less because it assumes the three misses occur before the seven hits; the binomial does not. In EXCEL, the syntax for the NBD is $=$NEGBINOM.DIST$[j,$ $r, (1 - \pi),$ cumulative $= 1,$ noncumulative $= 0]$. For example, the noncumulative probability of two failures before three successes when $\pi = 70\%$ is $=$NEGBINOM. DIST$(2, 3, .70, 0) \to 18.5\%$.

The NBD can be used for a totally different purpose—modeling overdispersed Poisson data. Unlike the Poisson distribution, the variance of the NBD is greater than the mean because of the additional parameter, r. Specifically, for the NBD, $Mean = \frac{pr}{1-\pi}$ and $Var = \frac{\pi r}{(1-\pi)^2}$. The denominator of the variance is smaller than the mean, making the total variance larger than the mean. The *ancillary parameter*

$(r = k^{-1})$ can be used as an index of overdispersion, an ancillary parameter of zero reflects no overdispersion (Poisson).

The Polya distribution is a special case of the NBD used to model "contagious" discrete events, like measles, tornadoes, or the snowballing of argument moves in an oral discussion (see R. C. Anderson et al., 2001). These violations of statistical independence among cases will cause Poisson processes to become overdispersed.

References

Agresti, A. (1996). *An introduction to categorical data analysis.* New York, NY: John Wiley & Sons.

Agresti, A. (2002). *Categorical data analysis* (2nd ed.). Hoboken, NJ: Wiley-Interscience.

Allison, P. D. (1999). *Logistic regression using the SAS system: Theory and application.* Cary, NC: SAS Institute.

Anderson, C. J., & Rutkowski, L. (2008). Multinomial logistic regression. In J. W. Osborne (Ed.), *Best practices in quantitative methods* (pp. 390–409). Thousand Oaks, CA: Sage.

Anderson, R. C., Nguyen-Jahiel, K., McNurlen, B., Archodidou, A., Kim, S., Reznitskaya, A., . . . Gilbert, L. (2001). The snowball phenomenon: Spread of ways of talking and ways of thinking across groups of children. *Cognition and Instruction, 19,* 1–46.

Antelman, G. (1997). *Elementary Bayesian statistics.* Lyme, NH: Edward Elgar.

Brockmann, H. J. (1996). Satellite male groups in horseshoe crabs, *Limulus polyphemus. Ethology, 102,* 1–21.

Cacioppo, J. T., Petty, R. E., & Kao, C. F. (1984). The efficient assessment of need for cognition. *Journal of Personality Assessment, 48,* 306–307.

Clogg, C. C., & Shihadeh, E. S. (1994). *Statistical models for ordinal variables.* Thousand Oaks, CA: Sage.

Coxe, S., West, S. G., & Aiken, L. S. (2009). The analysis of count data: A gentle introduction to Poisson regression and its alternatives. *Journal of Personality Assessment, 9,* 121–136.

DeHart, T., Tennen, H., Armeli, S., Todd, M., & Affleck, G. (2008). Drinking to regulate romantic relationship interactions: The moderating role of self-esteem. *Journal of Experimental Social Psychology, 44,* 537–538.

Denham, B. E. (2011). Alcohol and marijuana use among American high school seniors: Empirical associations with competitive sports participation. *Sociology of Sport Journal, 28,* 362–379.

Hosmer, D. W., & Lemeshow, S. (2013). *Applied logistic regression* (3rd ed.). New York, NY: John Wiley & Sons.

Johnson, N. L., Kotz, S., & Kemp, A. W. (1992). *Univariate discrete distributions* (2nd ed.). New York, NY: John Wiley & Sons.

Klein, J. P., & Moeschberger, M. L. (2003). *Survival analysis: Techniques for censored and truncated data.* New York, NY: Springer.

Kleinbaum, D. G., & Klein, M. (2012). Survival analysis: A self-learning test. New York, NY: Springer.

Kotler-Berkowitz, L. A. (2001). Religion and voting behavior in Great Britain: A reassessment. *British Journal of Political Science, 31,* 523–544.

Larsen, R. J., & Marx, M. L. (1990). *Statistics.* Englewood Cliffs, NJ: Prentice Hall.

Le, C. T. (1998). Applied categorical data analysis. New York, NY: John Wiley & Sons.

Nussbaum, E. M. (2008). Using argumentation vee diagrams (AVDs) for promoting argument/counterargument integration in reflective writing. *Journal of Educational Psychology, 100,* 549–565.

Nussbaum, E. M., Elsadat, S., & Khago, A. H. (2008). Best practices in analyzing count data: Poisson regression. In J. W. Osborne (Ed.), *Best practices in quantitative methods* (pp. 306–323). Thousand Oaks, CA: Sage.

Nussbaum, E. M., & Schraw, G. (2007). Promoting argument–counterargument integration in students' writing. *The Journal of Experimental Education, 76,* 59–92.

O'Connell, A. A. (2006). *Logistic regression models for ordinal response variables.* Thousand Oaks, CA: Sage.

Padilla, D. P. (2003). *A graphical approach for goodness-of-fit of Poisson models* (Unpublished master's thesis). University of Nevada, Las Vegas.

Peterson, B. L., & Harrell, F. E. (1990). Partial proportional odds models for ordinal response variables. *Applied Statistics, 39,* 205–217.

Rosenfeld, M. J. (2002). *Some notes on different families of distributions.* Retrieved from http://www.stanford.edu/~mrosenfe/soc_388_notes/soc_388_2002/pdf%20version%20of%20notes%20on%20diff%20distributions.pdf

Segal, S. P. (2012). Civil commitment law, mental health services, and US homicide rates. *Social Psychiatry and Psychiatric Epidemiology, 47,* 1449–1458.

Sherkat, D. E. (2008). Beyond belief: Atheism, agnosticism, and theistic certainty in the United States. *Sociological Spectrum, 28,* 438–459.

Simonoff, J. S. (2003). *Analyzing categorical data.* New York, NY: Springer.

Tsitsipis, G., Stamovlasis, D., & Papageorgiou, G. (2012). A probabilistic model for students' errors and misconceptions on the structure of matter in relation to three cognitive variables. *International Journal of Science and Mathematics Education, 10,* 777–802.

Williams, R. (2006). Generalized ordered logit/partial proportional odds models for ordinal dependent variables. *The Stata Journal, 6,* 58–82.

Wilson, G. (2012). Women's mobility into upper-tier occupations: Do determinants and timing differ by race? *The ANNALS of the American Academy of Political and Social Science, 639,* 131–148.

Zayeri, F., Kazemnejad, A., Khansfshar, N., & Nayeri, F. (2005). Modeling repeated ordinal responses using a family of power transformations: Applications to neonatal hypothermia data. *BMC Medical Research Methodology, 5,* 29. Retrieved from http://www.biomedcentral.com/471-2288/5/29

CHAPTER 12

LOG-LINEAR ANALYSIS

Log-linear methods are applicable when most variables are nominal (although a few covariates are allowed). These methods are more general and flexible than the ones presented in Chapters 4 and 5 for analyzing contingency tables. For example, although the Mantel-Haenszel (MH) test can be used with $2 \times 2 \times k$ tables to control for the confounding effect of a third variable, the test cannot be used with larger contingency tables (e.g., $3 \times 3 \times k$). In contrast, log-linear analysis can be used for any size contingency table and, unlike the MH test, does not necessarily assume the absence of interaction effects. Logistic regression can also handle situations where all the variables are nominal, as can Poisson regression, and log-linear analysis can be considered a special case of these techniques when one variable is being predicted by others. However, it is preferable to use log-linear analysis when one does not have clear hypotheses about causal directionality and/or when all the variables in the model could be considered response variables.

This chapter presents the basic theory underlying log-linear analysis, focusing primarily on 2×2 contingency tables. The theory is then extended to mulinomial and multiway tables. The chapter next considers contingency tables where there are ordered categories reflecting metric or ordinal levels of measurement. The final section of the chapter addresses sample consideration (e.g., minimum sample sizes, structural zeroes.)

Basic Theory

Statistical Models

A general model for two random variables (X and Y), as given by Goodman (cited in Mair, 2006), is

$$m_{ij} = \eta \tau_i^x \tau_j^y \tau_{ij}^{xy},$$

where m_{ij} is the expected value for the cell in the ith row and jth column of the contingency table, and η is the geometric mean of the cell counts. (See Technical Note 12.1 for an explanation of geometric means.) The τs represent the effect that the variables have on the frequencies (Mair, 2006). If we take the logarithms of both sides of the equation, we obtain:

$$ln\left(m_{ij}\right) = ln(\eta) + ln\left(\tau_i^x\right) + ln\left(\tau_j^y\right) + ln\left(\tau_{ij}^{xy}\right),$$

which can also be written as

$$ln\left(m_{ij}\right) = \lambda + \lambda_i^X + \lambda_j^Y + \lambda_{ij}^{XY}, \qquad \text{(Eq. 12.1)}$$

where the lambdas are the natural logarithms of the effects. If the two variables are statistically independent, $\lambda_{ij}^{XY} = 0$, so the independence model can be written as

$$ln\left(m_{ij}\right) = \lambda + \lambda_i^X + \lambda_j^Y. \qquad \text{(Eq. 12.2)}$$

Suppose, for the sake of illustration, that both X and Y are binary variables so that Eq. 12.2 applies to a 2×2 table. That means that five parameters need to be estimated $\left(\lambda, \lambda_1^X, \lambda_2^X, \lambda_1^Y, \lambda_2^Y\right)$, but there are only four observed cell frequencies. The model is therefore not *identifiable*; there are many values of the lambdas that could produce the expected values. To make the model identifiable, we have to impose constraints on their values. Different software programs use different constraints. If one uses the constraints $\sum \lambda_i^X = 0$ and $\sum \lambda_j^Y = 0$, then $\lambda_1^X, \lambda_2^X, \lambda_1^Y, \lambda_2^Y$ would represent main effects (differences from the means). This approach is known as *effects coding*. If one uses the constraint $\lambda_2^X = 0$ and $\lambda_2^Y = 0$, then λ_1^X and λ_1^Y would be dummy variable parameters in a regression equation (with the dummy variables being implicit); $X = 2$ and $Y = 2$ would signify the reference categories. This approach is known as *dummy coding*. There are thus alternative ways to *parameterize* the model, depending on the constraints chosen.

Consider the hypothetical contingency table (Table 12.1). Using the "General Log-Linear" procedure in SPSS (described later), the lambda estimates shown in Table 12.1 were generated for the independence model, and from these I calculated the expected log values shown in panel (c), and then, through exponentiation, the expected values shown in panel (d). For example, when X and Y both equaled zero, the expected log value was

$$ln(m_{11}) = 3.74 - 0.85 - 0.41 = 2.48, \qquad \text{(Eq. 12.3)}$$

(2.48491 with unrounded values), and $exp(2.48491) = 12.00$, the expected frequency count shown in Panel (d). This is the same value generated by Eq. 4.3 from

Table 12.1 Example 2 × 2 Log-Linear Analysis

(A) Sample contingency table

	$Y = 0$	$Y = 1$	Totals
$X = 0$	10 (F_{11})	20 (F_{12})	30
$X = 1$	30 (F_{21})	40 (F_{22})	70
Totals	40	60	100

(B) SPSS output: parameter estimates under independence

Parameter Estimates[b,c]

Parameter	Estimate	Std. Error	Z	Sig.	95% Confidence Interval	
					Lower Bound	Upper Bound
Constant	3.738	.145	25.822	.000	3.454	4.021
[X = .00]	−.847	.218	−3.883	.000	−1.275	−.420
[X = 1.00]	0[a]	.	.	.	.	.
[Y = .00]	−.405	.204	−1.986	.047	−.806	−.005
[Y = 1.00]	0[a]	.	.	.	.	.

[a]This parameter is set to zero because it is redundant.
[b]Model: Poisson
[c]Design: Constant + X + Y

(C) Expected log values

	$Y = 0$	$Y = 1$
$X = 0$	2.48491	2.89037
$X = 1$	3.33220	3.73767

(D) Expected values (rounded)

	$Y = 0$	$Y = 1$
$X = 0$	12	18
$X = 1$	28	42

Chapter 4, where the chi-square test for independence was discussed: $\frac{40*30}{100} = 12$. Taking logs we obtain $ln(40) + ln(30) − ln(100) = ln(12)$, or

$$3.689 + 3.401 − 4.605 = 2.48 \ (2.4849 \text{ with unrounded values}). \qquad \text{(Eq. 12.4)}$$

Eqs. 12.3 and 12.4 both yield the value 2.48, but the parameterizations differ. If we let the lambdas represent the differences from the marginal log means and the constant term λ represent the log of the geometric mean (Goodman, 1970), the log being 3.10, we would have yet a differently parameterized equation:

$$3.10 − 0.42 − 0.20 = 2.48. \qquad \text{(Eq.12.5)}$$

The equations differ because different constraints are imposed on the parameters.

EXPLORING THE CONCEPT

Rewrite Eqs. 12.3−12.5 for the cell where $X = 1$ and $Y = 0$. All equations should yield essentially the same values. (Hint: for Eq. 12.5, change -0.20 to positive because $\sum_{j=1}^{2} \lambda_j^Y = 0$.)

The model in Eq. 12.2 reflects statistical independence. Any statistical dependence between X and Y can be modeled with an interaction term. The full model can therefore be written as in Eq. 12.1. This model is considered *saturated*, meaning that all possible terms are included. A saturated model fits the data perfectly. If one of the terms on the right-hand side of the equation were omitted, there would be some deviance between the observed and predicted values. For notational convenience, Eq. 12.1 is often written without the dependent variable (DV) and subscripts:

$$\lambda + \lambda^X + \lambda^Y + \lambda^{XY}. \tag{Eq. 12.6}$$

Here, the DV and subscripts are implied. Constraints are also imposed on the interaction term (e.g., $\lambda_{12}^{XY} = \lambda_{21}^{XY} = \lambda_{22}^{XY} = 0$).

As with logistic regression, the lambdas can be estimated using maximum likelihood estimation (MLE) and a likelihood ratio test conducted to test their statistical significance. If $\hat{\lambda}^{XY}$ is not significant, it should be dropped from the model. If $\hat{\lambda}^X$ or $\hat{\lambda}^Y$ are not then significant, this would indicate, in the case of λ^X, no difference in $\text{Prob}(X = 1)$ and $\text{Prob}(X = 0)$, or, in other words, a 50%−50% split, but that usually is not of much theoretical interest. The same applies to λ^Y.

Next, consider the case where $\hat{\lambda}^{XY}$ is significantly different from zero. This result implies statistical dependence and the degree of dependence can be quantified with an odds ratio (OR). The predicted ORs can be derived from the lambdas through exponentiation. Following the notation in Table 12.1, when $X = 0$, the odds that $Y = 0$ are $\frac{F_{11}}{F_{12}}$. When $X = 1$, the odds are $\frac{F_{21}}{F_{22}}$. The observed OR is $\frac{F_{11}F_{22}}{F_{12}F_{21}}$. Substituting the predicted means for the observed frequencies, the estimated OR is equal to $\frac{m_{11}m_{22}}{m_{12}m_{21}}$. (It is shown in Appendix 12.1 that $\lambda_{ij}^{XY} = \ln\left(\frac{m_{11}m_{22}}{m_{12}m_{21}}\right)$, so $exp(\lambda_{ij}^{XY}) = \widehat{OR}$. However, this assumes a dummy variable parameterization, consistent with SPSS; in statistical programs using effects coding, multiply λ_{ij}^{XY} by 4 before exponentiating.)

It is important to note that if Y were metric and not nominal, one could conduct an ANOVA. Y would be the DV, and a statistical association would be reported as a main effect, not an interaction. However, in log-linear analysis, we have an additional variable—the log of the expected frequency counts $[\ln(m_{ij})]$—which serves as the DV, and Y is a predictor, so the analysis is of a higher order than in ANOVA (i.e., what would be treated as a main effect in ANOVA is treated as an interaction in log-linear analysis).

Table 12.2 2 × 2 Log-Linear Analysis: Hamilton High Example

	Frequencies[a]		
Completion Status	Gender		
	Male	Female	Total
Completer	35.5	46.5	82
Noncompleter	14.5	5.5	20
Total	50.0	52.0	102

	Probabilities		
Completion Status	Gender		
	Male	Female	Total
Completer	34.8%	45.6%	80.4%
Noncompleter	14.2%	5.4%	19.6%
Total	49.0%	51.0%	100.0

Parameter estimates (SPSS)			
Variable	Estimate	Exp(Estimate)	P-value
Constant	3.839	46.479	.000
Completion	−2.135[b]	0.118	.000
Gender	−0.270[b]	0.763	.226
Completion × Gender	1.239[c]	3.452	.024

[a]Frequencies include addition of 0.5 to avoid structural zeroes.
[b]Reflects log odds.
[c]Reflects *ln*(OR).

As an example of a log-linear analysis of a 2 × 2 contingency table, Table 12.2 displays the data for the Hamilton High example from Chapter 4, but with 0.5 added to each cell. (This practice prevents any cells from having a zero count, as that would result in a zero probability estimate, and $ln(0)$ is undefined; critiques of this practice will be discussed later in the chapter.) The lower part of the table shows the parameter estimates from SPSS. (In the "General Log-linear" Procedure, go to "Options" and check "Estimates" to obtain these. Do not add 0.5 to the cell counts; the program will do that for you.) In SPSS, the estimates are, except for the intercept, natural logs of odds or ORs. Therefore, the exponentiated values are also shown in Table 12.2. Of particular importance is that the interaction term between gender and completion status is significant. The parameter estimate is 1.239, and $exp(1.239) = 3.45$, which is the OR; the odds of females completing high school are about three and half times greater than males. We can verify that this is the correct value because the odds for males are $\frac{35.5}{14.5} = 2.448$; the odds for females are $\frac{46.5}{5.5} = 8.455$; the OR is therefore $\frac{8.455}{2.448} = 3.45$. (The estimated and observed ORs do not always perfectly match, but in this case the model is saturated and fits the data perfectly.)

What does the exponentiated parameter estimate for completion (0.118) represent? This is the odds of females not completing high school. (The reciprocal is 8.46, which is the odds of females completing high school, as noted previously.) Now, consider the exponentiated estimate for the gender variable (0.763); this value represents the odds of being male if you are a completer. (Note that from Table 12.2 the probability of being a male, if you are a completer, is $\pi_{male|completer} = \frac{35.5}{82} = 43.3\%$ and therefore the $odds_{male|completer} = \frac{43.3\%}{(1-43.4\%)} = 0.76$, which conforms to the regressions results.)

EXPLORING THE CONCEPT

Suppose we drop the negative sign from the parameter estimate for gender. Then the exponentiated estimate would be $exp(0.27) = 1.31$. What does this number represent? (Hint: Find the reciprocal.)

Extensions to Multinomial and Multiway Situations

The models described in the previous section apply to binary variables. In a multinomial situation, at least one of the variables (e.g., X or Y) has more than two categories. As was explained in Chapter 9, nominal variables can be included in regression models as dummy variables, which are binary, 0/1 variables. For a multinomial variable with three categories (e.g., Democrat, Republican, Other), one designates one category as the reference category and then two dummy variables are included in the model for the other two categories. Suppose the variable X has three categories designated as A, B, C. Then the saturated model is written as

$$\lambda_0 + \lambda_{i=B}^X + \lambda_{i=C}^X + \lambda_j^Y + \lambda_{i=B,j}^{XY} + \lambda_{i=C,j}^{XY}. \tag{Eq. 12.7}$$

EXPLORING THE CONCEPT

(a) What is the reference category in the previous model? (b) Why is there no interaction term between B and C (i.e., λ_{BC})? Are these categories mutually exclusive?

The other important situation to consider is when there is a third variable, Z, that needs to be controlled, and/or interactions between Z and X or Y examined. This situation is three-way rather than the two-way situation analyzed in the previous section. We shall assume for the moment that all the variables are binary. The real usefulness of log-linear analysis, however, is either when some of these variables are multinomial, or when there are even more independent variables (IVs) to include in the model. These situations are beyond the reach of the MH and Breslow-Day tests.

For the three-way (binary) situation, the saturated model is

$$\lambda_0 + \lambda_i^X + \lambda_j^Y + \lambda_k^Z + \lambda_{ij}^{XY} + \lambda_{ik}^{XZ} + \lambda_{jk}^{YZ} + \lambda_{ijk}^{XYZ}. \tag{Eq. 12.8}$$

Notice the inclusion of a three-way interaction term, λ^{XYZ}. If this term is significant, it means that there is a different population odds ratio (for X and Y) when $Z = 0$ than when $Z = 1$. In Chapter 5, the example was given of different ORs for males and females. The MH test assumes homogeneity of the ORs (i.e., there is no interaction effect). Therefore, if λ^{XYZ} is nonsignificant, an MH test could be performed instead. Alternatively, we could just drop λ^{XYZ} from the model. If λ^{XZ} is significant, it means that there is a significant association between X and Z (and likewise for λ^{YZ}). If both λ^{XZ} and λ^{YZ} are significant, this implies that Z would be a confounding variable if it were not included in the model. The term λ^{XY} reflects the association between X and Y while controlling for Z. The parameter is a measure of *conditional* association.

EXPLORING THE CONCEPT

Suppose λ^{YZ} is significant but λ^{XZ} is not. Is it then still necessary to retain Z in the model (and the associated interaction terms) to avoid confounding? If it is not a confounding variable, could Z be removed from the model and therefore the three-way table "collapsed" to a two-way table?

Table 12.3 summarizes many of these different situations (see also Technical Notes 12.2 regarding Simpson's paradox and 12.3 regarding collapsibility).

When there are nonsignificant terms in the model, it is usually best to drop them, as including irrelevant variables reduces statistical power and may make the other terms harder to interpret. Also, parsimony is considered a virtue in science. However, there is an important constraint on dropping terms that should be observed:

> Never drop a term involving a variable that is also involved in a higher-order interaction represented in the model.

So, for example, do not drop λ^X if λ^{XY} is retained in the model. Likewise, do not drop λ^{XY} if λ^{XYZ} is retained in the model. Doing so would make the higher-order term uninterpretable. Remember that the higher-order coefficients are beta estimates that reflect differences added to, or subtracted from, the lower-order terms.

Table 12.3 Types of Independence and Association in Three-Way Tables

Type	Description	Model term
Marginal independence	No association in the (overall) marginal table.	$\lambda_{xy} = 0$
Mutual independence	No association among three or more variables.	$\lambda_{xy} = 0$, $\lambda_{xz} = 0$, and $\lambda_{yz} = 0$
Joint independence	One variable (e.g., X) is independent of the other two.	$\lambda_{xy} = 0$, $\lambda_{xz} = 0$, but $\lambda_{yz} \neq 0$
Conditional independence	No association in a partial table (e.g., X and Y), controlling for the kth level of a third variable (Z).	$\lambda_{xy} = 0$, but $\lambda_{xz} \neq 0$, and $\lambda_{yz} \neq 0$
Marginal association	Association in the (overall) marginal table.	$\lambda_{xy} \neq 0$
Homogenous association	No three-way interaction.	$\lambda_{xyz} = 0$
Conditional (or partial) association	Association in a partial table, controlling for the kth level of a third variable.	$\lambda_{xy} \neq 0$ with λ_{xz} and λ_{yz} included in the model
Three-way interaction	The degree of partial association depends on the kth level of a third variable.	$\lambda_{xyz} \neq 0$

Without the lower-order terms, such an interpretation does not work (and also omitting them would create biased estimates). (For a discussion of nonhierarchical models, see Mair, 2006.)

In log-linear analysis, the likelihood ratio (LR) test is typically used to compare models and to assess the statistical significance of each term. It is clear from Eq. 12.8 that a saturated model will contain lots of terms, and the significance and retention of the higher-order terms should be assessed before the lower-order terms. The number of interaction terms grows exponentially as additional variables are added to the model, and even more so if any of these are multinomial. Stepwise regression is often used to help specify the model terms. It is typical to use some form of backward elimination, but forward selection algorithms can also be used.

In SPSS, to use stepwise regression in log-linear analysis, go to "Analyze" ➔ "Log-linear" ➔ "Model Selection." The default option is backward elimination beginning with the saturated model. Numeric coding of each variable is required. You must specify the range of each variable (e.g., 0 to 1 for binary variables, or 0 to 2 for three-category multinomial variables; do not interpret the numbering as an ordering). The procedure uses a backward elimination algorithm to select the best statistical model. Parameter estimates will be displayed for the final model; there is also an option to display estimates for the saturated model.

After model selection, one can use the "General Log-linear Analysis" procedure in SPSS to obtain further information about the model (or if one would like more control over how the final model is selected). The default is Poisson regression, but one can also choose multinomial. The latter is equivalent to performing a binary or multinomial logistic regression with the cell frequency counts as the DV. The multinomial option can be used if the total number of cases in the population is fixed (e.g., if one fixed the sample size in advance of collecting the data). On the other hand, if the sample size is determined by when data collection stopped (e.g., counting everyone in an emergency ward for 3 days), then the Poisson option should be used.

Finally, there is also a separate "Logit" procedure under the log-linear menu. This procedure performs *conditional* log-linear analysis, which is based on conditional probabilities. In general log-linear analysis, Y is not considered a DV, whereas in conditional log-linear analysis, it is. If we can assume that X is a fixed variable that has a causal influence on Y (or at least that Y has no causal influence on X or any of the other IVs), then we can use this information to make the model more parsimonious and reduce the number of statistical tests performed. The mathematical details of conditional log-linear analysis are explained in Appendix 12.2.

Power Girls **Example**

In the *Power Girls* example from Chapter 5, the effect of watching *Power Girls* on children's aggression was examined with gender as a third variable. For purposes of the present chapter, I also conducted a log-linear analysis. The data were recoded into dummy variables as follows: watch = 1, not watch = 0; female = 1, male = 0; aggressive = 1, not aggressive = 0. Using the "Log-linear" ➔ "Model Selection" procedure, the three-way interaction term was nonsignificant ($z = .257, p = .797$). Using backward elimination (and LR test), the best model was

Aggression * Gender, Aggression * TV, Gender * TV.

This best model is indicated on the SPSS output at the last step in the step summary box. Only the higher-order effects are shown; inclusion of the main effects for these variables is implied. The goodness-of-fit tests show that the $\chi_{LR}^2 (1) = 0.042, p = .838$ (Pearson X^2 is identical). The lack of significance indicates a good fit.

The results show that watching *Power Girls* significantly predicts whether a child will become aggressive, when gender is controlled. Gender is significantly related to both aggression and TV, so it is a relevant (confounding) variable. This conclusion conforms with the MH test presented in Chapter 5, where the common OR was found to significantly differ from 1.0.

One could also use the "General Log-linear" procedure to run the final model. Clicking on "Model" and selecting "Custom," specify main effects for "Aggression,"

"Gender," and "TV" and all two-way interaction terms. If one wishes to obtain parameter estimates, select "Estimates" under "Options." The results show that TV watching and aggression are significantly related (conditional on controlling for gender), Wald $z = 4.64$, $p < .001$, $B = 0.674$ for [Aggression = 0] * [TV = 0] and by implication for [Aggression = 1] * [TV = 1], because changing the sign of two dummy variables results in no net change in sign. The common odds ratio is $exp(0.674) = 1.96$, which conforms to what is produced by the MH test.

EXPLORING THE CONCEPT

Would the results have conformed to the MH test if the three-way interaction term had been significant? Why or why not?

For the final model, both log-linear procedures produce a list of residuals, one for each of the eight cells in the full contingency table. For the "General Log-linear" procedure, the largest raw residual is 0.646, adjusted residual (raw residual divided by its estimated standard error) is 0.205, standardized residual is 0.147, and deviance residual is 0.146. None of these are especially large, so there is no need for remedial action.

Conditional Log-Linear

However, for this example, one could have also used conditional log-linear analysis, because aggression was hypothesized to depend on gender and/or TV watching. (This assumption is a little risky, though, because theoretically aggression could influence TV watching.) To use conditional log-linear analysis in SPSS, go to "Log-linear" ➔ "Logit," specify "Aggression" as the DV and "TV" and "Gender" as factors, producing the model

Constant + Aggression + Aggression * Gender + Aggression * TV + Aggression * Gender * TV.

Note that because "TV" and "Gender" are fixed variables, no main effects are specified for these variables, and there are only interaction terms involving the random variable (RV) "Aggression." There are, however, four different constant terms estimated, one for each possible combination of "TV" and "Gender." The three-way interaction term is nonsignificant, and when it is dropped from the model we obtain the same results as with the "General Log-linear" procedure for "Gender," and somewhat more significant results for "TV." The conditional model is somewhat more statistically powerful; there are also fewer statistical tests performed, making a conditional model easier to present.

Generalized Linear Models

It is also possible to conduct log-linear analysis using the generalized linear model (GLM) procedure. GLM uses a conditional structure (one must specify a DV), so it is easier to see the parallels of this technique with conditional (logit) log-linear analysis than with general Poisson log-linear analysis.

In the social sciences, it is often the case that one has a mixture of metric and nominal predictors, and GLM has a greater ability for incorporating multiple metric covariates and modeling interactions with them. In IBM SPSS, go to the GLM "Type of Model" tab and choose "Binary logistic." For the *Power Girls* example, if one continues to weight the data by the frequencies, using "Gender" and "TV" (main effects only) to predict aggression, the analysis produces identical results as in the previous section. Note that binary logistic regression assumes the random component of the model is binomial or multinomial—which in turn assumes n is fixed. If n is random (e.g., the sample size was not fixed in advance), it is more appropriate to use Poisson regression for a conditional analysis, or to use general Poisson log-linear analysis.

For the *Power Girls* example, changing the model type to "Poisson log-linear" (and still weighting the data by frequencies) produced a "TV" coefficient of 0.267, which is much smaller than the previous coefficient of 0.674. The latter, when exponentiated, estimates the common OR whereas the former, when exponentiated, estimates the relative risk (RR) ratio, because Poisson regression is used. As explained in Chapter 11, it is more technically correct to use the unweighted data, the frequency counts (for Aggression = 1) as the DV, and the total counts to compute an offset; for example, the count for females who watch *Power Girls* is 150 and the offset would be $ln(150 + 400) = 6.31$. This method yields the same beta estimates as the weighted approach but gives a more valid estimate of the deviance and dispersion parameter.

An Extended Example of Log-Linear Analysis

Pugh (1983) conducted an experiment examining the likelihood of rape convictions. Participants were 358 volunteers obtained from an introductory sociology class. Each participant read a case file regarding a hypothetical date-rape case and then indicated whether they would likely find the defendant guilty. Pugh manipulated the content of court transcripts so that the victim either had low, high, or neutral moral character (this variable was called *Stigma*). If the victim had admitted to inviting other male friends met in a bar to her apartment to have sex, the prosecutor suggested in the transcript that she had low moral character (LMC). In the high moral character (HMC) condition she denied these accusations, and in the neutral condition this section of the court transcript was not presented. The researcher then measured on a 6-point scale the extent that each participant thought that the rape

Table 12.4 Rape Verdict (*V*) by Juror Gender (*J*), Stigma (*S*), and Fault (*F*)

Gender (*J*)	Stigma (*S*)	Fault (*F*)	Verdict (*V*)	
			Not guilty (*V* = 0)	Guilty (*V* = 1)
Male (*J* = 0)	HMC (*S* = 0)	Low (*F* = 0)	4	17
		High (*F* = 1)	7	11
	Neutral (*S* = 1)	Low (*F* = 0)	4	36
		High (*F* = 1)	18	23
	LMC (*S* = 2)	Low (*F* = 0)	6	10
		High (*F* = 1)	18	4
Female (*J* = 1)	HMC (*S* = 0)	Low (*F* = 0)	0	25
		High (*F* = 1)	4	12
	Neutral (*S* = 1)	Low (*F* = 0)	8	43
		High (*F* = 1)	23	42
	LMC (*S* = 2)	Low (*F* = 0)	2	22
		High (*F* = 1)	6	13

Data from Pugh (1983).

was the victim's own fault and used the median score to divide the participants into a high- and low-fault group. Finally, participants were asked whether or not they would find the defendant guilty if they were a juror. The dependent variable was therefore Verdict (V). The independent variables were the participant's gender (*J* for juror's gender), Stigma (S), and Fault (F). The data are shown in Table 12.4.

I entered these data into SPSS so as to create contingency tables (see Chapter 4); the first few rows of the data set appeared as follows:

Juror	Stigma	Fault	Verdict	Frequency
Male	HMC	Low	Not Guilty	4
Male	HMC	Low	Guilty	17

An easy way of entering data to create a complex contingency table is to copy and paste the rows already created to create new rows and to then change the applicable variable(s) and frequencies. The next two rows appeared therefore as:

Male	HMC	~~Low~~High	Not Guilty	~~04~~ 7
Male	HMC	~~Low~~High	Guilty	~~17~~ 11

The next step was to copy the four existing rows to create four new ones, and so forth. Next, because log-linear analysis requires numerical codes, I used "Transform" ➜ "Recode into Different Variables," using the numerical codes

Table 12.5 Numerical Codes (Design Matrix) for Rape Conviction Model

Juror2	Stigma2	Fault2	Verdict2	Frequency
0	0	0	0	4
0	0	0	1	17
0	0	1	0	7
0	0	1	1	11

shown in Table 12.4. The first four rows of the database therefore appeared as shown in Table 12.5.

The last step in creating the database was to weight the data by "Frequency."

The model selection procedure identified the best, most parsimonious model as

Verdict * Fault, Verdict * Stigma * Juror.

Although Verdict is hypothesized to be influenced by the other variables, it is possible that there could have been reciprocal effects of a juror's verdict leanings and perceptions of fault (and even stigma), so it would be safest to use a general log-linear model rather than a conditional one. We shall use a multinomial (rather than a Poisson) model to analyze the data, because the size of the sample (and the number of people in each factor, other than Verdict) was determined by the researcher. In the SPSS "General Log-linear" procedure, click on "multinomial" as well as on "model" to customize the model. Because of the three-way interaction term, the lower-order two-way interaction terms (Verdict * Stigma, Verdict * Juror, and Stigma * Juror) must be included, as well as main effect terms for each factor. Also, click on "Options" to request estimates.

Other than the main effects (which are usually not substantively interesting), there were three parameter estimates that were significant. It is best to interpret these in conjunction with generating a contingency table involving the applicable variables to help interpret the results.

1. [Fault = .00] * [Verdict = .00]: Estimate was −1.529 ($p < .001$). Recall that [Verdict = .00] represents a not guilty verdict. It follows that [Fault = .00] * [Verdict = 1.00] was 1.529 (dropping the equal sign) and, repeating this operation, that the [Fault = 1.00] * [Verdict = 1.00] estimate was −1.529. (When there are two terms involved, changing both values from 0 to 1 has no net effect on the sign of the coefficient.) The negative coefficient means that if the rape victim was perceived to be at fault, it was less likely that the defendant would have been judged guilty. Using the "Descriptive Statistics" ➜ "Crosstabs Option," the contingency table shown in the top half of Table 12.6 was generated. Note that $exp(−1.529) = 0.22$, which is the actual OR shown in Table 12.6.

EXPLORING THE CONCEPT

(a) In Table 12.6, was the percentage of guilty verdicts higher or lower in the high fault category? Were the odds of a conviction higher or lower when the victim was perceived to be at fault? (b) Table 12.6 was generated based on the observed values. Do you think an OR based on estimated values would differ by much? Would there be *any* difference if one was using a saturated model?

Table 12.6 Contingency Tables Associated with Significant Results in Rape Conviction Data

Fault Effect * Verdict

| Fault | Verdict | | Total | Guilty Verdicts | | OR |
	Guilty	Not guilty		Proportion	Odds	
High	105	76	181	58.0	1.38	0.22
Low	153	24	177	86.4	6.38	
Total	258	100	358			

Juror Gender * Verdict

| Gender | Verdict | | Total | Guilty Verdicts | | OR |
	Guilty	Not guilty		Proportion	Odds	
Female	157	43	200	78.5	3.65	2.06
Male	101	57	158	63.9	1.77	
Total	258	100	358			

Stigma (LMC vs. Neutral) * Verdict, for Females

| Stigma | Verdict | | Total | Guilty Verdicts | | OR |
	Guilty	Not guilty		Proportion	Odds	
LMC	35	8	43	81.4	4.38	1.60
Neutral	85	31	116	73.3	2.74	
Total	120	39	159			

Stigma (LMC vs. Neutral) * Verdict, for Males

| Stigma | Verdict | | Total | Guilty Verdicts | | OR |
	Guilty	Not guilty		Proportion	Odds	
LMC	14	24	38	36.8	0.58	0.22
Neutral	59	22	81	72.8	2.68	
Total	73	46	119			

2. [Juror = .00] * [Verdict = .00]: Estimate was 2.015 ($p < .001$). This figure would also be the estimate for [Juror = 1.00] * [Verdict = 1.00]. If a juror were a female, a guilty verdict would have been more likely, as demonstrated in Table 12.6. $Exp(2.015) = 7.50$, which is the actual OR shown in Table 12.6.

3. [Juror = .00][Verdict = .00][Stigma = 1.00], Estimate = -1.993, $p = .001$. The estimate was for the three-way interaction when [Stigma = 1.00], there was a separate, nonsignificant estimate when [Stigma = 0], corresponding to high moral character (HMC). The reference category in SPSS was [Stigma = 2], corresponding to low moral character (LMC). So the significant comparison was between LMC and neutral [Stigma = 1]. The term relates to males because [Juror = .00].

The estimated coefficient of a three-way interaction cannot be properly interpreted without consulting the coefficients for the two-way interactions. The coefficient for [Verdict = .00][Stigma = 1.00] was 0.467. This relates to females, because that was the reference category for juror gender (SPSS used the 1.0 value as the reference category). $Exp(0.467) = 1.60$, which is the actual OR shown in Table 12.6 for females.

To obtain the OR for males from the coefficients, take into account the three-way interaction term, which reflects how the ln(ORs) differ between genders (see the discussion of dummy variables in Chapter 9). Adding the Verdict * Stigma coefficient for females to the coefficient for [Juror = .00][Verdict = .00][Stigma = 1.00] yields $0.467 - 1.993 = -1.526$, $exp(-1.526) = 0.22$, which is the actual OR for males shown in Table 12.6. Males were much less likely to vote for a conviction when the victim had low moral character. The odds declined by 78% ($0.22 - 1.0 = 0.78$).

EXPLORING THE CONCEPT

It is important to correctly interpret the directionality of a relationship. (a) To verify the above conclusion, for males, are the odds of a guilty verdict shown in Table 12.6 less when the victim has low moral character (LMC)? (b) The term [Juror = .00][Verdict = .00][Stigma = 1.00] relates to acquittal, because [Verdict = .00], so to find the OR for conviction, reverse the sign of the coefficient to positive. Because the reference category is LMC [Stigma = 2.00], this means that the victim having a better (more neutral) character increases the chance of conviction. Is this conclusion consistent with that stated previously?

Ordered Contingency Tables

It is often the case that the columns or rows of a contingency table can be ordered (e.g., high, medium, or low). Such categories can be coded numerically (e.g., 3, 2, 1). Many times the variables are crude measurements of a latent continuous scale. If we further assume that the distance between any two adjacent category scores is about equal, then that variable can be considered metric. If both X and Y are metric, then a linear-by-linear (uniform) association model can be fit to the data. This assumption may be reasonable for a Likert scale (especially if there are many points on the scale so that the distances are not large) or for actual values of a discrete variable that has been binned into categories (e.g., midpoints of salary categories).

Linear-by-Linear Association Model

The form of this model (see Simonoff, 2003) is

$$\lambda_0 + \lambda_i^X + \lambda_j^Y + \theta(u_i - \bar{u})(v_j - \bar{v}). \tag{Eq. 12.9}$$

Typically, u_i and v_j are the numeric codes of the variables involved (e.g., 1, 2, 3). A new variable is formed consisting of the mean-deviated cross products with the coefficient θ. The cross-product term replaces the interaction term λ_{ij}^{XY} in the saturated model (however, Eq. 12.9 will not fit the data perfectly, unlike the saturated model). If θ is significantly greater than zero, then X and Y can be considered to interact. If X or Y has multiple categories, there will be only one regression estimate ($\hat{\theta}$) to interpret rather than multiple ones. Also, because more information from the data is being used by taking the ordering of the categories into account, there will be potentially more statistical power. Theta can be shown to be equal to the log of the OR for any contiguous 2×2 subtable in the larger contingency table estimated by the model (Simonoff, 2003). Because metric variables involve values that are equally spaced, the model assumes that the ORs for all subtables are the same.

As an example, Table 12.7 presents data from Ishi-Kuntz (1991) on the relationship between marital happiness and age based on a random sample of 750 noninstitutionalized adults. To fit a linear-by-linear association model to these

Table 12.7 Age and Marital Happiness Frequencies (and Mean Deviated Cross Products)

Age	Happiness			Total
	Not too happy (1)	Pretty happy (2)	Very happy (3)	
Young (1)	13 (1) [1.40]	129 (0) [1.30]	199 (−1)	341
Middle-aged (2)	7 (0) [0.97]	97 (0) [1.49]	195 (0)	299
Aged (3)	2 (−1)	27 (0)	81 (1)	110
Total	22	253	475	750

Figures in brackets are odd ratios for the 2×2 subtable including contiguous cells with higher values on Happiness or Age.
From Ishi-Kuntz (1991).

data, I coded the variables 1, 2, and 3 (as shown in the table) and computed the mean deviated cross products (these are shown in parentheses in the table). For example, the mean of both variables was 2.0. The cross-product for the cell where Age = 1 and Happiness = 1 was −1 ∗ −1 = 1. These values were entered into the database as a new variable, MDCP (for mean-deviated cross product). This variable could have also been created in SPSS using "Transform" → "Compute Variable": MDCP = (Age − 2) ∗ (Happiness − 2), as the averages for each variable were 2. Whereas Age and Happiness were coded as nominal variables, consistent with the log-linear model, MDCP was coded as a scale (i.e., metric) variable. The next step was to weight the data by frequency, and to use the general log-linear option to fit a custom model including main effects for Age and Happiness (as factors) and MDCP (as a covariate). (One can also use SPSS to perform a chi-square analysis on the contingency table, and the output will include the results of a linear-by-linear association analysis.) I also ran an independence model that did not include the MDCP term.

The independence model did not fit the data well $[\chi^2_{LR}(4) = 9.86, p = .04]$ whereas the linear-by-linear association model did $[\chi^2_{LR}(3) = .264, p = .967]$. The change in the chi-square statistic was significant $[\chi^2_{LR}(1) = 9.60, p = .002]$. The estimated coefficient for MDCP ($\hat{\theta}$) was 0.30, and $exp(0.30) = 1.35$. Therefore, the estimated odds of greater marital happiness (as measured by an additional point on the three-point happiness scale) increase by 35% with age. The actual ORs for each contiguous 2 × 2 subtable are shown in brackets in Table 12.7. These range from 0.97 to 1.49, so the 1.35 estimate seems reasonable.

Ishi-Kuntz (1991) also collected data on a third variable, social interaction with relatives (never, sometimes, frequently). Let R denote this variable. Including this variable in the model necessitates adding three additional terms to the model (R ∗ Age, R ∗ Happiness, and R ∗ Age ∗ Happiness), where each term has its own mean deviated cross-product. The three-way interaction term was nonsignificant and therefore dropped from the model. Age ∗ Happiness remained significant even when controlling for relatives (by including R ∗ Age and R ∗ Happiness).

These examples use *integer scoring*, as the categories were coded with the integers 1, 2, and 3. (Another common method is *centered* scoring; see Technical Note 12.4.) If treated as a metric variable, it is assumed that the distance between any two adjacent scores is the same. It could be argued, however, that this assumption might not hold for Happiness, because for many individuals the distance between *not too happy* (scored 1) and *pretty happy* (scored 2) might not be the same as that between *pretty happy* and *very happy* (scored 3). How critical is the equispaced assumption? For example, what would happen if one scored *not too happy* as 0 rather than 1, to make the distance between adjacent scores unequal (R was not included in the model for this simulation)? The results were that $\hat{\theta}$ dropped to 0.25, and the p value dropped in half (.004). These results were therefore sensitive to the distances assumed, so while some researchers do use association models for ordinal variables,

I do not recommend it unless it can be assumed that the scores represent approximately equal distances on a latent variable.

If only one of the variables can be considered metric, then it is more appropriate to fit a row or column effects model. If the column variable were metric, then one would fit a row effects model and vice versa. A row effects model takes the form

$$\lambda_0 + \lambda_i^X + \lambda_j^Y + \tau_i\left(v_j - \bar{v}\right), \tag{Eq. 12.10}$$

where $\Sigma\tau_i = 0$.

A column effects model takes the form

$$\lambda_0 + \lambda_i^X + \lambda_j^Y + \rho_j\left(u_i - \bar{u}\right), \tag{Eq. 12.11}$$

where $\Sigma\rho_j = 0$.

Considering the data in Table 12.7, suppose—for the sake of argument—one is not willing to consider the happiness variable as consisting of equispaced scores. The age variable is still metric, so one could fit a column effects model. The mean deviated values for age would be, as before, $-1, 0, 1$; call this MDAge. In SPSS, one would specify the model as

Age, Happiness, MDAge ∗ Happiness.

The variables Age and Happiness should be designated as factors and the MDAge term as a covariate. Unlike the linear-by-linear association model, where information relevant to any interaction would be contained in the newly created MDCP term, here it is necessary to specify an interaction term, MDAge ∗ Happiness. The model fits well $[\chi_{LR}^2(2) = .21, p = .90]$. The MDAge ∗ Happiness term is significant ($z = -2.78, p = .005$) for the second level of happiness (with the third level as the reference category).

The estimates are $\hat{\rho}_1 = -0.53$, $\hat{\rho}_2 = -0.31$ and $\hat{\rho}_3 = 0$ (as the third level of happiness is the reference category). However, the $\hat{\rho}_j$ estimates do not sum to 0, as required by the model. All we need to do here is to mean deviate the estimates so that they will. The mean is -0.28, and the mean deviated estimates are $\hat{\rho}_1' = -0.25$, $\hat{\rho}_2' = -0.03$, and $\hat{\rho}_3' = 0.28$. Although the estimates show a trend, only the comparison between the third and second levels of happiness was significant. The estimated local OR for those levels is given by $exp\left(\hat{\rho}_3' - \hat{\rho}_2'\right) = exp(0.31) = 1.36$. The odds of being *very happy* (as opposed to *pretty happy*) increase by 36% as a couple moves up one category of age.

Finally, suppose that neither variable is metric but there is some ordering of categories on both variables. If the variables reflect underlying continuous constructs, one could fit the following log-multiplicative row and column effects model (also known as an RC) model:

$$\lambda_0 + \lambda_i^X + \lambda_j^Y + \theta\tau_i\rho_j, \tag{Eq. 12.12}$$

where $\Sigma\tau_i = 0 = \Sigma\rho_j = 0$ and $\sum\tau_i^2 = \sum\rho_j^2 = 1$. This model is appropriate when the observed scores for any two adjacent levels cannot be assumed to be equispaced or even the categories correctly ordered. One can try to determine the correct ordering and spacing from the data by using special algorithms to estimate the row and column effects τ_i and ρ_j.

The RC model is similar in form to the linear-by-linear association model (Eq. 12.9), except that the mean-deviated scores are estimated from the data; in effect, one is searching for the best-fitting linear-by-linear association model by allowing the columns and rows to be reordered and moved closer to or farther apart from one another (Simonoff, 2003). Using data on the New York Public Library book collection, Simonoff (2003) gave the example of examining the association between book condition (intact, slight, moderate, and extreme deterioration) and preservation method (repair, microfilm, restore, and "no preservation needed"). The latter variable is ordered in terms of the most to least serious (in terms of cost and technical difficulty). The data are shown in Table 12.8. Although counterintuitive, Simonoff and Tsai (1991) did not find more intensive preservation methods to be used on the most deteriorated books. Instead, they found an inverse relationship between the two variables, because the more deteriorated a book was, the less likely the library was to spend money upgrading it. (Repairs and microfilming tended to be made to intact and slightly deteriorated books.) However, the best-fitting RC model made "no preservation needed" the first column category, rather than the last, because the vast majority of books were intact and needed no preservation. Reordering the preservation method variable in this way maximized the linear association. Technical Note 12.5 presents more detailed results.

RC models are not log-linear models but are known instead as *log-multiplicative* models and are more difficult to estimate. Estimating an RC model cannot be done in SPSS and requires special software. Free software that can be downloaded from the Internet includes LEM (Vermunt, 1997), CDAS (Eliason, 1997), and the *gnm* package in R (Turner & Firth, 2007). A good overview of the RC and other association models is provided by Wong (2010).

Table 12.8 Book Deterioration and Preservation Data

Book condition[a]	Preservation method			No preservation needed
	Repair	Microfilm	Restore	
Intact	27	50	7	676
Slight	1	6	0	22
Moderate	2	3	1	13
Extreme	0	34	30	3

[a]Degree of deterioration.
From DeCandido (cited in Simonoff & Tsai, 1991).

With metric variables, an alternative to the RC or linear-by-linear association models is to fit a R + C (row and column effects) model. This model is log-linear and can be estimated in SPSS (for a two-way table, include two interaction terms (e.g., book_condition*MDpreservation_method and MDbook_condition*preservation_method). Wong (2010) recommended comparing different association models and choosing a model (linear-by-linear, RC, C, R, or R + C) with the smallest, most negative Bayesian information criterion (BIC), assuming that all other aspects of these models have been correctly specified and the chosen model makes theoretical sense. According to Wong, BIC differences of 5 or more points should be considered meaningful (or use the guidelines presented in Chapter 10).

When both the IV and DV are ordinal, a simpler but less powerful alternative is to run a nonparametric $\hat{\tau}$ or partial $\hat{\tau}$ test (see Chapter 6), assuming one can determine the correct ordering of the categories for each variable. So for example, $\hat{\tau}_b$ for Table 12.8 is −.45, but if one correctly moves the "no preservation needed" to be the first column category, the $\hat{\tau}_b$ value is now +.52 ($p < .001$). However, these tests will be less sensitive to complex patterns in the data, such as those shown in Table 12.8. Ordinal regression is another possibility (see Chapter 11).

If one of the variables is nominal and the other ordinal, one can run a Kruskal-Wallis test (if one uses the nominal variable to predict the ordinal), as Kruskal-Wallis can be used for ordered categories and not just ranks (see Chapter 7).

Sample Considerations

Sample Size and Sparse Cells

Large sample sizes are desirable when one is using log-linear analysis. The approach uses maximum likelihood estimation, the models tend to contain a large number of terms, and stepwise regression is often employed. Large samples are therefore often needed to reduce MLE bias, achieve sufficient statistical power, and reduce error rates.

It is generally recommended that there be at least 5 times the number of cases as cells. For example, a 2 × 2 × 3 table would have 12 cells and therefore require 60 cases. Without a sufficient sample size, one may need to collect additional data or to drop one or more variables (but be careful not to drop potentially confounding, relevant variables). Furthermore, for all two-way associations, the expected cell frequencies should be greater than 1, and no more than 20% should be less than 5. If there is a problem with sparse cells, remedial options include (a) doing nothing (violation of the requirement has more an effect of Type II than Type I errors), (b) combining some categories with others if it makes theoretical sense, (c) deleting some sparse or empty cells from the database (Allison, 1999), or (d) with an unsaturated model, adding 0.5 to each cell so that there are no empty cells (although this can also adversely affect statistical power and distorts the sample size). Agresti (2002) noted that it is acceptable to have some empty cells.

Cell Structure and Structural Zeroes

In the SPSS "Log-Linear" procedure, there is a box called "Cell Structure." If using a Poisson model, the cell structure is used to indicate the offset, which must be greater than zero. "Cell Structure" can also be used to designate structural zeroes. If the value for a cell in the cell structure is zero or less, than the cell values will be treated as structural zeroes (as opposed to sampling zeroes) and will not be used to fit the model. (A structural zero is a cell value that must be zero theoretically or methodologically; see Chapter 11.) If one is just interested in specifying structural zeroes and not using an offset, then set the value for cells that are not structural zeroes equal to 1 and those that are equal to 0. In either case, one must create a variable with the appropriate information and insert the variable name into the "Cell Structure" box in SPSS.

Doubly Classified Data

When cases are classified twice on the same set of categories, a square contingency table is produced. For example, two different raters may classify the cases for the purpose of calculating an interrater reliability statistic. Another example is when individuals are classified into occupational categories twice—one based on a parent's occupation and one based on the child's. A third example is when matched pairs are formed. It is beyond the scope of this chapter to review the large number of techniques for analyzing square contingency tables; Agresti (2002) and Simonoff (2003) provide good coverage of this topic. A major concept in such analyses is testing for *quasiindependence* (statistical independence excluding the main diagonal of the contingency table) and *quasisymmetry* (where $\lambda_{ab} = \lambda_{ba}$, when $a < b$).

Summary

Log-linear techniques can in principle be used for any size contingency table and for any number of variables (and interaction effects), although the sample size has to be sufficiently large. Remedial actions must also be taken if there are too many sparse cells. Log-linear analysis is useful when one primarily has nominal variables, although there are log-linear models for ordered contingency tables as well.

Problems

1. Schwarz (as cited in Pomer, 1984) distributed a survey in 1946 to 1,040 Americans from which he evaluated whether the person was prejudiced against African Americans. The survey was readministered to a different

sample in 1964. Following are the data broken down by year and region of the country (North, South). Perform a log-linear analysis to determine whether prejudice declined over time, and whether it declined more in the South or North.

a) Enter the data into your statistical software program. Be sure to recode string codes into numeric ones and weight the data by frequency. Generate a contingency table to check the accuracy of your data entry.

b) Find the best-fitting model (in SPSS, use "Model Selection"). What are the final generating classes?

c) Because the researcher expects the amount of prejudice to be influenced by region and year, and the size of the survey sample was fixed in advance, run a conditional logit log-linear model with Prejudice as the DV and Region and Year as predictors.

 (Do not include a Year * Region term.) Report the B, *exp*(B), and *p* values for all interactions. Interpret the estimates and associated ORs. Was the amount of prejudice greater in the North or South? Did the amount of prejudice decline over time?

d) What is the largest standardized residual? Is there any reason for concern?

e) From the data, construct the contingency table for Year * Prejudice. What are the conditional probabilities (conditioned on Year)? Are they declining over time? What is the actual OR? How does it compare with the estimated OR?

f) Because one of the research questions was whether prejudice declined more in the North or South, rerun the model with the interaction term Year * Region. For that term, what is B, *exp*(B), and the *p* value. Interpret the results.

g) Rerun the model in the previous subproblem using the IBM SPSS GLM procedure. (Continue to weight the data by frequency.) Set the reference category to first ("lowest value.") Run a custom model (binomial with logit link). Set the reference category to first ("lowest value.") Then rerun the model with the negative log-log link. Is the interaction term now significant? Is there any rationale for using this model?

Year	North Prejudiced		South Prejudiced	
	Yes	No	Yes	No
1946	374	439	163	64
1963	56	410	31	126

2. Fennema and Sherman (1977; see also Burnett, 1983) collected data on the number of high school students in "on grade" math classes from four schools, broken down by grade and gender. The data are below (and in file *On_Grade.12. Problem_2*). "Gender" is coded with "Female" = 1. Be sure to weight the data by the frequency counts.

a) Generate a multiway contingency table. Treat "Grade" as a nominal variable.
b) Find the best-fitting model.
c) Run a general log-linear analysis using the best-fitting model from Part (b), requesting parameter estimates. Report the Bs, *exp*(Bs), z and p values for "On Grade" * Grade, "On-Grade" * Gender, and "On Grade" * Grade * Gender (there will be other significant effects besides these). Interpret the estimates and associated ORs. Is there a steeper decline in "On Grade" enrollments for males or females?

Grade	Gender	School 1		School 2		School 3		School 4	
		On	Off	On	Off	On	Off	On	Off
9	F	262	62	249	102	179	57	289	39
	M	280	62	231	104	180	63	249	74
10	F	149	95	121	116	141	86	100	149
	M	199	133	111	111	160	86	102	181
11	F	63	112	22	71	41	95	24	87
	M	94	124	43	90	50	98	39	131
12	F	8	60	1	35	8	63	7	27
	M	25	98	11	62	24	79	15	46

3. A researcher analyzes data similar to that in Barth et al. (2007) relating to the number of out-of-home placement moves of children with emotional and behavioral disorders (EBD). Data were obtained from a table in a published report that dichotomized all variables. All variables were therefore nominal: Less than four moves a year versus 4 or greater, whether the child was assessed as depressed versus not, lived with siblings versus not, or was initially placed before age 11 versus not. The data are below.
   a) Enter the data (using dummy variables) and weight by the frequency counts. Using a log-linear model selection algorithm (e.g., Log-Linear ➜ Model Selection in SPSS), find the generating classes.
   b) Because the sample size was not fixed in advance, should one use a Poisson or multinomial model?
   c) Run a general log-linear model using the terms identified in Part (a). In SPSS, use the "General Log-Linear" procedure and customize the model (be sure to include all the appropriate lower-order terms for any higher-order terms) and to request estimates under options. For each interaction, report the coefficient, z score, and p value.
   d) Are depressed children likely to be moved more? What is the estimated OR?
   e) Are children with siblings more likely to be moved? What is the estimated OR?

Placement Moves of EBD Children

Depressed	Lives with siblings	Age at baseline				Total
		<11 years Placements		11 years+ Placements		
		<4 moves	4+ moves	<4 moves	4+ moves	
Yes	Yes	7	8	8	5	28
	No	6	10	5	4	25
No	Yes	40	15	28	6	89
	No	20	8	10	15	53
Total		73	41	51	30	195

4. Le (1998) presented data, slightly modified from Korff, Taback, and Bean (1952), on the number of people who became ill after attending a company picnic in Baltimore (data follow). Variables included whether an employee ate crab meat and/or potato salad.

a) Using a model selection algorithm (e.g., "Log-Linear" → "Model Selection" in SPSS), find the generating classes.

b) Why should one use a Poisson model for this problem?

c) Run a general log-linear analysis for the selected model. Report the coefficient, z score, and p value for each interaction effect. (In SPSS, be sure to check "estimates" from the options box.)

d) Run a general log-linear analysis with "crabmeat," "ill," and "crabmeat * ill." Why does eating crabmeat have a significant association with illness when "potato salad" is omitted from the model? (Hint: Was there a significant relationship between eating "potato salad" and "crabmeat" from part (a)?) Is "crabmeat" conditionally or marginally independent from contracting food poisoning?

e) Generate a 2×2 contingency table between "ill" and "potato salad," along with the actual OR. Compare that value with what is estimated from the log-linear analysis [using $exp(B)$].

Ill	Potato salad eaten				Total
	Yes		No		
	Crabmeat eaten	No crabmeat eaten	Crabmeat eaten	No crabmeat eaten	
Yes	120	22	4	1	147
No	80	24	31	23	158
Total	200	46	35	24	305

5. Fergusson, Boden, and Horwood (2007) collected data on a large sample of young people in New Zealand. Individuals were interviewed about whether they ever thought about suicide (suicidal ideation, or SI) and their employment history. The data are below (and in file *Suicide.12.Problem_5*).

a) Conduct a general log-linear analysis of the data using a row or column model, given that age is a metric variable. Use the multinomial option for the distribution of cell counts. Treat the employment categories as nominal because the categories are not equally spaced (and the statistical models we will use do not address ordinal predictors). Be sure to treat Age as a factor and MDAge as a covariate. The model should have the following terms: SI, Age, Unemployment, SI * MDAge, SI * Unemployment, MDAge * Unemployment, and SI * MDAge * Unemployment. Report the z and p value for the three-way interaction term. Is it significant?

b) Run the model without the three-way interaction term. Report the B, *exp*(B), z, and p values for the two-way interaction terms. (Note: There will be two sets of statistics for terms involving Unemployment.)

c) Do the parameter estimates suggest that older workers experience more long-term unemployment (> 6 months)? Find the mean of the parameter estimates, and then mean deviate the parameter estimates. Does there appear to be a trend? What are the estimated local ORs?

d) Based on a calculation of Kendall's $\hat{\tau}_b$, is age positively or negatively associated with SI? What about unemployment and SI? How do the beta estimates support your conclusions?

e) Ferguson et al. (2007) did not find significant relationships between unemployment and SI when controlling for some other factors, such as drug use. Explain theoretically how drug use could be a confounding factor.

Age	Suicidal ideation present			Suicidal ideation absent		
	Unemployment			Unemployment		
	None	<6 mo.	+6 mo.	None	<6 mo.	+6 mo.
16	96	14	10	844	56	5
17	77	29	12	743	134	30
18	53	17	10	750	142	39
19	58	27	6	730	135	55
20	58	21	7	736	140	49
21	50	6	12	831	54	48
22	53	9	12	786	89	53
23	46	10	5	824	73	43
24	43	9	5	826	72	46

6. Kim, Yang, Kim, Kim, and Yoon (2007) collected data on 308 patients who had been treated for eye injuries at a Korean medical center. One variable was the physical zone of each injury: Zone 1 was isolated to the cornea, Zone 2 included injury to the sclera (i.e., white of the eye), and Zone 3 included a greater zone on the eyeball. The other variable was each patient's visual acuity after treatment, measured on 5-point scale: $1 = > \frac{20}{40}$, $2 = \frac{20}{50}$ to $\frac{20}{100}$, $3 = \frac{19}{100} - \frac{5}{200}$, $4 = \frac{4}{200} - LP$ (*some light perception*), $5 = $ *no light perception*. Both scales were considered reasonably metric.

    a) Perform a linear-by-linear association analysis. In SPSS, use the "General Log-Linear" procedure (Poisson option). Report the estimate of θ and the associated z and p values. Describe the direction of the association.

    b) Perform a chi-square test between zone and acuity; in SPSS, use the cross-tabs procedure. What is the χ^2 value for the linear-by-linear association? Is it the square of the z value computed in part (a)?

Visual acuity	Injury zone			Total
	1	2	3	
1	70	28	19	117
2	25	7	3	35
3	20	7	8	35
4	33	13	36	82
5	9	4	26	39
Total	157	59	92	308

Technical Notes

12.1 The geometric mean of n numbers is the nth root of the product of those numbers. For example, the geometric mean of 1, 3, and 5 is $\sqrt[3]{(1*3*5)} = 2.47$.

12.2 According to *Simpson's paradox*, marginal independence does not imply conditional independence and vice versa. The magnitude and even direction of the association between two variables can be quite different depending on whether one does or does not control for a third variable.

12.3 Contingency tables can be collapsed into lower-order tables if the control variable (Z) is conditionally independent from either X or Y (and there is no three-way interaction). In that case, the marginal and conditional ORs (for X and Y) will be the same.

12.4 An alternative to integer scoring is centered scoring, where $X_i = i - \left(\frac{I+1}{2}\right)$, and $Y_j = j - \left(\frac{J+1}{2}\right)$, where I and J are the total number of rows and categories. For a 2×2 table, the scores would be -0.5 and 0.5 for each variable; for a 3×3 table, the scores would be $-1, 0, 1$. Use of one scoring system over another should reflect how one wants to conceptualize and represent the variable in question.

12.5 For the data in Table 12.8, the estimated parameters for the RC model were $\hat{\theta} =$ 8.10, $\hat{\tau} = (-.38, -.26, -.23, .86)$, and $\hat{\rho} = (-.20, .36, .55, -.72)$; see Simonoff (2003). The $\hat{\rho}$ effects, which related to preservation method, are generally increasing, except when we come to the last category ("no preservation needed"), which should be placed first in the ordering rather than last.

APPENDIX 12.1

DERIVATION OF $\widehat{OR}$ FROM LOG-LINEAR DUMMY PARAMETERS

For a 2×2 contingency table, let $ln(m_{11}) = \beta_0 + \beta_1 R_i + \beta_2 C_j + \beta_3 R_i C_j$. Therefore,

$ln(m_{11}) = \beta_0 + \beta_1 + \beta_2 + \beta_3$, because $R_i = 1$ and $C_j = 1$.

$ln(m_{12}) = \beta_0 + \beta_1$, because $R_i = 1$ and $C_j = 0$. Therefore $\beta_1 = ln(m_{12}) - \beta_0$

$ln(m_{21}) = \beta_0 + \beta_2$, because $R_i = 0$ and $C_j = 1$. Therefore $\beta_2 = ln(m_{21}) - \beta_0$

$ln(m_{22}) = \beta_0$, because $R_i = 0$ and $C_j = 0$.

Solving for β_3:

$\beta_3 = ln(m_{11}) - \beta_0 - \beta_1 - \beta_2$ (from the second equation),

$\quad = ln(m_{11}) - ln(m_{22}) - (ln(m_{12}) - \beta_0) - (ln(m_{21}) - \beta_0)$,

$\quad = ln(m_{11}) - ln(m_{22}) - ln(m_{12}) + ln(m_{22}) - ln(m_{21}) + ln(m_{22})$,

$\quad = ln(m_{11}) - ln(m_{12}) - ln(m_{21}) + ln(m_{22}) = ln\left(\frac{m_{11}m_{22}}{m_{12}m_{21}}\right) = ln\ \widehat{OR}$.

β_3 is λ_{ij}^{XY}, so $exp(\lambda_{ij}^{XY}) = \widehat{OR}$.

Source: Allison (1999).

APPENDIX 12.2

CONDITIONAL LOG-LINEAR ANALYSIS: MATHEMATICAL FOUNDATIONS

Conditional (logit) log-linear analysis assumes (a) one variable depends on the others, and (b) that variable follows a multinomial distribution. The latter implies that n should be fixed. The random component of the model can be formalized as $(F_{ij} \dots F_{rj}) \sim Multinomial(n_j, \pi_{ij} \dots \pi_{rj})$, where r is the number of categories comprising the DV. The systematic component of the model involves the term $v_{ij} = \Sigma_{k=1}^{p} x_{ij} \beta_k$, where p is the number of nonredundant predictors (indexed by k). Assuming the cell structure $z_{ij} > 0$, then:

$$m_{ij} = z_{ij} * e^{\alpha_j + v_{ij}},$$

where α_j is the normalizing constant for the jth category of an IV. We define $\alpha_j = ln\left(\frac{n_j}{\Sigma_{i=1}^{\gamma} z_{ij} e^{v_{ij}}}\right)$. Normalizing constants ensures that the sum of the DV probabilities equal one. If the cell structure value $z_{ij} \leq 0$, then we set $m_{ij} = 0$.

Concentration is a measure of the degree of association between the DV and the set of IVs. It is a ratio of the amount of dispersion due to the model to the total dispersion, where total dispersion is defined as $S(B) = n\left(1 - \Sigma_{i=1}^{\gamma} \hat{\pi}_i^2\right)$; i is the index for the DV categories. An alternative measure of association is *entropy* (see Haberman, 1982).

References

Agresti, A. (2002). *Categorical data analysis* (2nd ed.). Hoboken, NJ: Wiley.

Allison, P. D. (1999). *Logistic regression using the SAS system: Theory and application.* Cary, NC: SAS Institute.

Barth, R. B., Lloyd, E. E., Green, R. L., James, S., Leslie, L. K., & Landsverk, J. (2007). Predictors of placement moves among children with and without emotional and behavioral disorders. *Journal of Emotional and Behavioral Disorders, 15,* 46−55.

Burnett, J. D. (1983). Loglinear analysis: A new tool for educational researchers. *Canadian Journal of Education, 8,* 139−154.

Eliason, S. R. (1997). *The categorical data analysis system supplemental user's manual for command line pre-release version 4.0 of PROG MLLSA for DOS and OS/2.* Retrieved from http://www.u.arizona.edu/~seliason/software.html

Fennema, E., & Sherman, J. (1977). Sex-related differences in mathematics achievement, spatial visualization and affective factors. *American Educational Research Journal, 14,* 51−71.

Fergusson, D. M., Boden, J. M., & Horwood, L. J. (2007). Unemployment and suicidal behavior in a New Zealand birth cohort: A fixed effects regression analysis. *Crisis, 28,* 95−101.

Goodman, L. A. (1970). The multivariate analysis is of qualitative data: Interactions among multiple classifications. *Journal of the American Statistical Association, 65,* 226−256.

Haberman, S. J. (1982). Analysis of dispersion of multinomial responses. *Journal of the American Statistical Association, 77*, 568–580.

Ishi-Kuntz, M. (1991). Association models in family research. *Journal of Marriage and the Family, 53*, 337–348.

Kim, J.-H., Yang, S. J., Kim, D. S., Kim, J.-G., & Yoon, Y. H. (2007). Fourteen-year review of open globe injuries in an urban Korean population. *The Journal of TRAUMA Injury, Infection, and Critical Care, 62*, 746–749.

Korff, F. A., Taback, M. A. M., & Beard, J. H. (1952). *Public Health Reports, 67*, 909–913.

Le, C. T. (1998). *Applied categorical data analysis.* New York: John Wiley & Sons.

Mair, P. (2006). *Interpreting standard and nonstandard log-linear models.* New York, NY: Waxmann.

Pomer, M. I. (1984). Demystifying log-linear analysis: Four ways to assess interaction in a $2 \times 2 \times 2$ table. *Sociological Perspectives, 27*, 111–135.

Pugh, M. D. (1983). Contributory fault and rape convictions: Loglinear models for blaming the victim. *Social Psychology Quarterly, 46*, 233–242.

Simonoff, J. S. (2003). *Analyzing categorical data.* New York, NY: Springer.

Simonoff, J. S. & Tsai, C.-L. (1991). Higher order effects in log-linear and log-non-linear models for contingency tables with ordered categories. *Journal of the Royal Statistical Society. Series C (Applied Statistics), 40*, 449–458.

Turner, H. L., & Firth, D. (2007). Generalized nonlinear models in R. *Statistical Computing & Graphics Newsletter, 18*, 11–16.

Vermunt, J. K. (1997). LEM 1.0: *A general program for the analysis of categorical data.* Tilburg University, Tilburg, The Netherlands. Retrieved from http://spitswww.uvt.nl/fsw/mto/lem/manual.pdf

Wong, R. S.-K. (2010). *Association models.* Thousand Oaks, CA: Sage.

CHAPTER 13

GENERAL ESTIMATING EQUATIONS

The general estimating equations (GEE) approach, initially developed by Zeger and Liang (1986), is used in lieu of generalized linear models (GLMs) when there is statistical dependence among observations. So, for example, use GEE if using ordinal, logistic, or Poisson regression for pre–post or repeated measure analysis. GEE can also be used for clustered data structures as an alternative to multilevel models, especially when there are many small groups (e.g., many small discussion groups), because approaches such as multilevel models are better used with larger groups of at least 20–30 students each so that individual regression equations can be estimated for each group (Hox, 2010). However, of the two situations (repeated measures and clustered observations), the repeated measures application is stressed in this chapter (and the IBM SPSS procedure is designed only for repeated measures).

In this chapter, I first discuss the foundations of GEE, including the various types of correlational structures between observations that can be assumed and how to estimate standard errors and perform statistical hypothesis tests. Sample size requirements, the approach for estimating missing values, and residual tests are then discussed, followed by step-by-step instructions on how to perform a GEE analysis in SPSS.

Foundations of General Estimating Equations

Example

I will present the GEE procedure using the data in the file *Arousal.13*. The data are from a hypothetical experiment in which 30 participants were asked to solve a complex mathematical problem on three different occasions (T_1, T_2, T_3) for Time 1, Time 2, and Time 3, respectively. The occasions were spaced one week apart, with a different but comparable problem randomly chosen for each occasion. The dependent variables (Y_1, Y_2, Y_3) were whether the participant was able to successfully solve the problem, and hence the dependent variables (DVs) were nominal. The independent variables (X_1, X_2, X_3) were measures of physiological arousal taken just prior to problem solving on each of the three different occasions. The research hypothesis was that higher levels of arousal will increase the chance of successfully solving a problem.

The research hypothesis will be tested with logistic regression. Recall that ordinary logistic regression is based on maximum likelihood estimation (MLE), which chooses parameter estimates that maximize the likelihood of obtaining the observed data. The likelihood function for logistic regression that was presented in Chapter 10 is based, in part, on the binomial distribution; specifically, the conditional distributions of Y_i are assumed to be distributed binomially in grouped data (and as Bernoulli when ungrouped). The binomial distribution is defined as a series of *independent* Bernoulli trials, so if the trials are no longer statistically independent, the likelihood function is no longer valid and therefore MLE cannot be used. Instead, one can use an approach known as *quasilikelihood estimation* (Wedderburn, 1974). This approach bases the estimates of the standard errors not on the empirical variances but rather as a function of the means. (This method will be illustrated in more detail in the next chapter.)

Quasilikelihood estimation is similar to MLE in that we attempt to find the maximum point of a function iteratively, by starting with initial guesses of the parameter estimates and then updating those guesses in several rounds of computation. In our example, there are two parameters, β_0 (the intercept) and β_1 (the slope). My initial guess for the parameters is that the intercept is –1.0 and the slope is 0.30. There are a number of ways to make the initial guesses. For example, an ordinary logistic regression could be performed assuming all the observations are independent (i.e., the GLM procedure could be used). This procedure is the one used by IBM SPSS. In our example, the initial guesses were chosen here more on pedagogical grounds, as these particular guesses make the example more dramatic and memorable.

Based on the initial guesses, we can generate predicted values of Y_0. The prediction equation is

$$\text{Predicted logit} = \text{predicted log odds} = ln\left(\frac{\hat{\pi}_{it}}{(1-\hat{\pi}_{it})}\right) = \hat{\beta}_0 + \hat{\beta}_1 X_{it}. \tag{Eq. 13.1}$$

In this example, $\hat{\pi}_{it}$ is the probability of a success for the ith individual (there are 30) at time t (T_1, T_2, T_3). Table 13.1 shows the predicted logits for the first individual in our data, using the initial parameter estimates. From these, the predicted probabilities were derived by exponentiating the logits (see Table 13.1). Recall that the logit is the log of the odds, so exponentiation produces the predicted odds. One can then move from the odds scale to the probability scale by using the formula.

Table 13.1 Predictions for the First Individual Based on Initial Parameter Estimates

Predicted logit calculations			
Time			
1	$-1.0 + 0.30 * 0.85 = -0.75$		
2	$-1.0 + 0.30 * 4.10 = 0.23$		
3	$-1.0 + 0.30 * 3.55 = 0.07$		

Predicted probabilities			
Time	*Logit*	*Odds[a]*	*Probabilities[b]*
1	-0.75	0.47	32.2%
2	0.23	1.26	55.7%
3	0.07	1.07	51.7%

Residuals			
Time	*Y*	*Predicted probabilities*	*Residual[c]*
1	1	32.2%	67.8%
2	1	55.7%	44.3%
3	1	51.6%	48.4%

Pearson standardized residuals				
Time	$\widehat{Var\ (\pi)}_{it}$ [d]	$\widehat{SD}$ [e]	*Raw residual[f]*	*Pearson residual[g]*
1	0.22	0.47	0.68	1.45
2	0.25	0.50	0.44	0.89
3	0.25	0.50	0.48	0.97

[a] Predicted odds $= exp$(logit).

[b] $\hat{\pi}_{it} = \frac{\widehat{odds}}{(1 + \widehat{odds})}$.

[c] Residual $= Y_{it} - \hat{\pi}_{it}$

[d] Predicted $Var = \hat{\pi}_{it}(1 - \hat{\pi}_{it})$. For any case, $n = 1$ so the n term is not needed.

[e] $\widehat{SD} = \sqrt{\widehat{Var}}$.

[f] Reexpressed in decimal form.

[g] Pearson residual $=$ raw residual/$\widehat{SD}$.

$$\hat{\pi}_{it} = \frac{\widehat{odds}}{\left(1 + \widehat{odds}\right)} . \qquad \text{(Eq. 13.2)}$$

This formula is based on the definition of odds as $\frac{\pi}{(1 - \pi)}$.

EXPLORING THE CONCEPT

See if you can derive Eq. 13.2 from the definition of odds with a little algebra.

Once we have calculated the predicted values of π_{it}, we can then calculate the residuals (i.e., errors) with the following formula:

$$\text{Residual} = Y_{it} - \hat{\pi}_{it}. \tag{Eq. 13.3}$$

This step is illustrated in Table 13.1. The last part of the table standardizes the residuals using the Pearson method. Note that the variances are computed as a function of the mean, which is part of the quasilikelihood approach.

Correlational Structure

In regular logistic regression, one assumes that in the population, the conditional values of Y follow a binomial distribution at each level of X. However, this assumption is problematic here because of statistical dependence (the binomial distribution assumes that the trials are independent). Fortunately, there is a saving grace! We do have empirical data on the degree that the observations for the different time periods are correlated with one another. One can therefore measure the degree of statistical dependence between each time period and control for it.

Table 13.2 presents the correlations between the Y_{it}s by time period. It also presents the correlations between the Pearson residuals. The estimates for both analyses are close to one another; however, it is better to use the Pearson

Table 13.2 Correlations Between Outcomes and Pearson Residuals

	Outcomes (Y_{it})		
	Y_1	Y_2	Y_3
Y_1	1.00	.36	.46
Y_2		1.00	.66
Y_3			1.00
	Pearson residuals (r_{it})		
	r_{i1}	r_{i2}	r_{i3}
r_{i1}	1.00	.44	.50
r_{i2}		1.00	.68
r_{i3}			1.00

residuals because the influence of variation in X is removed. Let the population parameters for the correlation be represented by $a_{tt'}$ where t is one time period and t' is the other. We are therefore estimating three parameters, α_{12}, α_{13}, and α_{23}. The specific estimates are $\hat{\alpha}_{12} = .44$, $\hat{\alpha}_{13} = .50$, and $\hat{\alpha}_{23} = .68$. The correlation matrix in the second part of Table 13.2 is known as the *unstructured initial working correlation matrix*.

One can potentially increase statistical power by imposing some structure on the correlation matrix, specifically by assuming how the correlations are related to one another. This approach allows one to estimate fewer parameters. (Estimating more parameters requires more information, thus consuming statistical power.) For repeated measures analyses, a common structure is AR(1), which is shorthand for an autoregressive model of Order 1. In an AR(1) model, it is assumed that each observation is causally affected by the preceding observation. (The spacing between time points must be the same for all individuals.) So, for example, the amount of cigarettes that a person smokes per day is influenced by the number smoked the preceding day, because smoking is a nasty drug habit. An AR(1) model would be as follows:

Number Smoked at Time t = Constant + Coefficient ∗ Number Smoked at Time $(t - 1) + \varepsilon$.

In our psychology experiment, the coefficient is denoted by α. There is only one parameter to estimate, not three. We can estimate α by finding the correlation between the residuals at time t with those one time period prior $(t - 1)$. It turns out that $\hat{\alpha}_{AR(1)} = .56$. We therefore estimate the correlation between the residuals at T_1 and T_2 to be .56 in the population, and the correlation between T_2 and T_3 also to be .56. By transitivity, the correlation between T_1 and T_3 is

Corr (T_1, T_3) = Corr (T_1, T_2) ∗ Corr (T_2, T_3) = 0.56 ∗ 0.56 = 0.31.

The resulting correlation matrix is shown in Table 13.3. The table also shows the correlation matrix for an *exchangeable* correlation structure, where the correlation between any two times is the same as all the others. It turns out that $\hat{\alpha}_{exchangable} = .54$.

EXPLORING THE CONCEPT

In an AR(1) structure, why do the correlations become increasingly smaller as the time periods become farther apart?

An AR makes sense for longitudinal data, although an exchangeable structure is also sometimes sensible. An AR does not make sense for clustered data (e.g., students

Table 13.3 Different Correlational Structures

	r_{i1}	r_{i2}	r_{i3}
		AR(1)	
r_{i1}	1.00	.56 (α)	.31 (α * α)[a]
r_{i2}		1.00	.56 (α)
r_{i3}			1.00
		Exchangeable	
r_{i1}	1.00	.54 (α)	.54 (α)
r_{i2}		1.00	.54 (α)
r_{i3}			1.00
		Independent	
r_{i1}	1.00	0.00	0.00
r_{i2}		1.00	0.00
r_{i3}			1.00

Note: Not shown are an unstructured correlational matrix (was shown in previous table) and an *M-dependent* structure. An example of an *M*-dependent structure with a 4 × 4 matrix would be where there are two correlation parameters (*M* = 2), one for adjacent time periods and one for time periods two periods apart. Time periods three periods apart would be independent.

[a]Because α is the correlation between residuals at times t and $t - 1$ and between times $t - 1$ and $t - 2$, the correlation between times t and $t - 2$ would be α * α.

clustered into small groups), because the statistical dependence is not between time periods. An exchangeable structure is more sensible here.

One other correlational structure shown is an independent structure, where there is no correlation between different time periods. Some GEE programs begin with an independence structure and then impose a structure on the next iteration. Independence structures sometimes fit best when "the number of experimental units is large and the cluster sizes are small" (Davis, 2002, p. 209) so that statistical dependencies make little difference to the results.

It is quite possible to specify an initial correlational structure that is not correct. Fortunately, we shall see that one can use robust estimators of the other parameters in the GEE procedure that are not affected much by specifying the wrong correlational structure. The estimates are *consistent,* meaning that any bias is negligible in large samples; however, one might lose some statistical power. Therefore, one should give some thought as to what type of correlational structure makes the most sense for one's data. For the psychology experiment, because students presumably learn how to better solve these complex math problems with experience, an AR structure would make the most sense. With only 30 participants, we should use an AR(1) structure. We could also consider an AR(2) structure if we had a larger sample. [An AR(2) structure assumes that problem solving two time periods previously has a causal effect on current performance in addition to problem solving one time period previously. This structure entails estimating two parameters so requires more data.]

We therefore settle on the first correlation matrix in Table 13.3 as the initial working correlation matrix. How does one use it to control for the statistical dependence inherent in the data? Recall from Chapter 1 that statistical dependence makes the probability of extreme events greater (assuming positive dependence). Therefore, there is higher probability that extreme results will be obtained under the null hypothesis—increasing the chance of a Type I error unless one makes appropriate increases to the standard errors. This is where the working correlation matrix comes in. The underlying variance-covariance matrices in the estimation process are multiplied by the working correlation matrix, just as was the case with ϕ when we corrected for overdispersion in Chapter 10. The correlation matrix and ϕ are called *nuisance parameters*, because we have to estimate them (and that's a bother!) but we are not directly interested in their values.

When using multiple predictors, let $\hat{\boldsymbol{\beta}}$ represent estimates of the various true (population) regression parameters, one for each predictor plus the intercept. The parameters are arranged in a vector (a vertical list). When a symbol is bold it represents a vector or matrix.

$$\hat{\boldsymbol{\beta}} = \begin{bmatrix} \hat{\beta}_0 \\ \hat{\beta}_1 \end{bmatrix}.$$

Covariance Matrix

We will be estimating $\hat{\boldsymbol{\beta}}$. Note that associated with $\hat{\boldsymbol{\beta}}$ is a $p \times p$ variance–covariance matrix, where p represents the number of predictors (in this case two). Therefore, the 2×2 variance–covariance matrix would be of the form:

	$\hat{\beta}_0$	$\hat{\beta}_1$
$\hat{\beta}_0$	$Var(\hat{\beta}_0)$	$Cov(\hat{\beta}_0, \hat{\beta}_1)$
$\hat{\beta}_1$	$Cov(\hat{\beta}_0, \hat{\beta}_1)$	$Var(\hat{\beta}_1)$

The square roots of the variances here are the standard errors of the estimates. The term *variance–covariance* matrix is somewhat long, and so the matrix is often referred to as the *covariance matrix* of $\hat{\boldsymbol{\beta}}$ even though we are typically more interested in the variances than in the covariances. However, a variance is just a covariance of a variable with itself, so the term *covariance matrix* is still apt.

Now, as part of estimating the covariance matrix for $\hat{\boldsymbol{\beta}}$, one first needs to calculate a covariance matrix for each individual. Because in our example there are 30 individuals, there will be 30 individual covariance matrices. Here, I will just illustrate the calculations for the first person. Let A_i be a diagonal matrix for the ith individual with $Var(Y_{it})$ on the diagonal (with all the other entries zero). Using the values of $Var(\pi_{1t})$ from Table 13.1 (the portion on Pearson residuals), where $Var(\pi_{1t}) = \pi_{1t} * [1 - (\pi_{1t})]$, we have

$$\mathbf{A}_1 = \begin{bmatrix} .22 & 0 & 0 \\ 0 & .25 & 0 \\ 0 & 0 & .25 \end{bmatrix}.$$

If one thinks of each observation as a random variable that can differ from sample to sample, then the sampling distribution will have a mean π_{it} and variance $\pi_{it}(1 - \pi_{it})$; the variances are reflected in this matrix. However, one next needs to correct for statistical dependence by multiplying the matrix by the working correlation matrix or, more specifically, multiplying by its inverse, using the following formula:

$$\mathbf{V}_i = \mathbf{A}_i^{1/2}\mathbf{R}(\alpha)\mathbf{A}_i^{1/2}\mathbf{\Phi}, \tag{Eq. 13.4}$$

where $\mathbf{A}_i^{1/2}$ is the square root of $\mathbf{A}_i$. The scale parameter matrix $\mathbf{\Phi}$ will, in our example, be made an identity matrix and ignored. The calculations of the component matrices are

$$\mathbf{A}_1^{1/2} = \begin{bmatrix} .47 & 0 & 0 \\ 0 & .50 & 0 \\ 0 & 0 & .50 \end{bmatrix}, \mathbf{R}(\alpha) = \begin{bmatrix} 1.00 & .56 & .31 \\ .56 & 1.00 & .56 \\ .31 & .56 & 1.00 \end{bmatrix}, \mathbf{A}_1^{1/2}\mathbf{R}(\alpha) = \begin{bmatrix} .47 & .26 & .15 \\ .28 & .50 & .28 \\ .16 & .28 & .50 \end{bmatrix}.$$

(The reader is encouraged to verify the matrix multiplication using EXCEL; see Chapter 6, e.g., Table 9.5, for a review of linear algebra.) Now, because we have only so far multiplied by the square root of $\mathbf{A}_1$, we need to multiply by that term again to complete the calculation:

$$\mathbf{V}_1 = \mathbf{A}_1^{1/2}\mathbf{R}(\alpha)\mathbf{A}_1^{1/2}\mathbf{\Phi} =$$

$$\begin{bmatrix} .47 & .26 & .15 \\ .28 & .50 & .28 \\ .16 & .28 & .50 \end{bmatrix} * \begin{bmatrix} .47 & 0 & 0 \\ 0 & .50 & 0 \\ 0 & 0 & .50 \end{bmatrix} = \begin{bmatrix} .22 & .13 & .07 \\ .13 & .25 & .14 \\ .07 & .14 & .25 \end{bmatrix}.$$

This is the working covariance matrix for the first individual. Comparing the main diagonals of $\mathbf{V}_1$, the estimated variances are the same as in $\mathbf{A}_1$, but there are also now covariances shown between time periods. The covariances reflect the associations between the sampling distributions of each observation.

The next step is to apply the Fisher scoring algorithm (or a related algorithm) to reestimate the betas. Fisher scoring is explained in the next chapter; for now, just note that the working covariance matrices are used in this process. After one iteration, the $\hat{\boldsymbol{\beta}}$ vector has been updated as follows:

$$\hat{\boldsymbol{\beta}}_0 = \begin{bmatrix} -1.0 \\ 0.3 \end{bmatrix} \text{(initial matrix)}; \ \hat{\boldsymbol{\beta}}_1 = \begin{bmatrix} -1.68 \\ 0.41 \end{bmatrix}.$$

Note that the subscripts here represent the number of the iteration. The first number in each vector represents the intercept ($\hat{\beta}_0$) and the subsequent number represents the slope ($\hat{\beta}_1$). If there were multiple predictors, there would be additional slope estimates for each predictor. Once $\hat{\boldsymbol{\beta}}$ has been updated, we can then make new predictions, which will generate a new set of residuals and a new estimate of the working correlation and covariance matrices, and in turn new estimates for $\hat{\boldsymbol{\beta}}_2$ (the matrix for the second iteration). The process is repeated until the $\hat{\boldsymbol{\beta}}$ estimates stop changing (or the changes are infinitesimal), which is known as *convergence*. Using the GEE function in IBM SPSS, I found the final beta estimates to be

$$\hat{\boldsymbol{\beta}}_{\text{convergence}} = \begin{bmatrix} -1.93 \\ 0.52 \end{bmatrix}.$$

On each iteration, a new set of beta parameters is used to update the residuals and in turn the working correlation matrix. Again using IBM SPSS, I found the final working correlation matrix to be

$$\mathbf{R}(\boldsymbol{\alpha}) = \begin{bmatrix} 1.000 & 0.614 & 0.377 \\ 0.614 & 1.000 & 0.614 \\ 0.377 & 0.614 & 1.000 \end{bmatrix}.$$

Estimating Standard Errors

Once one has the final beta estimates, one needs to compute a final set of individual covariance matrices and residuals. From these one can compute the standard errors of the betas. This process is described below. Be forewarned that this section is somewhat complex and may benefit from a second reading after the next chapter in this book is completed.

An important type of matrix in this process is one described by the great statistician Sir Ronald Fisher and is the cornerstone of information theory. It is the *information matrix* for $\hat{\boldsymbol{\beta}}$. The information matrix is the inverse of the covariance matrix. So once one has derived the information matrix, one can then calculate the covariance matrix and from that, the standard errors. Knowing that the information matrix is the inverse of the covariance matrix is also important for conceptual reasons. The intuition here is that estimates with small standard errors are more precise and therefore more informative; estimates with high standard errors are more uncertain and less informative (hence will have low information values).

EXPLORING THE CONCEPT

As explained in the next chapter, the information matrix is also a function of the second derivative of the log likelihood function; the "information" is a measure of the amount of curvature of this function at the mean. Will log likelihood functions that have a low amount of curvature (i.e., low information) be more spread out than those with high curvature? Will this make the variance lower or higher?

Estimation procedures usually do not calculate the second derivative directly from the log likelihood function but rely on various computational formulas that simplify the calculations. All of the formulas will involve some type of partial derivative. In the case of GEE, the relevant derivative is

$$d_i = \frac{\partial \hat{\mu}_i}{\partial \beta}. \qquad \text{(Eq. 13.5)}$$

Now, specifically in a logistic regression, and as illustrated in Table 13.1, the predicted value is a proportion, and $\hat{\mu}_i = \hat{\pi}_i$ (the mean of the binomial distribution). Substituting $\hat{\pi}_i$ for $\hat{\mu}_i$, it therefore follows that for a logistic regression

$$d_i = \frac{\partial \hat{\pi}_i}{\partial \beta}. \qquad \text{(Eq. 13.6)}$$

As is illustrated in Table 13.1, we use our beta estimates to predict the probability of a hit for any given individual. The d_i is a measure of how much our probability prediction would change if the beta estimate were to change a little bit. If the log likelihood function (L) is relatively flat (i.e., has high variance and low information), d_i will be low and, as will be seen shortly, the information will be low. This point will be explained more in the next chapter, but the explanation rests on the fact that L is a function $\hat{\pi}_i$, and both L and $\hat{\pi}_i$ are functions of $\hat{\beta}_j$.

Let D_i be a $t * p$ matrix of d_is (or more specifically, d_{itj}s) for the ith individual. In logistic regression, a computational formula for D_i is

$$\mathbf{D}_i = \mathbf{A}_i * \mathbf{X}_i. \qquad \text{(Eq. 13.7)}$$

Recall that $\mathbf{A}_i$ is the covariance matrix for the ith individual before being multiplied by the working correlation matrix. $\mathbf{X}_i$ is just the data matrix. One computational formula for the information matrix (specifically what we will call the *model-based information matrix*) is

$$\mathbf{M}_0 = \sum_{i=1}^{n}\left(\mathbf{D}_i^T * \mathbf{V}_i^{-1} * \mathbf{D}_i\right). \tag{Eq. 13.8}$$

To better understand this formula, note that the information should be high when d_{itj} is high and low when d_{itj} is low. It should also make sense that $\mathbf{M}_0$ should be an inverse function of the covariance matrices. The information matrix is after all the inverse of the beta covariance matrix, and that matrix, as shown in Eq. 13.6, is indirectly a function of the individual covariance matrices.

To make all this clearer, let us continue with our example for the first individual. Tables 13.4 and 13.5 reestimate all the relevant values and matrices for the first individual, this time using the *final* beta estimates. If one then sums the individual information matrices for all 30 individuals, one obtains $\mathbf{M}_0$, which according to my calculations, is

$$\mathbf{M}_0 = \begin{bmatrix} 9.21 & 27.70 \\ 27.64 & 129.08 \end{bmatrix}.$$

Inverting the matrix to obtain the covariance matrix for the beta estimates yields

$$\text{Cov. Matrix} = \mathbf{M}_0^{-1} = \begin{bmatrix} .31 & -.07 \\ -.07 & .02 \end{bmatrix}.$$

Examining the main diagonal, one can see that the variance of the sampling distribution for $\hat{\beta}_0 = 0.31$ and for $\hat{\beta}_1 = 0.02$. Take the square roots to obtain the standard errors: $SE(\hat{\beta}_0) = 0.56$ and for $SE(\hat{\beta}_1) = 0.14$. Recall that $\hat{\beta}_0 = -1.93$ and $\hat{\beta}_1 = 0.52$. Both beta estimates are well more than twice the standard errors and so appear significant.

Robust Standard Errors

Royall (1986) showed that $\mathbf{M}_0$ is only a consistent estimator if the working correlation matrix is correctly specified, and such may not be the case. In the example so far, we used an AR(1) structure, but perhaps the structure should be AR(2) or even exchangeable. ($\mathbf{M}_0$ is said to be "model based" because it depends on a particular model of the correlational structure.) Fortunately, Royall (1986; see also Liang & Zeger, 1986) developed a more robust estimator of the information matrix, one that is consistent even if the working correlation matrix is misspecified. The general idea is that there is information in the data on how the residuals are correlated with one another, and one can use this information to improve the estimate of the

Table 13.4 Predictions for the First Individual Based on Final Parameter Estimates

	Predicted logit calculations	
Time		
1	$-1.93 + 0.52 * 0.85 = -1.49$	
2	$-1.93 + 0.52 * 4.10 = 0.20$	
3	$-1.93 + 0.52 * 3.55 = -0.08$	

	Predicted probabilities	
Time	*Odds[a]*	*Probabilities[b]*
1	0.23	18.4%
2	1.22	55.0
3	0.92	47.9

	Residuals	
Time	*Y*	*Predicted probabilities* $(\hat{\pi}_t)$	*Residual[c]*
1	1	18.4%	81.6%
2	1	55.0	45.0
3	1	47.9	52.1

	Pearson standardized residuals			
Time	$\widehat{Var}$ [d]	$\widehat{SD}$ [e]	*Raw residual[f]*	*Pearson residual[g]*
1	0.15	0.39	0.82	2.10
2	0.25	0.50	0.45	0.90
3	0.25	0.50	0.52	1.04

Note: Figures in table based on unrounded values.

[a]Predicted odds $= exp(\text{logit})$.

[b] $\hat{\pi}_{it} = \frac{odds}{(1 + odds)}$.

[c]Residual $= Y_{it} - \hat{\pi}_{it}$.

[d]Predicted $Var = \hat{\pi}_{it}(1 - \hat{\pi}_{it})$. For any case, $n = 1$ so the n term is not needed.

[e] $\widehat{SD} = \sqrt{\widehat{Var}}$.

[f]Reexpressed in decimal form.

[g]Pearson residual = raw residual/$\widehat{SD}$.

information matrix. Let us calculate a more empirically driven information matrix, $\mathbf{M}_1$, with the following formula:

$$\mathbf{M}_1 = \sum_{i=1}^{n}\left(\mathbf{D}_i^{\mathrm{T}} * \mathbf{V}_i^{-1} * (\text{residual}_i) * (\text{residual}_i)^{\mathrm{T}} * \mathbf{V}_i^{-1} * \mathbf{D}_i\right), \quad \text{(Eq. 13.9)}$$

Table 13.5 Calculation of Covariance and Information Matrices for First Individual Based on Final Parameter Estimates

$$\mathbf{X}_1^T = \begin{bmatrix} 1 & 1 & 1 \\ 0.85 & 4.10 & 3.55 \end{bmatrix}, \ \mathbf{A}_1^{1/2a} = \begin{bmatrix} .39 & 0 & 0 \\ 0 & .50 & 0 \\ 0 & 0 & .50 \end{bmatrix}, \ \mathbf{R}^b = \begin{bmatrix} 1.00 & .614 & .377 \\ .614 & 1.00 & .614 \\ .377 & .614 & 1.00 \end{bmatrix}$$

$$\mathbf{A}_1^{1/2} * \mathbf{R} = \begin{bmatrix} .39 & .24 & .15 \\ .31 & .50 & .31 \\ .19 & .31 & .50 \end{bmatrix}, \ \phi = 1, \ \mathbf{V}_1^c = \phi * \mathbf{A}_1^{1/2} * \mathbf{R} * \mathbf{A}_1^{1/2} = \begin{bmatrix} .15 & .12 & .07 \\ .12 & .25 & .15 \\ .07 & .15 & .25 \end{bmatrix}$$

$$\mathbf{V}_1^{-1} = \begin{bmatrix} 10.55 & -5.05 & 0.00 \\ -5.05 & 8.84 & -3.94 \\ 0.00 & -3.94 & 6.42 \end{bmatrix}, \ \mathbf{D}_1^d = \mathbf{A}_1\mathbf{X}_1 = \begin{bmatrix} 0.15 & 0.13 \\ 0.25 & 1.03 \\ 0.25 & 0.89 \end{bmatrix}.$$

$$\mathbf{D}_1^T * \mathbf{V}_1^{-1} = \begin{bmatrix} 0.34 & 0.46 & 0.62 \\ -3.82 & 4.91 & 1.66 \end{bmatrix}.$$

$$\textbf{Information matrix}_1 = \mathbf{D}_1^T * \mathbf{V}_1^{-1} * \mathbf{D}_1 = \begin{bmatrix} 0.32 & 1.06 \\ 1.06 & 6.01 \end{bmatrix}$$

Note: All figures based on unrounded values. Explanation and calculation of the score matrix is presented in the next chapter. The information matrices for all participants are summed to obtain the overall information matrix, $\mathbf{M}_0$. The inverse of the information matrix is the covariance matrix for $\hat{\boldsymbol{\beta}}$, and the square roots of the diagonal entries are the model-based standard errors.
[a]Diagonal matrix with *SD*s from Table 13.4 on the main diagonal.
[b]Final working correlation matrix assuming AR(1), rounded to two decimal places.
[c]Covariance matrix.
[d]Derivative $\frac{\partial \hat{\mu}_1}{\partial \beta}$.

where the residual term is defined as $(\mathbf{Y}_i - \hat{\boldsymbol{\mu}}_i)$. The formula differs from Eq. 13.8 in that the residual terms have been inserted into the middle of the equation, and $\mathbf{V}_i^{-1}$ appears twice. This is necessary because each residual term requires a $\mathbf{V}_i^{-1}$ to standardize it. The bottom part of Table 13.5 shows the calculation for the first individual. Summing over all individuals, we find that $\mathbf{M}_1 = \begin{vmatrix} 9.67 & 29.03 \\ 28.51 & 148.83 \end{vmatrix}$. Liang and Zeger (1986) recommend combining $\mathbf{M}_0$ and $\mathbf{M}_1$ to produce a "robust" estimator of the standard error, specifically with the formula

$$\widehat{Var}(\hat{\boldsymbol{\beta}}) = \mathbf{M}_0^{-1}\mathbf{M}_1\mathbf{M}_0^{-1}.$$

This expression is also known as an *information sandwich* estimator (Royall, 1986), with $\mathbf{M}_1$ sandwiched between $\mathbf{M}_0^{-1}$. This rather complicated procedure is necessary, in part, to make the matrix dimensions conform to one another. Recall that the estimated $\widehat{Var}(\hat{\boldsymbol{\beta}}) = \mathbf{M}_0^{-1}$, because the covariance matrix is the inverse of the information matrix. We are now adjusting the estimated variance by $\mathbf{M_1M_0^{-1}}$. It is this information sandwich estimator $(\mathbf{M_0^{-1}M_1M_0^{-1}})$ that has been shown to be robust to incorrect specification of the correlation matrix.

EXPLORING THE CONCEPT

The expression $\mathbf{M_1M_0^{-1}}$ is analogous (in scalar arithmetic) to dividing $\mathbf{M_1}$ by $\mathbf{M_0}$. What would this adjustment ratio equal if the model-based and empirical information values were the same? (In matrix algebra, it would be the identity matrix; see Chapter 9, Table 9.5, for a review of linear algebra.)

To continue with our example, we have

$$\mathbf{M_1M_0^{-1}} = \begin{bmatrix} 1.05 & 0.00 \\ -1.02 & 1.37 \end{bmatrix}.$$

We then multiply $\mathbf{M_0^{-1}}$ by this adjustment factor to obtain the covariance matrix of

$$\hat{\boldsymbol{\beta}}: \widehat{Var}(\hat{\boldsymbol{\beta}}) = \begin{bmatrix} 0.39 & -0.09 \\ -0.09 & 0.03 \end{bmatrix}.$$

These variances are higher than the model-based ones; here, we are paying a small price for the insurance that our estimates are in fact consistent, regardless of the underlying correlational structure. Note that the variances will be higher with increasing values of $\mathbf{M_1}$, which occurs with larger residuals. (For some variations on the GEE procedure, see Technical Notes 13.1–13.3.)

Statistical Hypothesis Tests

We can now, at long last, perform the statistical hypothesis tests. The variance of the sampling distribution for $\hat{\beta}_0 = 0.39$ and for $\hat{\beta}_1 = 0.03$. Take the square roots to obtain the standard errors: $SE(\hat{\beta}_0) = 0.62$ and $SE(\hat{\beta}_1) = 0.17$. Recall that $\hat{\beta}_0 = -1.93$ and for $\hat{\beta}_1 = 0.53$. Using rounded figures, the squared ratio of $\hat{\beta}$ to its standard error yields the Wald $\chi^2(1)$ of 9.59 (intercept), 9.72 (slope). The associated p values are $p = .002$ for $\hat{\beta}_0$ and $p = .002$ for $\hat{\beta}_1$.

As discussed in Chapter 10, a limitation of the Wald test is that it assumes the model is correctly specified. If one has not yet settled on a final model, a different test should be used. We cannot use likelihood ratio tests because in GEE there are no valid likelihood functions, but we can used the *generalized score test*. Score tests are described in more detail in Chapter 14.

Although the standard error estimates are robust, there can still be some loss of statistical power if the wrong correlational structure is used. Therefore, it is good practice to try out several different structures to see which fits the data best. Because likelihood functions are not valid when statistical independence is violated, GEE does not use traditional fit statistics like the negative two log likelihood (−2LL). Instead, the quasilikelihood information criteria (QIC) is used (Pan, 2001), which is a statistic based on the information statistic in the Fisher scoring method discussed in the next chapter. Lower levels of QIC indicate a better fit. QIC is analogous to Akaike's information criterion (AIC), which penalizes one for using more parameters to obtain a better fit. QIC(V) is a function of the covariance matrix $\mathbf{V}$, so you should use this statistic for comparing models with different initial working correlation matrices. QIC_u (also written QICC) is used for comparing models that have different IVs. In our example, the QIC for the AR(1) structure is 116.48, but it is 116.23 for unstructured (see Table 13.6). Smaller is better in these comparisons although the difference here is very small.

According to Ballinger (2004), theoretical considerations should primarily drive the choice of correlational structure, but fit statistics can be used if there is some uncertainty among the two or three different models.

Notes on Model Specification

In the psychology experiment example, the predictor "Arousal" changed from one time period to another. If using time-dependent covariates, to avoid bias it is important that outcomes at that time point not depend on covariate values at other time points (Davis, 2002; Pepe & Anderson, 1994). AR and M-dependent correlational structures are likely to violate this assumption, whereas independent and exchangeable structures do not violate the assumption. Pepe and Anderson's (1994) simulations suggested that the beta estimates will be biased downward when the assumptions are not met. In the psychology experiment, because we adopted an AR(1) structure, we should perform a sensitivity analysis to determine how beta and chi-square estimates would change using other correlational structures. An independent structure is unlikely on theoretical grounds. However, an exchangeable structure is plausible if task performance depends mostly on arousal at a given time point (and a person's underlying ability, which is what causes correlations between time points). A counterargument is that task success might also be affected by successful performance at a previous time point (e.g., learning how to solve the problem), which would be affected by the covariate value of arousal at a previous time point. This possibility might explain why an autoregressive structure fits the data somewhat better than an exchangeable structure. An unstructured correlation matrix fits the data best and might be most theoretically defensible given that there are different types of psychological processes involved (some autoregressive and others not). However, an unstructured matrix might not meet the underlying GEE assumption for consistent estimation (i.e., unbiased in large samples) and the B = 0.49 estimate might therefore be biased downward. Fortunately, the B estimates are significant regardless of the correlational structure used (and the standard errors robust to choice of structure). The example shows, though, why caution is needed when using time-dependent covariates.

With GEE, one can also use predictors that do not change over time; for example IQ scores measured at the start of the experiment. If IQ scores were the sole predictor, the predicted logits would not change among time periods, but the actual logits would (because these are based on problem-solving performance at each time point). The residuals would also therefore change, as would the correlations among the residuals.

Note that in the example, "Time" is not used as a predictor. It appears in the model as a subscript, indexing various values. We could also choose to use "Time" as a predictor. In the example, however, doing so did not produce a significant result $[\chi^2_{GS}(1) = 0.63, p = .43]$.

Minimum Sample Sizes

It should be noted that our example data set is of minimal size to use GEE. Ziegler, Kastner, Grömping, and Blettner (1996) recommend that there should be at least 30 clusters when there are four or fewer time points. (See Technical Note 13.4 for more specific recommendations and results.) In small and medium-sized samples, one should use model-based estimates of the standard errors, but one needs some confidence that one is using the correct correlational structure. One can, however, rerun the analysis with several different correlational structures to determine the sensitivity of the results to the choice. Hanley, Negassa, Edwardes, and Forrester (2003) reported that robust standard errors are likely to be more trustworthy unless

Table 13.6 Comparison of Results for Selected Working Correlation Structures

Structure	B_x	χ^2_{GS}	P-value	QIC
AR(1)	0.52[a]	8.96	.003	116.476
Independent	0.50	5.59	.018	117.570
Exchangeable	0.56	8.46	.004	117.218
Unstructured	0.47	8.40	.004	116.228

[a]Differs slightly from the previous manual calculation due to rounding error.

the sample is sparse. However, some simulation studies show that when using robust standard errors, the Wald test is far too liberal in small samples (Firth, 1993). On the other hand, the generalized score test is too conservative when there are less than 30 clusters (Guo & Pan, 2002; Guo, Pan, Connett, Hannan, & French, 2005; see Technical Note 13.5).

Hanley et al. (2003) state that, unlike multilevel modeling techniques, there is not a minimum sample size for the clusters. With multilevel models, one attempts to estimate different regression lines for different clusters, which requires the clusters to be fairly large. In GEE, the goal is to model the correlations among the residuals by estimating one (or a couple) correlation parameters; the estimation requires there to be a lot of clusters, but the clusters do not have to be large. This makes GEE particularly useful for analyzing small-group attitudes or behaviors. For example, in argumentation research, arguments and counterarguments made by individuals in the same discussion group are not statistically independent; however, groups may consist of only two to six students.

General Estimating Procedures and Missing Values

Unlike repeated measures with ANOVA, where cases with missing values must be thrown out, one strong feature of GEE is that it can be used with some values missing on some variables; correlations can still be computed from all available pairs. You do not need to estimate the missing values. We simply define a separate working correlation matrix for each individual. So, in our example, suppose the fifth individual in our experiment ("Maurice") was absent at Time 2. Then his correlational and covariance matrices would be 2×2 (reflecting two observations) rather than 3×3. His information matrix would also be 2×2, but all the other subjects' information matrices are also 2×2 because we are estimating two beta parameters (see Table 13.5). Therefore, his values can still contribute to the sums in Eqs. 13.8 and 13.9 and to the overall information matrix. (The same logic can be applied to the score matrix, discussed in Chapter 14.)

GEE will still produce consistent estimates as long as the observations are missing completely at random, a condition known as MCAR for short (Stokes, Davis, & Koch, 2001). It would help to find out why Maurice was absent, as well as the reason for absences of the other individuals. If there is a pattern in the absence, then the MCAR assumption is violated. So, for example, if more of the absences are at Time 2 (because there was a snow storm that day), then the missing values are not MCAR. Alternatively, if there are more absences when the participant failed the problem in the previous session, then the MCAR assumption is violated. In cases such as these, more sophisticated techniques for handling missing values should be used (see Cole, 2008; Davis, 2002).

General Estimating Procedures Residual Tests

Some authors recommend examining the final residuals and testing whether the directions of the residuals is random. Chang (cited in Hardin & Hilbe, 2003) recommended use of the nonparametric Wald-Wolfowitz runs test for this purpose. Code residuals as 1 or −1, depending on whether they are positive or negative, then compute

$$W_z = \frac{T - E(T)}{\sqrt{Var(T)}},$$

where T is the count of the total number of runs (i.e., sequence of numbers that have the same sign). So, for example, the sequence <1, −1, 1, 1, −1, −1, −1> has a two positive run <1, 1> and a three negative run <−1, −1, −1>, producing $T = 5$. Let $n_p = 2$ (number of positive runs) and $n_n = 3$ (number of negative runs). It can be shown that

$$E(T) = \frac{2n_p n_n}{n_p + n_n} + 1,$$

$$Var(T) = \frac{2n_p n_n (2n_p n_n - n_p - n_n)}{(n_p + n_n)^2 (n_p + n_n - 1)}.$$

W_z will be normally distributed under the null hypothesis that the order is random. The Wald-Wolfowitz runs test is available in SPSS on the nonparametrics menu.

The runs test is especially useful when the errors are assumed to follow a normal distribution. We saw in Chapter 9 that in ordinary least squares the residuals should not display any pattern. However, when using nonnormal models, involving logistic or Poisson regression, for example, some patterning in the residuals is to be expected. Nevertheless, unusual patterns may signify potential problems. A valuable visual test is to plot the residuals against the fitted values for each time point (Ballinger, 2004). Look for any clustering around certain values or differences among time points. Problems may be caused by misspecification of the working correlation structure, distribution, or link function.

Performing General Estimating Procedures in SPSS

To run GEE in IBM SPSS, go to "Analyze" ➔ "Generalized Linear Models" ➔ "Generalized Estimating Equations" and specify the following:

1. On the "Repeated" tab, input the variable that designates the subject (or cluster). It is in this tab that one also specifies what assumption to make about the working correlation matrix (R). Also, specify "Time" or a comparable variable as a "Within-subject variable" even if it is not used as a predictor. (In the psychology experiment example, it was not necessary to put "Arousal" in this box even

though it varied within subjects. Doing so created a working correlation matrix with 6,400 entries and increased the QIC.)

2. On the "Type of Model" tab, one specifies the distribution and link function to use by choosing a particular function.

3. On the "Response" tab, specify the DV.

4. On the "Predictors" tab, specify the IVs. If using an offset variable (e.g., in Poisson regression, the size of the population in a given year), also specify that here as well.

5. On the "Model" tab, specify the specific variables and interactions that you wish to have included in the model.

6. On the "Estimation" tab, you can specify if you wish GEE to compute and include a dispersion parameter for an overdispersed model (e.g., by choosing "Deviance" from the "Scale Parameter Method" choices).

7. On the "Statistics" tab, you can specify whether or not to use the Wald method or generalized score method. The Wald method should only be used if one has finished specifying the model and is no longer comparing models; otherwise, use the generalized score method.

8. The "EM Means" tab allows one to conduct planned and pairwise comparisons between means.

9. The "Save" and "Export" tabs are self-explanatory.

Summary

The GEE methodology is used in lieu of GLMs when there are statistical dependencies among the observations. GEE has different sample size requirements than does multilevel modeling. The user can make some assumptions about how the observations (or errors) are correlated and estimate (if desired) robust standard errors that are not affected much if an incorrect correlational structure is used. The use of robust standard errors, however, may come at a cost of some statistical power.

Problems

1. One hundred married men were assessed by a clinician on their levels of defensiveness both before and after couples therapy. Defensiveness was categorized into one of five ordinal levels ranging from 0 (*low*) to 5 (*high*). The IV was the number of therapy sessions (1 to 9). The data are in the file *Defensiveness.13. Problem_1*. Use the GEE procedure to assess whether more therapy sessions reduces levels of defensiveness. Specify "Time" as a within-subject variable (on the Repeated Tab" if using IBM SPSS) and set the Working Correlation Matrix to AR(1), Type of Model: Ordinal Logistic, Response: DV = "Defensiveness,"

Predictors: "Sessions," "Time," Model: "Sessions," "Time," "Sessions × Time," Statistics: Generalized Score, check "Include exponential parameter estimates."

a) Other than the thresholds, which terms are statistically significant? For "Time," and "Sessions * Time," report the B, *exp*(B), Generalized Score (and *p* value) and Wald χ^2 and Wald *p* values (because the model is final).

b) Rerun the analysis with an independent correlational structure. Is the interaction now significant? What is the Wald *p* value? Which set of results is more defensible, and why?

2. Similar to Lin et al. (2012), a researcher measures the number of analogies generated in oral discussions in small groups of fourth graders engaged in collaborative reasoning discussions of short stories. Each small group consisted of six or seven students. Using the group as the level of analysis, the number of analogies generated during the first, fifth, and ninth discussion were counted to ascertain any trends over time. The data are in the file *Analogies.13. Problem_2*. Use "Time" as the sole independent variable. Fit a Poisson model with no offset.

a) Run GEE, setting the working correlation matrix to unstructured and using robust standard errors. Find the QIC.

b) Find the QIC also using an AR(1) and *M*-dependent(2) structure. Which returns the lowest QIC?

c) Run the analyses in parts (a) and (b) and request, under "Statistics," the working correlation matrices. Describe how they differ.

d) The *M*-dependent(2) structure should estimate the following correlations: Times 1 and 2 (.596), Times 2 and 3 (.596), and Times 1 and 3 (.261). Is the last correlation the product of the first two, as would be the case with the AR(1) structure? How many population parameters are actually estimated with the *M*-dependent(2) structure? With the AR(1) structure?

e) Using the *M*-dependent(2) structure, what is *exp*(B)? Does the number of analogies generated appear to increase over time?

3. Similar to Kasperski et al. (2011), a researcher conducts a study on college students' use of cocaine. She surveys a random sample of 100 undergraduates on their use of cocaine. The survey is administered four times. The first survey (Y1 for Year 1), administered during the freshman year, asks students whether they have ever used cocaine. The second (Y2) asks whether they have used cocaine during the past year, as do the Y3 and Y4 surveys. The data are in the file *Cocaine.13.Problem_3*. Gender is coded with "Female" = 1.

a) Perform separate logistic GEE analyses using an AR(1), exchangeable, and unstructured working correlation matrix, and report the QICs. Which QIC is lowest? (Use "Year," "Gender," and their interaction as predictors. Be sure to code "Year" as metric/scale.)

b) Using an AR(1) structure and robust standard errors, report the B and generalized score statistic (and *p* value) for each predictor. Is there an increase in cocaine use over time? Does gender have an effect? Is gender significant if the interaction term is dropped? (Be sure to make "Cocaine = 0" the reference category and to use descending for "category order for factors.")

c) Run the model with just "Year" as a predictor. Report and interpret exp(B). Also report the QICC.

d) Calculate the actual percentage of students who used cocaine for Y1 through Y4. Calculate the observed logits and plot these against year. Does the relationship look linear or nonlinear? Run a GEE using "*ln*(Year)" instead of "Year." Have the fit statistics declined? (Recall that lower values indicate a better fit.)

e) On the survey for Year 4, the researcher asked questions to assess whether the student was dependent on cocaine. For example, she asked whether the student spent a lot of time obtaining cocaine, gave up important activities, or continued to use despite health problems. Out of seven items, affirmative answers to at least three indicated dependency. Using a logistic generalized linear model (not GEE), assess whether gender predicts dependency. (To avoid quadruple counting, in IBM SPSS use "Data" ➔ "Select Cases" and set "If condition is satisfied" to Year = 4.) Are males or females more likely to become dependent? Report B, χ^2_{LR}, and the p value. Report and interpret exp(B).

4. Replicating Aretouli, Tsilidis, and Brandt (2013), a researcher conducts a study with 83 older adults with mild cognitive impairments to determine how many will later develop dementia. Participants were administered a clinical rating scale to assess dementia at baseline, 2 years later, and 4 years later; those with scores above 1.0 were considered to have dementia (which was treated as a nominal variable). Participants were also administered two cognitive tests at baselines: a maze test (which measured their ability to plan and sequence without running into dead ends) and the Hayling test (Burgess & Shallice, 1997). The Hayling test measures an individual's ability to inhibit a natural response. (The individual must first complete a series of sentences with one natural response and then complete them a second time with an artificial word.) The Hayling and maze scores, along with age at baseline and time (T1, T2, and T3) were used to predict dementia. The data are in the file *Dementia.13.Problem_4*.

a) Restructure the database so that there are separate rows for each time period. In SPSS, go to "Data" ➔ "Restructure" ➔ "Restructure Selected Variables into Cases," and on the following screens choose the following:
   - Number of Variable Groups: One.
   - Variables to Cases: Under "Variables to be Transposed," choose Baseline, Y2, Y4. Also rename "trans1" as "dementia."
   - Variables to Cases: Create Index Variables, One.

- Variables to Cases: Create One Index Variable; rename "Index 1" to "Time."
- Options: Under "Handling of Variables Not Selected," select "Treat and Keep as Fixed Variable(s)."
- Finish: Select "Restructure the Data Now."
- Also, on the variable view screen, set "Time" to "scale" in the measures column and erase or modify any label.

b) Including main effects only, perform separate logistic GEE analyses using AR(1) and exchangeable working correlation matrices, and report the QIC's. Which QIC is lowest?

c) Using an AR(1) structure and robust standard errors, report the B and generalized score statistic (and p value) for each predictor, and the overall QICC. (Only include main effects in the model.)

d) Why does it makes sense that the B coefficient for time and age would be positive (indicating an increased risk of dementia)? Calculate and interpret the estimated OR for age.

e) The construct measured by the Hayling test (inhibition) is associated with a greater number of brain regions than the maze test. Why might this fact make the former a better predictor of dementia than the latter? Are higher scores on the Hayling test positively or negatively associated with increased odds of dementia?

f) Calculate "ln(Time)" and substitute it for "Time" in the model. Does the revised model fit the data better than the previous one? Use the QICC to justify your answer.

Technical Notes

13.1 There is a version of GEE that can be used for modeling individual growth curves; Zeger, Liang, and Albert (1988) called it the "subject-specific approach."

13.2 Carey, Zeger, and Diggle (1993) developed an approach using odds ratios, rather than Pearson residuals, when fitting binomial model. For a given individual (or cluster), and two time points (t_1 and t_2), compute the odds that $Y_{t_2} = 1$ (a "hit") given that $Y_{t_1} = 1$, divided by the odds of a hit given that $Y_{t_1} = 0$. There are odds ratios for every pairwise combination of time points. To maximize efficiency, the procedure alternates between estimating the ORs with logistic regression and with GEE and is therefore known as *alternating logistic regression* (or ALR) because it alternates between GEE and logistic regression steps (Hardin & Hilbe, 2003).

13.3 A procedure known as GEE2 does not assume that the estimating equations for α and β are orthogonal. According to Hardin and Hilbe (2003, p. 105), the procedure is not often used because the link function and covariance functions have to be correctly specified.

13.4 Using continuous data following the gamma distribution and $t = 4$, Paik (1988) found that GEE performed satisfactorily when $n \geq 30$ with one predictor and when $n \geq 50$ with four predictors. However, Davis (2002) cautioned that bias and statistical power will also depend on the distribution of cluster sizes, the degree of correlations among repeated measures, and the amount of missing observations, so one should use the above results cautiously. Regarding missing observations, Park (1993) performed simulations with missing data probabilities of 10% to 30% using multivariate normal data, $t = 4$, $p = 3$, and $n = 30$ or 50, and found that only the latter simulation ($n = 50$) performed satisfactorily.

13.5 When there are less than 30 clusters, Guo et al. (2005) proposed remedying the conservativeness of the generalized score test by adjusting it by a factor of $\frac{J}{(J-1)}$, where J is the number of clusters.

References

Aretouli, E., Tsilidis, K. K., & Brandt, J. (2013). Four-year outcome of mild cognitive impairment: The contribution of executive dysfunction. *Neuropsychology, 27*, 95–106.

Ballinger, G. A. (2004). Using generalized estimating equations for longitudinal data analysis. *Organizational Research Methods, 7*, 127–150.

Burgess, P. W., & Shallice, T. (1997). *The Hayling and Brixton Tests manual*. Edmunds, England: Thames Valley Test Company.

Carey, V. J., Zeger, S. L., & Diggle, P. J. (1993). Modeling multivariate binary data with alternating logistic regressions. *Biometrika, 80*, 517–526.

Cole, J. C. (2008). How to deal with missing data: Conceptual overview and details for implementing two modern methods. In J. W. Osborne (Ed.), *Best practices in quantitative methods* (pp. 214–238). Thousand Oaks, CA: Sage.

Davis, C. S. (2002). *Statistical methods for the analysis of repeated measurements*. New York, NY: Springer.

Firth, D. (1993). Recent developments in quasi-likelihood methods. *Proceedings of the ISI 49th Session* (pp. 341–358). The Hague, The Netherlands: ISI.

Guo, X., & Pan, W. (2002). *Small-sample performance of the score test in GEE*. Minneapolis: University of Minnesota. Retrieved from http://www.sph.umn.edu/faculty1/wp-content/uploads/2012/11/rr2002-013.pdf

Guo, X., Pan, W., Connett, J. E., Hannan, P. J., & French, S. A. (2005). Small-sample performance of the robust score test and its modifications in generalized estimating equations. *Statistics in Medicine, 24*, 3479–3495.

Hanley, J. A., Negassa, A., Edwardes, M. D., & Forrester, J. E. (2003). Statistical analysis of correlated data using generalized estimating equations: An orientation. *American Journal of Epidemiology, 157*, 364–375.

Hardin, J. W., & Hilbe, J. M. (2003). *Generalized estimating equations*. Boca Raton, FL: Chapman & Hall/CRC.

Hox, J. (2010). *Multilevel analysis: Techniques and applications* (2nd ed.). New York, NY: Taylor & Francis.

Kasperski, S. J., Vincent, K. B., Caldeira, K. M., Garnier-Dykstra, L. M., O'Grady, K. E., & Arria, A. M. (2011). College students' use of cocaine: Results from a longitudinal study. *Addictive Behaviors, 36*, 408–411.

Liang, K. Y., & Zeger, S. L. (1986). Longitudinal data analysis using generalized linear models. *Biometika, 73,* 12–22.

Lin, T.-J., Anderson, R. C., Hummel, J. E., Jadallah, M., Miller, B., Nguyen-Jahiel, K., . . . Dong, T. (2012). Children's use of analogy during collaborative reasoning. *Child Development, 83,* 1429–1443.

Paik, M. C. (1988). Repeated measurement analysis for nonnormal data in small samples. *Communication in Statistics—Simulations and Computation, 17,* 1155–1171.

Pan, W. (2001). Akaike's information criterion in generalized estimating equations. *Biometrics, 57,* 120–125.

Park, T. (1993). A comparison of the generalized estimating equation approach with the maximum likelihood approach for repeated measurements. *Statistics in Medicine, 12,* 1723–1732.

Pepe, M. S., & Anderson, G. L. (1994). A cautionary note on inference for marginal regression models with longitudinal data and general correlated response data. *Communications in Statistics—Simulation and Computation, 23,* 939–951.

Royall, R. (1986). Model robust confidence intervals using maximum likelihood estimators. *International Statistical Review, 54,* 221–226.

Stokes, M .E., Davis, C. S., & Koch, G. G. (2001). *Categorical data analysis using the SAS system* (2nd ed.). Cary, NC: SAS Institute/Wiley.

Wedderburn, R. W. M. (1974). Quasi-likelihood functions, generalized linear models, and the Gauss-Newton method. *Biometrics, 61,* 439–447.

Zeger, S. L., & Liang, K. Y. (1986). Longitudinal data analysis for discrete and continuous outcomes. *Biometrics, 42,* 121–130.

Zeger, S. L., Liang, K. Y., & Albert, P S. (1988). Models for longitudinal data: A generalized estimating equation approach. *Biometrics, 44,* 1049–1060.

Ziegler, A., Kastner, C., Grömping, U., & Blettner, M. (1996). The generalized estimating equations in the past ten years: An overview and a biomedical application. *Sonderforschungsbereich, 386* [Discussion paper 24]. Munich, Germany: Ludwig-Maximilians University.

CHAPTER 14

ESTIMATION PROCEDURES

In prior chapters, I discussed how various generalized linear models (GLMs) and the general estimating equation (GEE) procedure work. I did not discuss much the algorithms for estimating the beta parameters. These algorithms are complex but are easily implemented in statistical computing packages, thus reducing the need for users to understand the algorithms. Nevertheless, some grasp of the algorithms is useful for understanding the underlying theory. It is difficult to read advanced articles and books on statistics without being somewhat literate in these estimation procedures. This chapter therefore presents an explanation of them.

Estimating a Binomial Parameter

Recall that in maximum likelihood estimation (MLE), one chooses an estimate of a given population parameter that maximizes a likelihood function. The likelihood function gives the probability of obtaining the observed data given various estimates; one then chooses the estimate that is most likely to have produced the data. Let us begin with a simple example. The example is so simple that readers may wonder why all the mathematical gyrations are necessary. The answer is twofold. First, the intent is to build the reader's intuitions as to what the MLE procedures are and why they work. Second, simulation studies (e.g., Brown, Cai, & DasGupta, 2001) show that the normal approximation method for computing binomial confidence intervals that is traditionally taught to students (presented in Chapter 2) often results in confidence intervals (CIs) that are inaccurate for many values of P (and not just for extreme values, as is often thought; see Technical Note 2.2 in Chapter 2). Therefore, more complicated methods for computing binomial CIs are necessary.

Our goals will be to estimate a population proportion, π. It turns out that P (the sample proportion) is the MLE of π (see Chapter 10). In the proof, the derivative of the log-likelihood equation is set to zero and then the equation is solved. In more complicated situations, the equation cannot be solved algebraically and so iterative methods are used. Iterative methods make an initial guess of the parameter being estimated and then update the guess at each subsequent step. On each step, we get closer and closer to the maximum of the likelihood function and continue until the estimates stop changing (or the changes are infinitesimal).

Iterative methods rely on three functions that are derived from the likelihood function:

1. *The log-likelihood function (L).* This is simply the likelihood function on the natural log scale. Logarithms simplify calculations because terms that need to be multiplied together can instead—on the log scale—be added together. The maximum point of *L* will also be the maximum point of the likelihood function, so we can restate the overall goal to be finding the maximum point of *L*.

2. *The score function (U).* This is the *first* derivative of the log-likelihood function and is a measure of the slope of *L*. The slope (i.e., score) will be zero at the maximum point, so our overall goal can be again restated as finding the value of the estimated parameter corresponding to a score of zero.

3. *The information functions (−H or I).* These functions are related to the second derivative of *L* and also to the first derivative of *U* (which itself is the first derivative of *L*). The information value at a particular point is a measure of how rapidly the slope of *L* changes and, as such, is a measure of curvature. When the curvature is low, larger changes in our current parameter estimates (of π) are needed to find the maximum point of *L* than when the curvature is high. The information functions are also used to derive the standard errors of the parameter estimates.

In this discussion, I use the term *derivative* to refer to the partial derivative of a function (typically *L* or *U*) in respect to the parameter being estimated, unless otherwise indicated. There are two main iterative estimation methods: the Newton-Raphson procedure and the Fisher scoring method. Each will be described in turn.

Newton-Raphson Procedure

The Newton-Raphson procedure has been the most commonly used one to obtain beta estimates. Appendix 14.1 gives a fuller account of the logic and mathematics of how the procedure was derived; here I focus on the actual procedure for estimating a binomial parameter.

Suppose our initial guess is that π is 50%. We collect 100 observations and find 58 hits and 42 failures. Therefore *P* is 58%. Let us first examine the shape of each of the key functions.

Likelihood and Log-Likelihood Functions

Figure 14.1 displays the shape of the *likelihood function* for different values of π. The figure was created in EXCEL using the binomial formula. For example, the likelihood when $\pi = 58\%$ is .02, because $\binom{100}{58} * 50\%^{58} * (1 - 50\%)^{(100-58)} = .02$. Sometimes *kernel* likelihoods are used; kernel likelihoods omit the binomial coefficient $\binom{100}{58}$. Because the binomial coefficient will not vary with different values of π, it is not necessary to include it when estimating.

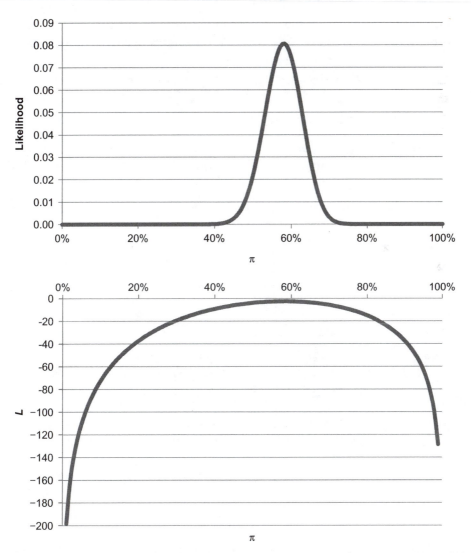

Figure 14.1 Likelihood and log-likelihood (*L*) functions for estimating a proportion (π). The observed proportion (*P*) is 58%.

It is clear from the figure that the maximum point of the likelihood function is 58%. Such graphical methods of finding the maximum likelihood estimate are unfeasible in more complicated situations, so let us focus on iterative, mathematical ways of estimating π. We first need to estimate the *log-likelihood function* (L) by taking the logs of all the likelihood values. So for example, $ln(0.02) = -3.91$, and this value corresponds to the maximum point of L, as can be seen approximately in Figure 14.1.

Note that the all the values of L are negative. That is because all the likelihoods are probabilities which, by definition, range from 0 to 1. The log of a number < 1 is negative.

EXPLORING THE CONCEPT

Although the values of L on the vertical axis in Figure 14.1(b) are negative, will the maximum point still corresponds to 58% on the horizontal axis? At the maximum point, will the slope and score be zero?

Score Function

Figure 14.2 displays the score function, which reflects the first derivatives. The slope of the score function is negative because the slope of L is initially positive and large (see Figure 14.1) and then the slopes of the L function decrease, eventually becoming negative. Figure 14.2 shows that at the maximum point of L, the score is 0. It can be shown that the mean of the scores is zero.

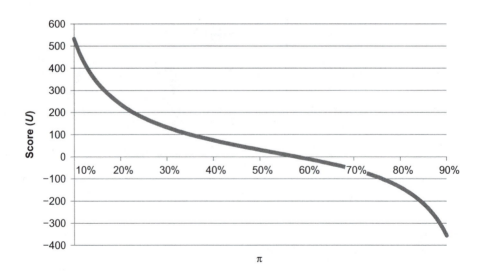

Figure 14.2 The score function for the log-likelihood function shown in Figure 14.1. The score is the slope of the L function. (This figure shows that the slopes are higher at the extreme and become negative for π values higher than 58%.)

It is shown in Appendix 14.2 that the score (U) in this example is given by the following formula:

$$U = \frac{Y - n\pi}{\pi(1 - \pi)},$$
(Eq. 14.1)

where Y is the number of hits.

This formula was used to make Figure 14.2. In our example, we will initially guess that $\pi = 50\%$, so $U = \frac{58 - 100 * 50\%}{50\% * 50\%} = 32$. The score is not 0, indicating that 50% is not the MLE.

Information Functions

In the Newton-Raphson procedure, we also need to calculate the information, which is a measure of the curvature of L at a specific point. We are specifically interested in the point $\pi = 50\%$ (our initial guess). The curvature is how rapidly the slope of the function is changing. We will use knowledge of the curvature to calculate how much to change our initial estimate.

The curvature can be calculated by finding the first derivative of the score function in respect to the parameter being estimated, π. Because at each point the score is the first derivative of L, the curvature is the second derivative of L. Let us call the latter value the *Hessian* (*H*), named after the 19th-century German mathematician, Otto Hesse. The Hessian at each point will be negative because, as shown in Figure 14.2, the slope of the score function at each point is negative. To make the Hessians positive, we multiply by -1, yielding a measure known as the *observed information*:

$$Observed\ information = -H(\pi) = -\frac{\partial Y}{\partial X}.$$
(Eq. 14.2)

For our example, taking the first derivative of U (Eq. 14.1) produces the following equation for the observed information (see Appendix 14.2 for proof):

$$-H(\pi) = \frac{Y}{\pi^2} + \frac{n - Y}{(1 - \pi)^2}.$$
(Eq. 14.3)

In our example, $\pi_0 = 50\%$, so $-H(50\%) = \frac{58}{50\%^2} + \frac{100 - 58}{(1 - 50\%)^2} = 400$. Figure 14.3 plots $-H(\pi)$ for all values of π; the curve is shaped like the letter U.

Updating the Estimate

Following is a key point. With the Newton-Raphson procedure, on each iteration we update the estimate of π by $\frac{Score}{Observed\ Information}$. Formally, this is written as

$$\pi_{(t+1)} = \pi_t + \frac{U_t}{-H_t},$$
(Eq. 14.4)

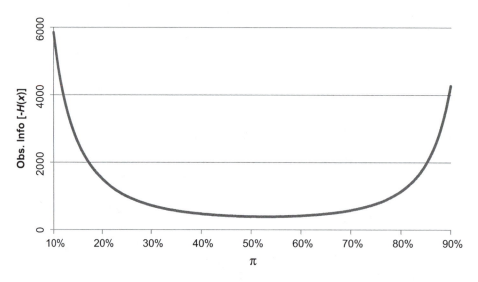

Figure 14.3 The observed information function for the log-likelihood function shown in Figure 14.1 and score function shown in Figure 14.2. The information is the slope of the score function. (This figure shows that the information values are generally higher the farther the points are from the ML estimate of $\pi = 58\%$, although the minimum $-H(x)$ value is actually at $\pi = 53\%$.)

where the subscript t denotes the number of the iteration (first, second, third, etc.). In our example, the value of the second term is $\frac{32}{400} = 8\%$, so Eq. 14.4 becomes $\pi_1 = 50\% + 8\% = 58\%$. Based on Eq. 14.1, the score at this point is 0 (because the residual $Y - n\pi = 0$), so there will be no further updating ($\pi_2 = 58\%$). We have reached convergence, meaning that our estimates stop changing or only change by an infinitesimal amount. So the MLE of $\hat{\pi} = 58\%$, which is what it should be (i.e., the observed proportion 58%)! The score also sets the direction of movement; the score is positive below the mean ($\pi = 58$) so the steps will be positive, increasing the estimated value of π. Likewise, above the mean, the score is negative, decreasing the estimated π, thereby moving our estimate of π closer to the ML value.

Now in this example we reached convergence in one iteration, but usually multiple iterations are necessary. Suppose our initial estimate of $\hat{\pi}$ was 10% rather than 50%. Table 14.1 illustrates the Newton-Raphson procedure and shows that six iterations are needed before convergence is established. In this second example, we started further away from 58% so more iterations were needed.

Ten percent is a poor initial guess of π. In more complicated situations, one might not reach convergence if the initial guess of the parameter is poor. The estimates might be attracted to a "local maximum" of L rather than the actual maximum and therefore take the calculations in the wrong direction. Or the estimates might end up in an endless loop. It is therefore important to start with a reasonably good initial estimate.

Why does the updating formula generally work? It does not guarantee that one will reach the maximum point of L in one step, but the equation is designed to take

Table 14.1 Illustration of MLE Methods for Estimating a Proportion

Givens: $n = 100$; $Y = 58$; $\hat{\pi}_0 = 10\%$[a]

Step 1

$\hat{\pi}_0 = 10\%$ (arbitrarily chosen for this example).

Score: $u_0 = \dfrac{Y - n\hat{\pi}_0}{\hat{\pi}_0(1 - \hat{\pi}_0)} = \dfrac{58 - 100(10\%)}{10\% * 90\%} = 533.33$

Observed Information: $-H_0 = \dfrac{Y}{\hat{\pi}_0^2} + \dfrac{n - Y}{(1 - \hat{\pi}_0)^2} = \dfrac{58}{10\%^2} + \dfrac{100 - 58}{(1 - 10\%)^2} = 5851.85$

Update: $\dfrac{u_0}{-H_0} = \dfrac{533.33}{5851.85} = 9.11\%$

$\hat{\pi}_1 = \hat{\pi}_0 + \dfrac{u_0}{-H_0} = 10\% + 9.11\% = 19.11\%.$

Additional Steps[b]

$\hat{\pi}_2 = 34.34\%$

$\hat{\pi}_3 = 52.15\%$

$\hat{\pi}_4 = 58.06\%$

$\hat{\pi}_5 = 58.00\%$

$\hat{\pi}_6 = 58.00\%$

Fisher Scoring Method

Score: $u_0 = \dfrac{Y - n\hat{\pi}_0}{\hat{\pi}_0(1 - \hat{\pi}_0)} = \dfrac{58 - 100(10\%)}{10\% * 90\%} = 533.33$.

Expected Information: $\dfrac{n\hat{\pi}_0}{\left(\hat{\pi}_0^2\right)} + \dfrac{n - n\hat{\pi}_0}{(1 - \hat{\pi}_0)^2} = \dfrac{n}{\hat{\pi}_0(1 - \hat{\pi}_0)} = \dfrac{100}{10\%(1 - 10\%)} = 1111.11$.

$\dfrac{Score}{Expected\ Information} = \dfrac{533.33}{1111.11} = 48\%.$

$\hat{\pi}_1 = \hat{\pi}_0 + \dfrac{Score}{Exp.\ Info} = 10\% + 48\% = 58\%.$

Note: Convergence reached after one step because the score is now 0.

[a]A low value of $\hat{\pi}_0$ was chosen for illustration purposes.

[b]Calculations not shown. [See answer to Problem 3(a).]

one closer to the maximum. For example, the smaller the score in absolute value, the closer we are to the maximum; the score will be zero at the maximum point. This fact is why the score is the numerator in the updating equation.

The denominator in the second term of Eq. 14.4 is the observed information. Examine Figure 14.3. When there is higher curvature to the L function, by definition the information values will be higher and therefore the updating steps are smaller. This fact helps to prevent overshooting the maximum point.

Fisher Scoring Procedure

The score method was invented by R. A. Fisher (1925) and his doctoral student, C. R. Rao (1948). The functions in Figures 14.1 through 14.3 are based on $P = 58\%$ (and on $Y = 58$). However, if we were to take another sample, we might obtain a different value of Y. There is a sampling distribution with a mean corresponding to the expected value of Y. The Fisher scoring method is based on the *expected value* of Y rather than the actual value, and in turn the expected variance and information.

Binomial Example Continued

To use the Fisher scoring method with the binomial example, we use the *expected information* in the equations rather than the observed information. For the case where $\pi_0 = 50\%$, we simply use the expected value of Y in Eq. 14.3 (50) rather than the observed value (58). Doing so produces the following:

$$\frac{50}{50\%^2} + \frac{100-50}{(1-50\%)^2} = 400 .$$

While the observed information is partially a function of Y, the expected information is not (it is only a function of the hypothesized π and n). In this example, the expected information (I) is 400, the same as the observed invformation. Here, the Fisher scoring procedure produces the same result as Newton-Raphson. However, identical outcomes are not always the case, especially when one method fails to reach convergence.

Let us return to the example where our initial estimate of π was 10%. The bottom portion of Table 14.1 shows that the Fisher scoring method reaches convergence after just one iteration, whereas the Newton-Raphson procedure required six iterations. (With Newton-Raphson, the observed information values are larger than the expected information value, producing smaller update steps.) The Fisher scoring procedure is therefore more computationally efficient in this situation, where we started with a poor initial guess.

Now examine Figure 14.4. Two curves are shown. The first is the log-likelihood function based on $\pi = 58\%$ (given $Y = 58$). The second log-likelihood function is based on the expected values of assuming $\pi_0 = 10\%$. Now at $\pi_0 = 10\%$, the curvature of the function based on the expected value of Y is much less than that of the other curve. In other words, the expected information is lower than the observed information. The Fisher scoring procedure will therefore make a larger update in Eq. 14.4 (when H_t is replaced by the expected information), producing faster convergence.

EXPLORING THE CONCEPT

In the example, why does the Fisher scoring procedure result in a larger update? At $\pi_0 = 10\%$, the expected information is smaller than the observed information. Do the information values enter into the numerator or denominator of the updating formulas?

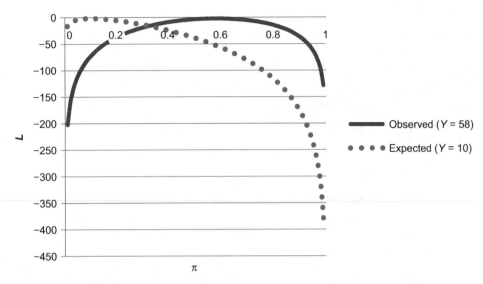

Figure 14.4 Log-likelihood functions for (a) observed value of $Y = 58$, and (b) expected value of $Y = 10$, given an initial guess of $\pi_0 = 10\%$. For the latter, the figure shows the curvature (i.e., expected information) to be lower at $\pi_0 = 10\%$ than for the observed L function, thereby producing larger update steps.

It is not always the case that the Fisher scoring method will produce convergence faster and, in fact, the Newton-Raphson method generally converges more quickly (Knight, 2000). However, the Fisher scoring method is more robust and will often converge in cases where the Newton-Raphson method does not. This strength is because the Fisher scoring method uses expected values and is therefore less dependent on the actual data points, some of which may cause convergence problems. The scoring method is also more robust against poor starting values (Jennrich & Sampson, 1976). It is also common to use a hybrid estimation method that alternates between Fisher scoring and the Newton-Raphson procedure; hybrid updating is the default option in IBM SPSS. Finally, it can be shown (see Technical Note 14.2) that when using a GLM with a canonical link function, the two methods are equivalent.

Variances and Standard Errors

Another important feature of the Fisher scoring procedure is that it can be used to calculate variances and standard errors. There is a simple formula for the variance of U, given different values of Y. One first calculates the deviations reflecting how each value of U differs from the mean, and then one squares the deviations. Because the mean is zero, one can just calculate U^2. However, many of the numbers become unmanageably large.

Fisher devised the alternative statistic, the *Information ($\mathcal{I}$)*, which is based on taking the derivative of the U function for different values of Y, as the derivative values are much smaller. Theoretically, the derivatives could then be squared, and the average (i.e., expected value) calculated. However, the purpose of squaring is to make all the values positive, and this goal can also be accomplished by multiplying the values by -1, because all the derivatives are negative. After averaging, the resulting number was termed the *information*, but denoted by I instead of $\mathcal{I}$.

$$I = -E\left[\frac{\partial^2 L(\theta)}{\partial \theta^2}\right],$$
(Eq. 14.5)

where θ is the parameter being estimated (π in our example). I is the square root of $\mathcal{I}$. It also turns out (and this point is especially important) that I is the inverse of the variance of the log-likelihood function, so a high information value is related to low variance (indicating certainty). This fact is why Fisher used the term *information*.

Because the information is the inverse of the variance, and the variance in estimating a proportion is $\frac{\pi(1-\pi)}{n}$, the formula for the (expected) information is just the reciprocal:

$$I = \frac{n}{\pi(1-\pi)}$$
(Eq. 14.6)

Eq. 14.6 can also be formally derived; it is the expected value of the observed information, $E\left(\frac{Y}{\pi^2} + \frac{n-Y}{(1-\pi)^2}\right)$; see Eq. 14.3. Because $E(Y) = n\pi$, the formula further reduces to Eq. 14.6 (see Appendix 14.2 for proof).

Because it was shown in Chapter 10 that P is the MLE of π, the Newton-Raphson and Fisher scoring methods of estimating ML values may seem like overkill in this example, but—as we shall see in the latter part of this chapter—the methods are very useful in more complicated situations when there are multiple predictors. The Fisher scoring method is also useful for testing statistical hypotheses and forming confidence intervals, as we shall see next.

Testing Hypotheses

Given some parameter θ, Fisher proved that the test statistic $S(\theta) = \left(\frac{U(\theta_0)^2}{I(\theta_0)}\right)$ follows a chi-square distribution (with 1 *df*) when the null hypothesis is true. The null hypothesis is that the parameter θ is equal to a certain value (say 50%). The hypothesized value is denoted by θ_0. $S(\theta)$ is just the score statistic squared divided by the information and is the test statistic for the *score test*. In the above example, $S(\theta) = \frac{32^2}{400} = 2.56$. So $\chi^2(1) = 2.56$, $p = .11$. We therefore cannot reject the null hypothesis that θ (i.e., π) is 50%.

One could also perform the Wald test. The score test uses the slope and curvature of the log-likelihood function at the hypothesized value (50%), whereas the

Wald test uses the slope and curvature at the ML value of 58%. Plugging 58% (rather than 50%) into Eq. 14.6 yields: $\frac{100}{58\%(1-58\%)} = 410.5$, a variance of $\frac{1}{410.5} = .0024$, and an SE of $\sqrt{.0024} = .049$. Wald's z is $\frac{\left|\hat{\theta} - \mathrm{E}(\hat{\theta})_{null}\right|}{SE} = \frac{58\% - 50\%}{.049} = 1.63, p = .10$ (two-sided). We again cannot reject the null hypothesis. Note that in this example, the Wald test boils down to a simple z test (normal approximation) without a continuity correction. Furthermore, of the two procedures, Agresti (2002) recommended using the score test because "it uses the actual null SE rather than an estimate" (p. 14). He argued against incorporating a continuity correction with the score test as it "makes the score test too conservative" (p. 77).

Finally, one could perform the test using the likelihood ratio test described in Chapter 10. The test is a compromise between the other two tests because it compares values at both the hypothesized null value of 50% and predicted value of 58%. In both cases, we can compute the likelihood of obtaining $P = 58\%$ given $\pi_0 = 50\%$ and $\pi_1 = 58\%$ using the binomial formula. We can then compute the difference in the likelihoods and the negative two log likelihood ($-2LL$) statistic. Specifically,

π:	$\pi_0 = 50\%$	$\pi_1 = 58\%$
Likelihood (l):	2.23%	8.06%
Log Likelihood (L):	-3.80 (L_0)	-2.52 (L_1)

$L_0 - L_1 = -3.80 - (-2.57) = -1.29.$
$-2(L_0 - L_1) = 2.57, \chi^2(1) = 2.57, p = .11.$

The likelihood ratio test, which is based on the $-2LL$, gives a p value closer to the score test.

In general, the score test is more versatile than the Wald test because it does not assume that one is at the maximum of the likelihood function. The score test is more complicated than the likelihood ratio (LR) test but is more versatile when the exact likelihood ratio function is not known. This situation is the case with such procedures as GEE, which was described in the previous chapter.

Building Confidence Intervals

One could create a 95% CI around P by finding the p values for different estimates of π and choosing those with a .025 or .975 cumulative probability as the end points of the interval. However, it is awkward forming CIs manually, as the entire procedure is not embodied in a formula. Clopper and Pearson (1934) developed formulas for computing exact CIs, but because binomial variables are discrete, the CIs produce larger than 95% coverage of the probability distribution. (Do not therefore be misled by the word "exact," it only refers to the use of the exact binomial distribution.) Other methods for forming CIs are preferable. According to Agresti (2002, p. 19), the score method is more accurate than the other methods of forming CIs except for extreme values of π close to 0 or 1. Other methods may be too liberal or conservative due to the discrete nature of the variables (see Chapter 2). I therefore focus first on the score method.

In performing the statistical hypothesis test, it was found that $\chi^2(1) = 2.57$, $p = .11$. The z value (for the score method) is the square root of 2.57, so $z_s = 1.60$, with a hypothesized null π_o of 50%. According to Agresti (2002, p. 15), "The score confidence interval contains π_o values for which $|z_s| < |z_{\alpha/2}|$." For a 95% CI, $\alpha/2$ is 2.5%, so $z_{\alpha/2} = -1.96$. The endpoints of the CI are the endpoint of the equation:

$$\frac{(\hat{\pi} - \pi_0)}{\sqrt{\dfrac{\pi_0(1 - \pi_0)}{n}}} = \pm\, Z_{\alpha/2}.$$

Bear in mind that here π_0 are the values of the end points. We want to find end points that yield a z value of ±1.96. The equation that solves this problem was found by Wilson (1927). The midpoint is

$$\tilde{\pi} = \hat{\pi}\left(\frac{n}{n + z_{\alpha/2}^2}\right) + 50\%\left(\frac{z_{\alpha/2}^2}{n + z_{\alpha/2}^2}\right), \qquad \text{(Eq. 14.7)}$$

which in our example turns out to be

$$58\%\left(\frac{100}{100 + 1.96^2}\right) + 50\%\left(\frac{1.96^2}{100 + 1.96^2}\right)$$

$$= 58\% * \frac{100}{103.84} + 50\%\left(\frac{3.84}{103.84}\right)$$

$$= 58\% * 0.96 + 50\% * 0.04$$

$$= 57.68\%.$$

Note that the figure reflects a weighted average between $\pi = 58\%$ (weighted at .96) and $\pi = 50\%$ (weighted at .04). The final n is 103.84 rather than 100; basically 3.84 cases (i.e., $z_{\alpha/2}^2$) have been added to the sample. It is these 3.84 cases that have a $\pi = 50\%$. They make up 4% of the sample, hence the weight of .04.

Why is this adjustment necessary? It is basically a small sample correction. The correction is much larger in small samples and becomes negligible in larger samples. For example, if $n = 1,000$, the weight given the 3.84 cases is $\frac{3.84}{1000 + 3.84} = 0.38\%$ rather than 4% with $n = 100$. If n were equal to 20, the weight would be $\frac{3.84}{20+3.84} = 16\%$. Brown et al. (2001) showed that unadjusted ML estimators (e.g., the sample proportion P) are biased in small samples. ML estimators are consistent, meaning that the amount of bias decreases as the sample size increases. One way of thinking about the bias is that extreme events (e.g., outliers) are more influential in small samples, creating an overestimation of the mean and standard error.

Now let us build the actual CI. The end points for a 95% CI are given by the expression

$$
1.96 * \sqrt{ \frac{1}{n+3.84} \left[\hat{\pi}(1-\hat{\pi}) \left(\frac{n}{n+3.84} \right) + 50\%(1-50\%) \frac{3.84}{n+3.84} \right] } \quad \text{(Eq. 14.8)}
$$

$$
= 1.96 * \sqrt{ \frac{1}{103.84} * \left[58\% * 42\% * \frac{100}{103.84} + 50\%(1-50\%) \frac{3.84}{103.84} \right] }
$$

$$
= 1.96 * \sqrt{ \frac{1}{103.84} * [24.36\% * .963 + 25\% * .037] }
$$

$$
= 1.96 * \sqrt{.0023}
$$

$$
= 9.40\%.
$$

Eliminating rounding error in the calculation produces a value of 9.50%. The CI therefore ranges from 48.18% to 67.18% (i.e., 57.68% ± 9.50%).

There are other methods for forming binomial CIs. In IBM SPSS, if one goes to "Nonparametric Tests" ➜ "One Sample" ➜ "Customize Tests" ➜ "Compare observed binary probabilities to hypothesized (binomial test)" ➜ "Options," one can then select from three different types of binomial CIs: Clopper-Pearson, Jeffreys, and likelihood ratio. (One can ignore the setting for the "hypothesized value," as we are forming confidence intervals and not performing hypothesis tests.) As noted previously, the Clopper-Pearson CI is directly based on the exact binomial distribution and, because of discreteness, typically has coverage greater than 95%. The likelihood ratio CI is based on inverting the acceptance region of the LR hypothesis test. The Jeffreys CI involves Bayesian statistics (see Technical Note 14.1) and is equivalent to the Clopper-Pearson CI with a continuity correction. The results for this example, along with the score method, which must be calculated by hand, are shown in Table 14.2. Although the differences are minor, the Clopper-Pearson exact CI is largest, whereas the CI for the score method is smallest. The differences are more dramatic in small samples, with $n \leq 40$. Therefore, the table also shows CIs under the scenario that the number of hits (Y)

Table 14.2 Comparison of Different Methods for Calculating Binomial Confidence Intervals

Method	Y = 58, n = 100		Y = 17, n = 30	
	95% CI	Range	95% CI	Range
Clopper-Pearson	.477 to .678	.201	.374 to .745	.371
Jeffreys	.482 to .673	.191	.390 to .731	.341
Likelihood ratio	.482 to .674	.192	.389 to .733	.344
Score method	.482 to .672	.190	.391 to .726	.335

is 17 and $n = 30$. The table shows the score CI is again shortest, by 0.6% compared to Jeffreys, 0.9% compared to LR, and by 3.6% compared to Clopper-Pearson.

Estimation Procedures for Generalized Linear Models and General Estimating Equations

The prior section discussed how to use MLE methods to estimate a population proportion π. Tackling this problem led to many theoretical developments, particularly the various hypothesis testing methods, which can be easily illustrated when using P to estimate π. However, in most applied statistical situations, the goal is to predict a DV (for example, π, a logit, or count) based on one or more independent variables (IVs). The independent variables make up a *design matrix* that looks something like the following (more simplified) example for two predictors and three cases):

$$\mathbf{X} = \begin{matrix} X_0 & X_1 & X_2 \\ \begin{bmatrix} 1 & x_{11} & x_{21} \\ 1 & x_{12} & x_{22} \\ 1 & x_{13} & x_{23} \end{bmatrix} \end{matrix}.$$

The first column consists of the data for the intercept term; all the observations are set equal to one because the intercept term will not vary from case to case (see Chapter 9). Recall that when a variable is in bold, it represents a matrix.

When using multiple predictors, let $\boldsymbol{\beta}$ represent the various true (population) regression parameters, one for each predictor. The parameters are arranged in a vector (a vertical or horizontal list). The vector should be written as follows:

$$\boldsymbol{\beta} = \begin{matrix} \beta_0 \\ \beta_1 \\ \beta_2 \end{matrix}$$

With a simple linear (OLS) regression, the statistical model is:

$$\mathbf{Y}_{n \times 1} = \mathbf{X}_{n \times p} * \boldsymbol{\beta}_{p \times 1} + \mathbf{e}_{n \times 1}. \tag{Eq. 14.9}$$

Eq. 14.9 reflects the general linear model but is only one of many *generalized* linear models. With GLMs such as logistic or Poisson regression, the linear model does not predict $\mathbf{Y}$ but something else, such as the logit or a log. We should therefore rewrite Eq. 14.9 as

$$\boldsymbol{\eta}_{n \times 1} = \mathbf{X}_{n \times p} * \boldsymbol{\beta}_{p \times 1} + \mathbf{e}_{n \times 1}, \tag{Eq. 14.10}$$

where $\eta_{n \times 1}$ is a vector consisting of the *linear predictions* for each case. We use the subscript i to index these cases, where $i = 1, 2, 3 \ldots n$. Therefore the entries of $\eta_{n \times 1}$ are η_i, the linear prediction for each case. For illustration purposes, I concentrate in this chapter on logistic regression, so η will represent the predicted logits (log odds). As was the case with the illustrations in Chapter 10, we predicted logits for each case, then transformed them to odds and then transformed those to predicted probabilities (π_i). These series of transformations, put together in one formula, constitute the *inverse link function*. In contrast, the *link function* transforms Y_i into η_i. The link function is sometimes denoted by $g(\mu)$, and by definition, $g(\mu) = \eta$.

The expression $g(\mu) = X * \beta + e$ is a generic definition of a link function; applied to logistic regression, $g(\pi)$ is the logit, and g is the logistic link function. The inverse link function is denoted generically by $g^{-1}(\eta)$.

Let $L(\beta_{p \times 1})$ represent a vector of log likelihoods, one for each parameter. Recall that the likelihood is the probability of obtaining the data, given a certain parameter value. One can then compute a score vector (**u**). (See Technical Note 14.3 for computational formulas.) Recall from the first part of this chapter that the scores represent the slope of the log-likelihood function at a certain point, specifically the one corresponding to $\hat{\beta}_p$. We want to choose the betas where the scores are zero, because that represents the maximum point of the log-likelihood function. However, the resulting equation must be solved iteratively; because of nonlinearities, algebraic solutions are not possible. For example, in the Fisher scoring method, the steps are defined by the scores divided by the expected information.

Computing the Scores

Because there is a different likelihood for each parameter ($\hat{\beta}_0, \hat{\beta}_1, \hat{\beta}_2, \ldots \hat{\beta}_k$), there is also a different score for each parameter (e.g., slope of L_0 at $\hat{\beta}_0$). There is therefore a *score matrix* $U_{p \times 1}$, which is a vector of scores, one for each parameter. From calculus, the slope (i.e., score) is the first derivative of L_i evaluated at $\hat{\beta}_j$ [j ranges from $0 \ldots k$; $p = (k + 1)$]. However, there are a variety of computational formulas for estimating the score matrix that simplify the calculations; these formulas are derived from the first derivative (the proofs are in Appendix 14.2).

Table 14.3 summarizes the different score equations. The first equation is the most general; it applies to any situation. The second is much simpler but applies only when a canonical link function is used. The third equation is used for quasilikelihood estimation, which is an approach used when the actual likelihood function is unknown. The fourth equation is used as part of the GEE procedure. I describe in more detail below the nature of each equation.

Table 14.3 Score and Quasiscore Formulas

(Eq. 14.11) General Score Equation

$$u_j = \sum_{i=1}^{n} \frac{Residual_i}{Var(Y_i)} * \frac{\partial \mu_i}{\partial \eta_i} X_{ij},$$

where $Residual_i = (Y_i - \mu_i)$

(Eq. 14.12) Score Equation for Canonical Links

$$u_j = \sum_{i=1}^{n} Residual_i * X_{ij}.$$

(Eq. 14.13) Quasi-likelihood Score Equation (Quasiscore)

$$\sum_{i=1}^{n} \frac{Residual_i}{\phi v(\mu_i)^{a}} * \frac{\partial \mu_i}{\partial \beta_j}$$

$\mathbf{D}^{\mathrm{T}} \, \mathbf{V}_i^{-1} \, (\mathbf{Y} - \boldsymbol{\mu})/\phi$ (matrix form)

(Eq. 14.14) GEE Quasiscore Equation

$$\sum_{i=1}^{N} \frac{Residual_i}{\phi v(\mu_i)} * \frac{\partial \mu_i}{\partial \beta_j}$$

$\sum_{i=1}^{n} \mathbf{D}_i^{\mathrm{T}} \, \mathbf{V}_i^{-1} \, [(\mathbf{Y}_i - \boldsymbol{\mu}_i)]$ (matrix form).

[a]The parameter ϕ is the dispersion factor, estimated by $\frac{x^2}{(n-p)}$. $v(\mu_i)$ is the variance function that expresses the relationship between the means μ_i and the variances. So for example, in the Bernoulli distribution, π_i is the mean and $\pi_i (1 - \pi_i)$ is the variance. See Table 14.4 for a list of variance functions.

The General Score Equation (Eq. 14.11)

It is clear from this equation that if there is a perfect fit, the residuals will be zero, as will the score. A perfect fit is rare. Nevertheless, when one has good estimates of the beta parameters, the positive residuals will cancel out the negative ones, and so when these are summed, the score will be close to zero. However, when the beta estimates are poor, there will not be a good fit, and the score will be some distance from zero. We do not want this measurement of this distance to be affected by the metrics we are using, so the sum is standardized by dividing by $Var(Y_i)$.

What is the role of the second term, $\frac{\partial \mu_i}{\partial \eta_i} X_{ij}$? It is a partial derivative. Recall that the score is the slope of the log-likelihood function, and because we are measuring a slope, we need some sort of derivative term in the equation. But why this particular term? The proof of Eq. 14.11 is in Appendix 14.2. A key step in the proof invokes the following equation:

$$L_i = f(\theta_i) = f(\mu_i) = f(\eta_i) = f(\beta_j). \tag{Eq. 14.15}$$

Let us examine each part of Eq. 14.15. The log-likelihood L_i is a function of θ_i, which is the parameter(s) being estimated, such as π_i or λ_i. I will concentrate on π_i because our extended example relates to that. The variable π_i is also the mean of the logistic regression line (predicted value) and so θ_i(or π_i) $= f(\mu_i)$. Now, $\mu_i = f(\eta_i)$, because η_i is the linear prediction (the logit in the logistic case), and from the logit we can deduce π_i. Pi may differ for different cases because the IVs may differ. We see from Eq.14.9 that $\eta_i = f(\beta_j)$, as the betas (and Xs) are used to predict η_i.

We can now address the question of the role of $\frac{\partial \mu_i}{\partial \eta_i} X_{ij}$ in Eq. 14.11. Recall that, by definition, the score function is the first derivative of the log-likelihood function in respect to β_j. In calculus, there is something known as *the chain rule*, which applies when taking partial derivatives of functions that are functions of other functions. (Specifically, $\frac{dz}{dx} = \frac{dz}{dy} * \frac{dy}{dx}$.) After applying the chain rule, all the partial derivatives are dissolved except for one: $\frac{\partial \mu_i}{\partial \eta_i}$. The formula needs to contain some sort of derivative, because it is measuring a slope of a function.

EXPLORING THE CONCEPT

(a) Although the score is the slope of L, why does L not appear in Eq. 14.11? Has L been deconstructed by the chain rule? (b) Recall from Chapter 10 that likelihood functions reflect a combination of the Y distribution and the link function. Are both pieces of information still in Eq. 14.11? Do the residuals and variance terms give information on the distribution of Y? Does the $\frac{\partial \mu_i}{\partial \eta_i} X_{ij}$ term give information on the link function?

Score Equation for Canonical Links (Eq. 14.12)

We have seen in previous chapters that for different distributions (binomial, Poisson, etc.), there is a particular link function that it is customary to use (e.g., the log-link function with a Poisson distribution). These usual and customary link functions are known as *canonical links* and are listed in Table 14.4. In performing a GLM analysis, one is not obligated to use the canonical link—for example, one

Table 14.4 Canonical Link and Variance Functions

Distribution	Canonical Link	Link Formula	Variance Function[a]
Normal	Identity	$\mu = X\beta$	$v(\mu) = 1$
Binomial/Multinomial	Logistic	$ln\left(\frac{\mu}{1-\mu}\right) = X\beta$	$v(\mu) = \mu(1-\mu)$
Poisson	Log	$ln(\mu) = X\beta$	$v(\mu) = \mu$
Exponential/Gamma	Inverse	$\mu^{-1} = X\beta$	$v(\mu) = \mu^2$
Inverse Gaussian	Inverse squared	$\mu^{-2} = X\beta$	$v(\mu) = \mu^3$

[a]Used in quasilikelihood estimation.

could use the identity link in a Poisson regression if one thought that means for different values of X form a straight line. Nevertheless, the canonical links have some special mathematical properties. For example, when one uses a canonical link, the score function Eq. 14.11 in Table 14.3 simplifies to Eq. 14.12. This result is proven in Appendix 14.2.

Quasilikelihood Estimation

The preceding discussion assumed we knew what the likelihood function was, and from this function we derived the other functions. We sometimes do not know what the likelihood function is. For example, if the observations are not statistically independent, then the empirical variance in the sample, $Var(Y_i)$, will produce biased estimates. Sometimes the distribution of Y is unknown or it is not a member of the natural exponential family (e.g., not a Poisson, binomial, gamma distribution, etc.; see Appendix 14.2—Proof of Eq. 14.11—for more explanation). Valid likelihood functions depend on the correct specification of the Y distribution, so in these situations, the likelihood functions will be unknown. However, an alternative approach is to predict the population variance from the means.

For example, with the binomial distribution, the mean is π_i and the variance is a function of the mean, $\pi_i(1 - \pi_i)$ if $n = 1$. More generally, if we know or can estimate the relationship of the mean to the variance, we can construct a variance function and a quasilikelihood function (Wedderburn, 1974). The variance function specifies the relationship between the mean and the variance. This function can be estimated from the data, or it can be derived theoretically. Let $v(\mu_i)$ be the variance function specifying the relationship between the mean and variance. It then follows, by definition, that

$$Var(Y_i) = v(\mu_i). \tag{Eq. 14.16}$$

As an example, for the binomial distribution, $v(\mu_i) = \mu_i(1 - \mu_i) = \pi(1 - \pi)$, which is the variance for a single observation (Bernoulli trial). For the Poisson distribution, $v(\mu_i) = \mu = \lambda$, because the variance is equal to the mean. Similar functions can be stated for other distributions used in GLMs, such as the gamma (μ_i^2), inverse Gaussian (μ_i^3), and negative binomial ($\mu_i + \mu_i^2/k$). However, it is not necessary to assume that the conditional distribution of Y follows one of these distributions. The quasilikelihood procedure can estimate the variance function from the data and then update it iteratively with the other functions, particularly the quasilikelihood function. The quasilikelihood function is basically Eq. 14.11 with the $Var(Y_i)$ term replaced by $v(\mu_i)$ and the derivative expressed as a function of β. More specifically, $\frac{\partial \mu_i}{\partial \eta_i} X_{ij}$ is replaced by $\frac{\partial \mu_i}{\partial \beta_j}$. One can also include a dispersion factor (ϕ) in the score and variance functions. (See Appendix 14.2 for a proof of Eq. 14.13.)

Quasilikelihood models are similar to GLMs in that there is a link function connecting Y with a set of IVs. There is no assumption, however, that Y_i has a distribution from those discussed above. It is also unnecessary to assume that the responses are discrete or continuous or that the variances are constant.

Generalized Estimating Equations

In the GEE procedure, we also relax the assumption that the observations are independent. As explained in the previous chapter, distributions such as the binomial assume that the observations are independent, and so if independence is compromised, the likelihood equations are invalid. However, one can still use quasilikelihood procedures with some modifications. Specifically, there will be separate covariance and related matrices for each individual.

Estimation With Multiple Predictors

Information Matrices

In addition to the score equation, the estimation procedures require computation of an information matrix. It can be shown that when one uses a canonical link function, the expected and observed information matrices are the same, so for now we will just focus on the expected information matrix. The formula for that matrix, with the dimensions of each matrix shown, is

$$\mathbf{I}_{j \times j} = \mathbf{X}_{j \times n}^{T} \mathbf{W}_{n \times n} \mathbf{X}_{n \times j}, \qquad \text{(Eq. 14.17)}$$

where j is the number of predictors. (The formula is derived in the Appendix 14.2.)

$\mathbf{W}_{n \times n}$ (known as the weight matrix) is a diagonal matrix, which is a square matrix with entries (one for each observation) along the main diagonal and zero entries elsewhere. For each observation, each entry in the matrix is the information value associated with the likelihood of obtaining that observation, given the parameter estimates.

Using linear algebra, the updating formula for Fisher scoring in matrix form is

$$\hat{\boldsymbol{\beta}}^{(t+1)} = \hat{\boldsymbol{\beta}}^{(t)} + \left(\hat{\boldsymbol{\vartheta}}^{t} \right)^{-1} \hat{u}^{t}. \qquad \text{(Eq. 14.18)}$$

Conceptually, the last term is the score matrix divided by the expected information matrix. The updating formula for Newton-Raphson scoring is

$$\hat{\boldsymbol{\beta}}^{(t+1)} = \hat{\boldsymbol{\beta}}^{(t)} + \left(-\mathbf{H}^{(t)} \right)^{-1} \hat{u}^{(t)}, \qquad \text{(Eq. 14.19)}$$

where $-\mathbf{H}^{(t)}$ is the observed information matrix. These matrices and procedures may make more sense when illustrated with actual examples. The Newton-Raphson procedure will be illustrated in the following examples (but remember that the Fisher scoring procedure is identical when canonical link functions are used, which is the case in these examples).

Logistic Regression Example

The first example concerns some synthetic logistic regression data that I created. I began by decreeing that the population intercept would be –2.0 and the slope 0.7, chose the X values, used the appropriate equations to derive mean values (π_i's), and the SPSS binomial random variable function to add the random component and generate the final data, which is shown in the first part of Table 14.5.

Note that the data are grouped; the subscript i therefore refers to the value of X and not to the different individuals. Y_i is the number of hits per value of X_i.

Next, I pretended that I did not know the actual population parameters. I made an initial guess of what they might be, starting with a slope (β_1) of zero and an intercept based on the overall proportion of cases that were hits ($\beta_0 = -0.167$). These values are shown in the first panel of Table 14.5. Following along this panel, we see that the predicted values $\hat{\eta}_i$ (predicted logits) were all the same because the slope was zero for the first iteration. I then exponentiated to obtain the odds, and from these derived the predicted probabilities ($\hat{\pi}_i$'s) and means ($\hat{\mu}_i$'s). The variance term used in the score equation, $Var(Y_i)$ is the actual proportion of hits for each level of Y, specifically it was $n_i P_i (1 - P_i)$.

The second panel shows a diagonal variance matrix, with the variances along the main diagonal. From this matrix one can derive the information matrix using the formula $\mathbf{I} = \mathbf{X}^\mathsf{T}(\mathbf{Diag})\mathbf{X}$ (see Appendix 14.2), and the inverse of this matrix is the covariance matrix. The third panel shows the residuals. According to Eq. 14.12, the scores are just $\mathbf{X}^\mathsf{T} * \mathbf{Residuals}$, producing two scores, one for the intercept and one for the slope. These are then multiplied by the covariance matrix (which is the same as dividing by the information matrix), to produce measures of how much each parameter estimate should be updated (–2.074 for the intercept, and 0.824 for the slope).

The fourth panel shows that the updated betas were $\hat{\beta}_0 = -2.241$ and $\hat{\beta}_1 = 0.824$. These were closer to the population values than the initial estimates but still somewhat off. Using these estimates as the inputs for the next iteration and repeating the whole process several times produced estimates that converged to $\hat{\beta}_0 = -1.799$ and $\hat{\beta}_1 = 0.634$, which were reasonably close to the population values. The calculations for the additional iterations are not shown in the table but are addressed in the problems at the end of the chapter.

Table 14.5 Initial Step of Maximum-Likelihood Estimation (Logistic Regression Example)

							Data and Design Matrix				
X_i	n_i	Y_i	P_i		X		$\hat{\beta}_{t=0}$	Logits $\hat{\eta}=X\hat{\beta}$	Odds Exp($\hat{\eta}$)	$\hat{\pi}_i=\frac{odds}{(1+odds)}$	$\hat{\mu}_1=n\hat{\pi}_i$	Var(Y_i)$=$ $n_iP_i(1-P_i)$
				X_0	X_1							
0	20	5	25%	1	0	$\beta_0=-0.167$[a]	-0.167	0.85	45.8%	9.17	3.75
1	20	3	15	1	1	$\beta_1=0.000$[b]	-0.167	0.85	45.8	9.17	2.55
2	20	6	30	1	2		-0.167	0.85	45.8	9.17	4.20
3	20	11	55	1	3		-0.167	0.85	45.8	9.17	4.95
4	20	13	65	1	4		-0.167	0.85	45.8	9.17	4.55
5	20	17	85	1	5		-0.167	0.85	45.8	9.17	2.55

Diagonal Var Matrix	Information: X^T (Diag)X $2\times6, 6\times6, 6\times2=2\times2$	Info Inverse: Cov Matrix

$$\begin{bmatrix} 3.75 & 0.00 & 0.00 & 0.00 & 0.00 & 0.00 \\ 0.00 & 2.55 & 0.00 & 0.00 & 0.00 & 0.00 \\ 0.00 & 0.00 & 4.20 & 0.00 & 0.00 & 0.00 \\ 0.00 & 0.00 & 0.00 & 4.95 & 0.00 & 0.00 \\ 0.00 & 0.00 & 0.00 & 0.00 & 4.55 & 0.00 \\ 0.00 & 0.00 & 0.00 & 0.00 & 0.00 & 2.55 \end{bmatrix} \quad \begin{bmatrix} 22.550 & 56.750 \\ 56.750 & 200.450 \end{bmatrix} \quad \begin{bmatrix} .154 & -.044 \\ -.044 & .017 \end{bmatrix}$$

X_i	Residual $Y_i-\hat{\mu}_1$	Scores $X^T*Resid$ $2\times6, 6\times1=2\times1$	$\hat{\beta}$ update Cov$*Score$[b] $2\times6, 6\times1=2\times1$
0	-4.17		
1	-6.17	$\begin{bmatrix} 0.00 \\ 47.50 \end{bmatrix}$	$\begin{bmatrix} -2.074 \\ 0.824 \end{bmatrix}$
2	-3.17		
3	1.83		
4	3.83		
5	7.83		

				Beta update(s)				
	$\hat{\beta}_{t=0}$	$\hat{\beta}$ update	$\hat{\beta}_{t=1}$	$\hat{\beta}_{t=2}$[c]	$\hat{\beta}_{t=3}$[c]	$\hat{\beta}_{t=4}$[c]	$\hat{\beta}_{t=5}$[c]	$\hat{\beta}_{t=6}$[c]
B_0	-0.167	-2.074	-2.241	-1.885	-1.808	-1.800	-1.799	-1.799
B_1	0.000	0.824	0.824	0.660	0.637	0.634	0.634	0.634

[a] Initial guess of intercept based on: $P=\frac{\sum Y_i}{n}=\frac{55}{120}=45.8\%; odds=\frac{45.8\%}{54.2\%}=0.846, logit=\ln(.846)=-0.167$. Initial guess of slope is 0.

[b] The covariance matrix is the inverse of the information matrix, so this operation is equivalent to dividing by the information matrix.

[c] Calculations not shown.

As shown in the second panel of the table, the covariance matrix was:

	$\hat{\beta}_0$	$\hat{\beta}_1$
$\hat{\beta}_0$	0.15	−0.04
$\hat{\beta}_1$	−0.04	0.02

Here, the matrix is a function of the actual data and so is unaffected by updates to the beta estimates. The matrix shows that $Var(\hat{\beta}_0) = 0.15$ and $Var(\hat{\beta}_1) = 0.02$. The standard errors are just the square root of these: $SE(\hat{\beta}_0) = 0.39$ and $SE(\hat{\beta}_1) = 0.14$. Wald z for the estimates is the betas divided by the SEs; for example, for the slope, Wald z is 4.5 and highly significant.

EXPLORING THE CONCEPT

(a) As shown in the tables, the beta update matrix is equal to the covariance matrix times a score matrix. If the scores all are zero, what will the beta update values be?
(b) The inverse of the covariance matrix is the information matrix. Conceptually, is multiplying by the covariance matrix like dividing by the information matrix? Is doing so consistent with the Fisher scoring updating formula, Eq. 14.18?

General Estimating Equations

In illustrating the GEE procedure in the previous chapter, I deferred to this chapter the description of the steps for calculating quasiscores and updating the beta estimates. These steps are illustrated below.

In Chapter 13, we saw that the derivative D_i for the first individual in the example was a function of the covariance matrix (A_1), specifically, $D_i = A_i * X_i$. Using the values in Table 13.1, we can therefore calculate D_1 as follows:

$$A_1 = \begin{bmatrix} .22 & 0 & 0 \\ 0 & .25 & 0 \\ 0 & 0 & .25 \end{bmatrix}, X_1 = \begin{bmatrix} 1 & 0.85 \\ 1 & 4.10 \\ 1 & 3.55 \end{bmatrix}, D_1 = \begin{bmatrix} 0.22 & 0.19 \\ 0.25 & 1.03 \\ 0.25 & 0.89 \end{bmatrix}.$$

After adjusting the A matrix by $R(\alpha)$ per Eq. 13.4, the adjusted covariance matrix is

$$V_1 = \begin{bmatrix} .22 & .13 & .07 \\ .13 & .25 & .14 \\ .07 & .14 & .25 \end{bmatrix}.$$

To apply Eq. 14.14, we next need to invert the matrix:

$$\mathbf{V}_1^{-1} = \begin{bmatrix} 6.56 & -3.47 & 0.11 \\ -3.47 & 7.67 & -3.32 \\ 0.11 & -3.32 & 5.83 \end{bmatrix}.$$

Multiplying $\mathbf{V}_1^{-1}$ by the raw residuals in Table 13.1 yields $\begin{bmatrix} 2.99 \\ -0.58 \\ 1.41 \end{bmatrix}$.

To derive the quasiscore matrix, $\mathbf{u}_1$, multiply this matrix by the transpose of the derivative matrix $(\mathbf{D}_i^T)$ and then the residuals:

$$\mathbf{u}_1 = \mathbf{D}_1^T \mathbf{V}_1^{-1} (\mathbf{Residuals})_1 = \begin{bmatrix} 0.86 \\ 1.21 \end{bmatrix}.$$

The quasiscore matrices for all individuals are then summed together and divided by the information matrix to produce the updates to the beta estimates. It should be noted that I rounded all the numbers above to two decimal places, creating rounding error. More precise calculations in EXCEL yielded the following quasis-core matrix: $\begin{bmatrix} 0.93 \\ 1.32 \end{bmatrix}$. Recall that this was the quasiscore matrix for the first iteration and was used to calculate the beta update matrix. On the last iteration, we want the scores to be zero.

Summary

The big idea in estimation is to find parameter estimates that maximize a likelihood function and a corresponding log-likelihood function (L). The maximization is performed iteratively in the Newton-Raphson procedure, by starting with a good initial guess and then updating the guess by dividing the score (first derivative of L) by the information (the negative of the second derivative of L). Because one wants the score to be zero, low scores require smaller update steps; however, even with a large score, the steps can be small if the curvature of L is high, as measured by the information.

Fisher scoring, which is more robust, uses the expected information, based on the expected value of Y ($n\pi$), and the corresponding expected variance. Because the expected information is the reciprocal of the variance of the sampling distribution, it can be used to derive standard errors and to perform hypothesis testing (score tests). For statistical hypothesis testing, alternatives to Fisher scoring are the Wald method, which assumes one is at the maximum of the "true" log-likelihood function, and the LR test, which is a compromise between Fisher and Wald.

The Fisher scoring procedure is also the basis of quasilikelihood estimation, which uses expected variances. More specifically, quasilikelihood estimation uses a variance function v to determine the expected variances. The variance function expresses the relationship between the mean and variance in the data and can be derived analytically or empirically. If the conditional distribution of Y is from the natural exponential family, the variance function can be derived analytically (see Table 14.4). However, if this is not the case, or the log-likelihood function is unknown, the variance function can be estimated empirically, by examining the ratio of the variance to the mean in the data. Once the value of v is estimated, it is used in place of the empirical variance in the score equations (see Table 14.3).

Quasilikelihood estimation is the basis of the GEE procedure, where, because of violations of statistical independence, the empirical variances will lead to biased estimates of the standard errors. The true log-likelihood functions are therefore unknown. It is fitting that this chapter ends where this book began. The axiom of probability theory holds that $\text{Prob}(A \,\&\, B) = \text{Prob}(A) * \text{Prob}(B)$ only if events A and B are statistically independent. If that is not the case, the whole statistical machinery built on this axiom breaks down. The GEE procedure (described in more detail in Chapter 13) is one attempt to address this problem.

Problems

1. Compute and compare 95% binomial confidence intervals (CIs) if $n = 30$ and $P = 30\%$. Compare Clopper-Pearson, Jeffreys, likelihood ratio, and the score CI. You may use statistical software. If using IBM SPSS, you will need to compute the score CI by hand using Eqs. 14.7 and 14.8. Report the end points and CI range, and rank order the CIs as in Table 14.2. Also, for the score interval, report the midpoint ($\hat{\pi}$).

2. Repeat Problem 1 with $n = 20$ and $P = 90\%$.

3. Using an EXCEL worksheet, find the ML estimates for the following scenarios, using both the Newton-Raphson and Fisher scoring methods, using Table 14.1 as a guide.

   a) $n = 100$, $\hat{\pi}_0 = 10\%$, $P = 58\%$ (and therefore $Y = 58$). In the EXCEL worksheet, have step number (t) for the rows, and for the columns, $\hat{\pi}_t$, score, information, and update. Use six steps. Use the F4 key for absolute cell addresses (for n, Y). Your answers should reproduce the values in Table 14.1.

   b) Using the worksheet from Part (a), change the parameter values to those in Problem 1 and to $\hat{\pi}_0 = 40\%$.

4. Using the worksheet from the previous problem, change the parameter values to those in Problem 2 and $\hat{\pi}_0 = 80\%$. Use six steps.

5. The following problems involve linear algebra.

a) What are the dimensions of the following two matrices? Can they be mul-

tiplied together? $\begin{bmatrix} 1 & 6 \\ 2 & 8 \\ 5 & 1 \\ 3 & 1 \end{bmatrix}$ $\begin{bmatrix} 1 & 8 \\ 3 & 1 \\ 5 & 5 \\ 7 & 2 \end{bmatrix}$

b) If one were to transpose the second matrix, what would be its dimensions? Could one then multiply the two matrices together? What would be the dimensions of the resulting matrix?

c) Enter the two matrices into an EXCEL spreadsheet. Transpose the second matrix by highlighting a 2 × 4 set of blank cells and type =TRANSPOSE, select the second matrix, and press shift-ctrl-enter. Report the transposed matrix.

d) Multiply the first matrix by the second matrix transposed. Highlight a 4 × 4 set of blank cells and type =MMULT, select the two matrices as the arrays, and press shift-control-enter. Report the product.

e) Invert (and report) the following square matrix: $\mathbf{M} = \begin{bmatrix} 1 & 3 \\ 2 & 4 \end{bmatrix}$. Highlight a 2 × 2 set of blank cells, type =MINVERSE, select the matrix as the array, and press shift-ctrl-enter.

f) Multiply $\mathbf{M}$ by $\mathbf{M}^{-1}$. The resulting matrix should be the identity matrix, $\mathbf{I}$. Calculate $\mathbf{MI}$. What is the result?

6. Repeat Problem 5, parts (a)–(e), using the following two matrices:

$\begin{bmatrix} -4 & -2 \\ -7 & -2 \\ 10 & -3 \\ -8 & 7 \end{bmatrix}$, $[-7\ 8]$. For Part (e), $\mathbf{M} = \begin{bmatrix} 7 & -9 & -5 \\ 9 & -7 & -7 \\ -4 & -1 & -9 \end{bmatrix}$.

7. Using the data and beta estimates in Table 14.5 (and following the steps shown there), find the following matrices for $\hat{\boldsymbol{\beta}}_{t=1}$ using EXCEL: (a) residuals, (b) scores, (c) $\hat{\boldsymbol{\beta}}$ updates (to three decimal places to minimize subsequent rounding error), and (d) $\hat{\boldsymbol{\beta}}_{t=2}$ (to three decimal places). You may use the following covariance matrix (which is a more precise version of the one shown in the table): $\begin{bmatrix} 0.1542 & -0.0437 \\ -0.0437 & 0.0174 \end{bmatrix}$. This matrix does not change with each iteration.

8. Repeat Problem 7 for $\hat{\boldsymbol{\beta}}_{t=2}$. (Note: Use the $\hat{\boldsymbol{\beta}}_{t=2}$ values shown in Table 14.5, as the ones from Problem 7 may contain more rounding error.)

9. Find the score matrix (to four decimal places) for the data in Problem 7 for $\hat{\boldsymbol{\beta}}_{t=5}$. Why are the scores close to zero?

10. What does the information matrix in Table 14.5 represent? Is the curvature of the log-likelihood function greater for the intercept or the slope? Is the standard error greater for the intercept or the slope? (Hint: the standard errors are the square root of the variances.)

11. Use the X and Y data in Table 14.5 but assume the predicted values are generated using Poisson regression with a canonical log-link function. For the initial values ($\hat{\boldsymbol{\beta}}_{t=0}$), use 1.26 for the intercept and 0.32 for the slope (these values are based on an OLS regression of $ln(Y_i)$ on (X_i)). Use the predicted means to estimate the variances (as the variance of a Poisson distribution equals the mean), these will be needed to derive the covariance matrix and then the beta update matrix. Report the following matrices for the initial step ($\hat{\boldsymbol{\beta}}_{t=0}$), to three decimal places: (a) $\boldsymbol{\eta}$ (linear predictions), (b) variances, (c) information, (d) covariance, (e) residuals, (f) scores, (g) beta updates, and (h) $\hat{\boldsymbol{\beta}}_{t=1}$.
12. Resimulate the previous problem using the actual frequencies (not the predicted ones) for the variances. To three decimal places, report the matrices for (a) information, (b) covariance, (c) beta updates, and (d) $\hat{\boldsymbol{\beta}}_{t=1}$.

Technical Notes

14.1 The Jeffreys CI is based on Bayesian statistics. The Bayesian approach (see Nussbaum, 2011) considers probability estimates as reflecting degrees of beliefs that a proposition is true. Prior beliefs are updated based on Bayes's theorem when new evidence is observed. A prior probability distribution for a variable X reflects prior beliefs of the probabilities that X is equal to different values. As is explained below, the Jeffreys CI assumes a prior beta distribution with parameters (0.5, 0.5). The shape of this distribution (which can be simulated in EXCEL using =BETA.DIST) is relatively flat except at the extremes; however, after P is observed, the "posterior" distribution resembles the typical binomial shape. The posterior beta distribution has parameters $T + \frac{1}{2}, n - T + \frac{1}{2}$, where T is the number of hits. For the example discussed in the text, $P = 58\%$ and $n = 100$; the posterior probability parameters are therefore 58.5 and 42.5. The end points of the 95% CI are found by finding the points with .025 and .975 probabilities. In EXCEL, the end points can be found using the inverse beta command, for example = BETAINV(0.025, 58.5, 42.5) ➔ .482, which is the lower point of the CI (see Table 14.2). For intervals that span values $\geq 95\%$ or $\leq 5\%$, see Brown et al. (2001) for an appropriate adjustment. Brown et al. (2001) also show that the Jeffreys CI is equivalent to making a continuity correction using mid-p values to the exact Clopper-Pearson CI. As noted above, the beta distribution is the prior for the Jeffreys CI. A beta distribution is defined by a probability density function with the following form: $P(q) = \frac{q^{(\alpha-1)}(1-q)^{(\beta-1)}}{B(\alpha, \beta)}$. B is the *beta function*, which gives the entire area under the curve, thus ensuring that all the probabilities sum to 1.0. Note that the numerator of the beta distribution is somewhat similar in form to the formula for the binomial distribution (Eq. 1.10). This feature allows Bayesian updating to produce a binomial distribution as a posterior distribution. The beta and binomial distributions are therefore called *conjugate distributions* and are the reason that the beta distribution is used as the prior.

14.2 When canonical links are used, the observed and expected information matrices are the same (which also makes the Newton-Raphson and Fisher scoring procedures identical). When a canonical link is used, the score function (see Eq. 14.12) is $u_j = \sum_{i=1}^{n}(Y_i - \mu_i) * X_{ij} = \sum(Y_i X_{ij} - \mu_i X_{ij})$. (With a dispersion parameter, the score function is $u_j = \frac{\sum_{i=1}^{n}(Y_i - \mu_i) * X_{ij}}{a(\phi)}$.) We take the derivative to obtain the Hessian: $\frac{\partial u_j}{\partial \beta}$. The term $Y_i X_{ij}$ is not a function of β so can be ignored. However, for the second term, μ_i is a function of beta, and the derivative can be expressed as $\frac{\partial \mu_i}{\partial \beta}$. We therefore have: *Hessian* $= \frac{\partial u_j}{\partial \beta} = -\frac{\partial u_i}{\partial \beta} X_{ij}$, and therefore the *Observed Information* $= \frac{\partial \mu_i}{\partial \beta} X_{ij}$. Because the random variable Y does not appear in this equation, then applying an expectation operator will make no difference: $E\left(\frac{\partial \mu_i}{\partial \beta} X_{ij}\right) = \frac{\partial \mu_i}{\partial \beta} X_{ij}$. For this reason, the observed and expected information matrices will be identical when canonical links are used.

14.3 For logistic regression, it can be shown that the score function is $u_j^{(t)} = \left(\sum_i Y_i - \sum_i n_i \hat{\pi}_i^{(t)}\right) X_{ij}$ and the observed information matrix, which is the negative of the Hessian, is $-(\mathbf{H}^{(t)}) = \left\{ \mathbf{X}' \text{diag}\left[n_i \pi_i^{(t)} \left(1 - \pi_i^{(t)}\right) \right] \mathbf{X} \right\}$.

APPENDIX 14.1

LOGIC BEHIND THE NEWTON-RAPHSON PROCEDURE

In MLE, we set the log-likelihood function (L) equal to zero. The equation may be hard to solve algebraically, so, in the Newton-Raphson procedure, we attempt to simplify the equation using the first few terms of a Taylor expansion. Taylor (1715/1969) showed that any function evaluated at a given point (call it point a) could be rewritten as a polynomial $f(a) + T + T^2 + T^3 \cdots$ with an infinite number of terms. Specifically,

$$\sum_{n=0}^{\infty} \frac{f^{(n)}(a)}{n!}(x-a)^n .$$

(Eq. 14.20)

So, for example, in the case of T^2, $T^2 = \frac{f''(a)}{2!}(x-a)^2$. Now, it is not practical to compute an infinite number of terms, but one can approximate a function by taking the first few terms of the Taylor expansion. For example, an exponential function can be well approximated by the first seven terms. For our purposes, we can try to approximate a log-likelihood function by taking the first two terms of the Taylor expansion. This will not produce a very good approximation but will produce a U-shaped curve similar in shape to the log-likelihood function. Finding the maximum of the approximation will not usually be the maximum of the log-likelihood function, but it will take us in the right direction.

Using this methodology, we can approximate L as:

$$L(\boldsymbol{\beta}) \approx L(\boldsymbol{\beta}_t) + \mathbf{u}_t^{\mathbf{T}}(\boldsymbol{\beta} - \boldsymbol{\beta}_t) + \frac{1}{2}(\boldsymbol{\beta} - \boldsymbol{\beta}_t)^{\mathbf{T}} \mathbf{H}_t(\boldsymbol{\beta} - \boldsymbol{\beta}_t),$$

where $\boldsymbol{\beta}$ reflects the maximum of the simplified function (but not yet the MLEs), $\mathbf{u}_t^{\mathbf{T}}$ is the score matrix (transposed) and $\mathbf{H}_t$ is the matrix of second derivatives (Hessian matrix), divided by 2! Both of these are evaluated at $\boldsymbol{\beta}_t$. The explanation to come is easier to follow using nonmatrix notation, so let us rewrite the equation as

$$L(\beta) \approx L(\beta_t) + u_t(\beta - \beta_t) + \frac{1}{2}(\beta - \beta_t)^2 H_t.$$

(Eq. 14.21)

Now, let us map Eq. 14.20 onto Eq. 14.21, starting with the more complex second order term, $\frac{f''(a)}{2!}(x-a)^2$. The second derivative of the L function is the Hessian, H_t. The matrix is divided by 2 ($= 2!$) in Eq. 14.20, which is the same as multiplying by $\frac{1}{2}$ in Eq. 14.21. Finally, $(x-a)$ in Eq. 14.20 corresponds to $(\beta - \beta_t)$ in Eq. 14.21. See if you can perform the mapping for the first order term, $\mathbf{u}_t^{\mathbf{T}}(\boldsymbol{\beta} - \boldsymbol{\beta}_t)$.

The next step is to find the maximum point of the simplified quadratic function, Eq. 14.21. To do this, we need to take the first derivative of the equation in respect to the β. The function is a sum of these terms, and in calculus, the sum rule states that we can just take the derivative of each term and sum them together. Because β_t is a constant (our estimate on the tth iteration), so are $L(\beta_t)$, u_t, and H_t. So in taking the derivative, the first term, $L(\beta_t)$, disappears because it is a constant. The second term expands to $u_t\beta - u_t\beta_t$, and the derivative is u_t. The derivative of the third term is $(\beta - \beta_t)H_t$. Setting the entire derivative to zero (to find the maximum point) yields $u_t + (\beta - \beta_t)H_t = 0$. Rearranging terms yields $u_t + \beta H_t - \beta_t H_t = 0$. Therefore, $\beta H_t = \beta_t H_t - u_t$ and $\beta = \beta_t - \frac{u_t}{H_t}$. β is the maximum point of the simplified function, though not of the log-likelihood function prior to convergence. However, it becomes our estimate on the next iteration, so we can write

$$\beta_{(t+1)} = \beta_t - \frac{u_t}{H_t}. \tag{Eq. 14.22}$$

Conceptually, Eq. 14.22 asserts that we update our estimates by the score divided by the negative of the Hessian, which is the observed information. In matrix terms, this is

$$\boldsymbol{\beta}_{(t+1)} = \boldsymbol{\beta}_t - \mathbf{H}_t^{-1}\mathbf{u}_t, \tag{Eq. 14.23}$$

which is the Newton-Raphson updating equation.

APPENDIX 14.2

PROOFS OF IMPORTANT RESULTS

Proof of Eq. 14.1 $U = \frac{Y - n\pi}{\pi(1-\pi)}$, Where Y Is the Number of Hits

Start with the definition of the log likelihood of π:

$$L(\pi) = \ln\,[\pi^Y\,(1-\pi)^{n-Y}] = Y\ln(\pi) + (n-Y)\,\ln\,(1-\pi).$$

Then to derive U, take the partial derivative of L in respect to π:

$$U(\pi) = \frac{\partial L(\pi)}{\partial \pi} = \frac{Y}{\pi} - \frac{(n-Y)}{(1-\pi)}.$$

Form common denominators:

$$U(\pi) = \frac{Y(1-\pi)}{\pi(1-\pi)} - \frac{\pi(n-Y)}{\pi(1-\pi)} = \frac{Y - Y\pi - n\pi + Y\pi}{\pi(1-\pi)} = \frac{Y - n\pi}{\pi(1-\pi)}. \text{ QED}$$

Proof of Eq. 14.3 for Binomial Distribution, $-H(\pi) = \frac{Y}{\pi^2} + \frac{n-Y}{(1-\pi)^2}$

This is the negative of the Hessian (i.e., derivative of the score function). From the proof of Eq. 14.1, $U(\pi) = \frac{Y}{\pi} - \frac{(n-Y)}{(1-\pi)} = Y\pi^{-1} - (n-Y)(1-\pi)^{-1}$. $H(\pi) = \frac{\partial U}{\partial \pi} = -Y\pi^{-2} - (n-Y)(1-\pi)^{-2}$. (The sign of the last term is negative because of the chain rule.) Therefore, $-H(\pi) = \frac{Y}{\pi^2} + \frac{n-Y}{(1-\pi)^2}$. QED.

Proof of Eq. 14.6 for Binomial Distribution, $I = \frac{n}{\pi(1-\pi)}$

Start with the definition of the log likelihood of π:

$$L(\pi) = \ln\,[\pi^Y\,(1-\pi)^{n-Y}] = Y\ln(\pi) + (n-Y)\,\ln\,(1-\pi).$$

Then take the expectation (multiplied by -1) of the second partial derivative of L in respect to π: $I = -E\left[\frac{\partial^2 L(\pi)}{\partial \pi^2}\right] = E\left[\frac{Y}{\pi^2} + \frac{(n-Y)}{(1-\pi)^2}\right]$. As explained in Chapter 3, one can move an expectation operator right to the random variable Y. Because Y has a binomial distribution, $E(Y) = n\pi$. Therefore, $I = \frac{n\pi}{\pi^2} + \frac{(n-n\pi)}{(1-\pi)^2}$. Forming common denominators yields $I = \frac{n\pi(1-\pi)^2}{\pi^2(1-\pi)^2} + \frac{(n-n\pi)\pi^2}{(1-\pi)^2\pi^2} = \frac{n\pi(1-\pi)^2+(n-n\pi)\pi^2}{(1-\pi)^2\pi^2}$. Factoring out from the numerator π,

n, and then $(1-\pi)$ produces: $I = \frac{\pi\left(n(1-\pi)^2+(n-n\pi)\pi\right)}{(1-\pi)^2\pi^2} = \frac{n\pi[(1-\pi)^2+[(1-\pi)\pi]}{(1-\pi)^2\pi^2} = \frac{n\pi(1-\pi)[(1-\pi)+\pi)]}{(1-\pi)^2\pi^2}$.

Cancelling terms produces: $I = \frac{n[1-\pi+\pi]}{(1-\pi)\pi} = \frac{n}{\pi(1-\pi)}$, QED.

Proof of Eq.14.11: $u_j = \sum_{i=1}^{n} \frac{(Y_i - \mu_i)x_{ij}}{Var(Y_i)} * \frac{\partial \mu_i}{\partial \eta_i}$.

The random component of a GLM refers to the conditional distribution of the Y variable. Most of the random distributions used (normal, binomial, Poisson, gamma) are from the *natural exponential family*, because the function defining the distribution contains some sort of exponentiation. If the DV has independent observations from the natural exponential family with one parameter θ, its PDF will be of the form

$$l_i = f(Y_i;\theta_i;\phi) = exp\left[\frac{Y_i\theta_i - b(\theta_i)}{a(\phi)} + c(Y_i,\phi)\right],$$

where the letters a, b, and c refer to functions. The l_i is the likelihood for each observation. The symbol ϕ is the dispersion parameter, and when ϕ is known, the equation reduces to

$$f(Y_i;\theta_i) = a(\theta_i)b(Y_i)exp(Y_i[Q(\theta_i)]),$$

where $Q(\theta_i)$ is the *natural parameter*.

This formula is very general, and it may be more meaningful when applied to an actual distribution like the Bernoulli. There, the parameter θ estimated is π. The PDF is $\pi^Y(1-\pi)^{(1-Y)}$, where Y is either 0 or 1. Now it can be shown that the probability mass function is equal to:

$$(1-\pi)exp\left[Y\,ln\left(\frac{\pi}{(1-\pi)}\right)\right].$$

Comparing the last two equations, we see that the very last term, $Q(\theta_i)$, is the log odds, and $a(\theta_i) = (1-\pi)$. Finally, there is no term for $b(Y_i)$; implicitly, $b(Y_i) = 1$.

L_i is the log-likelihood for each observation. Taking the logs of both sides, we obtain

$$L_i = \frac{[Y_i\theta_i - b(\theta_i)]}{a(\phi)} + c(Y_i,\phi). \qquad \text{(Eq. 14.24)}$$

The parameter ϕ is the dispersion factor. Assuming no overdispersion, $\phi = 1$ and so we can ignore the c term of Eq. 14.24.

To find the score for each observation, take the first derivative of Eq. 14.24:

$$u_i = \frac{\partial L_i}{\partial \theta_i} = \frac{Y_i - b'(\theta_i)}{a(\phi)},$$

(Eq. 14.25)

where $b'(\theta_i)$ is the first derivative of $b(.)$ at θ_i. The mean score is zero, as shown in Figure 14.2. Therefore, if we take the expected value of each side of Eq. 14.25, we obtain

$$E(u_i) = E\left[\frac{Y_i - b'(\theta_i)}{a(\phi)}\right] = \frac{E(Y_i) - b'(\theta_i)}{a(\phi)} = 0.$$

Solving for $E(Y_i)$, we find that $E(Y_i)$, $= b'(\theta_i)$ and because $E(Y_i)$ is the mean (μ_i):

$$\mu_i = b'(\theta_i).$$

(Eq. 14.26)

We shall use this result in the main proof to come.

The negative of the second derivative of Eq. 14.24 (i.e., the negative of the first derivative of Eq. 14.25) gives the observed information: $\frac{\partial^2 L_i}{\partial \theta_i^2} = \frac{-b''(\theta_i)}{a(\phi)}$. We take the expected value of both sides to obtain the expected information: $E\left[\frac{\partial^2 L_i}{\partial \theta_i^2}\right] = E\left[\frac{-b''(\theta_i)}{a(\phi)}\right]$. It can be shown that $E\left[\frac{\partial L_i}{\partial \theta_i}\right]^2 = -E\left[\frac{\partial^2 L_i}{\partial \theta_i^2}\right]$. Therefore:

$$E\left[\frac{b''(\theta_i)}{a(\phi)}\right] = E\left[\frac{\partial L_i}{\partial \theta_i}\right]^2.$$

(Eq. 14.27)

Recall from Eq. 14.25 that $\frac{\partial L_i}{\partial \theta_i} = \frac{Y_i - b'(\theta_i)}{a(\phi)}$. Substituting into Eq. 14.27, we find that the formula for the expected information is $E\left[\frac{b''(\theta_i)}{a(\phi)}\right] = E\left[\frac{Y_i - b'(\theta_i)}{a(\phi)}\right]^2$. Because from Eq. 14.26, $b'(\theta_i)$ is the mean, the numerator is simply the variance of Y_i, so we have

$$E\left[\frac{b''(\theta_i)}{a(\phi)}\right] = \frac{Var(Y_i)}{[a(\phi)]^2}.$$

(Eq. 14.28)

We can drop the expectation sign from the left-hand side of the equation because the random variable Y_i does not appear there. Solving for $Var(Y_i)$ yields

$$Var(Y_i) = b''(\theta_i)a(\phi).$$

(Eq. 14.29)

It follows that $b''(\theta_i) = \frac{Var(Y_i)}{a(\phi)}$. Because $b'(\theta_i) = (\mu_i)$ from Eq. 14.26, $b''(\theta_i) = \frac{\partial \mu_i}{\partial \theta_i} = \frac{Var(Y_i)}{a(\phi)}$. Flipping the numerators with denominators, it follows that

$$\frac{\partial \theta_i}{\partial \mu_i} = \frac{a(\phi)}{Var(\phi)}.$$

(Eq. 14.30)

We shall use this result in the main proof to come.

Having derived the mean and variance of the distribution Y_i (and the random component), we now return attention to the log-likelihood function. Eq. 14.24 gave the log-likelihood (L_i) for each observation. The probability of obtaining all the observations is a joint, multiplicative probability of independent events, or a sum on the log scale. So the overall likelihood is the sum of the L_is. There will be a different log likelihood for each parameter β_j, because the log likelihood is the probability (on the log scale) of obtaining your results given a certain parameter value. These log likelihoods will form the vector, $\mathbf{L(\beta)}$. Based on Eq. 14.24, we therefore have $\mathbf{L(\beta)} = \sum_{i=1}^n \frac{[Y_i\theta_i - b(\theta_i)]}{a(\phi)} + \sum_i c(Y_i, \phi)$. We want to maximize the function by taking the first partial derivative and setting it to zero: $\frac{\partial L(\beta)}{\partial \beta_j} = \sum_i \frac{\partial L_i}{\partial \beta_j} = 0$.

According to Eq. 14.24, the L_is for each observation are a function of both Y_i (the observations on the DV) and θ_i (which represent parameters such as π or λ). The subscript i indicates that these parameters can differ for different observations. (This fact is the case because X_{ij} will very well differ for different observations, altering the value of π_i or λ_i.) Now the θ_is are a function of the μ_is, which are the predicted/expected values of Y_i, and the μ_is are in turn a function of η_is (which are the predicted values using the link function) and these are a function of the betas. In summary, we have

$$L_i = f(\theta_i) = f(\mu_i) = f(\eta_i) = f(\beta_j).$$ (Eq. 14.31)

As an example, using the betas and the logistic link function, we might predict for a particular case a logit of 0.405 (η_i), which through exponentiation produces odds of 1.5 (μ_i), which corresponds to a probability of 60% ($\pi_i = \theta_i$). From the estimate of π_i and the observed value, we can predict the likelihood (see Eq. 10.3). This sequence follows the one shown in Eq. 14.31, from right to left.

Using the chain rule from calculus on Eq. 14.31, we have $\frac{\partial L_i}{\partial \beta_j} = \frac{\partial L_i}{\partial \theta_i} \frac{\partial \theta_i}{\partial \mu_i} \frac{\partial \mu_i}{\partial \eta_i} \frac{\partial \eta_i}{\partial \beta_j}$. Recall from Eq. 14.25 that $\frac{\partial L_i}{\partial \theta_i} = \frac{Y_i - b'(\theta_i)}{a(\phi)}$. Therefore, $\frac{\partial L_i}{\partial \beta_j} = \frac{Y_i - b'(\theta_i)}{a(\phi)} \frac{\partial \theta_i}{\partial \mu_i} \frac{\partial \mu_i}{\partial \eta_i} \frac{\partial \eta_i}{\partial \beta_j}$. Since $\mu_i = b'(\theta_i)$ (from Eq. 14.26), we have $\frac{\partial L_i}{\partial \beta_j} = \frac{Y_i - \mu_i}{a(\phi)} \frac{\partial \theta_i}{\partial \mu_i} \frac{\partial \mu_i}{\partial \eta_i} \frac{\partial \eta_i}{\partial \beta_j}$. Because $\frac{\partial \theta_i}{\partial \mu_i} = \frac{a(\phi)}{Var(Y_i)}$ (from Eq.14.30), we have $\frac{\partial L_i}{\partial \beta_j} = \frac{Y_i - \mu_i}{a(\phi)} \frac{a(\phi)}{Var(Y_i)} \frac{\partial \mu_i}{\partial \eta_i} \frac{\partial \eta_i}{\partial \beta_j}$. Because $\eta_i = \beta_j X_{ij}$ and therefore $\frac{\partial \eta_i}{\partial \beta_j} = X_{ij}$, we have $\frac{\partial L_i}{\partial \beta_j} = \frac{Y_i - \mu_i}{a(\phi)} \frac{a(\phi)}{Var(Y_i)} \frac{\partial \mu_i}{\partial \eta_i} X_{ij}$. Rearranging terms and cancelling $a(\phi)$:

$$\frac{\partial L_i}{\partial \beta_j} = \frac{(Y_i - \mu_i)}{Var(Y_i)} \frac{\partial \mu_i}{\partial \eta_i} X_{ij}.$$ (Eq. 14.32)

The above equation produces the score for each observation. To obtain the overall score, we sum the individual scores, yielding

$$u_j = \sum_{i=1}^n \frac{(Y_i - \mu_i)}{Var(Y_i)} \frac{\partial \mu_i}{\partial \eta_i} X_{ij}. \text{ QED.}$$ (Eq. 14.33)

We then set the score equal to zero in order to minimize the score and maximize the likelihood. Further information can be found in Agresti (2002).

Proof of Eq.14.12 Score Equation for Canonical Links

I will demonstrate how the general score equation, Eq. 14.11, reduces to Eq. 14.12 when a canonical link is used. Recall that the two equations were

$$u_j = \sum_{i=1}^{n} \frac{Residual_i}{Var(Y_i)} * \frac{\partial \mu_i}{\partial \eta_i} X_{ij} = 0 \text{ (general)} \qquad \text{(Eq. 14.11)}$$

$$u_j = \sum_{i=1}^{n} Residual_i * X_{ij} = 0 \text{ (with canonical link).} \qquad \text{(Eq. 14.12)}$$

When a canonical link is used, I will demonstrate how $\frac{\partial \mu_i}{\partial \eta_i} = Var(Y_i)$, so that these terms cancel one another out in Eq. 14.11, producing Eq. 14.12. This result can be demonstrated most easily for the case of Poisson regression. There, because the canonical link is the log function, we have: $\eta_i = ln(\mu_i) = X\beta$. To find $\frac{\partial \mu_i}{\partial \eta_i}$, we need to express μ_i as a function of η_i by exponentiating both sides of the equation: $exp(\eta_i) = \mu_i$. Reversing the two sides yields $\mu_i = exp(\eta_i)$. Also, $\frac{\partial \mu_i}{\partial \eta_i} = exp(\eta_i) = e^{\eta_i}$, because e is defined as the point where its derivative is equal to itself. Putting the last two expressions together, it follows that $\frac{\partial \mu_i}{\partial \eta_i} = \mu_i$, and because the variance of a Poisson distribution is equal to the mean (μ_i), the derivative is equal to the variance. The derivative and variance terms cancel one another out in Eq. 14.11, producing Eq. 14.12.

Now consider the case of logistic regression. The canonical link is the logistic link function: $\eta_i = ln\left(\frac{\pi_i}{1-\pi_i}\right) = X\beta$. A key relationship in logistic regression is that, because of the binomial distribution, $E(Y_i) = n_i\pi_i = \mu_i$, so it follows that $\pi_i = \frac{\mu_i}{n_i}$. Replacing π_i in the link function produces: $\eta_i = ln\left(\frac{\frac{\mu_i}{n_i}}{1-\frac{\mu_i}{n_i}}\right) = X\beta$. The previous step is needed because in order to find $\frac{\partial \mu_i}{\partial \eta_i}$, we need to express μ_i as a function of η_i. However to do that, we need to simplify the expression in the parentheses. For the denominator, it follows that $1 - \frac{\mu_i}{n_i} = \frac{n_i}{n_i} - \frac{\mu_i}{n_i} = \frac{n_i-\mu_i}{n_i}$. Dividing the last term into the numerator produces $\frac{\mu_i}{n_i-\mu_i}$. The logarithm is $ln(\mu_i) - ln(n_i - \mu_i)$. So it follows that

$$\eta_i = ln(\mu_i) - ln(n_i - \mu_i). \qquad \text{(Eq. 14.34)}$$

Let us first find $\frac{\partial \eta_i}{\partial \mu_i}$ and then we can invert the result to find $\frac{\partial \mu_i}{\partial \eta_i}$ (which is the term we are trying to eliminate in the score equation 14.11). Focus first on the first term in Eq. 14.34, $ln(\mu_i)$. According to the rules of calculus, the derivative of a natural log is the reciprocal. So $\frac{\partial ln(\mu_i)}{\partial \mu_i} = \frac{1}{\mu_i}$. Likewise, for the second term of Eq. 14.34, $\frac{\partial ln(n_i-\mu_i)}{\partial (n_i-\mu_i)} = \frac{1}{(n_i-\mu_i)}$. However, we are ultimately interested in the derivative in respect to μ_i, not $(n_i - \mu_i)$. This requires application of the chain rule; $\frac{\partial ln(n_i-\mu_i)}{\partial \mu_i} = \frac{1}{(n_i-\mu_i)} * \frac{\partial (n_i-\mu_i)}{\partial \mu_i}$.

The last term is simply -1, reversing the sign of the last term. Putting these results together yields $\frac{\partial \eta_i}{\partial \mu_i} = \frac{1}{\mu_i} + \frac{1}{(n_i - \mu_i)} = \frac{n_i}{\mu_i(n_i - \mu_i)}$. Inverting the result yields $\frac{\partial \mu_i}{\partial \eta_i} = \frac{\mu_i(n_i - \mu_i)}{n_i}$.

Our goal is to show that this derivative term is equal to the variance, $n_i \pi_i (1 - \pi_i)$. To accomplish this goal, we will express the derivative in terms of π_i. Because $\mu_i = n_i \pi_i$, it follows through substitution that $\frac{\partial \mu_i}{\partial \eta_i} = \frac{n_i \pi_i (n_i - n_i \pi_i)}{n_i} = \pi_i(n_i - n_i \pi_i) = n_i \pi_i (1 - \pi_i)$, which is what we set out to prove. The variance and derivative terms in Eq. 14.11 again cancel one another out, producing the simplified formula in 14.12.

Similar results can be demonstrated for the other GLMs; for a more general proof of these results covering all GLMs; see Myers, Montgomery, and Vining (2002, p. 163). Further information can be found in Lawal (2003, p. 35).

Proof of Eq. 14.13. Quasilikelihood Score Equation

To prove: Quasiscore $= \sum_{i=1}^{n} \frac{Residual_i}{v(\mu_i)} * \frac{\partial \mu_i}{\partial \beta_j}$.

Let $v(\mu_i)$ be the variance function relating the variance to the mean. Let K be the quasilikelihood function. Wedderburn (1974) defined K so that the slope in respect to μ_i would equal the residual divided by $v(\mu_i)$:

$$\frac{\partial K(Y_i, \mu_i)}{\partial \mu_i} = \frac{Y_i - \mu_i}{v(\mu_i)}. \qquad \text{(Eq. 14.35)}$$

In MLE, we defined the score as the slope of L in respect to β_j. Eq. 14.35 is similar except that $\frac{\partial K(Y_i, \mu_i)}{\partial \mu_i}$ is the slope of the quasilikelihood function, and is in respect to μ_i, not β_j. (β_j will be introduced into the equations momentarily so that we can estimate the betas.) In Eq. 14.35, we can eliminate the derivative sign by taking the integral (which is differentiation in reverse). K can therefore be formally defined as

$$K(Y_i, \mu_i) = \int^{\mu_i} \frac{Y_i - \mu'_i}{v(\mu'_i)} d\mu'_i + function\ of\ Y_i.$$

To find the quasiscore, we need to find the slope of the function in respect to β_j. Recall that μ_i is a function of the betas; we use the betas to predict the means. By the chain rule in calculus, $\frac{\partial K}{\partial \beta_j} = \frac{\partial K}{\partial \mu_i} * \frac{\partial \mu_i}{\partial \beta_j}$. Using Eq. 14.35 to substitute for $\frac{\partial K}{\partial \mu_i}$ produces $\sum_{i=1}^{n} \frac{Y_i - \mu_i}{v(\mu_i)} * \frac{\partial \mu_i}{\partial \beta_j}$, which completes the proof.

It can be shown that the mean of the quasiscore function is zero (see Wedderburn, 1974, p. 440), just like a score function. One sets the equation equal to zero to find the maximum point of K. Furthermore, Wedderburn (1974, p. 441) showed that the quasilikelihood function (K) reduces to the log-likelihood function if and only if the distribution is from the natural exponential family (e.g., normal, Poisson, binomial, gamma); however, quasilikelihood estimation can be used if we do not know the shape of the distribution but can estimate the variance as some proportional function of the mean.

Proof of Eq. 14.14 GEE Score Equation

Eq. 14.14 is the same as Eq. 14.13 except every individual (or cluster) has its own unique matrix of derivatives, so the immediately preceding proof applies.

Proof of Eq. 14.17 Information Matrix Formula: $I_{j \times j} = X_{j \times n}^T W_{n \times n} X_{n \times j}$

The information matrix **I** has elements $E(-\frac{\partial^2 L(\beta)}{\partial \beta_h \beta_j})$, where β_h and β_j are two different predictors in the matrix. It can be shown that $E\left(\frac{\partial^2 L_i}{\partial \beta_h \beta_j}\right) = -E\left(\frac{\partial L_i}{\partial \beta_h}\right)\left(\frac{\partial L_i}{\partial \beta_j}\right)$. Substituting the term in Eq. 14.32 yields $E\left(\frac{\partial^2 L_i}{\partial \beta_h \beta_j}\right) = -E\left[\frac{(Y_i - \mu_i)X_{ih}}{Var(Y_i)}\frac{\partial \mu_i}{\partial \eta_i}\frac{(Y_i - \mu_i)X_{ij}}{Var(Y_i)}\frac{\partial \mu_i}{\partial \eta_i}\right] = -E\left[\frac{(Y_i - \mu_i)^2 X_{ih}}{Var(Y_i)}\left(\frac{\partial \mu_i}{\partial \eta_i}\right)^2 \frac{X_{ij}}{Var(Y_i)}\right]$.

Because we are taking the expected value, $E(Y - \mu_i)^2$ is just $Var(Y_i)$, which cancels out one of the $Var(Y_i)$ terms in the denominator, producing

$$E\left(\frac{\partial^2 L_i}{\partial \beta_h \beta_j}\right) = \frac{-X_{ih}X_{ij}}{Var(Y_i)}\left(\frac{\partial \mu_i}{\partial \eta_i}\right)^2.$$

Moving the negative sign to the other side of the equation produces $E\left(-\frac{\partial^2 L_i}{\partial \beta_h \beta_j}\right) = \frac{X_{ih}X_{ij}}{Var(Y_i)}\left(\frac{\partial \mu_i}{\partial \eta_i}\right)^2$. The next step is to take the sum of each side. For the left-hand side of the equation, note that $L(\beta) = \sum_{i=1}^n L_i$. Therefore, because $L(\beta)$ expresses a sum, taking the sums of each side produces $E\left(-\frac{\partial^2 L(\beta)}{\partial \beta_h \beta_j}\right) = \sum_{i=1}^n \frac{X_{ih}X_{ij}}{Var(Y_i)}\left(\frac{\partial \mu_i}{\partial \eta_i}\right)^2$.

Let **W** be a diagonal matrix with the following elements on the diagonal: $w_i = \frac{\left(\frac{\partial \mu_i}{\partial \eta_i}\right)^2}{Var(Y_i)}$. It follows that the information matrix **I** is equal to the following: $I = X^T W X$. Note that the terms for $X_{ih}X_{ij}$ have been absorbed into the X matrices. It is important to note that the inverse of the information matrix is the covariance matrix of $\hat{\beta}$, estimated asymptotically. Therefore, $\widehat{cov}(\hat{\beta}) = \hat{I}^{-1} = \left(X^T \widehat{W} X\right)^{-1}$.

References

Agresti, A. (2002). *Categorical data analysis* (2nd ed.). Hoboken, NJ: Wiley-Interscience.
Brown, L. D., Cai, T. T., & DasGupta, A. (2001). Interval estimation for a binomial proportion. *Statistical Science, 16,* 101–133.

Clopper, C. J., & Pearson, E. S. (1934). The use of confidence or fiducial limits illustrated in the case of the binomial. *Biometrika, 26,* 404–413.

Fisher, R. A. (1925). Theory of statistical estimation. *Proceedings of Cambridge Philosophical Society, 22,* 700–725.

Jennrich, R. I., & Sampson, P. F. (1976). Newton-Raphson and related algorithms for maximum likelihood variance component estimation. *Technometrics, 18,* 11–17.

Knight, K. (2000). *Mathematical statistics.* Boca Raton, FL: Chapman & Hall/CRC.

Lawal, B. (2003). *Categorical data analysis with SAS® and SPSS applications.* Mahwah, NJ: Erlbaum.

Myers, R. H., Montgomery, D. C., & Vining, G. G. (2002). *Generalized linear models with applications in engineering and the sciences.* New York, NY: John Wiley & Sons.

Nussbaum, E. M. (2011). Argumentation, dialogue theory, and probability modeling: Alternative frameworks for argumentation research in education. *Educational Psychologist, 46,* 84–106. doi:10.1080/00461520.2011.558816

Rao, C. R. (1948). Large sample tests of statistical hypotheses concerning several parameters with applications to problems of estimation. *Proceedings of Cambridge Philosophical Society, 44,* 50–57.

Taylor, B. (1969). Direct and reverse methods of incrimination. (pp. 21–23, Proposition VII, Theorem 3, Corollary 2). In *A source book in mathematics 1200–1800* (pp. 329–332; D. J. Struik, Trans.). Cambridge, MA: Harvard University Press. (Original work published 1715.)

Wedderburn, R. W. M. (1974). Quasi-likelihood functions, generalized linear models, and the Gauss-Newton method. *Biometrika, 61,* 439–447.

Wilson, E. B. (1927). Probable inference, the law of succession, and statistical inference. *Journal of the American Statistical Association, 22,* 209–212.

CHAPTER 15

CHOOSING THE BEST STATISTICAL TECHNIQUE

This chapter summarizes the major considerations that should be taken into account when choosing the best statistical technique. I have presented in this book a variety of statistical techniques related to categorical and nonparametric data analysis, techniques that can be used as alternatives to traditional parametric technique (e.g., ANOVA, ordinary least squares regression). In choosing the best statistical technique for a given situation, one needs to consider two major questions: (a) Is use of the technique valid in the situation; that is, are all assumptions met? and (b) Is the technique more statistically powerful than other ones? In my opinion, the first consideration is a bit more important than the second, because it makes little sense to maximize power if the resulting estimates of p values are biased or inconsistent. (I grant that a researcher may sometimes find it permissible to allow a minor violation of assumptions and a small amount of bias for the sake of power, but the burden of proof should be on the researcher to justify such a trade-off.) It should be noted that violation of assumptions can increase Type II errors as well as Type I errors, so considerations of validity and power are not always contrary.

Table 15.1 is a test matrix organized around levels of measurement as a tool for helping to choose the best statistical technique. The matrix should only be considered a starting point in the decision process, as use of the matrix may identify more than one technique that could potentially be used. Furthermore, when the dependent variable (DV) has excessive skew or kurtosis, reducing the level of measurement can sometimes increase statistical power, but doing so expands the number of alternatives to be considered. In using the matrix and evaluating alternative techniques, one should next consider sample size, whether independent or related samples are used, and whether one desires to control for "third variables." These considerations will help eliminate some of the potential alternatives. Third, one needs to consider the statistical power of different techniques. One can search for asymptotic relative efficiency studies (or consult Table 6.3) or use a rule of thumb

Table 15.1 Matrix of Nonparametric and Categorical Procedures and Tests for All Levels of Measurement

| | X variable(s) | | |
	Nominal	Ordinal	Metric
Metric Y variable	(A) Two matched samples: Wilcoxon signed ranks Interactions: ART ANOVA/Regression with dummy variables; Poisson regression for count data; (repeated measures or use GEE w/matched samples)	(D) Reduce X to nominal or Y to ordinal	(G) Correlation and OLS Poisson regression (count data) (use GEE with matched samples)
Ordinal	(B) Two samples: Wilcoxon-Mann-Whitney Multiple samples: Kruskal-Wallis, Median Two matched samples: Sign test Multiple matched samples: Friedman test Interactions/confounding variables: ANOVA-type statistics (ATS) Moses extreme reactions Ordered categories on Y: Ordinal regression (use GEE with matched samples)	(E) Spearman's $\hat{\rho}$ Many ties/ordered categories: Kendall's $\hat{\tau}$, Somer's d (directional), Ordinal regression (3rd variables). Ranks (Y) and ordered categories (X): Jonckheere-Terpstra test (Page test with matched samples)	(H) Ordered categories on Y: Ordinal regression (use GEE with matched samples)
Nominal	(C) Chi-square Small sample: Fisher's exact test Matched samples: McNemar (2×2; Cochran ($r \times c$)) Confounding variables: Mantel-Haenszel ($2 \times 2 \times k$) or Log linear (use GEE with matched samples)	(F) Kendall's $\hat{\tau}$ ($2 \times k$)	(I) Logistic regression (use GEE with matched samples)

Note: If one is analyzing an ordered contingency table, see Chapter 12 for more guidance.

(e.g., use a nonparametric or categorical technique if the skew or kurtosis of the DV is greater than 1.0). If one chooses to use more than one technique to analyze the data (and doing so does help one to learn the alternative techniques), one needs to decide, preferably in advance, which analysis will be binding in the event that a result is significant in one analysis but not another. The decision needs to be supported by a clear rationale that can be presented in a journal article or whatever medium will be used to present the research.

Finally, one needs to consider whether all the assumptions of a given technique are satisfied, over and above those mentioned above. Oftentimes one will not be able to make this determination until one has begun using the technique, but the assumptions one needs to check should be noted in advance. The main assumption in many nonparametric techniques is that the population distributions are identically shaped, but, as argued in Chapter 6, this assumption can be relaxed if one uses a more robust null hypothesis (of identically shaped distributions) and supplements the analysis with some measure of effect size. For some techniques, such as the aligned rank transform, the likely shape of the underlying DV population distribution also needs to be considered (in conjunction with the results of computer simulation studies). In categorical data analysis, one needs to check specification assumptions (e.g., Are all relevant predictors included? Is the most appropriate link function used?) and distributional assumptions (e.g., Is the distribution of the DV modeled correctly? Is overdispersion an issue?). Choosing an appropriate technique is only part of what needs to be done; the researcher also needs to develop an appropriate statistical model. That requires both labor and thought. I presented in Chapter 10 some steps for developing such models. It is good practice to consider and evaluate two or three competing statistical models.

In checking assumptions, researchers often use statistical hypothesis tests—for example, testing a null hypothesis that curves are parallel (in ordinal regression) or that the differences between observed and expected values are zero (as in the Hosmer-Lemeshow test for logistic regression). Such statistical tests are themselves subject to Type I and II errors. For example, in a small sample, one might not have enough statistical power to detect a major violation and in a large sample, one might have enough power to detect even minor violations. It is recommended that such tests be considered as only one piece of evidence in deciding whether or not to take remedial action. These diagnostic tests should be used along with other approaches, such as graphical inspection. Doing so, however, introduces a greater subjective element into the process. Given these issues regarding assumption checking, some authors (e.g., Garcia-Peréz, 2012) recommend instead using robust statistics, which by definition make few or no assumptions. This point is debatable, because robust statistical techniques are not always the most statistically powerful ones, and assumptions can often be reasonably checked (in my opinion), but robustness is one desirable criterion that should weigh into one's ultimate decision.

For any technique used, one should always check for outliers. As explained in Chapter 6, outliers can produce Type I and II errors even when nonparametric statistics are used. Outliers can result in a violation of assumptions as well as in a reduction in statistical power.

Summary

In closing, I note that this textbook is one of only a few to cover both nonparametric and categorical data analysis. I argue that knowledge of both is essential in choosing and using the best statistical approach. Sometimes one may need to choose between a nonparametric and categorical technique. Nonparametric techniques are generally simpler than categorical techniques but are also less flexible—for example, categorical techniques more easily accommodate "third variables."

Psychologists who study human reasoning differentiate between "routine problems" that can be solved algorithmically, through an application of a series of steps, and "nonroutine problems," which require thought and the evaluation of alternatives. Choosing the best statistical technique is sometimes a routine problem, when only one technique is clearly appropriate, but is often a nonroutine problem when two or more techniques appear potentially useful. Ultimately, one will need to make an argument for why one should be used over another. Abelson (1995) characterized the task of using statistics as making principled arguments in contrast to the stereotype of statistics as merely the application of algorithms. Of course, one can try to avoid the problem of choosing a technique entirely by just sticking to traditional parametric statistics. Doing so, however, is not always appropriate for one's data and can often sacrifice statistical power. I therefore urge researchers to consider a greater number of techniques and to choose one or two alternative techniques to master at a time. One learns these techniques by using them and reflecting on their use (e.g., by rereading the corresponding chapter of this text and applying the ideas to one's own data). I hope you will find them as useful as I have in the course of conducting research.

References

Abelson, R. P. (1995). *Statistics as principled argument*. Hillsdale, NJ: Erlbaum.
Garcia-Peréz, M. A. (2012). Statistical conclusion validity: Common threats and simple remedies. *Frontiers in Psychology, 3*, 325. doi:10.3389/fpsyg.2012.00325

ANSWERS TO ODD-NUMBERED PROBLEMS

Chapter 1: Levels of Measurement, Probability, and the Binomial Formula

1. a) metric (ratio)
    b) nominal
    c) metric (interval)

3. a) The probability of each permutation is $1/6 * 1/6 = 2.8\%$. There are six mutually exclusive permutations, $.0218 * 6 = 17\%$.
    b) 11% ($.028 * 4$ permutations).

5. It is 1 minus the probability of obtaining no tails (all heads). The probability of obtaining three heads out of three is the joint probability of three independent events, so the multiplication rule yields $\frac{1}{2} * \frac{1}{2} * \frac{1}{2} = \frac{1}{8}$. Subtracting from 1 yields $\frac{7}{8}$ or 88%.

7. The probability that the first card will be clubs is 13/52. If the first card is a club, the probability that the second card will be a club is 12/51, because the first card is not replaced. The probability that the third card will be clubs is 11/50. Multiplying the three probabilities together yields 0.13%.

9. $\binom{4}{3} = \frac{4!}{3!(4-3)!} = \frac{4 * 3 * 2 * 1}{3 * 2 * 1(*1)} = 4$.

11. $n = 4$, $k = 2$, $\pi = 0.5$, and the probability is $\binom{4}{2} * 0.5^2 (0.5)^2 = 6 * 0.5^4 = 37.5\%$.

13. In EXCEL, =BINOMDIST(2, 4, 0.75, 0) → 0.21.

15. Find $1 - \text{Prob}(k < 2) = \text{Prob}(k \le 1)$. In EXCEL, $=1 - \text{BINOMDIST}(1,4,0.75,1)$ → 95%.

17. a) In EXCEL, $=1 - \text{BINOMDIST}(29,100,0.25,1) = 15\%$.
    b) 18%. The mean is $n\pi = 100 * 25\% = 25$. $SD = \sqrt{n\pi(1-\pi)} = \sqrt{100(25\%)(75\%)} = 4.33$. $z = \frac{X - \bar{X}}{SD} = \frac{29 - 25}{4.33} = 0.92$. $\text{Prob}(z \ge 0.92) = 18\%$. [In EXCEL, $= 1 - \text{NORMSDIST}(0.92)$ → 18%.]
    c) 15%: $z = \frac{X - \bar{X}}{SD} = \frac{29.5 - 25}{4.33} = 1.04$ $\text{Prob}(z \ge 1.04) = 15\%$. [In EXCEL, $= 1 - \text{NORMSDIST}(1.04)$ → 15%.]

19. .00031%. The mean is $n\pi = 500 * .40 = 200$. $SD = \sqrt{n\pi(1-\pi)} = \sqrt{500(.40)(.60)} = 10.95$. For a continuity correction (and because we will be subtracting the

probability from one), set $X = 249.5$. $z = \frac{X - \bar{X}}{SD} = \frac{249.5 - 200}{10.95} = 4.53$. $\text{Prob}(z \geq 4.53) = .00031\%$. [In EXCEL, $= 1 - \text{NORMSDIST}(4.53) \rightarrow .00031\%$.]

Chapter 2: Estimation and Hypothesis Testing

1. Yes. $n = 20$; $X = 17$. For one-tailed test, H_0: $\pi = .60$; H_a: $\pi > .60$. Under null hypothesis, $\text{Prob}(X \geq 17) = 1 - \text{Prob}(X \leq 16) = .016$. [In EXCEL, $= \text{BINOMDIST}(16, 20, .60, 1) \rightarrow .016$.] $p = .016$ (one-tailed). For two-tailed test, probability of the other tail is based on $X \leq 7$ (because the deviation from the mean ($60\% * 20 = 12$) is 5 units in absolute value, as is the case with the upper tail). Under null hypothesis, $\text{Prob}(X \leq 7)$ is .021. The two-sided p value is therefore $.016 + .021 = .037$.

3. $X = 2,718$ ($2,718.5$ with continuity correction), $P = \frac{2718.5}{6,405} = .424$. For one-tailed test, H_0: $\pi = .454$; H_a: $\pi < .454$. $SE = \sqrt{\frac{.454 * (1 - .454)}{6,405}} = .006$. $z = \frac{.424 - .454}{.006} = -4.82$. $\text{Prob}(z \leq 4.82) = 7.10 * 10^{-7} = .000$. So one-tailed $p < .001$. Because the normal distribution is symmetrical, double the one-tailed value, yielding $\simeq 1.42 * 10^{-6}$, $p < .001$.

5. a) It is significantly below the statewide average. $n = 890$; $X = 30$ (30.5 with continuity correction), $P = \frac{30.5}{890} = .034$. For one-tailed test, H_0: $\pi = .094$; H_a: $\pi < .094$. $SE = \sqrt{\frac{.094 * (1 - .094)}{890}} = .01$. $z = \frac{.034 - .094}{.01} = -6.0$. $\text{Prob}(z \leq -6.0) = 9.87 * 10^{-10} = .000$. So one-tailed $p < .001$. Because the normal distribution is symmetrical, double the one-tailed value, yielding $\simeq 1.97 * 10^{-9}$, $p < .001$.

   b) It is significantly above the statewide average. $n = 890$; $X = 148$ (147.5 with continuity correction), $P = \frac{147.5}{1,197} = .123$. $SE = .01$, $z = \frac{.123 - .094}{.01} = 3.13$. $\text{Prob}(z \geq 3.13) = .001$ (one-sided), $.002$ (two-sided). Therefore, $p < .01$ (Note: using rounded figures, $z = 2.9$.).

7. There is a significant upward trend. H_0: $\text{Prob}(+) = .50$; H_a: $\text{Prob}(+) > .50$. Pair each observation with the one two years later (e.g., January, 2007 with January 2009). All pairs show an increase (+). Therefore, $T = 24$, $n = 24$, $p \simeq 5.96 * 10^{-8}$ (one-tailed binomial test), $p < .001$.

9. There is a cyclic pattern. First, list figures for the presidential election years followed by the figures for the nonpresidential election years. Second, test for a linear downward trend. Splitting the data in half, each presidential election year becomes paired with the following nonpresidential election year. There are a total of 13 pairs, all are decreases. So $T = 13$, $n = 13$. Under the null hypothesis, the probability of a decrease (π) $= .50$. $\text{Prob}(T = 13) = 1 - \text{Prob}(T \leq 12) = .0001$. [In EXCEL, $= 1 - \text{BINOMDIST}(12,13,.50,1) \rightarrow .0001$. Equivalently, $= \text{BINOMDIST}(13,13,.50,0) = .0001$.] Because the binomial distribution is symmetrical when $\pi = 50\%$, the two-tailed p value is .0002.

Chapter 3: Random Variables and Probability Distributions

1. $48,710 (or $48,711, depending on rounding)

Ethnicity	Income (in $)	No. Households	Proportional Weight	Income * Weight
White	51,861	95,489	.74487	38,630
Black	32,584	14,730	.11490	3,744
Asian/PI	65,469	4,687	.03656	2,393
Hispanic	38,039	13,289	.10366	3,943
Total		128,195		48,710

3. a) $3 + 7E(X)$
   b) $49\,Var(X)$
   c) $49\,Var(X)\ [= (-7)^2(X)]$
   d) $3 + 7E(X) + 2E(Y)$
   e) $49\,Var(X) + 4\,Var(Y)$
   f) $49\,Var(X) + 4\,Var(Y) + 2$

5. Consider the following random variable X:

X	Prob(X)	X * Prob(X)	$(X-2.55)^2$	$(X-2.55)^2$ * Prob(X)
0	.19	0.00	6.50	1.24
1	.08	0.08	2.40	0.19
2	.23	0.46	0.30	0.07
3	.08	0.24	0.20	0.02
4	.33	1.34	2.10	0.69
5	.09	0.45	6.00	0.54
Sum		2.55		2.75

(excludes rounding error)
Total 1.00

a) $\bar{X} = 2.55$, $Var(X) = 2.75$.
b) $E(2X) = 2E(X) = 2 * 2.55 = 5.10$. $Var(2X) = 4 * Var(X) = 4 * 2.75 = 11$.
c)

Y	Prob(Y)	Y * Prob(Y)	$(Y - 2.42)^2$	$(Y - 2.42)^2$ * Prob(Y)
0	.12	0.00	5.86	0.70
1	.17	0.17	2.02	0.34
2	.20	0.40	0.18	0.04
3	.25	0.75	0.34	0.08
4	.20	0.80	2.50	0.50
5	.06	0.30	6.66	0.40
Sum	1.00	2.42	17.56	2.06

$E(Z) = E(X) + E(Y) = 2.55 + 2.42 = 4.97$.
$Var(Z) = Var(X) + Var(Y) = 2.75 + 2.06 = 4.81$

7. a) 20%. (Expected value is π.)
    b) Variance: $\frac{\pi}{80} = 0.0025$. SE: $\sqrt{.0025} = .05$.
    c) Mean: $n\pi = 80 * .20 = 16$. SE: $\sqrt{n\pi(1-\pi)} = 3.58$.
    d) 25 hits. [In EXCEL, $=$ NORM.S.INV(.995) → 2.58 (upper critical value). $16 + 2.58 * 3.58 = 25.2$.]

9. a) z distribution. Due to the CLT, the test statistic will be normally distributed in a large sample. Dividing by the standard error standardizes the statistic, producing a z distribution. Because the population variance is a function of the population mean, we do not need to use the t distribution.
    b) Chi-square distribution. (The difference in proportions test is obsolete, because a two-sample chi-square test can be used instead, but the former is sometimes used for pairwise comparisons.)

11. a) Yes (KS $z = 1.31$, $p = .064$). The null hypothesis is not rejected, so we conclude that the data fit an exponential function.
    b) The data roughly follow an exponential function.

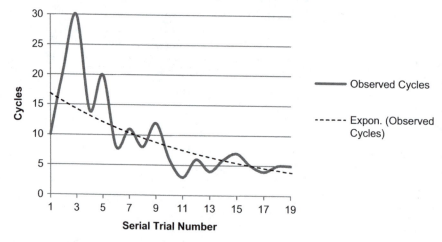

13. KS $z = 1.30$, $p = .068$, do not reject the null hypothesis, conclude that the data fit a normal distribution.

15. a) $\chi^2 (9) = 12.8$, $p = .172$. Do not reject the null hypothesis that the distribution is uniform.
    b) =CHIDIST(12.8,9) → .172.
    c) The KS test is statistically more powerful, meaning that it is more able to reject the null hypothesis, as was the case with these data. The p value was lower with the KS test.

Chapter 4: Contingency Tables: The Chi-Square Test and Associated Effect Sizes

1.

	Observed			Expected*		Deviations		$\frac{(Deviation)^2}{Expected}$	
	Japanese	US	Total	Japanese	US	Japanese	US	Japanese	US
Yes	15	5	20	8.2	11.8	6.8	−6.8	5.6	3.9
No	10	31	41	16.8	24.2	−6.8	6.8	2.8	1.9
Total	25		36	61	*Per Eq. 4.3	$\Sigma = 14.2$			

a) 60.0% $(= \frac{15}{25})$; 13.9%.

b) See table above: $\chi^2 (1) = 14.2$, $p < .001 (= .00016)$.

c) OR $= \frac{15/10}{5/31} = 9.3$(or 0.11 inverted). $\phi = \sqrt{\frac{14.2}{61}} = 0.48.$

d) $\Delta P = 46.1\%$. It is appropriate: Nationality is a fixed variable that is predetermined and used to predict a teacher's goal. The DV does not affect the IV.

e) $\chi^2(1) = 12.2$, $p < .001$. (Note: $p = .0005$, which $> .00016$; continuity corrections always increase p values. If there were a decrease, this would signify an error and I would have added one unit to first cell instead.)

f) SPSS returns the same values, except $\phi = -0.48$. The negative sign is an artifact of coding. Data file should look something like this (be sure to weight by frequency):

Country	Response	Frequency
Japanese	Yes	15
US	Yes	5
Japanese	No	10
US	No	31

3. a) $\chi^2 (9) = 2{,}751.56$, $p < .001$. $(df = 3 \times 3 = 9)$.

b) $V = \sqrt{\frac{2751.56}{17{,}733(4-1)}} = .23.$ $V^2 = 5\%$, which is the ratio of observed to maximum chi-square.

c)

	Father's 1880 occupation			
Son's 1910 occupation:	White collar	Farmer	Skilled/semiskilled	Unskilled
White collar	8.7%*	9.1%	6.8%	3.0%
Farmer	3.1%	19.0%	2.0%	2.3%
Skilled/semiskilled	5.1%	8.4%	9.8%	4.8%
Unskilled	2.8%	8.1%	3.5%	3.4%

*8.7% $= \frac{1{,}538}{17{,}773}$.

The farmer-farmer cell has the highest probability (19%), and the farmer column has the highest unconditional probabilities. Many male children of farmers were moving into other occupations, reflecting a growing urbanization of America.

d)

	White collar	Farmer	Skilled/semiskilled	Unskilled
Conditional Prob.	44.0%	42.6%	44.2%	25.4%
Cell Frequency	1,538%	3,371	1,736	611
Sum (Not Unskilled)		6,645		
n		15,326		2,407
Conditional Prob.		6,645/15,326 = 43.4%		25.4%

	Father's occupation		
Son same occup.:	Not unskilled	Unskilled	Total
Yes	6,645	611	7,256
No	8681	1,796	10,477
Total	15,326	2,407	17,733

The percentage of unskilled workers in 1910 with unskilled fathers was $\frac{611}{2,407} = 25.4\%$; the value is significantly lower than the percentage for other workers (43.4%): $\chi^2(1) = 277$ (with continuity correction), $p < .001$, $\Delta P = 18\%$, OR $= 2.25$ (or 0.44 inverted). This may reflect social mobility.

e) Father's occupation is a fixed variable, because in principle one could know the father's occupation before knowing the son's and it can be used to predict the latter. Also, son's 1910 occupation could not have a causal effect on the father's 1880 occupation. Therefore, it is appropriate to report ΔP conditionalized on father's occupation.

f) The odds that the son has the same occupation as the father increase by 125% ($2.25 - 1 = 125\%$) if the father is not unskilled.

Chapter 5: Contingency Tables: Special Situations

1. a) No, $p = .056$.
   b) Yes, $p = .037$. [In EXCEL,=HYPGEOM.DIST(sample_successes, number_sample, population_successes, population_number, cumulative) = HYPGEOM.DIST(14,20,17,30,0) = .039; $\frac{.039}{2} = .019$; $.056 - .019 = .037$.]

3. a) Yes, $(b + c) = (22 + 9) = 31$, is greater than 25.
   b) $T = \frac{(b-c)^2}{b+c} = \frac{(22-9)^2}{22+9} = \frac{13^2}{31} = \frac{169}{31} = 5.45$, $\chi^2(1) = 5.45$, $p = .02$.

c) $n = 77$, $b\% = 29\%$, $c\% = 12\%$, $29\% - 12\% = 17\%$. Course is effective, a net of 17% more students became proficient after taking course. ($n = 77$, $b\% = 29\%$, $c\% = 12\%$, $29\% - 12\% = 17\%$.)

d) $p = .03$. (The figures may differ slightly from those done by hand due to application of a continuity correction.)

5. a) Breslow-Day (BD) χ^2 (1)= .03, $p = .85$; MH test (χ^2 (1) $= 47.34$, $p < .001$, common OR 2.26.

b) $OR_{boys} = \frac{(111*1499)}{(718*104)} = 2.23. OR_{girls} = \frac{(46*1571)}{(736*42)} = 2.34.$

$$OR_{marginal} = \frac{(157*3070)}{(1454*146)} = 2.27.$$

c) The partial ORs are not significantly different according to the BD test, so one can interpret MH test. Controlling for gender, the odds of having a disability are 2.24 times greater if the father has had 12 or fewer years of education (or having 12 or fewer years of education increases the odds by 124% relative to having more education.)

7. a) BD χ^2 (1) $= 29.14$, $p < .001$. MH χ^2 (1) $= 5.29$, $p < .021$. Common OR $= 1.32$. The partial ORs are significantly different (BD test significant), so the partial tables must be analyzed with separate chi-square tests for statistical independence (χ^2_{boys} (1) $= 0.72$, $p = .396$, χ^2_{girls} (1) $= 34.75$, $p < .001$).

b) $OR_{boys} = 0.88$, $OR_{girls} = 3.53$, $OR_{marginal} = 1.47$.

c) Low birth weight only predicts RD in girls; low weight increases the odds by 253%.

9. a)

			Professor			
			claims	counter	null	rebuttal
GA	claims	Count	39	8	1	12
		% of Total	26.0%	5.3%	0.7%	8.0%
	counter	Count	3	22	0	0
		% of Total	2.0%	14.7%	0%	0%
	null	Count	8	5	18	0
		% of Total	5.3%	3.3%	12.0%	0%
	rebuttal	Count	20	0	1	13
		% of Total	13.3%	0%	0.7%	8.7%

Summing the diagonal, the agreement rate is 61.4%. The agreement rate is not quite satisfactory (70% is the usual minimum standard). The percentage disagreement rate is highest for claims/rebuttals, so more training and practice should be provided to the raters in how to recognize rebuttals.

c) $\kappa = 45.5\%$. According to guidelines by Landis and Koch (1977), this value reflects "moderate" (i.e., mediocre) agreement (60% is considered good), which is consistent with the interpretation for the unadjusted agreement rate. (The Landis & Koch guidelines have been criticized because κ is affected by the number of categories—it tends to be smaller with more categories—but here the number of categories is relatively small.

Chapter 6: Basic Nonparametric Tests for Ordinal Data

1. Skew $= 0$; kurtosis $= -1.15$ (reflecting heavier tails and less peakedness).

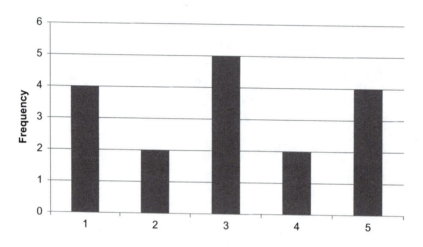

3. a) $\hat{\rho} = .73, p = .017$
   b) $\hat{\tau}_b = .63, p = .012$. Association is significant.
5. a) $\hat{\tau}$ is more appropriate for ordered categories, where there are a lot of ties.
   b) The table is not square.
   c) $\hat{\tau}_c = .26, p = .026$. (d) Somers' $d = .35, p = .026$. (Given the theory, the DV is attendance.)
7. $z = 2.48$ (or $\hat{\tau}_b = .264$), $p = .013$.
9. a) Sums: 62.5, 28.5. Averages: 8.9, 4.8.
   b) $W = 28.5$.
   c) $n = 6, m = 7$.
   d) 28.
   e) 56.
   f) W is not significant because it is not less than 28 or greater than 56.
   g) $p = .051$. The p value for the normal approximation should not be used because neither n nor $m \geq 10$.
   h) 79, 68.5. ΔHL $= 9$, CI: 0 to 16.

i) The difference between medians is $79 - 68.5 = 10.5$, which is slightly larger than ΔHL.

j) $PS_A = 82.1\%$ (based on a U of 34.5 or 7.5). The effect may be large, but the practical significance should not really be interpreted without statistical significance. [Calculation: $\frac{34.5}{42} = 82.1\%$. For $U = 7.5$, $\frac{7.5}{7*6} = 17.9\%$; but because this is below 50%, it is the statistic for the "inferior" group with the lower mean rank (PS_B); $1 - 17.9 = 82.1$.].

11. a) Sum of Ranks: 162 (CT), 91 (RC). Average Ranks: 14.7 (CT), 8.3 (RC). $W = 91$ (using group with lowest sum of ranks).

b) $E(W) = \frac{n(N+1)}{2} = \frac{11(22+1)}{2} = 126.5$.

c) $Var(W) = \frac{n(N+1)m}{12} = \frac{11(22+1)11}{12} = 231.9$.

d) $z = \frac{91-126.5}{\sqrt{231.9}} = -2.33$, $p = .02$.

e) Asymptotic $p = .02$. (Note: In newer versions of SPSS, W is based on the larger sum of rank and therefore the sign of z is sometimes reversed, but this will not affect the p value.) $PS_{CT} = 79.3\%$ (strong effect), based on $U = 25$ or 96.

f) Mean: 14.7 (CT), 8.2 (RC), Median: 17 (CT), 9 (RC), Var.: 46.8 (CT), 18.8 (RC), Range: 19 (CT), 13 (RC), Skew: -0.92 (CT), -0.46 (RC), Kurtosis: -0.71 (CT), -0.90 (RC). There appears to be more dispersion and skew with cognitive therapy. The assumption that the population distributions are identical is violated. One cannot therefore conclude that the population medians are different.

g) PS, because ΔHL assumes equality of variances.

h) $PS_{CT} = 79.3\%$ (strong effect), based on $U = 25$ or 96. One can therefore with high confidence conclude that CT tends to produce higher scores than RT.

i) The p value for the median test is .09. Note that the WMW test produces a lower p value.

13. a) 36 (exp.), 43 (control).

b) 37.

c) 35.

e) $p = .023$ (untrimmed), $p = .14$ (trimmed).

15. a) The value of 2 is an outlier. For that group, Mdn $= 15.5$, MAD $= 2.5$, ratio $\frac{|2-15.5|}{2.5} = 5.4$, which is greater than 5.0, indicating an outlier.

b) $W = 57$, $p < .001$.

c) The distributions do not look similar. The second group is more dispersed ($SD = 5.3$ vs. 3.5) and has more negative skew (-1.6 vs. -0.05) and kurtosis (-1.2 vs. 0.8). Given the significant W, one can conclude that there are significant distributional differences but should focus interpretation of the results on the probability of superiority.

d) $U = 2$, so $PS_{exp} = \frac{2}{10*9} = .02$. $PS_{control} = .98$. The control group is much more likely to have higher values (i.e., lower ranks.)

Chapter 7: Nonparametric Tests for Multiple or Related Samples

1. a) $H = 21.14$, $df = 3$, $p < .001$.
   b) Canadian females and Indian males. (If using the WMW test, the mean rank of Indian females are all significantly different from the other groups, but Canadian males are not quite significant.)
   c) Examining the means suggests that locational differences are involved in all the significant comparisons, but that for comparisons involving Group 4 (Indian females) there are also differences in the variances (i.e., scale), skew, and kurtosis. The latter two are caused by the one individual scoring a "1," and the mean differences would be even higher if this female was removed. However, because there are likely differences in both location and scale, it would be safest and easiest to interpret the differences in terms of the probability of superiority (PS).
   d) For the comparisons of Group 4 to both 2 and 3, $U = 17$, so $PS = \frac{17}{10*10} = 17\%$. There is only a 17% change that an Indian female would be more coercive than a Canadian female or Indian male. Compared to Group 1 (Canadian males), the chance is almost zero.
3. a) Tests based on tau ($\hat{\tau}$) assume that both variables are ordinal. In this problem, the DV is ordinal but the IV is not. The Kruskal-Wallis test can be used when the DV reflects ordered categories and the IV is nominal.
   b) $H = 7.62$, 2 df, $p = .02$
   c) =CHISQ.DIST.RT(7.62,2) ➔ .022.
   d) There are two homogenous subsets: Programs B/C and Programs C/A. Therefore, Programs B and A are statistically different from one another. For WMW, the Bonferroni-adjusted $\alpha = \frac{.05}{3} = .017$. The comparison between Programs A and B is significant ($p = .006$), other comparisons are not (A/C $p = .254$; B/C $p = .168$).
   e) For A/B, PS $= 57\%$., based on $U = 4740.5$. $\left(\frac{4740.5}{102*108} = 43.1\%, 1 - 43\% = 56.9\% \right)$
5. a) H_0: $\pi = 50\%$; H_a: $\pi \neq 50\%$. $T = 5$ (5 pluses). $p = .11$ (one-sided), $p = .22$ (two-sided). [In EXCEL, $=1 - $ BINOM.DIST(4, 6,.5,1) = .11. Double this value for a two-sided value because the sampling distribution is symmetrical when $\pi = 50\%$.]
   b) In SPSS, $p = .22$.
7. Disregard ties, so $n = 27$, $T = 20$. $z = \frac{2T - n - 1}{\sqrt{n}} = \frac{40 - 27 - 1}{5.20} = 2.31$, p-value (2 sided) $= .021$. [In EXCEL, $=1 - $ NORM.S.DIST(2.31,1) = .01.]

9. a) Sum of signed ranks $= 19$ (see table below), and $T+ = 20$. From Table 7.8, 20 corresponds to $w_{.975} = .025$, so $p = .025$ (one-tailed), $p = .05$ (two-tailed). (Note: Because the distribution shown in Table 7.8 is symmetrical, one can double the p value.) Only the one-tailed test is significant; as noted in the table, $T+$ needs to exceed the critical value of 20 to be significant.

City	Before	After	D_i	ABS(D_i)	Os	Rank	Remove rank (R_i^s)	Signed (R_i^s)^2
1	65	74	9	9	9	6	6	36
2	86	90	4	4	4	2	2	4
3	43	49	6	6	6	4	4	16
4	40	37	-3	3	3	1	-1	1
5	11	16	5	5	5	3	3	9
6	41	48	7	7	7	5	5	25
							$\Sigma = 19$	$\Sigma = 91$

The p values are much smaller than with those for a simple sign test.

b) $T = 19$, $z = \frac{19}{\sqrt{91}} = 1.99$, $p = .023$ (one-tailed), $p = .046$ (two-tailed). However, the asymptotic test is not valid because $n < 10$.

c) $\bar{D}_i = 4.67$, Mdn(D_i) $= 5.5$, Skew $= -1.51$. The assumption is not confirmed: The mean is much less than the median, and the skew exceeds 1.0 in absolute value. The difference scores are negatively skewed by an outlier (Subject 4). The skew may have caused some loss of statistical power.

d) $|z| = 1.99$, $p = .046$. (SPSS only reports the asymptotic p value, which should not be used with $n < 10$).

e) ΔHL $= 5.5$, CI: [0.5, 8.0], it does not bracket zero. Because this test is more robust to outliers, and the standard deviations of the two groups are similar, the result can be used as evidence of a significant increase over time. One could also use a robust ATS test (described in the next chapter) to test for significance.

11. a) 37.5.

b) -8.5, 10.5, 2.5, -4.5.

c) $T_1 = 8.36$, $p = .039.[-8.5^2 + 10.5^2 + 2.5^2 + -4.5^2] * .04 = 8.36$, $\sim \chi^2(3)$.

d) $T_3 = F(3,42) = 3.19$. $p = .033$. [In EXCEL, $=$F.DIST.RT(3.19,3,42) $= .033$.]

e) There are two homogenous subgroups: A/C/D and B/C/D, so definitions A and B are significantly different from one another. In Marascuilo and Agenais (1974), Definition A reflected free association and Definition B forced mixing. Using Eq. 7.13, the only significant comparison is A/B ($p = .007$, adjusted $p = .043$).

f) $W = 18.6\%$. W is not especially high; there is not that much agreement in the rankings (see, e.g., those for Definition A). .

13. a) $L_1 = 248$, one-sided critical values (from Table 7.11) are 248 ($\alpha = .001$), 243 ($\alpha = .01$), and 239 ($\alpha = .05$). So the one-sided p value is .001 and the two-sided p value is .002. However, the test is conservative in the presence of ties, so it would be best not to report exact p values (e.g., report instead $p < .01$).

   b) $T_1 = 11.765$ (2 df), $p = .003$; $T_2 = 8.07$ (2 df numerator, 36 df denominator, $p = .001$).

   c) In this problem, the p value for the Page statistic is only lower than the one for T_1 (however, it does provide evidence of an increasing trend).

   d) With the stepwise, stepdown procedure, all comparisons are significant. When all pairwise comparisons are made, only the Baseline/6-months comparison is significant (Bonferroni-adjusted $p = .004$).

Chapter 8: Advanced Rank Tests (for Interactions and Robust ANOVA)

1. a) Grand mean: 4.24, Factor means: 6.18 (A = 1), 4.96 (B = 1), Main effects ($\hat{\alpha}_1 = 1.94$, $\hat{\alpha}_0 = -1.94$; $\hat{\beta}_1 = 0.72$, $\hat{\beta}_0 = -0.72$).

   b)

Student	Y'	Rank(Y')	Student	Y'	Rank(Y')
1	0.20	19	15	0.83	24
2	−1.04	3	16	0.60	22
3	−0.25	11	17	−0.41	8
4	−0.04	13	18	1.28	28
5	−0.34	9	19	−0.11	12
6	0.17	17.5	20	0.76	23
7	−1.38	2	21	−0.32	10
8	0.02	14	22	−0.44	7
9	−0.85	4	23	1.25	27
10	0.27	21	24	0.11	16
11	0.07	15	25	0.90	25
12	−0.68	5	26	0.17	17.5
13	0.24	20	27	−0.55	6
14	−1.76	1	28	1.20	26

   c) The interaction is significant ($F(1, 24) = 5.42$, $p = .029$). [Note: If numbers are not rounded before ranking, $F(1, 24) = 5.28$, $p = .031$.]

   d) No, in general, a nonparametric approach would only be more powerful with excess *positive* kurtosis.

3. a) Grand mean: 1.87; column means: 1.63, 1.73, 2.23; experimental row means: 7.33, 1.0, 9.66, 7.67, 2.67, 1.33, 1.0, 0.33, 1.0, 0, 0, 0.33, 0, 1.0, 0; control row means: 6.67, 2.33, 7.67, 3.67, 0.67, 0, 1.0, 0.33, 0, 0, 0, 0, 0, 0, 0.33.

   b) Y' (Rank)

	Experimental group				Control group		
Patient	Prescore	3 months	6 months	Patient	Prescore	3 months	6 months
1	−.10 (44)	−1.20 (6)	1.30 (82)	16	−1.43 (4)	−.53 (20.5)	1.97 (87)
2	−.77 (13)	.13 (52.5)	.63 (74)	17	−.10 (39)	.80 (75)	−.70 (17.5)
3	−.43 (24)	−1.53 (3)	1.97 (88)	18	2.57 (89)	1.47 (83)	−4.03 (1)
4	.57 (74)	−.53 (20.5)	−.03 (45)	19	2.57 (90)	−.53 (22.5)	−2.03 (2)
5	−1.43 (5)	.47 (71)	.97 (79)	20	.57 (73)	−.53 (22.5)	−.03 (46)
6	−1.10 (7)	−0.20 (38)	1.30 (81)	21	.23 (64)	.13 (52.5)	−.37 (30.5)
7	−.77 (13)	−.87 (9)	1.63 (85)	22	−.77 (13)	−.87 (9)	1.63 (85)
8	−.10 (41.5)	.80 (77)	−.70 (17.5)	23	−.10 (41.5)	.80 (77)	−.70 (17.5)
9	−.77 (13)	1.13 (80)	−.37 (30.5)	24	.23 (64)	.13 (52.5)	−.37 (30.5)
10	.23 (64)	.13 (52.5)	−.37 (30.5)	25	.23 (64)	.13 (52.5)	−.37 (30.5)
11	.23 (64)	.13 (52.5)	−.37 (30.5)	26	.23 (64)	.13 (52.5)	−.37 (30.5)
12	−.10 (41.5)	.80 (77)	−.70 (17.5)	27	.23 (64)	.13 (52.5)	−.37 (30.5)
13	.23 (64)	.13 (52.5)	−.37 (30.5)	28	.23 (64)	.13 (52.5)	−.37 (30.5)
14	−.77 (13)	−.87 (9)	1.63 (85)	29	.23 (64)	.13 (52.5)	−.37 (30.5)
15	.23 (64)	.13 (52.5)	−.37 (30.5)	30	−.10 (41.5)	−.20 (37)	.30 (70.0)

   c) $H_{(A)} = 0.18$, $F(2, 27) = 2.435$, $p = .11$.

5. a) $\bar{R}_{pre} = 22.4063$, $\bar{R}_{post} = 10.5938$, $n = 16$. $N = 32$.

   b) $\hat{p}_{pre,} = .68$ and $\hat{p}_{post} = .32$ (from Eq. 8.3) and $\hat{p} = .87$ (or .13) from Eq. 8.4 ($\hat{p} = .86$ if based on rounded values).

   c) $SE = 9.8419$ [Calculation: for first case, [(Change score) $+ \bar{R}_{pre} - \bar{R}_{post}]^2 = (7.5 - 29.5 + 22.4063 - 10.5938)^2 = 103.7852$, $\Sigma 1452.9375$, $SE = \sqrt{1452.9375 / (16-1)} = 9.8419$].

   d) $T_n^F = -4.80$.

   e) $p = .0002$.

   f) No, because these are tests of a shift in location (e.g., medians), whereas T_n^F is a test of distributional shape. T_n^p is a test of locational shift but is more complicated to calculate in EXCEL (see Technical Note 8.2).

7. a) $\bar{R}_{placebo*time1} = \bar{R}_{placebo*time2} = \bar{R}_{ketamine*time1} = 38.6667, \bar{R}_{ketamine*time2} = 78.0000;$

$n_{placebo} = n_{ketamine} = 24, N = 96;$

b) $\hat{P}_{placebo*time1} = \hat{P}_{placebo*time2} = \hat{P}_{ketamine*time1} = .40; \hat{P}_{ketamine*time2} = .81.$

c) $\hat{\tau}^2_{placebo} = 125.5652, \hat{\tau}^2_{ketamine} = 449.2754.$

d) $U_n^{AT} = 8.04, p < .001.$

Chapter 9: Linear Regression and Generalized Linear Models

1. a) Two Parents: $\hat{Y} = 2.8 - 0.1 * Parent\,Involvement.$ One Parent: $\hat{Y} = 3.3 - 0.4 * Parent\,Involvement.$ EXCEL calculations:

Intercept (two parents)	2.8
Intercept change	0.5
Intercept (one parent)	3.3
Slope (two parents)	−0.1
Slope change	−0.3
Slope (one parent)	−0.4

b)

	$\hat{Y}$ = Externalizing behaviors	
Parental involvement	2 parents	1 parent
0	2.8	3.3
1	2.7	2.9
2	2.6	2.5
3	2.5	2.1
4	2.4	1.7
5	2.3	1.3
6	2.2	0.9
7	2.1	0.5

c)

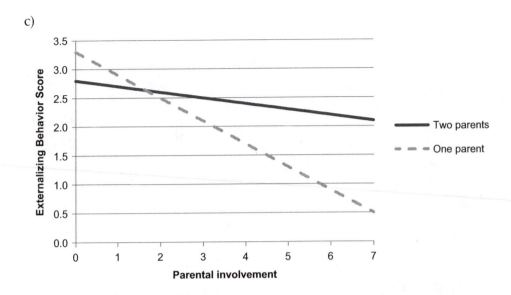

d) Greater parental involvement is associated with fewer behavior problems, especially for students in single-parent households.

3. a) Low Relevance: $\hat{Y} = 4.0 - 2.0 * Fear\_Appeal.$
    High Relevance: $\hat{Y} = 6.8 - 2.0 * Fear\_Appeal.$
    (Note that when *Text* = 0, both terms involving the variable disappear.)
    EXCEL calculations:

Intercept (low relevance)	4.0
Intercept change	2.8
Intercept (high relevance)	6.8
Slope (low relevance)	−2.0
Slope change	4.0
Slope (high relevance)	2.0

b)

	$\hat{Y}$ = Mean number of pages	
	No fear appeal	*Fear appeal*
High relevance	6.8	8.8
Low relevance	4.0	2.0

c)

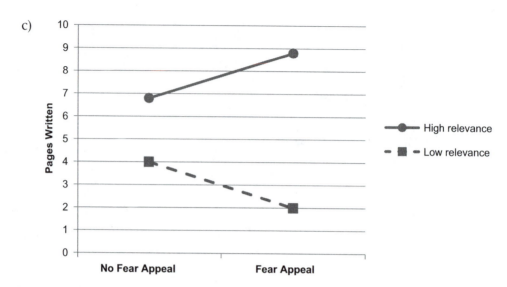

d) A fear appeal increases engagement on the topic with high personal relevance but decreased engagement on the topic of low relevance. Relevance has a positive effect on engagement; the effect is larger when combined with a fear appeal.

5. a) $\hat{Y}_i = 47.22 + -0.23 * X_i$, $p = 0.27$, $R^2 = 4.4\%$. No.
   b) The graph is curvilinear (U-shaped).
   c) $\hat{Y}_i = 55.5 - 2.49 * X_i + 0.11 * X_i^2$, $R^2 = 27.8\%$, $p = .004$ ($\hat{\beta}_1$), $p = .006$ ($\hat{\beta}_2$). Yes.
   d) With the simple linear regression, the functional form was misspecified, producing bias and a poor fit to the data. There are high values of Y_i when X_i is both very high and low, producing a relatively flat line.
   e) There is a discernible curvilinear, U-shaped pattern in the residuals for the linear but not the quadratic regression.
   f) The cubic, with a R^2 of 35.9%. However, adding a cubic term (X_i^3) will always increase the fit. If one runs the actual regression, the cubic term is not significant ($p = .08$).

7. a) 10.52%.
   b) 1.1052.
   c) $ln(1.1052) = 10\%$.
   d) $ln(Y) = 9.21 + 0.10 * Time$.

e)

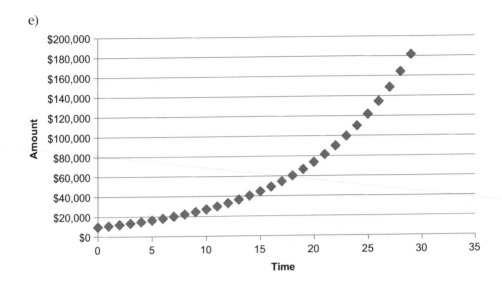

f) $R^2 = 87\%$ (linear), $R^2 = 99.2\%$ (quadratic), $R^2 = 100\%$ (exponential; highest).

9. a) *Skew* = 1.04, *kurt* = 1.41.
   b) 1, 37%; 2, 27%; 3, 15%; 4, 7%; 5, 3%; 6, 1%; 7, 1%; 8, 0%; 9, 0%; 10, 0%.
   c) 2,857.82 (linear model), 6.76 (gamma with log link), 6.67 (gamma with identity link). The last model fits best.
   d) There are more positive and larger residual than negative ones (skew = 1.15). This pattern is consistent with the errors being gamma distributed.
   e) Skew = 1.12 (the model does not assume, however, normally distributed errors).

Chapter 10: Binary Logistic Regression

1. a) skew = 0, kurt = −1.2. Minimum $z = -1.71$, maximum $z = 1.71$. None of the descriptive stats are problematic.
   b) $L = -7.68$, AIC = 19.36, BIC = 24.57, −2LL = 15.36.
   c) There is a .05% probability of obtaining these data given the estimated parameters. All other estimates produce a lower probability.
   d) AIC = $2p + 2LL = 2 * 2 + 15.36 = 19.36$. BIC = $p * ln(n) + 2LL = 2 * 4.605 + 15.36 = 24.57$.
   e) $\chi^2_{LR}(1) = 94.86$, $p < .001$. There is 1 *df* because the model has one additional estimated parameter (the B for E-mail) than the intercept-only model (with −2LL = 15.36 + 94.86 = 110.22).
   f) Yes: $\chi^2_{LR}(1) = 94.86$, $p < .001$; Wald χ^2 (1) = 7.67, $p = .006$. The LR test is more statistically powerful and does not assume the model contains all relevant variables.

g) B = 0.428, SE = 0.1547, z = 2.77 (the square root of the Wald χ^2).

h) The odds of having an account increase by 53% with each additional e-mail note received.

i) τ = .60 (positive).

j) The graph is linear; Box-Tidwell test for nonlinearity is nonsignificant, as the e-mail * ln(e-mail) B = .20 (p = .74).

k) $\chi^2_{LR}(2)$ = 0.28, p = 0.87. The logistic link is satisfactory.

l) Predicted logit = −10.069, predicated probability and category, 0.00. Leverage = .001; standardized Pearson res. = −.007; standardized deviance res. = −.009; likelihood res. = −.009, Cook's D = 0.

m) $\hat\pi_{24}$ = 48%. Value < 50% so predicted category is 0.

n) Cook's D [0, 0.58], leverage [0, 0.11], st. Pearson resids. [−2.24, 3.41], st. deviance resids. [−1.95, 2.32], likelihood resids. [−1.99, 2.44].

o) Cases with e-mail = 19 and 28 have outlying diagnostic statistics. These are the only two cases the model did not successfully predict. The cases should be retained because removing them would cause complete separation of the data (the model would fit perfectly with a zero −2LL) and no parameters could be estimated.

p) The second graph.

3. a) B = −.009 (using the lowest value Y category as the reference category), $\chi^2_{LR}(1)$ = 0.59, p = .44

b) $B_{x*ln(x)}$ = −0.135, $\chi^2_{LR}(1)$ = 8.52, p = .004. The model contains a significant nonlinear component.

c) Proportions: .6, .8, .9, .7, .5; logits: 0.41, 1.39, 2.20, 0.85, 0.00. The relationship is curvilinear (i.e., quadratic).

d) B_{x^2} = −0.001, $\chi^2_{LR}(1)$ = 7.14 p = .008, yes.

Chapter 11: Multinomial Logistic, Ordinal, and Poisson Regression

1. a) AIC = 57.95, BIC = 73.58, −2LL = 45.95, omnibus test χ^2_{LR} (4) = 63.32, p < .001, Nagelkerke pseudo-R^2 = 53.6%, 68.0% correctly classified.

b) For ED: χ^2_{LR} (2) = 55.88, p < .001. For AA: χ^2_{LR} (2) = 6.32, p = .04. There are 2 df because omitting an effect from the fuller model (to obtain the reduced model) involves omitting two parameter estimates, one for Managerial and one for Professional.

c) Each additional year of college education increases the odds of obtaining a managerial position by 79% and the odds of a obtaining a professional position by 330%. Being AA decreases the odds of obtaining a professional/technical position by 81% [exp(−1.64) = .19, .19 − 1 = −.81].

d) χ^2_{LR} (2) = 2.21, p = .33.

e) χ^2_{HL} (7) $= 1.77$, $p = .97$. The model fits well.

f) Largest positive deviance residual $= 2.16$, largest negative $= -2.25$, largest Cook's $D = .80$.

g) For cases 11 and 12, deviance residuals $= -2.20$, Cook's $D = 0.34$.

h) The magnitude of the AA estimate is larger ($\hat{\beta} = -1.65$) and p value smaller ($p = .02$ vs. $p = .09$).

3. a) $D = 78.15$, pseudo-$R^2 = .012$, B $= .013$, exp(B) $= 1.01$. The exp(B) estimate means that each additional 5 years of age increases the odds of believing more in god (i.e., being in a higher category) by 1%. The direction is positive ($\hat{\tau}_b = .08$).

b) No, $\chi^2(4) = 19.47$, $p = .001$.

c) Yes, the column (category) totals are not roughly equal but higher values are more likely. A cloglog model will likely fit best.

d) For negative log-log, $D = 106.22$, pseudo-$R^2 = .008$. For cloglog, $D = 71.62$, pseudo-$R^2 = .013$. Cloglog fits best, verifying hypothesis in part (c).

e) B $= .011$, exp(B) $= 1.01$, the estimate implies that as age increases by one category (approximately 5 years), the probability of being one category higher (in belief in god at any particular time point; see Technical Note 10.9) increases by 1%. However, the parallel lines test is still significant χ^2 (4) $= 13.23$, $p = .01$.

f) Yes, the B values only range from .007 to .01 (both round to 1%). Even if the values are multiplied by 3.38 (the standard deviation of age divided by 5 years, the span of each category), the exponentiated values range from 1.02 to 1.03. The individual regressions could be reported if higher precision is needed.

g) B $= 0.008$, χ^2_{LR} (1) $= 3.53$, $p = .06$; Wald χ^2 (1) $= 3.44$, $p = .063$. The largest likelihood residual is $|2.84|$. The residual value and other diagnostics are not different from those for neighboring points and the case should not be considered an outlier. (The statistical model does not predict atheism well, additional variables are likely needed.)

5. a) Mean: 0.52. Var: 0.49. Skew: 1.23. Kurt: 1.03. The DV is a count and is positively skewed, there are a lot of zeroes, and the variance is almost the same as the mean.

b) For Poisson, $z = .10$, $p \approx 1.0$. For normal, $z = 4.38$, $p < .001$. The empirical distribution differs significantly from the normal distribution but not the Poisson.

c) The loess line is flat.

d) $L = -133.81$. $D = 134.27$. $-2LL = 267.62$.

e) As an individual writes more, there is more opportunity to make a rebuttal. One may therefore wish to control for this by analyzing the number of rebuttals per page, which is equivalent to using ln(pages) as an offset.

f) 0.91 for D; 0.92 for Pearson. Both are less than 1.0, so overdispersion is not indicated.

g) $D = 134.27$. $-2LL = 267.62$. There is no change in D between the Poisson and negative binomial model. The ancillary parameter is practically zero, providing no evidence of overdispersion.

h) GO: 0.54 ($p = .02$); NC: -0.007 ($p = .46$). $Exp(0.54) = 1.72$. The chance of generating a rebuttal increases by 72% from using the organizer. There is no discernible effect of NC.

i) .39, .23, RR $= 1.72$ (without rounding the means). The actual RR is the same as the estimated one.

j) There is little evidence of outliers. The most extreme case (case 1) had a likelihood residual of 2.37, which is not excessively large; also, it is some- what close to the next largest residual, 1.86. Omitting the case made no substantive difference in the results. Omitting the case with the largest Cook's D (0.138 for case 5) also made no difference.

k) Identity Link: $D = 151.41$, AIC $= 290.76$, BIC $= 299.79$. Log link: $D = 134.27$, AIC $= 273.62$, BIC $= 282.65$. The log link fits better.

Chapter 12: Log-Linear Analysis

1. a) Should match the data table in the problem.

    b) Year * Region, Year * Prejudice, Region * Prejudice. (Intercept and main effect terms are implied.)

    c) Prejudice * Year: if Year $= 1963$ and Prejudice $=$ Yes, B $= -1.99$, exp(B) $=$ 0.14 (7.32 if inverted), $p < .001$. Prejudice * Region: if Region $=$ North, Prejudice $=$ Yes, B $= -0.94$, exp(B) $= 0.39$ (2.56 if inverted), $p < .001$. Prejudice was greater in the South and has decreased over time.

    d) 1.12. Because in log-linear analysis, standardized residuals tend to be nor- mally distributed, and $1.12 < 3.29$, there is no reason for concern.

    e) 51.6% were prejudiced in 1946; 14.0% in 1963, showing a decline over time. Actual OR $= 0.15$ (6.58 inverted). B $= -1.99$, exp(B) $= 0.15$ (7.32 inverted). The estimated and actual ORs are close (note that the former controls for region).

    f) B $= 0.497$, exp(B) $= 1.64$, $p = .09$. The decline in prejudice over time was not significantly greater in one region than the other.

    g) The interaction is significant with the negative log-log link. χ^2_{LR} (1) $=$ 11.21, $p = .001$. If Region $=$ South and Prejudice $=$ yes, B $= -0.59$, exp(B) $= 0.55$ (1.80 inverted). The decline in prejudice has been greater in the South than in the North. The negative log-log (NLL) link does not produce a better fit because both models are saturated (so the fits are perfect), but use of the NLL link is justified since some of the conditional probabilities are low (e.g., 12% for 1963 prejudice in the North). See Pomer (1984) for additional justification for alternative links.

3. a) Depression ∗ High_Moves, Siblings ∗ High_Moves, Age.
   b) Poisson.
   c) Depressed ∗ High_Moves: B = 0.84, z = 2.55, p = .011. Siblings ∗ High_ Moves: B = −0.79, z = −2.59, p = .01.
   d) Yes. The estimated OR is $exp(0.84) = 2.32$.
   e) No, they are likely to be moved *less*. The estimated OR is $exp(−0.79) = 0.45$.

5. a) z = −1.33, p = .18, no.
   b) Following SPSS, the table below assumes that the reference category for SI = 1 and for Unemployment is 6+ months.

Term	B	Exp(B)	z	P
SI ∗ MDAge	0.10	1.11	6.44	<.001
SI ∗ [Unemployment = None]	1.09	2.97	8.24	<.001
SI ∗ [Unemployment = < 6 months]	0.39	1.48	2.50	.013
MDAge ∗ [Unemployment = None]	−0.08	0.93	3.93	<.001
MDAge ∗ [Unemployment = < 6 months]	−0.13	0.88	−5.89	<.001

   c) Yes, because the coefficients are negative, older workers are less likely to fall into the [Unemployment = None] and [Unemployment = "< 6 months"] categories and therefore more likely to fall into the long-term unemployment category. Mean coefficient −0.07; mean deviated coefficients are −0.01, −0.06, and 0.07 respectively; a trend is not indicated. The local ORs are $exp(0.07 − (−0.01)) = 1.08$ and $exp(0.07 − (−0.06)) = 1.14$. Each additional year of age increases the odds of long-term unemployment by 8% (relative to no unemployment) and 14% (relative to short-term unemployment).
   d) Age is negatively associated with SI ($\tau = −.059$). The B (0.10) is positive when the reference category for SI = 0, so is negative when SI = 1. Unemployment is positively associated with SI ($\hat{\tau} = .105$). The Bs are positive when SI = 0 so are negative when SI = 1. Lower levels of unemployment are associated with less suicidal ideation.
   e) Higher levels of unemployment might increase suicidal ideation, but it is also possible that higher levels of drug use cause greater SI as well as more unemployment.

Chapter 13: General Estimating Equations

1. a) Only Time. Time: B = −1.59, $exp(B) = 0.20$, $\chi^2_{GS}(1) = 6.84$ and p = .009, Wald $\chi^2 (1) = 10.05$ and p = .002. Sessions ∗ Time: B = −0.14, $exp(B) = 0.87$, $\chi^2_{GS}(1) = 1.62$ and p = .20, Wald $\chi^2 (1) = 2.13$ p = .14.

b) Yes, $p = .02$. However, there is no reason to believe that the pre- and post-scores would be statistically independent, so the first set of results is more defensible. Violations of assumptions of statistical independence can create Type I errors.

3. a) QIC: 357.15 (AR-1), 358.27 (exchangeable), 360.56 (unstructured). AR(1) is lowest.

b) For gender (G = 1), B = -0.19, generalized score $\chi^2(1) = 0.07$, $p = .80$. For year, B = 0.35, $\chi^2_{GS}(1) = 12.27$, $p < .001$. For (G = 1) * Y, B = 0.06, $\chi^2_{GS}(1) = .11$, $p = .74$. The use of cocaine increased over time (because the coefficient is positive). Gender does not have a significant effect on use, even if the interaction term is dropped.

c) *Exp*(B) = 1.46. The odds of using cocaine increase by 46% per year. QICC = 346.371.

d) Y1 = 8%, Y2 = 14%, Y3 = 19%, Y4 = 23%. The logits are Y1 = -2.44 $\left[= ln\left(\frac{.08}{1-.08}\right)\right]$, Y2 = 1.82, Y3 = -1.45, Y4 = -1.21. The relationship appears slightly nonlinear. For the model using *ln*(Year), QICC = 345.849, which is smaller than for the previous model.

e) Females are more likely to develop dependency: B = 1.27, $\chi^2_{LR}(1) = 3.92$, $p = .048$. *Exp*(B) = 3.57. Even though use rates are similar, the odds of females developing dependency are 3.57 times those for males.

Chapter 14: Estimation Procedures

1. CP: [.147, .494], range = .347. Jeffreys: [.160, .477], range = .317. LR: [.157, .476], range = .319. Score: [.167, .479], range .312, $\tilde{\pi} = .323$. Rank order from high to low is CP, LR, Jeffreys, and score. (Note: of the four methods, the score method also contains an adjustment for small-sample bias.)

3. a)

	Newton-Raphson				Fisher Scoring			
Step	$\hat{\pi}_t$	U	Info (–H_t)	Update	Step	$\hat{\pi}_t$	U	Info (I)	Update
0	10.00%	533.33	5851.85	0.09	0	10.00%	533.33	1111.11	0.48
1	19.11%	251.52	1651.75	0.15	1	58.00%	0.00	410.51	0.00
2	34.34%	104.93	589.23	0.18	2	58.00%	0.00	410.51	0.00
3	52.15%	23.45	396.70	0.06	3	58.00%	0.00	410.51	0.00
4	58.06%	–0.24	410.83	–0.00	4	58.00%	0.00	410.51	0.00
5	58.00%	0.00	410.51	–0.00	5	58.00%	0.00	410.51	0.00
6	58.00%	0.00	410.51	–0.00	6	58.00%	0.00	410.51	0.00

b)

		Newton-Raphson					Fisher Scoring		
Step	$\hat{\pi}_t$	U	Info ($-H_t$)	Update	Step	$\hat{\pi}_t$	U	Info (I)	Update
0	40.00%	−12.5	114.58	−0.11	0	40.00%	−12.50	125.00	−0.10
1	29.09%	1.32	148.11	0.01	1	30.00%	0.00	142.86	0.00
2	29.98%	0.02	142.95	0.00	2	30.00%	0.00	142.86	0.00
3	30.00%	0.00	142.86	0.00	3	30.00%	0.00	142.86	0.00
4	30.00%	0.00	142.86	0.00	4	30.00%	0.00	142.86	0.00

5. a) 4×2, 4×2, no, the inner dimensions are not the same.

b) 2×4, yes, 4×4 (these are the outer dimensions of the two matrices, 4×2 and 2×4).

c) $\begin{bmatrix} 1 & 3 & 5 & 7 \\ 8 & 1 & 5 & 2 \end{bmatrix}$.

d) $\begin{bmatrix} 49 & 9 & 35 & 19 \\ 66 & 14 & 50 & 30 \\ 13 & 16 & 30 & 37 \\ 11 & 10 & 20 & 23 \end{bmatrix}$.

e) $\mathbf{M}^{-1} = \begin{bmatrix} -2 & 1.5 \\ 1 & -0.5 \end{bmatrix}$.

f) $\mathbf{I} = \begin{bmatrix} 1 & 0 \\ 0 & 1 \end{bmatrix}$. $\mathbf{MI} = \mathbf{M} = \begin{bmatrix} 1 & 3 \\ 2 & 4 \end{bmatrix}$.

7. a) Residuals: $\begin{bmatrix} 3.08 \\ -0.90 \\ -1.12 \\ -0.15 \\ -1.83 \\ -0.35 \end{bmatrix}$.

b) Scores: $\begin{bmatrix} -1.28 \\ -12.68 \end{bmatrix}$.

c) Beta updates: $\begin{bmatrix} 0.357 \\ -0.165 \end{bmatrix}$.

d) $\hat{\boldsymbol{\beta}}_{t=2}$: $\begin{bmatrix} -1.884 \\ 0.659 \end{bmatrix}$.

9. $\begin{bmatrix} 0.0075 \\ 0.0285 \end{bmatrix}$. The scores are close to zero because the beta estimates are converging onto the maximum-likelihood values. At the maximum, the slope of the likelihood and log-likelihood functions are zero, therefore the scores (which reflect the first derivative/slopes of the log-likelihood function) will be zero.

11. a) $\boldsymbol{\eta}$: $\begin{bmatrix} 1.260 \\ 1.580 \\ 1.900 \\ 2.200 \\ 2.540 \\ 2.860 \end{bmatrix}$.

b) The variances are the same as the means and can be found by exponentiating the predicted values. Variances: $\begin{bmatrix} 3.525 \\ 4.855 \\ 6.686 \\ 9.207 \\ 12.680 \\ 17.462 \end{bmatrix}$.

c) Information: $\begin{bmatrix} 54.415 & 183.875 \\ 183.875 & 753.877 \end{bmatrix}$.

d) Covariance: $\begin{bmatrix} 0.105 & -0.025 \\ -0.025 & 0.008 \end{bmatrix}$.

e) Residuals: $\begin{bmatrix} 1.475 \\ -1.855 \\ -0.686 \\ 1.793 \\ 0.320 \\ -0.462 \end{bmatrix}$.

f) Scores $\begin{bmatrix} 0.585 \\ 1.125 \end{bmatrix}$.

g) Beta updates: $\begin{bmatrix} 0.032 \\ -0.006 \end{bmatrix}$.

h) $\hat{\boldsymbol{\beta}}_{t=1}$: $\begin{bmatrix} 1.292 \\ 0.314 \end{bmatrix}$.

INDEX

Page numbers in italic indicate figures and tables.

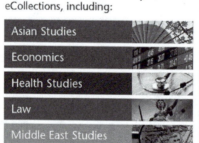